T0248825

Artificial Intelligence: Concepts and Applications

Artificial Intelligence: Concepts and Applications

Edited by
Mick Benson

www.willfordpress.com

Published by Willford Press,
118-35 Queens Blvd., Suite 400,
Forest Hills, NY 11375, USA

ISBN: 978-1-68285-409-9

Cataloging-in-Publication Data

Artificial intelligence : concepts and applications / edited by Mick Benson.
p. cm.
Includes bibliographical references and index.
ISBN 978-1-68285-409-9
1. Artificial intelligence. I. Benson, Mick.
TA347.A78 A78 2018
006.3--dc23

For information on all Willford Press publications
visit our website at www.willfordpress.com

Contents

Preface

The main aim of this book is to educate learners and enhance their research focus by presenting diverse topics covering this vast field. This is an advanced book which compiles significant studies by distinguished experts in the area of analysis. This book addresses successive solutions to the challenges arising in the area of application, along with it; the book provides scope for future developments.

The branch of computer science which studies the intelligence demonstrated by machines is known as artificial intelligence. It has helped in developing solutions in the field of computer science. Some of the tools used in the process are search and optimization, probabilistic methods for uncertain reasoning, classifiers and statistical learning methods, etc. Artificial intelligence is mainly used in autonomous vehicles, medical diagnosis, video games, security surveillance, search engines, etc. This book unravels the recent studies in the field of artificial intelligence. For all those who are interested in artificial intelligence, this book can prove to be an essential guide. It will serve as a valuable source of reference for graduate and postgraduate students.

It was a great honour to edit this book, though there were challenges, as it involved a lot of communication and networking between me and the editorial team. However, the end result was this all-inclusive book covering diverse themes in the field.

Finally, it is important to acknowledge the efforts of the contributors for their excellent chapters, through which a wide variety of issues have been addressed. I would also like to thank my colleagues for their valuable feedback during the making of this book.

Editor

1

A Registration Method for Multimodal Medical Images Using Contours Mutual Information

Ying Qian
The lab of Graphics and Multimedia
Chongqing University of Posts and Telecommunications
Chongqing, China

Meng Li, Qingjie Wei, Xuemei Ren
The lab of Graphics and Multimedia
Chongqing University of Posts and Telecommunications
Chongqing, China

Abstract—**In recent years, mutual information has developed as a popular image registration measure especially in multimodality image registration. For different modality medical images, the contour of tissues or organs is similarity. In this paper, an effective new registration method of the multimodal medical images based on the contour mutual information is proposed. Firstly, get the contour through variational level set method. Secondly, within the contour pixels are assigned the same grayscale value, obtain two contour images. Finally, two contour images using mutual information as similarity measure for image registration. The experiment results show that the registration algorithm proposed in this paper can do more effectively and more accurately work than normalized mutual information and gradient mutual information.** ***(Abstract)***

Keywords—contour mutual information; mutual information; multimodal medical image; image registration; variational level set method

I. INTRODUCTION

In actual clinical diagnosis, a single modal of image usually unable to provide enough information to doctors. So we need to have the step of different modal image fusion, thus providing more information to doctor, and then they can understand the comprehensive information from pathological tissue or organs, make better medical diagnosis. Moreover, image registration is a premise key in image fusion, so the multimodal medical image registration becomes a research hotspot in the field of medical image processing. Medical image registration is the ascertaining process of spatial mapping between two images that differs in image acquisition time, image properties, or viewpoint and subsequently producing a result image that is informative. At present, medical image registration methods can be classified as feature-based and intensity-based [1]. In intensity-based registration, mutual information (MI) is one of the intensity based measure, and does not require the definition of landmarks or features such as surfaces [2]. It is an automatic method and it applies to a variety of modal image registration [3].

The multimodal image registration method based on mutual information of image gray intensity, which is no need for us to preprocess. In order to improve the accuracy of the registration, so many scholars made deep researches and improved the registration methods.

Studholme [4] proposed a method based on normalized mutual information. But the method ignores the spatial information. Pluim [5] proposed to include spatial information by multiplying MI measure with a gradient term based on the magnitude and the orientation of the gradients. But gradient is sensitive to noise, thus the success rate become lower when images contains noise. Butz and Thiran [6] used mutual information in edge computing. But the edge information is sparse and the mutual information is sensitive to the number of sampling point, thus rendering appear local maximum and result in mismatching. Loeckx [2] proposed a method based on conditional mutual information, spread gray joint histogram from two-dimension to three-dimension. But the method has high computational cost. Ruecker [7] combined the mutual information with image's texture information which be described as gray level co-occurrence matrix. But this method assume that image's texture information have similarity when organ in different modal images.

This paper uses the characteristic that the contours of different modal medical images have similarity, proposed a registration method based on contour mutual information(CMI), extracts contours effectively from images which to be registered, and within the contour pixels are assigned the same gray-scale, That is, contour image. Finally, two contour images using mutual information as similarity measure for image registration. CMI efficiently solves problems of poor robustness and low accuracy in image registration.

II. MUTUAL INFORMATION

A. Definition

MI is selected as the similar measure. MI is an important concept of information theory which is used to represent the statistics correlation of two sets of data which has been widely used in the image registration. Starting from a reference image X and floating image Y with intensity bins x and y, the calculation of MI of two images X and Y is as follows:

$$I(X,Y)=H(X)+H(Y)-H(X,Y) \quad (1)$$

$H(X,Y)=-\sum_{x,y}p(x,y)\log(p(x,y))$ and $H(X)=-\sum_{x}p(x)\log(p(x))$ the joint and marginal entropy of random variables X and Y.

B. Problems

Mutual information registration method based on gray value is a statistical method for registration, has nothing to do with the pixel location. Constructs a size for the test image A (256 * 256), pixels of each column from top to bottom, with gray value from 0 to 255, as shown in Fig.1(a), A rotate 180 °, obtain test image B, as shown in Fig.1(b). A is the reference image, B is the floating image. Floating image 360 ° rotation around itself(from -180° to 180°), Observe the reference image and the floating image of the mutual information curve, Respectively, at 0 ° and 180 °, the maximum mutual information appears, as shown in Fig.1(c).

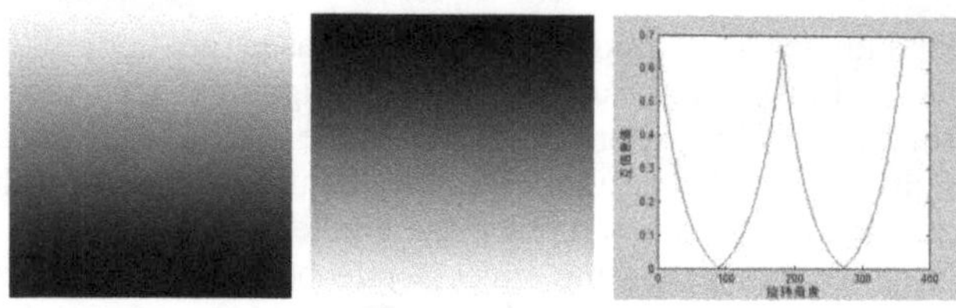

(a)reference A (b)floating image B (c)mutual information curve

Fig. 1. Test images and mutual information curve.

The research direction of many researchers is how the spatial information of image is introduced to the mutual information to produce a new similarity measure. But the robust is not very good; the main reasons are the following:

- As shown in Fig.2(a), image with noise, weaken the correlation between images, easy to fall into local extreme value leads to wrong registration;
- As shown in Fig.2(b), if the image appears to be missing which make registration less effective, the registration effect will be worse;
- As shown in Fig.2(c), due to less spatial resolution and image information content, the image of low resolution images can lead to the problem of robust ;
- Images to be registered have less effective information, such as the top of the skull, as shown in Fig.2(d).

In this paper, analyzes the robustness of registration according to mutual information measure curve. In Fig.1, the abscissa of mutual information is the rotation angle (-25° to 25°), the vertical axis is the amount of the mutual information. As can be seen from the figure, there are have different degrees of local maxima in the four cases mentioned above.

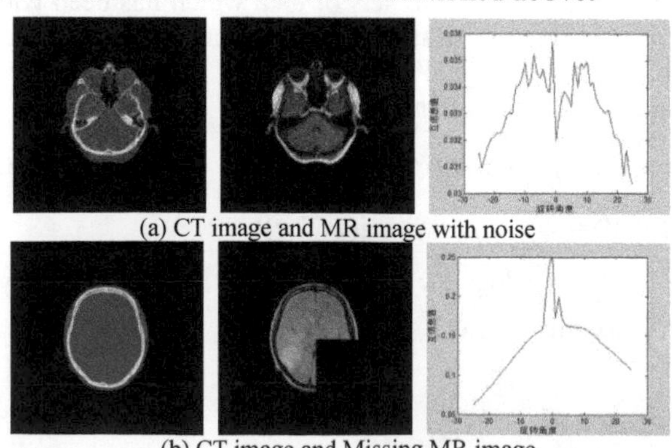

(a) CT image and MR image with noise

(b) CT image and Missing MR image

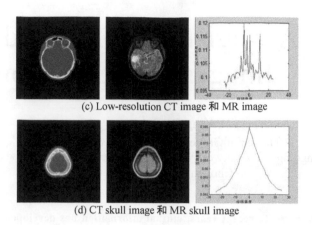

(c) Low-resolution CT image 和 MR image

(d) CT skull image 和 MR skull image

Fig. 2. Mutual information curves under different circumstances.

Although some of the improved algorithm considers the spatial information of the image, but have a major impact on the registration is still gray values, its spatial information is essentially the use of the gray information, because of the limitations are not widely used.

III. METHOD

Contour as a stable characteristic of image, when the image translating or rotating, which can keep better corresponding relation. So before get contour image, we should extract contours of image.

The contour extraction method will choose an operator to detect the image edge which based on the image edge features. Then we will connect the edge point into a closed curve to obtain the target contour.

Commonly used methods of edge detection include: Roberts operator, Sobel operator, Prewitt operator, etc. But these edge detection methods are sensitive to noise, and will enhance the noise in the edge detection. This paper mainly use feature of contour, so effectively getting contour of image is the key of CMI.

In this paper, we adopt a variational level set method proposed by Li [8], which have no need to reinitialize. The method bring penalty term in energy function, thus ensuring level set function keep become sign function in evolutionary process ,and it avoids reinitialize in the process. After get contours of two images, filling contour unified gray value and then get contour image.

CMI has two advantages, firstly, ensures the strict corresponding relations exist between two contours images, add gray-scale correlation between images. Secondly, it used spatial information that contours in different modal images have similarity.

The following describes specific steps which about a way of image registration based on contour mutual information:

1) Two images are given to be registered, and they are reference image R and floating image F.

2) Get contours of two images by variational level set method, and then fill unified gray value to get contour image(R'、F').

3) Mutual information be used as measure function, and Powell optimization algorithm execute optimization methodology of parameters for image R' and F' .The process of optimization uses the nearest neighbor interpolation algorithm.

4) Optimal spatial transform parameters reform the floating image F according to the Powell algorithm, obtain registration results.

IV. EXPERIMENT

In this paper, the experiment data are come from Retrospective Registration Evaluation Project of the Vanderbilt University, and the images resolution is 256×256. The experiment 1 validates the robustness of CMI, the rest experiments were compared with Gradient Mutual Information (GMI, from reference [3]) and Normalized Mutual Information (NMI, from reference [4]).

A. Experiment 1 Validate the Robustness

According to mutual information curve from Fig.3, using image mutual information method can lead peak value become not so sharp. However, the method proposed in this paper has a sharp peak, and it is easy to detect the best position of registration. Meanwhile, it is not easy to appear the phenomenon of local maximum because of the smooth curve.

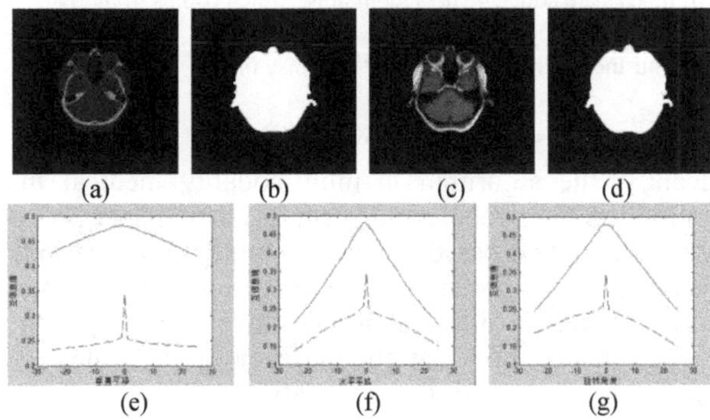

Fig. 3. The results of robustness of experiments.

(a) is CT original registration image and (b) is MR original registration image, (c)is CT contour image and (d) is MR contour image. (e)is mutual information curve based on image gray value and the method of paper to rotate the images (from -25° to 25°). (f) is mutual information curve based on image gray value and the method of paper for vertical translation (from -25 To 25 pixels). (g) is mutual information curve based on image gray value and the method of paper for horizontal translation. The solid line is the measure of the curve based on image gray value and the dashed line is the measure of the curve based on the contour mutual information.

In order to validate the robustness of the registration algorithm based on contour mutual information, using the method of this thesis to create mutual information curve, as Fig.4. Compared the curve of Fig.2, the mutual information curve created by CMI has good robustness.

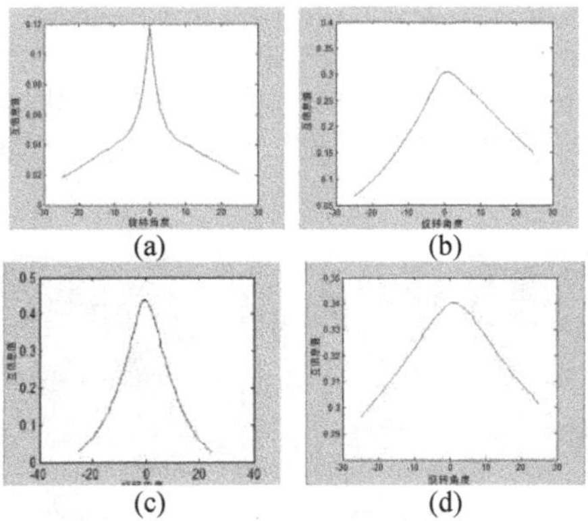

Fig. 4. Measure curve based on contour mutual information.

(a)(b)(c)(d)Corresponding to CMI of test image in Fig.1 mutual information measure curve.

B. Experiment 2 Validate the Accuracy

As shown in Fig.5, the edge images of CT are applied to the MR images, by this to detect the effect of registration. The bone tissue in MR images are close to the background gray-scale value, However, it have large different with the background gray-scale value of CT image, thus leading to registration errors which use the method of mutual information. As shown in Fig.5 (d); The gradient mutual information, due to considerate the information of space and orientation, the registration results have more advantages compare to the method which based on normalized mutual information; From the Fig.5 (e), there are many errors at edges, but effects obtained by CMI show that the image has been substantially aligned at edges after registration.

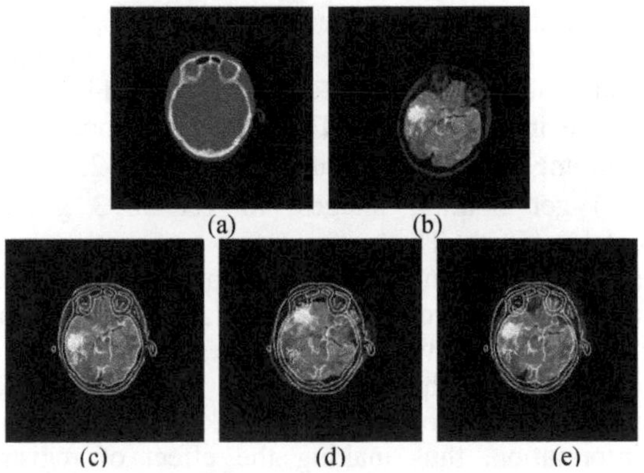

Fig. 5. Experimental results verify the accuracy of CMI.

C. Experiment 3 Validate the robustness when Image have Missing information

From Fig.6 (a), CT image are mainly reflects the skeletal structure, its gray-level changes small.

We mainly use edges information to register with MR image. When Image has large missing information, it will mismatch easier. From experimental results, CMI has obvious advantages compared with other two methods in gray-scale difference.

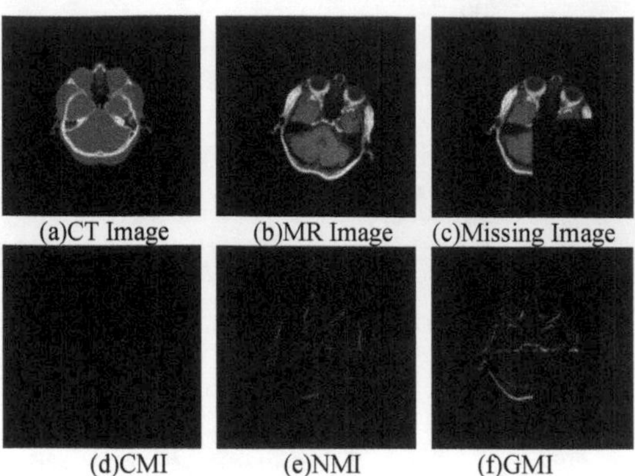

Fig. 6. The results on the robustness of image deletion.

(a)(b)(c)are CT images and MR image in the same layer from the head of a patient. (a) is original CT image (b) is original MR (T2-weighted) image, (c) have missing (Setting the region of the image's gray value to 0 to simulate the effect of information lacking). (d)(e)(f)corresponding to the result of the registration image obtained by CMI, NMI and GMI which subtract registration images.

D. Experiment 4 validate the robustness when image has noise

The gradient is sensitive to noise, when image have a little of noise, the registration method which based on gradient mutual information will easily appear local maximum, it will results in mismatching. As shown in Fig.7 (a), this paper choose three groups of head images in CT modal and MR modal (T1 weighted, T2 weighted, PD weighted), which have been registered and in the corresponding layer. As shown in Fig.7 (a), add noise in CT and MR images. Noised image of CT as a reference image, let the MR images horizontal, vertical position and angle transformation from (-25,-25,-25°) to (25, 25, 25°) get floating images, to get 50×3 groups of experimental subjects. Three parameters among them are the horizontal of translation (pixel), vertical translation (pixel), rotation angle. Results from three methods are shown in Table1. From data in the Table1we can see the registration of the CMI is more accurate than other two methods. In the CT image and PD image registration, the noise will make impact on gradient mutual information, thus making the effect of registration unsatisfactory.

TABLE I. THE EXPERIMENTAL RESULTS

Experimental object	**Robustness verification method for noisy image**		
	CMI	***NMI***	***GMI***
CT-T1	86%	80%	80%
CT-T2	86%	82%	78%
CT-PD	90%	84%	82%

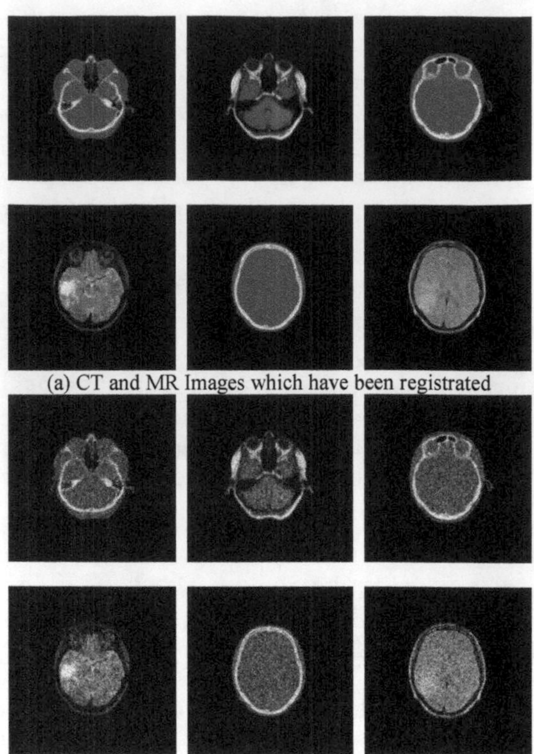

(a) CT and MR Images which have been registrated
(b) Images are added 40% random noise which comes from (a)

Fig. 7. Robustness verification method for noisy image.

V. CONCLUSIONS

Recently, the algorithm in multimodality medical image registration has poor robust and accuracy. This paper proposed a registration method based on contour mutual information. On the one hand, it uses a spatial characteristic, that the contours of tissue and organs in different modal have similarity. On the other hand, the method that fill the same gray values to the contour, which can add the correlation information among contour image and reduce the interference in image registration when image have missing information, thus improving precision of registration. The experiments show that the method of this paper can achieve more accurate registration results. But this paper has not considered nonlinear registration in multimodality medical image. We will make more research in further work.

REFERENCES

[1] Zitova B, Flusser J, Image registration methods a survey, Image and vision computing, pp: 977-1000, 2003.

[2] Loeckx D, Slagmolen P, Maes F, Vandermeulen D, Suetens P, Non-rigid image registration using conditional mutual information, IEEE transactions on medical imaging, pp:19-29, 2010.

[3] Nigris D, Collins D, Arbel T, Multi-modal image registration based on gradient orientations of minimal uncertainty, IEEE transactions on medical imaging, pp:2343-2354, 2012.

[4] Studholme C, Hill D L G, Hawkes D J, An overlap invariant entropy measure of 3D medical image alignment, Pattern recognition, pp: 71 -86, 1999.

[5] Pluim J P, Maintz J B, Viergever M A, Image registration by maximization of combined mutual information and gradient information, IEEE transactions on medical imaging, pp: 809-814,2000

[6] Butz T, Thiran J P, Affine registration with feature space mutual information, MICCAI 2001, LNCS 2208. 549-556.

[7] Rueckert D, Clarkson M J,Hill D L G, Hawkes D J, Non-grid registration using higher order mutual information, Proceeding of SPIE Medical Imaging 2000 : Image Processing, SanDiego, pp:438-447, 2000.

[8] Li CM, Xu CY, Gui CF, et,al. Level set evolution without re-initialization: A new variational formulation. In: Schmid C, Soatto S, Tomasi C, eds. Proc. of the 2005 IEEE Computer Society Conference. on Computer Vision and Pattern Recognition. Washington: IEEE Computer Society Press, pp: 430 – 436, 2005.

Prediction of New Student Numbers using Least Square Method

Dwi Mulyani
College of Informatics And Computer Management (STMIK) Banjarbaru
Banjarbaru Kalsel, Indonesia

***Abstract*—STMIK BANJARBARU has acquired less number of new students for the last three years compared to the previous years. The numbers of new student acquisition are not always the same every year. The unstable number of new student acquisition made the difficulty in designing classes, lecturers, and other charges. Knowing the prediction number of new student acquisition for the coming period is very important as a basis for further decision making. Least Square method as the method of calculation to determine the scores prediction is often used to have a prediction, because the calculation is more accurate then moving average.**

The study was aimed to help the private colleges or universities, especially STMIK BANJARBARU, in predicting the number of new students who are accepted, so it will be easier to make decisions in determining the next steps and estimating the financial matters.

The prediction of the number of new student acquisition will facilitates STMIK BANJARBARU to determine the number of classes, scheduling, etc.

From the results of the study, it can be concluded that prediction analysis by using Least Square Method can be used to predict the number of new students acquisition for the coming period based on the student data in the previous years, because it produces valid results or closer to the truth. From the test results in the last 3 years, the validity shows 97.8%, so it can be said valid.

Keywords—Prediction of New Students; Least Square method

I. INTRODUCTION

STMIK BANJARBARU as one of the colleges in the field of computer becomes one destination for new student candidates to continue their education. In the first year, the number of student candidates could be predicted since STMIK BANJARBARU was the only one computer college in South Kalimantan. However, in the last few years there are many other universities in South Kalimantan which provide department of computer science so it is assumed that the number of student candidates are divided into some computer universities in South Kalimantan. This causes the regression of the number of new students in the last three years. Based on the student data of STMIK BANJARBARU, the numbers of New Students accepted were 407 in 2009, 516 in 2010, 528 in 2011, 374 in 2012, 375 in 2013 and 386 in 2014. In 2012 to 2014, the number of new students accepted in STMIK BANJARBARU has decreased to 29.16%.

The problem faced by STMIK BANJARBARU is estimating the number of new students due to the regression of the number of new student acquisition in the last three years.

The study was aimed to help the private universities, especially STMIK BANJARBARU, in predicting the number of new students who are accepted, so it will be easier to make decisions in determining the next steps and estimating the financial matters. The prediction of the number of new student acquisition will facilitates STMIK BANJARBARU in determining the number of classes, setting schedules and others.

II. THEORITICAL BASIS

A. *Prediction or forecasting*

Prediction or forecasting is an important tool in an effective and efficient plan, especially in the economic field. In modern organizations, knowing the coming state is very important to look at the good or bad and aimed to prepare for the next activities (Rambe,2012).

According Heizer and Render (2009: 162), forecasting is the art and science to predict future events. This can be done by involving the retrieval of historical data and projected into the future with a form of mathematical models or predictions are subjective intuition, or using a combination of mathematical models that are tailored to the good judgment of a manager.

According Prasetya and Lukiastuti (2009: 43), forecasting is an attempt to predict the future state through state testing in the past. Forecasting relates to attempt to predict what happens in the future, based on the scientific method (science and technology) and carried out mathematically. However, forecasting activities are not solely based on scientific procedures or organized, because there is activity forecasting that uses intuition (feeling) or through informal discussions in a group.

According to Yamit, Forecasting is a prediction, projection or estimate of the level of an uncertain event in the future (Rambe 2012). According Makridakis, Forecasting is predictive values of a variable based on the known value of the variable or variables related (Rambe 2012).

According Pangestu Subagyo (1991: 1), forecasting is an activity / business to know (event) will happen in the future regarding a particular object by using experience / historical data. According to T Hani Handoko (1994; 260) Forecasting is an attempt to predict the future state through testing in the past.

From some of these explanations, it can be concluded that forecasting is a process or method of predicting an event that will occur in the future by basing it self on certain variables.

B. *Least Square method*

Least Square is a method for determining the approach polynomial function y = f (x) closest to the data (x1, y1) to (xn, yn). (Basuki 2014)

In the Collins English Dictionary says that the Least Square method is the best method to determine the value of an unknown quantity related to one or more sets of observations or measurements. (Harper Collins, 1991, 1994, 1998, 2000, 2003).

According to Dr. Setijo in his module entitled "Linear Regression with Least Squares Method" said that Least Square method is an approach method which is widely used for:

1) *Regression modeling based on the equation of the discrete data points*
2) *Analysis of measurement error (model validation) (Kristalina, 2015).*

This method is most often used to predict (Y), because the calculation is more accurate. The equation of the trend line to be searched is

$$Y = a0 + bx \quad a = (\Sigma Y) / n \qquad (1)$$
$$b = (\Sigma XY) / \Sigma x2 \qquad (2)$$

with:

Y = periodic data (time series) = estimated trend value.

a0 = trend value in the base year.
b = the annual average growth of value trend.
x = time variable (days, weeks, months or years).

To perform a calculation, it will require a specific value on the time variable (x), so that the amount of the time variable value is zero or $\Sigma x = 0$.

In this case,it will be devoted to discuss the analysis of time series with Least Square method which is divided into two cases, namely the even data case and the odd data case.

For odd n, then:

1) *The distance between the two time is rated one unit.*
2) *Above 0 is marked negative*
3) *Under 0is marked positive.*

For even n, then:

1) *The distance between the two time is rated two units.*
2) *Above 0 is marked negative*
3) *Under 0 is marked positive.*

In the data processing of odd data, the registration data of new student candidates for the past five years is required, which are the registration data in 2010 until the registration data in 2014. In the data processing of even data, the registration data of new student candidates for the past five years is required, which are the registration data in 2009 until the registration data in 2014.

In general, linear line equation of time series analysis is:

$$Y = a + b X. \qquad (3)$$

Information:
Y is the variable whichits trendis sought

X is the variable of time (years).

While, to find constant value (a) and parameter (b) is:

$$a = \Sigma Y / N \qquad (4)$$

and

$$b = \Sigma XY / \Sigma X2 \qquad (5)$$

III. System Analysis and Design

A. *Literature Review*

The previous study, conducted by Muhammad Ihsan Fauzi Rambe in 2012, examined the prediction of medicine supply using least square method which took the case study at Mutiara Hati Pharmacy, Medan. The study found that Least Square Method can be used to predict the medicine sales in the coming period based on the sales data in the previous year. Further, the analysis application can yield predictions and has minimized the forecast errors of the level of medicine sale in Pharmacies.

B. *Data Requirements*

The data required in the study is the data of Students accepted in STMIK BANJARBARU in 2009 to 2014.

TABLE I. Data of New Students of STMIK BANJARBARU

No.	Year	Total
1	2009	407
2	2010	513
3	2011	528
4	2012	374
5	2013	375
6	2014	385

(Source: PMB STMIK BANJARBARU)

IV. Results and Discussion

A. *Odd Data Case*

Before calculating the prediction of new student acquisition in 2015, some trialswere conducted in calculating the number of new student acquisition in 2012, 2013 and 2014 to determine the validity of the Least Square Method formula.

In calculating the prediction result of the number of students in 2012, the researcher used the student data in 2007 to 2011.

TABLE II. DATA OF NEW STUDENTS IN 2007 TO 2011

No.	Year	Total
1	2007	350
2	2008	512
3	2009	407
4	2010	513
5	2011	528

(Source: PMB STMIK BANJARBARU)

The next step is determining the values of variable X, XY and X^2.

TABLE III. DATA OF NEW STUDENTS IN 2007 TO 2011

No.	Year	Amount(Y)	X	XY	X^2
1	2007	350	-2	-700	4
2	2008	512	-1	-512	1
3	2009	407	0	0	0
4	2010	513	1	513	1
5	2011	528	2	1056	4
Total		2310		357	10

(Source: PMB STMIK BANJARBARU)

It is known that:

ΣY=2310

N=5

ΣXY=357

ΣX2=10

Then, to find the value of a:

a =ΣY/N

a =2310/5

a =462

And to find the value of b:

b =ΣXY/ΣX2

b=357/10

b=35.7

After the values of a and b are obtained, then the linear line equation is:

Y= a+bX

Y=462+(35.7) X

With the calculated equation of linear line, the number of new student in 2012 can be predicted:

Y=462+(35.7) X(For the year of 2012, the value of X is 3)

so that:

Y=462+(35.7 x3)

Y=462+107.1

Y=569.1 (6)

It means that the number of new student candidates who registered in 2012 was 569 people.

The next was calculating the result of the number of new student acquisition in 2013. The data used was the data of new student in 2008 to 2012.

TABLE IV. NEW STUDENT DATA IN 2008 TO 2012

No.	Year	Total
1	2008	512
2	2009	407
3	2010	513
4	2011	528
5	2012	374

(Source:PMB STMIK BANJARBARU)

Next was determining the values of X, XY and X^2.

TABLE V. NEW STUDENT DATA IN 2008 TO 2012

No.	Year	Amount	X	XY	X^2
1	2008	512	-2	-1024	4
2	2009	407	-1	-407	1
3	2010	513	0	0	0
4	2011	528	1	528	1
5	2012	374	2	748	4
Total		2334		-155	10

(Source: PMB STMIK BANJARBARU)

It is known that:

ΣY = 2334

N = 5

ΣXY = -155

ΣX^2 = 10

Then, to find the value of a:

a = ΣY / N

a= 2334/5

a= 466.8

And to find the value of b:

b= ΣXY/ ΣX^2

b= -155/10

b= -15.5

After the values of a and b are obtained, then the equation of linear line is:

Y = a + b X

Y = 466.8 + (-15.5)X

With the calculated linear line, it can be predicted that the number of new students in 2013 is:

Y = 466.8 + (-15.5) X (For the year of 2013, the value of X is 3)

Thus:

Y = 466.8– (15.5 x 3)

Y = 466.8 – 46.5

Y = 420 (7)

It means the number of candidates who registered in 2013was420. Next was calculating the number of new student acquisition in 2014. The data used was the student data in 2009 to 2013.

TABLE VI. DATA OF NEW STUDENT IN 2009 TO 2013

Year	Amount
2009	407
2010	513
2011	528
2012	374
2013	375
Total	2197

(Source:PMB STMIK BANJARBARU)

The next step is determining the variable values of X, XY and X^2.

TABLE VII. DATA OF NEW STUDENTS IN 2009 TO 2013

Year	Amount	X	XY	X^2
2009	407	-2	-814	4
2010	513	-1	-513	1
2011	528	0	0	0
2012	374	1	374	1
2013	375	2	750	4
Total	2197		-203	10

(Source: PMB STMIK BANJARBARU)

It is known that:

ΣY = 2197

N = 5

ΣXY = -203

ΣX^2 = 10

Then, to find the value of a:

a = ΣY / N

a= 2197/5

a= 439.4

And to find the value of b:

b= ΣXY/ ΣX^2

b= -203/10

b= -20.3

After the values of a and b are obtained, then the equation of linear line is:

Y = a + b X

Y = 439.4 + (-20.3)

With the calculated linear line, it can be predicted that the number of new students in 2014 is:

Y = 439.4+ (-20.3) X (For the year of 2014, the value of X is 3)

Thus:

Y = 439.4 – (20.3 x 3)

Y = 439.4 – 60.9

Y = 378.5 (8)

It means that the number of new student candidates who registered in 2014 was 378.

From the results of calculations in predicting the acquisition of the number of new students (6) (7) (8), it was found that in 2012 there was 569, in 2013 there was 420 and in 2014 there was 378. It is determined that if the deviation between the fact and the calculation with Least Square Method is >50 people, then the result is invalid. Compared to the tangible result obtained in 2012, the deviation is 34.2% (195 people), meaning that the result is invalid. In 2013, the deviation is 10% (45 people), meaning that the result is valid. In 2014, the deviation is 2.07% (8 people), meaning that the result is valid. From the three comparisons, it is found that two results are valid and one result is invalid. This means that the formula of Least Square Method is valid or closer to the truth. Next, the calculation would be performed to predict the number of new student acquisition in 2015. In the data processing of Odd Data case, the data of new students needed is the data from the last 5 years, from 2010 to 2014.

TABLE VIII. DATA OF NEW STUDENT REGISTRATION IN 2010 TO 2014

Year	Amount
2010	513
2011	528
2012	374
2013	375
2014	386
Total	2176

(Source: PMB STMIK BANJARBARU)

Then, determining the variable values of X, XY dan X^2.

TABLE IX. DATA OF NEW STUDENT REGISTRATION IN 2010 TO 2014

Year	Amount(Y)	X	XY	X^2
2010	513	-2	-1026	4
2011	528	-1	-528	1
2012	374	0	0	0
2013	375	1	375	1
2014	386	2	772	4
Total	2176		-407	10

(Source: PMB STMIK BANJARBARU)

Based on Table 3, the values of a and b will be discovered. To find the values of a and b:

It is known, that:

ΣY = 2176

N = 5

$\Sigma XY = -407$

$\Sigma X^2 = 10$

Then, to find the value of a:

$a = \Sigma Y / N$

a= 2176/5

a= 435,2

And to find the value of b:

$b = \Sigma XY / \Sigma X^2$

b= -407/10

b= -40.7

After the values of a and b are obtained, then to find the equation of linear line:

Y = a + b X

Y = 435.2 + (-40.7)

With the calculated equation of linear line, the number of new student in 2015 can be calculated:

Y = 435.2 + (-40.7) X (For the year of 2015, the value of X is 3)

So that:

Y = 435.2 – (40.7 x 3)

Y = 435.2 – 122.1

Y = 313.1 (9)

It means that the numbers of new student candidates who register are 313 people.

B. Even Data Case

TABLE X. DATA OF NEW STUDENT REGISTRATION IN 2009 TO 2014

Year	Amount
2009	407
2010	513
2011	528
2012	374
2013	375
2014	386
Total	2583

(Source: PMB STMIK BANJARBARU)

The next step is determining the variable values of X, XY and X^2.

TABLE XI. DATA OF NEW STUDENT REGISTRATION IN 2009 TO 2014

Year	Amount (Y)	X	XY	X^2
2009	407	-5	-2035	25
2010	513	-3	-1539	9
2011	528	-1	-528	1
2012	374	1	374	1
2013	375	3	1125	9
2014	386	5	1930	25
Total	2583		-673	70

(Source: PMB STMIK BANJARBARU)

Based on Table 4, the values of a and b will be discovered. To find the values of a and b:

It is known, that:

$\Sigma Y = 2583$

N = 6

$\Sigma XY = -673$

$\Sigma X^2 = 70$

Then, to find the value of a:

$a = \Sigma Y / N$

a= 2583 / 6

a= 430.5

And to find the value of b:

$b = \Sigma XY / \Sigma X^2$

b= -673/70

b= -9.6

After the values of a and b are obtained, then to find the equation of linear line:

Y = a + b X

Y = 430.5 + (-9.6)

With the calculated equation of linear line, the number of new student in 2015 can be calculated:

Y = 430.5 + (-9.6) X (For the year of 2015, the value of X is 7)

Thus:

Y = 430.5 – (9.6 x 7)

Y = 430.5 – 67.2

Y = 363 (10)

It means that the numbers of new student candidates who register are 363 people.

From the calculation using Least Square formula (9) (10), the results showed that the prediction of the number of new student acquisition in 2015 for Odd Data is 313 people and for Even Data is 363 people. But the calculation result of the numbers of new student prediction can be damaged or fell due to several reasons, for example because of changes in government regulations, regression of high school graduates,or other reasons.

V. CONCLUSIONS

Based on the results of the study, it can be concluded that:

1) Prediction or forecasting analysis using Least Square method can be used to predict the number of new students acquisition for the coming period based on the data of the previous years, because the results are valid or closer to the truth.

2) From the results of calculations in predicting the acquisition of the number of new students, it was found that in 2012 there was 569, in 2013 there was 420 and in 2014 there was 378. It is determined that if the deviation between the fact and the calculation with Least Square Method is >50 people, then the result is invalid. Compared to the tangible result obtained in 2012, the deviation is 34.2% (195 people), meaning that the result is invalid. In 2013, the deviation is 10% (45 people), meaning that the result is valid. In 2014, the deviation is 2.07% (8 people), meaning that the result is valid. From the 3 comparisons, it is found that 2 results are valid and 1 result is invalid. This means that the formula of Least Square Method is valid or closer to the truth.

REFERENCES

[1] Basuki, A. (2014). Metode Least Square. taken from i http://basuki.lecturer.pens.ac.id/lecture/numerik5.pdf

[2] Beny Mulyandi, Y. I. (2010). Sales Forecasting Analysis of Fuel type Premium at the pump Heroes Bandung Asri. National Conference: Design And Application Of Technology .

[3] Cahyo Adi Basuki, I. A. (2008). Analysis of Fuel Consumption In Steam Power Plant.

[4] Geer, S. A. (2005). Least Squares Estimation. Encyclopedia of Statistics in Behavioral Science, Volume 2, pp. 1041–1045 .

[5] Handoko, T. H. (1994). Dasar – Dasar Manajemen Produksi dan Operasi. Yogyakarta: BPFE.

[6] HarperCollins. (1991, 1994, 1998, 2000, 2003). Dictionary / Thesaurus. Taken from The Free Dictionary By Farlex: http://www.thefreedictionary.com/Least-squares+method

[7] Heizer, J. d. (2009). Manajemen Operasi, Edisi 9.

[8] Kosasih, S. (2009). Manajemen Operasi - Bagian Pertama. Edisi 1. 74.

[9] Kristalina, P. (2015). Metode Least Square.

[10] Mia Savira, Nadya N.K. Moeliono, S.SOS, MBA. (2014). Sales Forecasting Analysis of generic drugs bearing (OGB) At PT. Indonesia Farma.

[11] Rambe, M. I. (2012). Perancangan Aplikasi Peramalan Persediaan Obat Obatan Menggunakan Metode Least Square. Pelita Informatika Budi Darma, Volume : VI, Nomor: 1, Maret 2014

[12] Subagyo, P. (1999). Forecasting (Konsep dan Aplikasi). Yogyakarta.

[13] Sahara, Afni. (2013). Sistem Peramalan Persediaan Unit Mobil Mitsubishi Pada PT. Sardana Indah Berlian Motor Dengan Menggunakan Metode xponential Smoothing. Informasi dan Teknologi Ilmiah(INTI), Volume : I, Nomor : 1, Oktober 2013

[14] Tanojo, E. (2007.). DeriVation Of Moving Least Squares Approximation Shape Functions And ITS Derivatives Using The Exponential Weight Function. Civil Engineering Dimension, Vol 9, No 1, 19-24, March 2007 .

[15] Widodo, J. (2008). Ramalan Penjualan Sepeda Motor Honda pada pada CV. Mitra Roda Lestari. (xii + 32 + Lampiran).

3

Gram–Schmidt Process in Different Parallel Platforms

(Control Flow versus Data Flow)

Genci Berati
Tirana University, Department of Mathematics
Tirane, Albania

Abstract—**Important operations in numerical computing are vector orthogonalization. One of the well-known algorithms for vector orthogonalisation is Gram–Schmidt algorithm. This is a method for constructing a set of orthogonal vectors in an inner product space, most commonly the Euclidean space Rn. This process takes a finite, linearly independent set S = {b1, b2, ···, bk} vectors for k ⩽ n and generates an orthogonal set S1 = {o1, o2, ···, ok}. Like the most of the dense operations and big data processing problems, the Gram–Schmidt process steps can be performed by using parallel algorithms and can be implemented in parallel programming platforms. The parallelized algorithm is dependent to the platform used and needs to be adapted for the optimum performance for each parallel platform. The paper shows the algorithms and the implementation process of the Gram –Schmidt vector orthogonalosation in three different parallel platforms. The three platforms are: a) control flow shared memory hardware systems with OpenMP, b) control flow distributed memory hardware systems with MPI and c) dataflow architecture systems using Maxeler Data Flow Engines hardware. Using as single running example a parallel implementation of the computation of the Gram –Schmidt vector orthogonalosation, this paper describes how the fundamentals of parallel programming, are dealt in these platforms. The paper puts into evidence the Maxeler implementation of the Gram–Schmidt algorithms compare to the traditional platforms. Paper treats the speedup and the overall performance of the three platforms versus sequential execution for 50-dimensional Euclidian space.**

Keywords—Gram-Schmidt Algorithm; Parallel programming model; OpenMP; MPI; Control Flow architecture

I. Introduction

Classifications of parallel programming paradigms are mostly related to the hardware architectures.

The paradigms of parallel programming can be divided generally into two categories: process communicates [1] and problem decomposition [2].

Process communication is correlated to the instruments by which parallel processes communicate and share sources to each other. The most common forms of process interaction are shared memory and message passing between processes. Shared memory is an efficient instrument for passing data between programs by accessing that same shared memory. Algorithms may run on a single processor in sequential or on multiple separate processors in sequential way or in parallel. In shared memory model, parallel tasks share a global address space which they read and write to asynchronously. In shared memory systems the code can create threads each of them can access the same variable in parallel.

Message passing is a concept from computer science related mostly with distributed memory architectures for the parallel programming platforms that is used extensively in the design and implementation of modern software applications; it is very important for some models of concurrency and object-oriented programming. In a message passing model, parallel tasks exchange data and communicate through passing messages to one another. Either shared or distributed can be based Control Flow [3] Von Newman traditional architecture.

The paper deals with three different programming platforms (OpenMP, MPI and Maxeler). These three platforms can be grouped in two different architectures, in Control Flow (OpenMP and MPI) and Data Flow (Maxeler) architectures. These two different computing architectures are compared and analyzed in this paper by choosing a typical dense operations and big data problem which is the Gram – Schmidt process.

Is chosen Gram Schmidt classic algorithm for a 50-dimensional inner product space. The algorithm has operations rising in a significant progression from step to step. If we have a set S1={o1, o2, …, on} of orthogonal vectors as basis for the inner product space L, then we can express any vector of space L as a linear combination of the vectors in S1:

Let as have an arbitrary basis {b1, b2, … , bn} for an n-dimensional inner product space L. The Gram-Schmidt algorithm constructs an orthogonal basis {o1, o2, … , on} for L. In our paper we take the arbitrary basis {b1, b2, … , b50} for an 50-dimensional inner product pace L and after performing the Gram-Schmidt algorithm into a sequential machine platform, OpenMP platform, MPI platform and Maxeler controlfolw machine we than constructs an orthogonal basis {o1, o2, … , on} for L each time. The paper intends to compare the performance of the parallel platforms and to measure the speedup for each platform. The characteristics of the algorithms regards to the number of nested loops and the numbers of operations for iteration will define the best platform to recommend.

The reason why is selected the Gram – Schmidt algorithm is the time complexity. This algorithm complexity is $O(n^3)$. The operations in each iteration of the process rise progressively, so it is of large interest to study the behavior in different parallel programming platforms.

II. GRAM–SCHMIDT ALGORITHM

To obtain an orthonormal basis for an inner product space L, we use the Gram-Schmidt algorithm to construct an orthogonal basis. For R^n with the Euklidean inner product (dot product), we of course already know of the orthonormal basis {(1, 0, 0, …, 0), (0, 1, 0, … , 0), … , (0, … , 0, 1)}. For more abstract spaces, however, the existence of an orthonormal basis is not obvious. The Gram-Schmidt algorithm is powerful in that it not only guarantees the existence of an orthonormal basis for any inner product space, but actually gives the way of construction of such a basis.

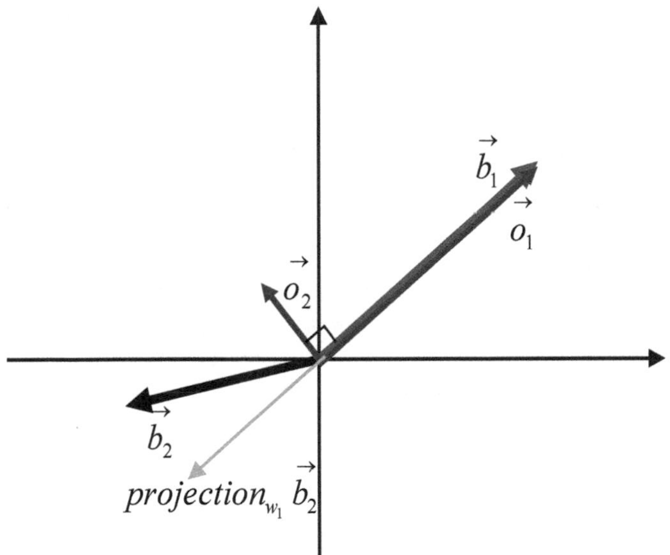

Fig. 1. Graphic representation of the Gram – Schmidt orthogonalisation

The Gram – Schmidt algorithm can be expressed in n steps to be performed. The algorithm steps are:

```
for i = 1 to n
vi = ai
for i = 1 to n
rii = ||vi||
qi = vi/rii
for j = i + 1 to n
rij = qi vj
vj = vj – rijqi
```

This algorithm is implemented in C++ code using Code::Blocks programming platform. This platform is chosen because it is portable to the parallel programming platforms.

III. GRAM – SCHMIDT VECTOR ORTHOGONALISATION ALGORITHM (SEQUENTIAL IMPLEMENTATION)

We implemented the steps mentioned in the previous section in the Code::Blocks [1] with C++ compiler. In our implementation, we take k=n=50, where k is the number of the linear independent vectors and n is the dimension of the Euclidian space. The C++ program code of Gram – Schmidt algorithm for a 50 dimensional inner product space, in our example named space L, for k=50, looks like:

```
#include <cstdlib>
#include <iostream>
#include <math.h>
using namespace std;
double b[50][50];
double r[50][50], q[50][50];
int main(int argc, char *argv[]) {
int i, j;
for (int i=0; i<50; i++)
for (int j=0; j<50; j++)
{b [i][j]=rand() % 10;}
int k;
for (k=0; k<50; k++){
r[k][k]=0; // equivalent to sum = 0
for (i=0; i<50; i++)
r[k][k] = r[k][k] + b[i][k] * b[i][k]; //rkk = sqr(a0k) + sqr(a1k) + sqr(a2k)
r[k][k] = sqrt(r[k][k]);  //
cout << endl << "R"<<k<<k<<": " << r[k][k];
for (i=0; i<3; i++)
{q[i][k] = b[i][k]/r[k][k];
cout << "q" <<i<<k<<": "<<q[i][k] << ", ";}
for(j=k+1; j<50; j++)
{r[k][j]=0; for(i=0; i<50; i++) r[k][j] += q[i][k] * b[i][j];
cout << endl << "r"<< k <<j<<": " <<r[k][j] <<endl;
for (i=0; i<50; i++) b[i][j] = a[i][j] - r[k][j]*q[i][k];
for (i=0; i<50; i++) cout << "b"<<j<<": " << b[i][j]<< ", "; }}
system("PAUSE");
return EXIT_SUCCESS;}
```

Fig. 2. Sequential Gram – Schmidt vector orthogonalisation. Program code in C++ (Code::Blocks)

The average execution time of this sequential algorithm is around 110 seconds. Now let's see in the next session the parallel implementation of this algorithm in OpenMP [2] parallel platform for C++.

IV. THE GRAM–SCHMIDT VECTOR ORTHOGONALISATION ALGORITHM FOR OPENMP PLATFORM

A parallel program is composed of parallel executing processes. A task-parallel model [4] focuses on processes , or threads of execution . These processes sometimes share the same sources, which emphasizes the need for communication between those processes. Task parallelism is a natural way to express message-passing communication between processing units. It is usually classified as MIMD/MPMD or MISD [5].

A parallel model consists of performing operations on a data set which usually regularly structures in an array. A set of tasks will operate on this data, but independently on separate partitions. In a shared memory system, the data will be accessible to all tasks, but in a distributed-memory system it will divide between memories.

Parallelism is usually classified as SIMD/SPMD (Single Instruction-Multiple Data)/(Single Program-Multiple Data) [6].

[1] Code::Blocks, "A free C, C++ and Fortran IDE". http://www.codeblocks.org/

[2] OpenMP Specifications. http://www.openmp.org/blog/specifications

The systems are categorized into two categories. [7] The systems of the first category were characterized by the isolation of the abstract design space seen by the programmer from the parallel, distributed implementation. In this, all processes are presented with equal access to some kind of shared memory space. In its loosest form, any process may attempt to access any item at any time. The second category considers machines in which the two levels are closer together and in particular, those in which the programmer's world includes explicit parallelism [8]. This category discards shared memory based cooperation in favor of some form of explicit message passing.

A classical shared memory parallel platform is OpenMP. OpenMP (Open Multiprocessing) is an API that supports multi-platform shared memory multiprocessing programming in C, C++ and Fortran programming language. OpenMP [3] is an application program interface providing a multi-threaded programming model for shared memory parallelism; it uses directives to extend sequential languages It consists of a set of compiler directives, library routines, and environment variables that influence run-time behavior.

OpenMP uses a portable, scalable model that gives programmers a simple and flexible interface for developing parallel applications for platforms like a standard desktop computer or supercomputer. After the configuration of the Code:: Blocks for OpenMP, we implemented on this platform our algorithm. The code is parallelized for our nested loops as below:

```
#include <cstdlib>
#include <iostream>
#include <math.h>
#include <omp.h>
using namespace std;
double b[50][50];
double r[50][50], q[50][50];
int main(int argc, char *argv[]) {
int i, j;
#pragma omp parallel for
for (int i=0; i<50; i++)
for (int j=0; j<50; j++)
{b [i][j]=rand() % 10;}
int k;
#pragma omp parallel for
for (k=0; k<50; k++){
r[k][k]=0; // equivalent to sum = 0
for (i=0; i<50; i++)
r[k][k] = r[k][k] + b[i][k] * b[i][k]; //rkk = sqr(a0k) + sqr(a1k) + sqr(a2k)
r[k][k] = sqrt(r[k][k]); //
cout << endl << "R"<<k<<k<<": " << r[k][k];
#pragma omp parallel for
for (i=0; i<3; i++)
{q[i][k] = b[i][k]/r[k][k];
cout << "q" <<i<<k<<": "<<q[i][k] << ", ";}
for(j=k+1; j<50; j++)
{r[k][j]=0; for(i=0; i<50; i++) r[k][j] += q[i][k] * b[i][j];
cout << endl << "r"<< k <<j<<": " <<r[k][j] <<endl;
for (i=0; i<50; i++) b[i][j] = b[i][j] - r[k][j]*q[i][k];
for (i=0; i<50; i++) cout << "b"<<j<<": " << b[i][j]<< ", "; }}
system("PAUSE");
return EXIT_SUCCESS;}
```

Fig. 3. OpenMP Gram – Schmidt vector orthogonalisation. Program code in C++ (Code::Blocks)

For directive in the code above splits the for-loop, so each thread in the current team handles a different portion of the loop. The main directive used is "#pragma omp parallel for". This statement is used to open the switch of OpenMP in this algorithm code. Only small changes in C++ sources code are required in order to use OpenMP. So each thread gets a different section of the loop, and they execute their own sections in parallel. We executed this code in the quad core computer where we before executed the sequential algorithm. The average execution time is 30 seconds. Significant speedup is reached for the Gram Schmidt algorithm when we use OpenMP parallel features. By trying this C++ code for OpenMP in PC with different number of cores and the execution time is shown in the table 1, is made a speedup analysis.

V. The Gram–Schmidt Vector Orthogonalisation in MPI

Message Passing Interface (MPI) is a standardized and portable message-passing system designed. The standard defines the syntax and semantics of a core of library routines useful to a wide range of users writing portable message-passing programs [9] in different programming languages such as Fortran, C, C++ and Java. Message passing this model uses communication libraries to allow efficient parallel programs to be written for distributed memory systems. These libraries provide routines to initiate and configure the messaging environment as well as sending and receiving data packets. Currently, the most popular high-level message-passing system for scientific and engineering applications is MPI (Message Passing Interface)[4].

We executed the code in a cluster with four computers with the same parameters like the quad core in which we executed the sequential algorithm and OpenMP code. The average time of execution is 22 seconds. The C++ code adopted for using the MPI library for parallelization of our Gram Schmidt vector orthogonalisation algorithm. Both platforms OpenMP and MPI are control flow based architectures. The figure 2 below show the control flow design architecture.

[3] OpenMP Specifications. http://www.openmp.org/blog/specifications/

[4] Message Passing Interface. http://www-unix.mcs.anl.gov/mpi/index.html

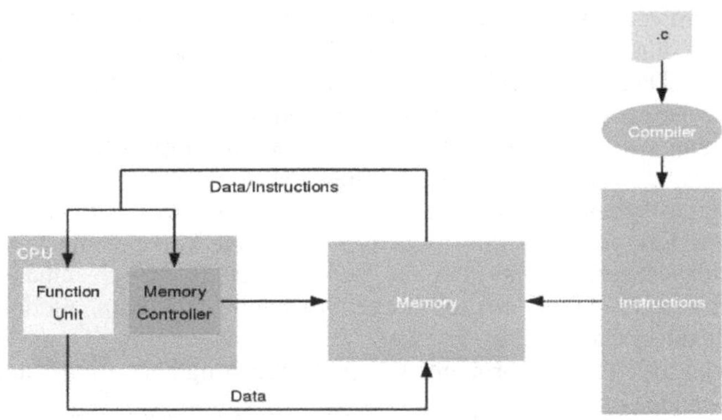

Fig. 4. Control Flow architecture design (source Maxeler)

```
#include <cstdlib>
#include <iostream>
#include <math.h>
#include <mpi.h>
using namespace std;
double b[50][50];
double r[50][50], q[50][50];
int main(int argc, char *argv[]) {
int i, j, rank, nrprocs, count, start, stop;
MPI_Init(&argc, &argv);
// get the number of processes, and the id of this process
MPI_Comm_rank(MPI_COMM_WORLD, &rank);
MPI_Comm_size(MPI_COMM_WORLD, &nprocs);
// we want to perform 50 iterations in total. Work out the
// number of iterations to perform per process...
for (int i=0; i<50; i++)
for (int j=0; j<50; j++)
{b [i][j]=rand() % 10;}
int k;
for (k=0; k<50; k++){
r[k][k]=0; // equivalent to sum = 0
for (i=0; i<50; i++)
r[k][k] = r[k][k] + b[i][k] * b[i][k]; //rkk = sqr(a0k) + sqr(a1k) + sqr(a2k)
r[k][k] = sqrt(r[k][k]); //
cout << endl << "R"<<k<<k<<": " << r[k][k];
for (i=0; i<3; i++)
{q[i][k] = b[i][k]/r[k][k];
cout << "q" <<i<<k<<": "<<q[i][k] << ", ";}
for(j=k+1; j<50; j++)
{r[k][j]=0; for(i=0; i<50; i++) r[k][j] += q[i][k] * b[i][j];
cout << endl << "r"<< k <<j<<": " <<r[k][j] <<endl;
for (i=0; i<50; i++) b[i][j] = a[i][j] - r[k][j]*q[i][k];
for (i=0; i<50; i++) cout << "b"<<j<<": " << b[i][j]<< ", "; MPI_Finalize();}}
system("PAUSE");
return EXIT_SUCCESS;}
```

Fig. 5. MPI Gram – Schmidt vector orthogonalisation. Program code in C++ (Code::Blocks)

VI. The Gram–Schmidt Vector Orthogonalisation in Maxeler

Dataflow architecture [10] is a computer architecture that differs in significant contrasts to the traditional Control Flow - Von Neumann architecture. Dataflow architectures do not have a program counter, or (at least conceptually) the executability and execution of instructions is based on the availability of input arguments to the instructions, so that the order of instruction execution is unpredictable: I. e. behavior is undetermined [5].

Dataflow machines have been around for more than two decades. Implementation challenges left the technology hidden for many years, but last five years the data flow parallel programming is becoming more and more a technological reality. One of the dataflow machines is the Manchester Dataflow Machine (MDFM) using single-assignment language SISAL. Another successful dataflow machine is Maxeler machine. The Gram – Schmidt algorithm is implemented to the MPC-X Series [6] machine.

The implementation process of our algorithm includes the adaption of the C++ host code for export to the MPC module. The Data Flow Engine (DFE) part of an accelerated solution itself contains two components: one or more Kernels, responsible for the data computations; and a single Manager, which orchestrates global data movement for the CPUs, DFEs and Kernels+Memory inside. Hence, accelerating an application requires the user to write three program parts: Kernel(s), A Manager, and a CPU application. The Kernel and the Manager is created by writing programs in MaxJ: an extended form of Java adding operator overloading. Using MaxCompiler requires only minimal familiarity with Java. A developer executes a MaxCompiler-based program to produce a ".max file" containing the DFE configuration, meta-data and SLiC functions. The CPU application is compiled and linked with the .max file, SLiC and MaxelerOS, to create the application executable.

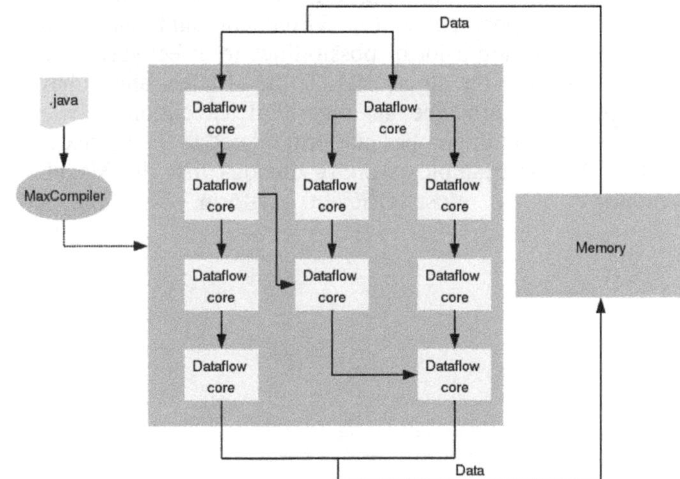

Fig. 6. Data Flow architecture design (source Maxeler)

[5] http://en.wikipedia.org/wiki/Dataflow_architecture, Retrieved on 21 April 2015

[6] https://www.maxeler.com/products/mpc-xseries/, Retrieved on 20 January 2015.

VII. RESULTS

This paper was dealing the behavior of parallel Gram – Schmidt vector orthogonalization algorithms with respect to OpenMP, MPI and Maxeler platform. The results we found are satisfactory. The number of input data size increased Maxeler gives very good performance. Nevertheless, the performance factor presented here is the execution times and speedup of the implementations for same input data size realized in the parallel programming models.

The speedup achieved by a parallel application varies for different programming models. The models chosen in this paper are only considered from the speedup perspective.

The results of the execution time of the algorithms in three machines are shown in figure 7. In this figure is shown the time in seconds in sequential (column 2) in OpenMP (column 3), MPI (column 4) which are Control Flow based architecture. In the figure 7 is shown also the speedup reached in OpenMP (column 5), speedup reached in MPI (column 6) and the speedup reached in the Maxeler machine (column 7). In Maxeler machine, the speedup is high, but limited and independent to the memory performance.

Memory	(Sequential Program) Execution Time	(OpenMP Program) Execution Time	(MPI Program) Execution Time	Speed up Seq/OpenMP	Speed up Seq/MPI	Speed up Seq/Maxeler
1MB	120	60	55	2.00	2.18	20
2MB	90	28	30	3.21	3.00	20
4MB	70	20	25	3.50	2.80	20
8MB	60	15	23	4.00	2.61	20

Fig. 7. Execution time analysis

VIII. CONCLUSIONS

For Gram – Schmidt vector orthogonalisation the parallel approach demands rethinking algorithms, adaption of the programming approach and environment and underlying hardware. There are a lot of possibilities to effectively create parallel version of the algorithms. To be efficient and to have the optimal performance in algorithm execution is very important to select the proper platform related to the contextual problem. In our example is pretty obvious that the Maxeler technology is the most efficient platform. The Maxeler machine spends some initial time for transfer from host to DFE, but control time is extremely slow compare to processing time. Data Flow architecture offers significant capabilities to accelerate scientifically numerical computations, such as the Gram – Schmidt vector orthogonalisation. Improvements in Maxeler and bus technology indicate that Data Flow will increase their lead over general purpose processors over the next few years. In this paper is shown that Maxeler machine with its software system are not wedded to von Neumann architectures nor to the von Neumann execution model. Maxeler platform works very well for calculating Gram – Schmidt vector orthogonalisation reaching a significant speedup.

This paper is addressed to the programmers, by providing taxonomy of parallel language designs. They can decide which language to use for contest of their project.

REFERENCES

[1] Gregory r. Andrews, ACM Computing Surveys, "Paradigms for Process Interaction in Distributed Programs", Vol 23, No 1, March 1991, page 50-52

[2] Stephen Boyd, Lin Xiao, and Almir Mutapcic, Notes for EE392o, Stanford University, Autumn, 2003: , "Notes on Decomposition Methods", page 1

[3] Micah Beck, Keshav Pingali, Department of Computer Science Cornell University, Ithaca, NY 14853, "From Control Flow To Dataflow", page 2

[4] Dounia Khaldi, Pierre Jouvelot, Corinne Ancourt and Franc¸ois Irigoin, " Task Parallelism and Data Distribution: An Overview of Explicit Parallel Programming Languages", page 10-16

[5] W, F, McColl, "A General Model of Parallel Computing", programming research group, Oxford University materials, page 6

[6] Mark A. Nichols, Howard Jay Siegel, Henry G. Dietz, "Data Management and Control-Flow Aspects of an SIMD/SPMD Parallel Language/Compiler", page 224-225

[7] Christoph Kessler, Jörg Keller, "Models for Parallel Computing: Review and Perspectives", PARS, Mitteilungen, December 2007, ISBN 0177-0454, page 3

[8] Manar Qamhieh, Serge Midonnet, "Experimental Analysis of the Tardiness of Parallel Tasks in Soft Real-time Systems", 18th Workshop on Job Scheduling Strategies for Parallel Processing (JSSPP), May 2014, page 5-6

[9] Board of Trustees of the University of Illinois, "Introduction to MPI", 2001 page 16

[10] ARTHUR H. VEEN, Center for Mathematics and Computer Science "Dataflow Machine Architecture", ACM Computing Surveys, December 1986, Vol. 18, No. 4, December 1986, page 2

Methods for Wild Pig Identifications from Moving Pictures and Discrimination of Female Wild Pigs based on Feature Matching Methods

Kohei Arai 1
Graduate School of Science and Engineering
Saga University
Saga City, Japan

Indra Nugraha Abdullah 2
2 Jakarta Office, Yamaha Co. Ltd.
Jakarta, Indonesia

Kensuke Kubo 3
3 Fujitsu Kyushu Network Technologies, Ltd.
Fukuoka Japan

Katsumi Sugawa 3
3 Fujitsu Kyushu Network Technologies, Ltd.
Fukuoka Japan

***Abstract*—Methods for wild pig identifications and discrimination of female wild pigs based on feature matching methods with acquired Near Infrared: NIR moving pictures are proposed. Trials and errors are repeated for identifying wild pigs and for discrimination of female wild pigs through experiments. As a conclusion, feature matching methods with the target nipple features show a better performance. Feature matching method of FLANN shows the best performance in terms of feature extraction and tracking capabilities.**

Keywords—OpenCV; Canny filter; Template matching; Feature matching

I. INTRODUCTION

Wildlife damage in Japan is around 23 Billion Japanese Yen a year in accordance with the report from the Ministry of Agriculture, Japan. In particular, wildlife damages by deer and wild pigs are dominant (10 times much greater than the others) in comparison to the damage due to monkeys, bulbuls (birds), rats. Therefore, there are strong demands to mitigate the wildlife damage as much as we could. It, however, is not so easy to find and capture the wildlife due to lack of information about behavior. For instance, their routes, lurk locations are unknown and not easy to find. Therefore, it is difficult to determine the appropriate location of launch a trap.

The purpose of this research work is to identify the wildlife, in particular, wild pigs for mitigation of wildlife damage. In particular, it is effective to capture female wild pigs (wild boar lays the child) for mitigation of wildlife damage. Therefore, there are very strong demands of capturing female wild pigs.

In order to identify the wild pigs and discriminate female wild pigs from the moving pictures acquired with Near Infrared: NIR camera, computer vision of technologies are utilized. First, target of wild pigs is attempted to extract from the moving pictures. Contour extraction and edge extraction are attempted. Secondly, background and target are attempted to separate. Using a template of nipple image (a small portion of image), discrimination of female wild pigs is attempted. Then feature matching methods are used for female wild pig discriminations with nipple features acquired from the moving pictures.

The following section describes research background followed by the proposed methods for wild pig identification and discrimination of female wild pigs. Then experiments are described followed by conclusion with some discussions.

II. RESEARCH BACKGROUND

According to the West, B. C., A. L. Cooper, and J. B. Armstrong. 2009. Managing wild pigs: A technical guide. Human-Wildlife Interactions Monograph 1:1–55[1], there are the following wild pig damages,

Ecological

Impacts to ecosystems can take the form of decreased water quality, increased propagation of exotic plant species, increased soil erosion, modification of nutrient cycles, and damage to native plant species [1]-[5].

Agricultural Crops

Wild pigs can damage timber, pastures, and, especially, agricultural crops [6]-[9].

Forest Restoration

Seedlings of both hardwoods and pines, especially longleaf pines, are very susceptible to pig damage through direct consumption, rooting, and trampling [10]-[12].

Disease Threats to Humans and Livestock

Wild pigs carry numerous parasites and diseases that potentially threaten the health of humans, livestock, and wildlife [13]-[15].

Humans can be infected by several of these, including diseases such as brucellosis, leptospirosis, salmonellosis, toxoplasmosis, sarcoptic mange, and trichinosis. Diseases of

[1] www.berrymaninstitute.org/publications,

significance to livestock and other animals include pseudorabies, swine brucellosis, tuberculosis, vesicular stomatis, and classical swine fever [14], [16]-[18].

There also are some lethal techniques for damage managements. One of these is trapping. It is reported that an intense trapping program can reduce populations by 80 to 90% [19]. Some individuals, however, are resistant to trapping; thus, trapping alone is unlikely to be successful in entirely eradicating populations. In general, cage traps, including both large corral traps and portable drop-gate traps, are most popular and effective, but success varies seasonally with the availability of natural food sources [20]. Cage or pen traps are based on a holding container with some type of a gate or door [21].

III. Proposed Method and System

A. Proposed System

Fig.1 shows an example of the system for trapping and capturing of wild pigs which consists of the trap cage and the video camera.

Fig. 1. Proposed system for trapping and acquiring moving picture of wild pigs

In the trap cage, there is bait. When wild pigs get in the trap cage, ultrasonic sensor sensed them. Then the entrance doors are shut downed. These processes are monitored and captured with the near infrared video camera with near infrared Light Emission Diode: LED. Because wild pigs are active in nighttime, Near Infrared: NIR camera with NIR LED is used.

The proposed system for trapping of wild pigs and for capturing their moving pictures is illustrated in Fig.2.

There are two ultrasonic sensors which are attached at the front and the back ends of the cage. When wild pigs get in the cage, then they are sensed with the ultrasonic sensors. Meantime, trap obstruction is activated. Drop-gates are then shut downed immediately after they are sensed with the ultrasonic sensors. Thus the wild pigs are trapped in the cage. These processes are monitored and captured with NIR camera with NIR LED. The captured moving pictures are transmitted through Bluetooth and then the transmitted moving pictures are transferred to the data collection center through WiFi networks or LAN. There are sensor control and battery box as well as solar panel for electricity supply.

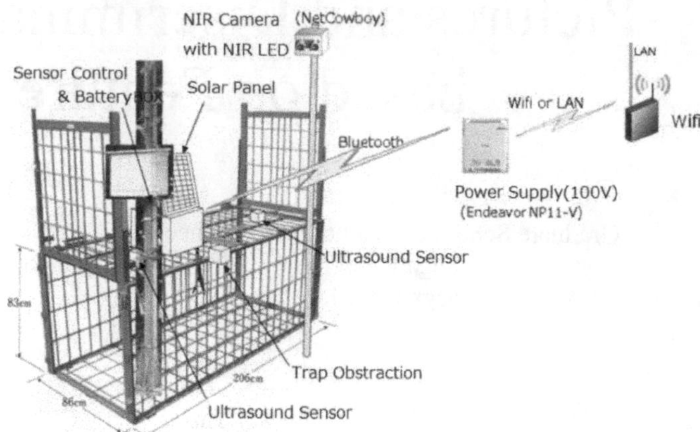

Fig. 2. Proposed system for trapping of wild pigs and for capturing their moving pictures

Outlooks of the NIR camera (NetCowboy) with NIR LED and ultrasonic sensors are shown in Fig. 3 (a) and (b), respectively. Meanwhile, specifications of these camera and sensor are shown in Table 1 and 2, respectively

TABLE I. Specification of NIR Camera (NetCowboy)

Pixel	1.3 M
Resolution	1280×1024
Frame rate	1280 x 1024 : 7.5fps, 640 x 480 : 30fps
Dimension	52mm (W) × 65mm (D) × 70mm (H)
Weight	85g
Operating condition	0 - 40deg.C
Interface	USB 2.0
IR Illumination	7 NIR LED

TABLE II. Features of Ultrasonic Range Finder[2]

Supply Voltage	5 V (DC)
Supply Current	30 mA (Typ), 35 mA (Max)
Range	3cm to 3m
Input Trigger	Positive TTL pulse, 2μS min, 5μS (Typ)
Echo Pulse	Positive TTL pulse, 115 μS to 18.5 mS
Echo Hold-off	750 μS from fall of Trigger pulse
Burst Frequency	40 kHz for 200 μS
Delay before next measurement	200 μS
Dimension	22 mm H x 46 mm W x 16 mm D

(a)NIR camera

(b)Ultrasonic sensor

Fig. 3. Outlook of NIR camera with NIR LED and ultrasonic sensor used in the proposed system for trapping and capturing of wild pigs

[2] Parallax ultrasonic sensor, http://www.parallax.com/dl/docs/prod/acc/PingDocs.pdf

B. *Proposed Moving Picture Analysis Methods*

Moving pictures are acquired with high resolution mode of 1280 by 1024 pixels. Therefore, frame rate is 7.5 fps. OpenCV is used for acquisition, processing, and analysis because it is totally easy to use. OpenCV is an open source computer vision library which is written in C and C++ and runs under Linux, Windows, and Mac OS X. It can be downloaded from *http://sourceforge.net/projects/opencvlibrary*

There so many library software for image processing and analysis. First, object has to be extracted from the moving picture. Then object contour has to be extracted. For the contour extraction and tracing, Canny filter related spatial filters are attempted. After that, it would be better to remove the background. The following background removals is attempted,

cv2.createBackgroundSubtractorMOG()

In order to discriminate female wild pigs, template matching method is applied with a template of small portion of nipple images. The following correlation functions are attempted for template matching,

CV_TM_SQDIFF , CV_TM_SQDIFF_NORMED , CV_TM_CCORR , CV_TM_CCORR_NORMED , CV_TM_CCOEFF, CV_TM_CCOEFF_NORMED

Also feature matching methods are applied for discrimination of female wild pigs. There are many feature matching methods in the OpenCV library. A couple of feature matching methods are attempted for the discriminations. The followings are typical feature matching methods which are provided from OpenCV,

- BruteForce
- BruteForce-L1
- BruteForce-SL2
- BruteForce-Hamming
- BruteForce-Hamming(2)
- FlannBased

The FlannBasedMatcher interface is used in the proposed method in order to perform a quick and efficient matching by using the FLANN (*Fast Approximate Nearest Neighbor Search Library*). Also Brute-Force matcher which is simple matching method is used in the proposed method. It takes the descriptor of one feature in first set and is matched with all other features in second set using some distance calculation. For both, feature descriptor is needed. Speeded-up Robust Feature: SURF is used in the proposed method.

IV. EXPERIMENTS

A. *Preliminary Image Processing*

One shot image of the acquired moving pictures is shown in Fig.4 as an example. This is a female wild pig on the route from habitat area to go to the calms feed. Wild boar children are followed by the female wild pig. By using the difference between the current and the previous frame of wild pig (targeted object), it is possible to extract the female wild pig. Also, it is possible to remove the background by frame by frame. Fig.5 shows the resultant image of the background removals.

Edge and contour extractions are attempted with Canny and sharp Canny filters. Fig.6 (a) shows the resultant image of Canny filter while Fig.6 (b) shows that of the sharp Canny filter. In the process, lower and higher thresholds are adequately set obviously. Through a comparison between Fig.6 (a) and (b), sharp Canny filter seems superior to Canny filter. It, however, is not sufficient for extraction.

Fig. 4. Portion of original image of the targeted object of female wild pig in concern

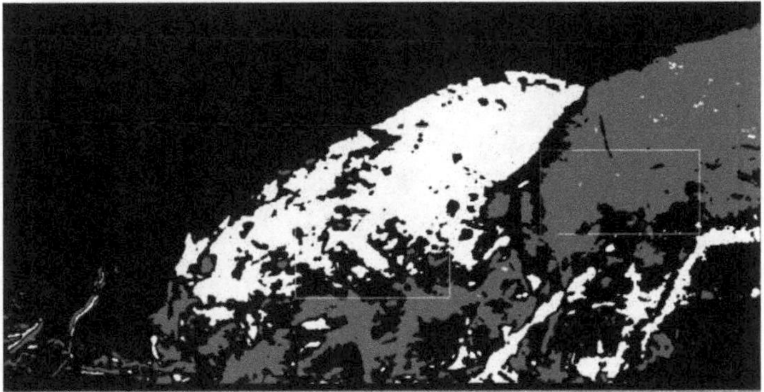

Fig. 5. Resultant image of background removal from the original image in frame by frame basis

(a)Canny filter

(b)Sharp Canny filter

Fig. 6. resultant images of edge and contour extractions

B. Descrimination of Female Wild pigs

Secondly, discrimination of female wild pigs is attempted with template matching and feature matching. Fig.7 shows an example of template image of nipple which indicates female wild pigs.

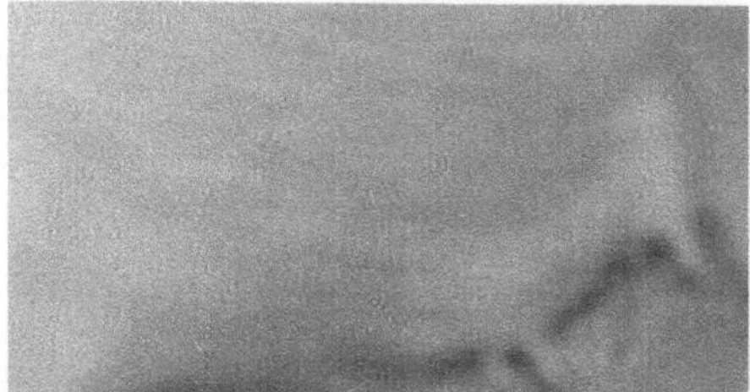

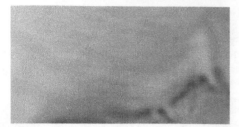

Fig. 7. Template image of nipple which is an indicator of female wild pigs

By using template matching software which is provided by OpenCV, nipple feature is matched and tracked. An example of template matching image with template image is shown in Fig.8. It seems does work well for female wild pig discrimination and tracking. It, however, does not work so well when the wild pig moves so fast and the portion of nipple is occluded and disappeared which are shown in Fig.9 (a) and (b), respectively. Also influence due to the both target moving speed and occlusion of the different object behind the other object is shown in Fig.9 (c).

Fig. 8. Example of resultant image of template matching with the template image of nipple portion of image

(a)Move so fast

(b)Nipple portion cannot be seen

(c)Cub behind male adult wild pig

Fig. 9. Influences due to target speed and occlusion in template matching

Other than these, feature matching methods are attempted for discrimination of female wild pigs. In order to describe the feature of female wild pigs, Scale-Invariant Feature Transform: SIFT and SURF based feature descriptors are used for representation of nipple features. SURF descriptor based feature matching and tracking is attempted. Fig.10 shows an example of the resultant image of SURF feature matching.

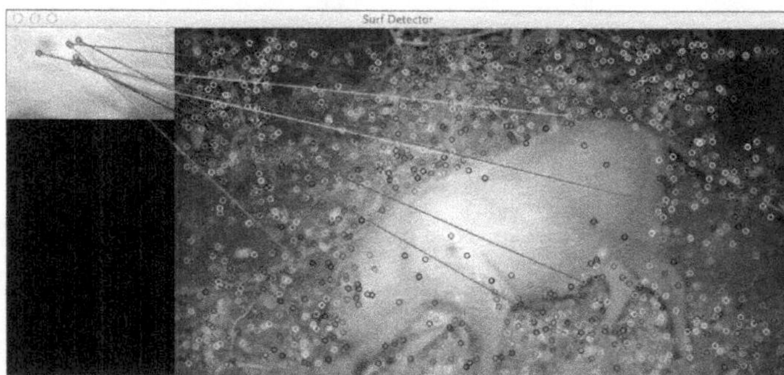

Fig. 10. Resultant image of SURF based feature description and feature matching

The SURF based feature description and feature matching is not good enough for feature tracking. Sometime it works well, but it does not work well as shown in Fig.10. Therefore, another feature matching methods are attempted after that.

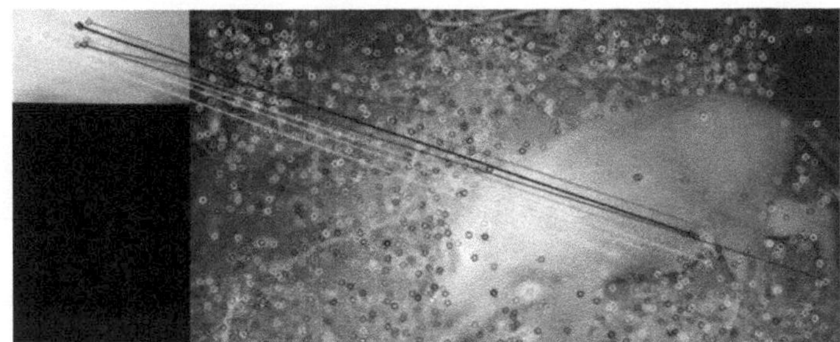

Fig. 11. Example of the resultant image of FLANN

Those are SURF Brute-Force and FLANN. The performance of discrimination is almost similar between both. Fig.11 shows an example of the resultant image of FLANN.

Nipple of features is matched so well. In particular, tracking capability of the FLANN feature matching is superior to the other template matching and SURF matching as well as SURF Brute-Force matching.

C. *Proposed System for Wild Pig Montoring Hardware System*

One of the issues for damage management due to wild pigs is how to count the number of female wild pigs in the area in concern. Although the proposed methods and systems above work well, the hardware system is costly. The hardware system proposed here is cheap version of the system for monitoring the number of female wild pigs. Because the areas where suffers from wild pig damage are situated almost all over the Japanese island. Such situation is common to the countries in the world. Therefore, the cheap version of hardware system for monitoring is required.

Android tablet terminal which equipped communication capability (Bluetooth, WiFi) and camera is not so expensive. For instance, Android tablet terminal of KEIAN M716S V2 with 7 inches display does cost about 7280 Japanese Yen. Major specification and outlook of the Android tablet terminal is shown in Table 3.

TABLE III. SPECIFICATION OF ANDROID TABLET TERMINAL OF KEIAN M716S V2

Dimension	OS	weight
800x480	Android 4.4.2	305g

NIR LED is also cheap. Therefore, one set of wild pig monitor does cost about 10000 Japanese Yen. The length of the route in the area in concern is a couple of hundred meters. Therefore, 30 sets of the monitoring system would cover the entire route of wild pigs. The total cost of the hardware system an area is around 300000 Japanese Yen. Obviously it is cheaper than damage cost.

Event driven application software is installed in the Android tablet terminal. When relatively large changes are detected in the current frame compared to the previous frame, the event driven software is activated. Then target object is detected with target extraction and contour extraction. Then size of target object is measured from the contour. Discrimination between adult and cub is done depending on the measured size. After that discrimination between male and female is done depending on presence or absence of nipple.

Along with the suspicious route of wild pigs, the monitoring hardware systems are set every 10 meters. The number of incoming and outgoing wild pigs is counted for each block with 10 meters long. Thus total number of wild pigs can be estimated.

V. CONCLUSION

Methods for wild pig identifications and discrimination of female wild pigs based on feature matching methods with acquired Near Infrared: NIR moving pictures are proposed. Trials and errors are repeated for identifying wild pigs and for discrimination of female wild pigs through experiments. As a conclusion, feature matching methods with the target nipple features show a better performance. Feature matching method of FLANN shows the best performance in terms of feature extraction and tracking capabilities.

Further study is required for wide area of spatial distribution of wild pigs. Spatial distribution of wild pigs in a relatively small size of area in concern can be estimated by the proposed system and method. Kiriging can be used for a much wide area in concern using estimated the number of wild pigs of the small size of areas.

ACKNOWLEDGMENT

The authors would like to thank Mr. Kenji Egashira, Dr. Herman Tolle of Arai's laboratory members for their useful comments and suggestions during this research works.

REFERENCES

[1] Patten, D. C. 1974. Feral hogs — boon or burden. Proceedings of the Sixth Vertebrate Pest Conference 6:210–234.

[2] Singer, F. J., W. T. Swank, and E. E. C. Clebsch. 1984. Effects of wild pig rooting in a deciduous forest. Journal of Wildlife Management. 48:464–473.

[3] Stone, C. P., and J. O. Keith. 1987. Control of feral ungulates and small mammals in Hawaii's national parks: research and management strategies. Pages 277–287 *in* C. G. J. Richards and T. Y. Ku, editors. Control of mammal pests. Taylor and Francis, London, England, and New York and Philadelphia, USA.

[4] Cushman, J. H., T. A. Tierney, and J. M. Hinds. 2004. Variable effects of feral pig disturbances on native and exotic plants in a California grassland. Ecological Applications 14:1746–1756.

[5] Kaller, M. D., and W. E. Kelso. 2006. Swine activity alters invertebrate and microbial communities in a coastal plain watershed. American Midland Naturalist 156:163–177.

[6] Bratton, S. P. 1977. The effect of European wild boar on the flora of the Great Smoky Mountains National Park. Pages 47–52 *in* G. W. Wood, editor. Research and management of wild hog populations. Belle W. Baruch Forest Science Institute, Clemson University, Georgetown, South Carolina, USA.

[7] Lucas, E. G. 1977. Feral hogs — problems and control on National Forest lands. Pages 17–22 *in* G. W. Wood, editor. Research and management of wild hog populations. Belle Baruch Forest Science Institute, Clemson University, Georgetown, South Carolina, USA.

[8] Thompson, R. L. 1977. Feral hogs on National Wildlife Refuges. Pages 11–15 *in* G. W. Wood, editor. Research and management of wild hog populations. Belle W. Baruch Forest Science Institute, Clemson University, Georgetown, South Carolina, USA.

[9] Schley, L, and T. J. Roper. 2003. Diet of wild boar *Sus scrofa* in Western Europe, with particular reference to consumption of agricultural crops. Mammal Review 33:43–56.

[10] Whitehouse, D. B. 1999. Impacts of feral hogs on corporate timberlands in the southeastern United States. Pages 108–110 *in* Proceedings of the Feral Swine Symposium, June 2–3, 1999, Ft. Worth, Texas, USA.

[11] Mayer, J. J., E. A. Nelson, and L. D. Wike. 2000. Selective depredation of planted hardwood seedlings by wild pigs in a wetland restoration area. Ecological Engineering, 15(Supplement 1): S79–S85.

[12] Campbell, T. A., and D. B. Long. 2009. Feral swine damage and damage management in forested ecosystems. Forest Ecology and Management 257:2319–2326

[13] Forrester, D. J. 1991. Parasites and diseases of wild mammals in Florida. University of Florida Press, Gainesville, Florida, USA.

[14] Williams, E. S., and I. K. Barker. 2001. Infectious diseases of wild mammals. Iowa State University Press, Ames, Iowa, USA.

[15] Sweeney, J. R., J. M. Sweeney, and S. W. Sweeney. 2003. Feral hog. Pages 1164–1179 *in* G. A. Feldhamer, B. C. Thompson, and J. A. Chapman, editors. Wild mammals of North America. Johns Hopkins University Press, Baltimore, Maryland, USA.

[16] Nettles, V.F., J. L. Corn, G. A. Erickson, and D. A. Jessup. 1989. A survey of wild swine in the United States for evidence of hog cholera. Journal of Wildlife Diseases 25:61–65.

[17] Davidson, W. R., and V. F. Nettles, editors. 1997. Wild swine. Pages 104–133 *in* Field manual of wildlife diseases in the southeastern United States. Second edition. Southeastern Cooperative Wildlife Disease Study, Athens, Georgia, USA.

[18] Davidson, W. R., editor. 2006. Wild swine. Pages 105–134 *in* Field manual of wildlife diseases in the southeastern United States. Third edition. Southeastern Cooperative Wildlife Disease Study, Athens, Georgia, USA.

[19] Choquenot, D. J., R. J. Kilgour, and B. S. Lukins. 1993. An evaluation of feral pig trapping. Wildlife Research, 20:15– 22.

[20] Barrett, R. H., and G. H. Birmingham. 1994. Wild pigs. Pages D65–D70 *in* S. Hyngstrom, R. Timm, and G. Larsen, editors. Prevention and control of wildlife damage. Cooperative Extension Service, University of Nebraska, Lincoln, Nebraska, USA.

[21] Mapston, M. E. 1999. Feral hog control methods. Pages 117–120 *in* Proceedings of the Feral Swine Symposium, June 2–3, 1999, Fort Worth, Texas, USA.

Evaluation of Reception Facilities for Ship-generated Waste

Sylvia Encheva
Stord/Haugesund University College
Bjørnsonsg. 45,
5528 Haugesund,
Norway

Abstract—**Waste management plans usually address all types of ship-generated waste and cargo residues originating from ships calling at ports. Well developed waste management plan is a serious step towards reduction of the environmental impact of ship-generated waste. Such important and at the same time complex considerations can be supported by application of modern mathematical theories. Evaluation of waste management plans based on application of grey theory is presented in this work.**

Keywords—Waste management, grey theory, assessments

I. Introduction

Waste management plans usually address all types of ship-generated waste including sewage, and cargo residues originating from ships. The volume of waste produces pressure on the environment, particularly with respect to ship-generated waste disposal at home ports and ports of call. Well developed waste management plan is a serious step towards reduction of the environmental impact of ship-generated waste. Obviously, there is a serious need for research on developing intelligent tools for evaluating such plans considering their importance and complexity.

Boolean logic is often used in the process of decision making, [4] and [16]. Thus if a response does not appear to be necessarily true, the system selects false. While Boolean logic appears to be sufficient for most everyday reasoning, it is certainly unable to provide meaningful conclusions in presence of inconsistent and/or incomplete input [5], [6]. This problem can be resolved by applying many-valued logic.

In real life situations qualities are often assessed by using linguistic terms. In order to facilitate such a process which is usually based on incomplete information we propose Grey theory, [7]. This theory is particularly useful with respect to working in situations where the information about elements (or parameters) is incomplete, the information about structure is incomplete, the information about boundaries is incomplete, and the behavior information of movement is incomplete. Occurrence of incomplete information is the main reason of being grey.

The rest of the paper is organized as follows. Basic terms and concepts are presented in Section II. The main results are described in Section III. Section IV contains the conclusion of this work.

II. Background

Grey theory is an effective method used to solve uncertainty problems with discrete data and incomplete information. The theory includes five major parts: grey prediction, grey relational analysis, grey decision, grey programming and grey control, [2], [3], and [8]. A quantitative approach for assessing the qualitative nature of organizational visions is presented in [10].

The Grey theory in this work follows [7].

Definition 1: A grey system is defined as a system containing uncertain information presented by a grey number and grey variables.

Definition 2: Let X be the universal set. Then a grey set G of X is defined by its two mappings $\overline{\mu}_G(x)$ and $\underline{\mu}_G(x)$.

$$\begin{cases} \overline{\mu}_G(x) : x \to [0,1] \\ \underline{\mu}_G(x) : x \to [0,1] \end{cases}$$

$\overline{\mu}_G(x) \geq \underline{\mu}_G(x), x \in X, X = R$, $\overline{\mu}_G(x)$ and $\underline{\mu}_G(x)$ are the upper and lower membership functions in G respectively.

When $\overline{\mu}_G(x) = \underline{\mu}_G(x)$, the grey set G becomes a fuzzy set. It shows that grey theory considers the condition of the fuzziness and can deal flexibly with the fuzziness situation.

The grey number can be defined as a number with uncertain information. For example, the ratings of attributes are described by the linguistic variables; there will be a numerical interval expressing it. This numerical interval will contain uncertain information. A grey number is often written as $\otimes G$, $(\otimes G = G|_{\underline{\mu}}^{\overline{\mu}})$.

Definition 3: Lower-limit, upper-limit, and interval grey numbers.

$\otimes G = [\underline{G}, \infty]$ - if only the lower limit of G can be possibly estimated and G is defined as a lower-limit grey number.

$\otimes G = [-\infty, \overline{G}]$ - if only the upper limit of G can be possibly estimated and G is defined as a upper-limit grey number.

$\otimes G = [\underline{G}, \overline{G}]$ - the lower and upper limits of G can be estimated and G is defined as an interval grey number.

Grey number operation is an operation defined on sets of intervals, rather than real numbers. Some basic operation laws of grey numbers $\otimes G_1 = [\underline{G}_1, \overline{G}_1]$ and $\otimes G_2 = [\underline{G}_2, \overline{G}_2]$ on intervals where the four basic grey number operations on the interval are the exact range of the corresponding real operation follow:

$$\otimes G_1 + \otimes G_2 = [\underline{G}_1 + \underline{G}_2, \overline{G}_1 + \overline{G}_2]$$

$$\otimes G_1 - \otimes G_2 = [\underline{G}_1 - \overline{G}_2, \overline{G}_1 - \underline{G}_2]$$

$$\otimes G_1 \times \otimes G_2 = [\min(\underline{G}_1\underline{G}_2, \underline{G}_1\overline{G}_2, \overline{G}_1\underline{G}_2, \overline{G}_1\overline{G}_2), \max(\underline{G}_1\underline{G}_2, \underline{G}_1\overline{G}_2, \overline{G}_1\underline{G}_2, \overline{G}_1\overline{G}_2)]$$

$$\otimes G_1 \div \otimes G_2 = [\underline{G}_1, \overline{G}_1] \times \left[\frac{1}{\underline{G}_2}, \frac{1}{\overline{G}_2}\right]$$

The length of a grey number $\otimes G$ is defined as

$$L(\otimes G) = [\overline{G} - \underline{G}].$$

Definition 4: [7] For two grey numbers $\otimes G_1 = [\underline{G}_1, \overline{G}_1]$ and $\otimes G_2 = [\underline{G}_2, \overline{G}_2]$, the possibility degree of $\otimes G_1 \leq \otimes G_2$ can be expressed as follows:

$$P\{\otimes G_1 \leq \otimes G_2\} = \frac{\max(0, L^\star - \max(0, \overline{G}_1 - \underline{G}_2))}{L^\star}$$

where $L^\star = L(\otimes G_1) + L(\otimes G_2)$.

For the position relationship between $\otimes G_1$ and $\otimes G_2$, there exist four possible cases on the real number axis. The relationship between $\otimes G_1$ and $\otimes G_2$ is determined as follows:

- If $\underline{G}_1 = \underline{G}_2$ and $\overline{G}_1 = \overline{G}_2$, we say that $\otimes G_1$ is equal to $\otimes G_2$, denoted as $\otimes G_1 = \otimes G_2$. Then $P\{\otimes G_1 \leq \otimes G_2\} = 0.5$.
- If $\underline{G}_2 > \overline{G}_1$, we say that $\otimes G_2$ is larger than $\overline{G}_1$, denoted as $\otimes G_2 > \otimes G_1$. Then $P\{\otimes G_1 \leq \otimes G_2\} = 1$.
- If $\overline{G}_2 > \overline{G}_1$, we say that $\otimes G_2$ is smaller than $\overline{G}_1$, denoted as $\otimes G_2 < \otimes G_1$. Then $P\{\otimes G_1 \leq \otimes G_2\} = 0$.
- If there is an inter-crossing part in them, when $P\{\otimes G_1 \leq \otimes G_2\} > 0.5$, we say that $\otimes G_2$ is larger than $\overline{G}_1$, denoted as $\otimes G_2 > \otimes G_1$. When $P\{\otimes G_1 \leq \otimes G_2\} < 0.5$, we say that $\otimes G_2$ is smaller than $\overline{G}_1$, denoted as $\otimes G_2 < \otimes G_1$.

Suppose a decision group has K persons, then the attribute weight of attribute Q_j can be calculated as

$$\otimes w_j = \frac{1}{K}\left[\otimes w_j^1 + \otimes w_j^2 + + \otimes w_j^K\right]$$

where $\otimes w_j^K, j = 1, 2, ..., n$ is the attribute weight of K-th decision maker and can be described by grey number $\otimes w_j^K = [\underline{w}_j^K, \overline{w}_j^K]$.

The rating values are

$$\otimes G_{ij} = \frac{1}{K}\left[\otimes G_{ij}^1 + \otimes G_{ij}^2 + + \otimes G_{ij}^K\right]$$

where $\otimes G_{ij}^K, i = 1, 2, ..., m, j = 1, 2, ..., n$ is the attribute rating value of K-th decision maker and can be described by the grey number $\otimes G_{ij}^K = [\underline{G}_{ij}^K, \overline{G}_{ij}^K]$.

The weighted normalized grey decision matrix can be established as

$$D^\star = \begin{bmatrix} \otimes V_{11} & \otimes V_{12} & \dots & \otimes V_{1n} \\ \otimes V_{21} & \otimes V_{22} & \dots & \otimes V_{2n} \\ & & \dots & \\ \otimes V_{m1} & \otimes V_{m2} & \dots & \otimes V_{mn} \end{bmatrix}$$

where $\otimes V_{ij} = \otimes G_{ij} \times \otimes w_j$.

$$S^{\max} = \left\{ \begin{array}{l} \left[\max_{1\leq i\leq m} \underline{V}_{i1}, \max_{1\leq i\leq m} \overline{V}_{i1}\right], \\ \left[\max_{1\leq i\leq m} \underline{V}_{i2}, \max_{1\leq i\leq m} \overline{V}_{i2}\right], \\ ..., \\ \left[\max_{1\leq i\leq m} \underline{V}_{in}, \max_{1\leq i\leq m} \overline{V}_{in}\right] \end{array} \right\}$$

The grey possibility degree between plan alternatives in set $Pl = \{Pl_1, Pl_2, ..., Pl_m\}$ and ideal referential plan alternative $Pl^{\max}$.

$$P\{Pl_i \leq Pl^{\max}\} = \frac{1}{n} \sum_{1\leq j\leq n} P\{\otimes V_{ij} \leq \otimes G_j^{\max}\}$$

A smaller $P\{Pl_i \leq Pl^{\max}\}$ implies worse ranking order of Pl_i.

A historical review and bibliometric analysis of grey system theory is presented in [14]. Application of grey theory for predication of electric power demand can be seen in [13].

The Analytic Hierarchy Process (AHP), [11] facilitates development of a hierarchical structure of complex evaluation problems. This way subjective judgment errors can be avoided and an increase of the likelihood for obtaining reliable results can be achieved. AHP employs paired comparisons in order to obtain ratio scales. Both actual measurements and subjective opinions can be used in the process. Grey relational analysis method and analytic network process [12] approach were used in [9].

III. DECISION-MAKING

The regulation on waste delivered to shore enforced by 2004 requires vessels entering ports within the European Economic Community (EEC) to report current status of waste on-board. This includes the amount of waste being produced, delivered in port, and planned to be delivered in next port of call. Prior to arrival all the required data has to be delivered to port authorities. Such information is of special interest regarding environmental reporting.

Cruise ships often generate waste that prevails their maximum storage capacity long before they have access to shore

TABLE I. ATTRIBUTE WEIGHTS

Scale	
Not satisfactory	[0.0, 0.2]
Somewhat atisfactory	[0.21, 0.4]
Average	[0.41, 0.5]
Good	[0.51, 0.8]
Very good	[0.81, 0.9]
Exellent	[0.91, 1.0]

waste disposal facilities. According to MARPOL 73/78 discharge of treated waste water is allowed 4 Mi off shore, while some regional port policies require discharge treated waste water 12 Mi off shore.

Establishment and operation of reception facilities for ship-generated waste is a very important regarding protection of the external environment. The reception facilities can be fixed, floating and mobile units. Examples of ship-generated waste include oily waste, sewage, cargo residues and garbage.

While large oil spills in the see are followed by marine pollution authorities, small spills cased by pumping oily bilge water overboard and refueling receive considerably less attention. They however have also negative effects on the marine environment. Marine bilge pump out services are to be used instead of pumping oily bilge water overboard. The latter may contain diesel and petrol, as well as lubricant and hydraulic oils.

Reception facilities are often ranked according to availability of the pre-treatment equipment and processes, methods of recording actual use of the port reception facilities, methods of recording amounts of received and disposed ship-generated waste and cargo residues. The impact of cruise ship generated waste on home ports and ports of call is studied in [1].

Actual use of reception facilities is sometimes stimulated by the so-called "no-special-fee" system. Thus 'fees covering the cost of the reception, handling and final disposal of ship-generated wastes are included in the harbor fee or otherwise charged to the ship, irrespective of whether any wastes are actually delivered', [17]. Recent port waste management plan for ship-generated waste can be found in [18].

In this work five plan alternatives $Pl1, Pl2, Pl3, Pl4$ and $Pl5$ are ranked with respect to four attributes. These attributes address provision of facilities to receive $A1$, treat $A2$, safely dispose of ship-generated waste $A3$, and use of green energy $A4$. We however are not to provide all numerical details due to agreed upon anonymity restrictions. Instead we present a graphical illustration for comparing the plan alternatives.

Further on we use the rule: the grater values the better. The calculations are done based on real data and according to the theory presented in [7]. This means calculating of linguistic ratings for weights and linguistic ratings for attributes, first and building a weighted normalized decision table afterwards. Linguistic ratings for applied weights presented in Table I and weighted normalized decision table is shown in Table II.

The possibility degrees are $P\{Pl_i \leq Pl^{\max}\} = \{0.65, 0.78, 0.72, 0.61, 0.67\}$. In other words $Pl2 > Pl3 > Pl5 > Pl1 > Pl4$. According to the performed calculations the second alternative should be chosen.

TABLE II. WEIGHTED NORMALIZED DECISION TABLE

	A1	A2	A3	A4
Pl1	[6.41,6.83]	[3.46,3.84]	[6.25, 6.81]	[3.52, 3.86]
Pl2	[5.46,5.72]	[5.34,5.75]	[7.32, 7.57]	[4.14, 4.76]
Pl3	[4.34,4.66]	[5.11,0.23]	[5.75, 6.13]	[6.25, 6.78]
Pl4	[4.93,5.28]	[4.33,4.62]	[4.32, 4.74]	[4.93, 5.08]
Pl5	[3.75,4.12]	[4.83,4.91]	[5.14, 5.63]	[5.55, 6.12]

Fig. 1. Plan alternative $Pl2$ and attributes

Applying grey theory in our case results in ordering proposed plan alternatives when all attributes are taken in consideration. In order to make decisions some authorities are interested to see how different plan alternatives are ranked with respect to each attribute. We answer that question in Fig. 1. Since the second alternative $Pl2$ is listed as the best according the executed calculations, we show where $Pl2$ is placed with respect to each attribute. $Pl2$ appears to be the best alternative according to $A2$ and $A3$, and is number two according to $A1$, and number four according to $A4$. Alternative $Pl3$ is highlighted Fig. 2.

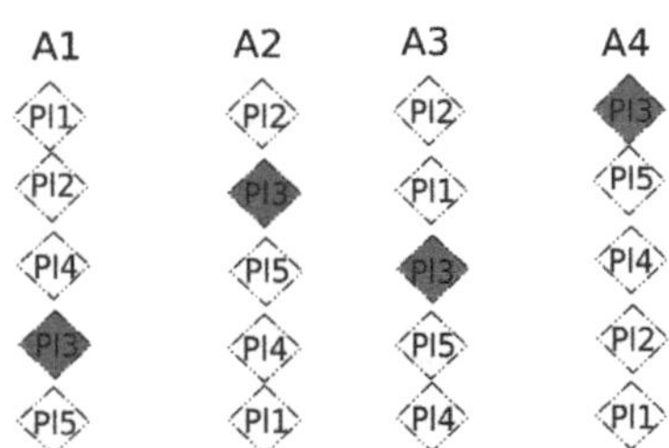

Fig. 2. Plan alternative $Pl3$ and attributes

Similar figures can be made for alternative $Pl5$ if for some reasons alternatives $Pl2$ and $Pl3$ cannot be accepted.

Instead of developing graphical representations one can study the weighted normalized decision table. The latter however proves to be quite difficult when populated with larger amount of data.

IV. CONCLUSION

Quite often different elements in waste management plans are evaluated independently of each other, which leads to multi-criteria decision inconsistencies. The presented approach can be used to evaluate all elements in such plans and compare those plans in order to make an optimal decision.

We have also compared the five plans applying Analytic Hierarchy Process, [15]. The outcomes conforms what has already been obtained with the Grey theory approach.

References

[1] N. Butt, The impact of cruise ship generated waste on home ports and ports of call: A study of Southampton, Marine Policy, vol.31(5), pp. 591–598, 2007.

[2] J. L. Deng, *Control problems of grey systems*, System and control letters, vol. 5, 288-294, 1982.

[3] J. L. Deng, *Introduction to grey system theory*, Journal of grey systems, 1, vol. 1-24, 1989.

[4] R. L. Goodstein, *Boolean Algebra*, Dover Publications, 2007.

[5] E. Gradel, M. Otto and E. Rosen, *Undecidability results on two- variable logics*, Archive of Mathematical Logic, vol. 38, pp. 313–354, 1999.

[6] N. Immerman, A. Rabinovich, T. Reps, M. Sagiv and G. Yorsh, *The boundery between decidability and undecidability of transitive closure logics*, In: CSL'04, 2004.

[7] S. Liu and Y. Lin, *Grey information*, Springer, 2006

[8] Y.C. Hu, *Grey relational analysis and radical basis function network for determining costs in learning sequences*, Applied mathematics and computation, vol. 184, pp. 291-299, 2007.

[9] C. Mi and W. Xia, *Prioritizing technical requirements in QFD by integrating grey relational analysis method and analytic network process approach*, Grey Systems: Theory and Application, vol. 5(1), pp. 117 – 126, 2015

[10] F. Rahimnia, M. Moghadasian and E. Mashreghi, *Application of grey theory organizational approach to evaluation of organizational vision*, Grey Systems: Theory and Application, vol. 1, (1), pp. 33–46, 2011.

[11] T. L. Saaty, *Principia Mathematica Decernendi: Mathematical Principles of Decision Making*, Pittsburgh, Pennsylvania: RWS Publications, 2010.

[12] T. L. Saaty, M. S. Ozdemir and J. S. Shang, *The Rationality of Punishment - Measuring the Severity of Crimes: An AHP-Based Orders-of-Magnitude Approach*, International Journal of Information Technology and Decision Making, vol. 14(1), pp. 5–16, 2015.

[13] X. Shen and Z. Lu, *The Application of Grey Theory Model in the Predication of Jiangsu Province's Electric Power Demand*, 2nd AASRI Conference on Power and Energy Systems (PES2013), AASRI Procedia, vol. 7, pp. 81–87, 2014.

[14] M-S. Yin, *Fifteen years of grey system theory research: A historical review and bibliometric analysis*, Expert Systems with Applications, vol. 40(7), pp. 2767–2775, 2013.

[15] J. Wang, *Multi-criteria decision-making approach with incomplete certain information based on ternary AHP*, Journal of Systems Engineering and Electronics, vol. 17(1), pp. 109–114, 2006.

[16] J. E. Whitesitt, Boolean Algebra and Its Applications, Dover Publications, 1995.

[17] http://www.helcom.fi/shipping/waste/en_GB/waste/

[18] http://www.bristolport.co.uk/sites/default/pfiles/files/bpc-waste-management-plan-2013-issue-3-revision-2-final.pdf

6

Scale-Based Local Feature Selection for Scene Text Recognition

Boyu Zhang, Jia Feng Liu, Xiang Long Tang
School of Computer Science and Technology,
Harbin Institute of Technology,
Harbin 150001, China

Abstract—**Scene text recognition has drawn increasing concerns from the OCR community in recent years. Among numerous methods that have been proposed, local feature based methods represented by bag-of-features (BoFs) show notable robustness and efficiency. However, as the existing detectors are based on assumptions about local saliency, a vast number of non-informative local features would be detected in the feature detection stage. In this paper, we propose to remove non-informative local features by integrating feature scales with stroke width information. Experiments taken both on synthetic data and real scene data show that the proposed feature selection method could effectively filter non-informative features and improve the recognition accuracy.**

Keywords—Scene Text Recognition; Local Feature; Stroke Width

I. Introduction

In recent years, scene text recognition (STR) [1] technologies have got increasing concerns from OCR community and other related fields. Compared with surrounding text, scene text is more connected to image contents in most cases. Thus the rich semantic information contained in scene text often plays vital roles in a host of computer vision applications, including impaired people assist, visual land-mark robot navigation and intelligent traffic system.

Even numerous potential applications exist, the STR is still challenging due to the following disadvantages: (1) The scales of scene text, even in same sentences, vary a lot; (2) The shapes and styles may be different since scene text are specially designed to fit different requirements; (3) Scene images always contain illumination changes, viewpoints variations and other disadvantages such as a non-flatness surface; (4) In most cases, no context information is provided.

During the past decades, a number of methods are proposed in response to these disadvantages. The existing methods inSTR area could be divided into two categories according toothier basic ideas. One of which is to achieve accurate STR by developing traditional OCR methods. Most approaches under this idea contain three procedures, which are, text detection, segmentation and character recognition. For instance, Chenand Yuille [2] train strong classifier which contains multiple features by integrating weak classifiers with AdaBoost to extract text regions, then text are recognized by employing commercial OCR software. Coates et al. [3] apply scalable leaning algorithm to feature extraction, text detector and classifier to produce high accurate STR system. Kai et al. [4]designed an end-to-end system for scene text recognition, in which Random Fern [5] is utilized as raw character detector as well as classifier. Moreover, they proposed to improve the accuracy of STR by introducing pre-defined vocabulary.

Another idea is to treat scene text as objects. Thus researchers can transplant objcct rccognition methods that are proposed mostly against image degradations and uncontrolled environments into STR area. For example, De Campus et al.[6] build up a STR framework by following classic BoFsmethods in which sample images are described by frequency histogram of local features. They also compare the effectiveness of different local descriptors by taking experiments one representative benchmark. Zheng et.al [7] recognize scene characters by matching detected SIFT [8] features between input samples and pre-build template images. Different from BoFs method that totally omits position information, they consider the relative position of local features by using MPLSH[9]. Diem and Sablatnig [10] build a historical document analysis system based on local descriptors and achieve a state-of-art accuracy for ancient character recognition.

Among these methods, the ones based on local features [6],[7], [10] show notable robustness and effectiveness, especially when in small sample size situations and situations containing image degradations [11]. They are more robust because they represent sample images using sets of local features and omitting other highly variable factors. It is obvious that their accuracy largely depends on the effectiveness of detected local features. However, even most local feature detectors assume that salient image patches are informative, the meanings of effective are different in different applications. Specific to our problem, not all detected saliency image patches reflect local structures of characters. Thus, for improving the accuracy, criteria are needed to filter features which are not related to the text.

In this paper, we focus on local feature based STR and propose a novel criterion which integrate stroke width information with local feature scales to remove non-informative local features and achieve higher accuracy. Our idea is based on the fact that text is constituted by strokes with specific width. Thus there should be an appropriate proportion between local feature scale and the corresponding stroke width if these features reflect local text structures such as corner and cross. Experiments taken on both natural and synthetic text images show that the proposed approach could effectively improve the accuracy of local feature based STR.

II. Related Works

Many techniques are developed for filtering redundancies and noises from original features set. In this paper, we make the specific consideration about methods based on codebook model. A classical codebook method includes local feature detection, codebook generation, quantization, and finally classification. Most efforts for feature selection are taken on codebook generation stage and code-word selection stages. In this section, we briefly introduce typical existing methods according to their categories and discuss differences between these methods and proposed method in the end.

A. Compact Codebook Generation

In codebook generation stage, the algorithm seeks for a group of code words (also referred as 'codebook'), which could describe the feature space effectively. A vast number of methods are proposed to generate effective codebook. For instance, Tuytelaars and Schmid [12] extract high-dimensional descriptors for sample images by partitioning feature space using lattices with regular sizes and then combine similar dimensions to make the descriptors more compact. The most widely applied idea is to get codebook utilizing unsurprised cluster algorithms such as K-means [13], which get the most descriptive k centers by minimizing the variance between k centers and the training data. Different from k-means that is dense sensitive, Jurie and Triggs [14] proposed a radius-based clustering which clusters all features within a fixed radius of similarity radius to one cluster.

B. Code-word Selection

Besides generating a compact codebook, a host of algorithms are proposed for picking the most effective subset from the original codebook. Code-word selection is equal to feature selection problem since sample images are rep- resented by frequency histograms of code-words and each bin corresponding to a feature dimension. Distinguishing by whether class labels are given existing methods could be divided into supervised and unsupervised ones.

Supervised methods analyze the relationship between the class labels and code words and then pick more discriminate subset based on pre-defined criteria. Literature [14] gives a performance evaluation for three typical methods including MI [15], OR [16] and Linear SVM weights [17] on representative datasets. Moosmann et al. [18] proposed to build supervised indexing trees using an ERC-Forest that considers semantic labels as stopping tests. The work in [19] aims to find the Descriptive Visual Words (DVWs) and Descriptive Visual Phrases (DVPs) for each image category.

For unsupervised situations, Zhang et al. [20] proposed to pick out the most discriminative code words which lead to minimal fitting errors between data matrix and indicator matrix. Maximum variance selects features with the largest variances and unsupervised feature selection for PCA selects a subset of features that can best reconstruct other features. Laplacian score [21] selects features tat preserve the local geometrical structure best. Q-α [22] measures the cluster coherence by analyzing the spectral properties of the affinity matrix.

C. Proposed Method

Different from the above methods, the proposed method in this paper filters non-informative features by per- forming a pre-selection based on analyzing both feature scale and stroke width information. Its advantage is that the algorithm effects before codebook generation stage and thus could avoid errors that occur in the following process. This means the proposed methods could be more effective when facing small sample size problems, which are common in STR and historical document analysis.

III. Scale-Based Local Feature Selection

The fundamental assumption of designing most local feature detectors is that salient image patches are informative. In fact, the concepts of 'informative' are different in different situations. Specifically, in STR process, it is not promised each salient patches indeed reflects character structure. Thus criteria are needed to remove features that are not effective.

According to whether they are helpful for distinguishing different characters, we divide detected local features into informative and non-informative. Features belong to the first category always localize in character bounding-boxes and they are salient since they contain character structures such as corners and stroke crosses. In contrast, most features that belong to the second category are generated by cluttered background and noises, thus do not provide information forSTR. It is worthwhile to emphasize that large local features that cover the majority of a character should be categorized into the second type since these features are not robust enough when numerous variations are included.

However, it is difficult to remove non-informative local features automatically as it is difficult to give a formally definition for non-informative features. The target can be achieved by training a binary classifier that could distinguish on-informative features from informative ones, however, a large number of training samples are needed to train such a classifier and the existence of varies fonts makes sample collecting rather difficult. Moreover, labeling all features manually is labor expensive and hardly objective. Another idea is to optimize learned codebook according to class label as we discussed in section II, which is under sophisticated mathematical model. These methods that select features by analyzing the relationship between code words and class labels also need large training dataset.

In this paper, we propose a novel local feature selection criterion that selects effective local features based on the ratio between character stroke width and local feature scale.

A. Feature Scale and Stroke Width

Our idea is based on the observation that it is impossible to write small character with wide strokes and large characters with thin strokes. Thus the ratio between character size s_c and stroke width w in the text area should keep within a reasonable range to ensure the character is recognizable. At the same time, for each detected local feature which reflects a local structure on character, its scale s_f should also be indirect

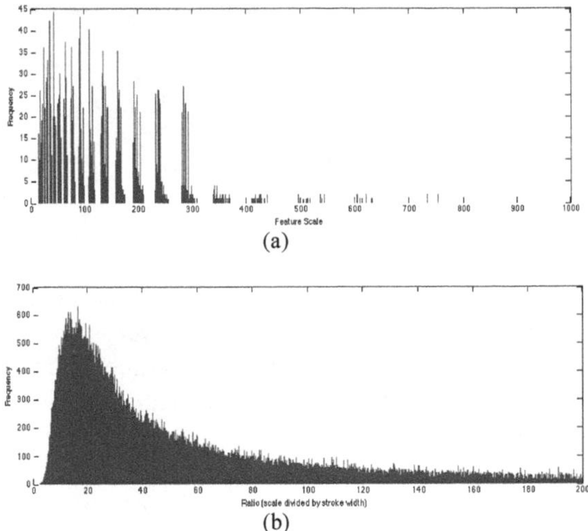

Fig. 1. Diagram (a) shows the frequency histogram of feature scales local features extracted form 'char 74k' dataset. (b) is the frequency histogram of ratio parameters that is calculate by feature scales divided by corresponding stroke width.

Proportion to character scale s_c abided by commonsense. This means that for a reasonable character, the scale of a representative local feature should have a stable ratio r with stroke width w .Based on this idea, we can filter non-effective features by checking whether the ratio r is in an interval $[\mathrm{r_{min}}, \mathrm{r_{max}}]$.

The reason we do not directly apply character size for feature selection is that local structures are directly instituted by strokes and thus the ratio between stroke width and feature scale is more stable than the ratio between character size and feature scale. Moreover, stroke width is more accurate then character size in two reasons. Firstly, the segmentation in scene images is difficult which would lead to inaccurate character size. Secondly, characters in the same size have different stroke width because of the existence of multi-font.

To prove this, we count the frequency histograms of the detected local features according to their feature scales and ratio parameters respectively. The definition of stroke width and the calculation of ratio parameters are described in detail in section IV. Fig 1(a) shows the frequency of local feature scale and Fig 1(b) gives the frequency of the ratio between feature scale and corresponding stroke width. We find that the ratio parameter depends on a uniform long-tail distribution which certify that a relationship exists between local feature scales and stroke width.

B. Scale-based Local Feature Selection

Typical local feature detectors such as SIFT and Multi-Scale Harris contain three stages. In the first stage, for each pixel $I(\mathrm{i},\mathrm{j})$ in an image I, its local saliency H corresponding to scale s is evaluated by using measurement function F. By noting the neighborhood of point $I(\mathrm{i},\mathrm{j})$ as $r(\mathrm{i},\mathrm{j})$, we have:

$$H(i,j,s) = F(\mathrm{r}(i,j,s)) \tag{1}$$

Then the algorithm searches local extreme through both spatial and scale space to find local maximums as candidate feature points, which we note as C. At last, a global thresholding process is taken on C abide by following equation:

$$L_{i,j} = \begin{cases} 1, if\ H(i,j,s) > th_s \\ 0, else \end{cases} \tag{2}$$

Where $L_{i,j}$ indictors whether pixel $r(\mathrm{i},\mathrm{j})$ is the center of an acceptable local feature and th_s is the threshold of feature saliency. Different from the above process considering the local saliency only, in our work, the relationship between the feature scale s and the stroke width w is also considered. Thus the probability that a local region is effective could be described as $P(H,\mathrm{s},\mathrm{w})$. According to Bayes formula, we have

$$P(H,\mathrm{s},\mathrm{w}) = P(H \mid \mathrm{s},\mathrm{w})P(\mathrm{s},\mathrm{w}) \tag{3}$$

Noticing that the calculation of local saliency H is independent to stroke width w, the probability $P(H,\mathrm{s},\mathrm{w})$ could be simplified into $P(H \mid s)$. Furthermore, in this paper, we describe the relationship $P(\mathrm{s},\mathrm{w})$ between s and w by a sign function of ratio r and use another sign function to describe $P(H \mid s)$, we get

$$L_{i,j} = P(H)\mathrm{P}(\mathrm{r}_{i,j}) \tag{4}$$

where

$$P(r) = \begin{cases} 1, r \in [r_{\min}, r_{\max}] \\ 0, else \end{cases} \tag{5}$$

and

$$P(H) = \begin{cases} 1, H > th_s \\ 0, \mathrm{else} \end{cases} \tag{6}$$

Thus we could give the feature selection algorithm based on the above analysis. According to Algorithm 1, we can improve the accuracy and efficiency by removing non-informative local features. Section IV demonstrates the effect of the proposed algorithm.

Algorithm 1 Scale-based Local Feature Selection

O P Q R S T U V
w X Y Z a b C d
e f g h i j k l

(a)

(b)

(c)

Fig. 2. (a) and (b) shows character samples from Fnt and NS data, respectively. (c) shows Chinese word samples form CH data

IV. USING THE TEMPLATE

In this section, we verify the effect of the proposed scale-based feature selection algorithm with experiments on representative benchmarks. Section IV demonstrate the effect of the proposed algorithm.

A. Experiment Setup

*1) Experimental Data:*To prove that local features with proper scales are more effective, we conduct experiments on a representative benchmark which is referred as 'char74k' [6]. The 'char 74k' dataset contains both synthetic and natural samples. Synthetic samples include 52 classes of English characters (capital letters and lower case letters) and 10 classes of numbers (0~9). For each class, 1016 character samples are generated according to 256 different system fonts with 4 different styles. For natural samples, characters are cropped manually from scene images. Fig 2(a) and Fig 2(b) shows some typical samples of 'Fnt' data and 'NS' data in this benchmark. This dataset is selected for two reasons. Firstly, it contains typical scene character samples which are segmented manually and labeled in detail. Secondly, synthetic data could be used as baseline in our experiment since these samples certify accurate stroke width information and all detected local features are useful for character recognition. Moreover, we collect our own Chinese words dataset (the dataset will be referred as 'CH' in the following parts of this paper) beside the above benchmark using Internet searching engine according to 12 different key words. For each text image we get, accurate text regions are cropped and labeled manually. Examples of CH data are shown in Fig 2(c).

*2) Local Feature Detection:*We employ two typical detectors, which are, Hessian-affine and difference of Gaussian (DoG). According to the literature [6], the combination of DoG detector and SIFT descriptor performs much better than others.

*3) Stroke Width Extraction:*In this paper, stroke width information is extracted by utilizing stroke width transform [23]. For each pixel in a text image, if it is localized between two edges pixels with opposite gradient directions, its stroke width value is defined as the distance between these two edge pixels. If more than one pair of edge pixels are found, the stroke width value is set as the minimum one. On the contrary, stroke width value is set as infinite when the algorithm cannot find pixels like that. For more details about stroke width extraction, readers could refer to the original paper by Epstein et.al.[23]. Two factors should be considered for extracting precise stroke width. The first one is the thresholds for edge detector (Canny here) should be selected very carefully since the precision of SWT heavily depends on the results of edge detection. The second is that the algorithm needs to know whether the character pixels are darker than the background or opposite. In practice, it is without any difficulties to assign parameters of edge detector for synthetic data as these images have high contrast (binary images, actually). Moreover, all synthetic samples have darker pixels compared to the background. For natural images, thresholds of Canny operator are assigned much lower by considering the image contrast and the contrast between text and background are assigned manually.

Based on detected local features and extracted stroke width value, we can calculate the ratio r for each local feature.

B. Character and Word Recognition

Text recognition is achieved based on classic bag-of-features framework, which is similar to literature [6]. In our experiments, 30 training samples and 15 testing samples are selected randomly for each class. Then local features are detected and described as mentioned above. The visual word vocabulary is generated by using k-means cluster algorithm, and the number P of visual words for each class is assigned equally (varies from 2 to 10 in the following experiments).

Finally, each sample is quantized into feature vector according to the vocabulary and thus each sample image is described bya $P \times C$ dimension vector where C is the number of classes. Support vector machine (SVM) with RBF kernel is chosen as classifier due to its effectiveness and representativeness and '1 VS all 'strategy is employed to solve multi-class problem.

Besides, we perform recognition separately for numbers, lowercase letters and capital letters to avoid the influence of similar symbols such as 'o'and '0', 'p' and 'P'. Thus the accuracy for NS and Fnt data is calculated by using the weighted average according the following equation

$$Acc = \frac{C_c}{C} Acc_c + \frac{C_l}{C} Acc_l + \frac{C_n}{C} Acc_n$$

where C_c, C_l and C_n is the class number of capital letters, lowercase letters and numbers and $C = C_c + C_l + C_n$.

In the feature selection stage, a group of samples for each class are selected to find the best threshold for filtering especially large or small features. For each training process, we

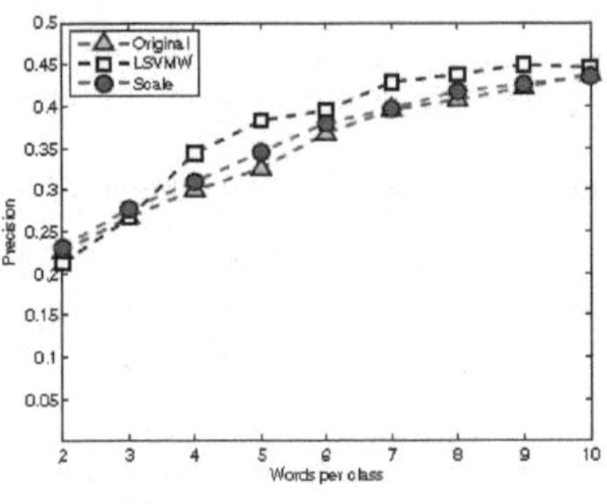

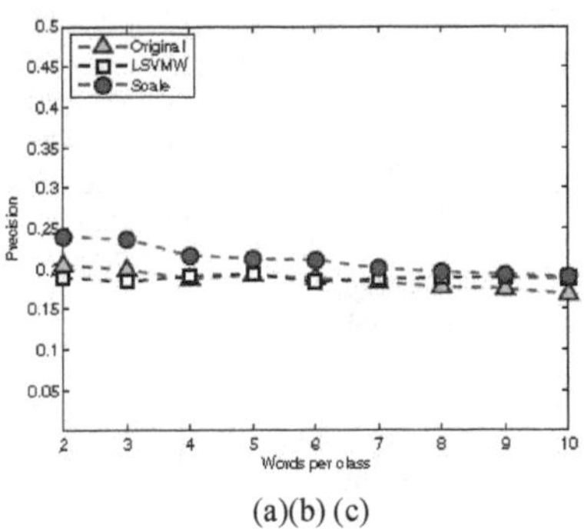

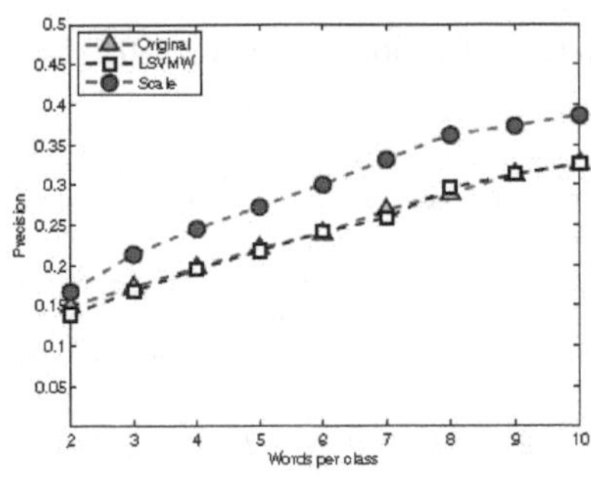

(a)(b) (c)

Fig. 3. Diagram (a) shows the recognition accuracy of LSVMW, proposed method and original BoFs based method on Fnt data. Diagram (b) and (c) shows the corresponding results on NS and CH data, respectively.

TABLE I. IMPROVEMENTS BROUGHT BY SCALE-BASED LOCAL FEATURE SELECTION

Words per Class		*2*	*3*	*4*	*5*	*6*	*7*	*8*	*9*	*10*
Rate of Improvement (%)	**Fnt**	2.20	3.84	3.27	6.13	3.39	0.39	2.51	1.07	-0.42
	NS	17.17	19.34	15.61	10.31	11.94	9.61	10.59	10.19	13.04
	CH	12.05	23.61	24.70	23.64	25.56	23.43	25.31	19.58	18.57

remove features that have extremely large or small ratio parameter as percentage. The best filter threshold is found by employing grid search. For Fnt data, the algorithm search the best threshold from 1% to 10% for both large and small sides. The reason for limiting the searching range is that very few non-informative features are detected for Fnt data. The experimental results also show that the best thresholds in the neighborhood of 1% in most cases for Fnt data. We can find that the selection slightly improve the accuracy of Fnt data. Besides, the results of feature selection using linear SVM weight is also shown in the Fig 3. The results of MI and IG are not attached as LSVMW over-performs them. We can discover that both Scale-based feature selection and LSVMW-based method can improve the accuracy of Fntdata. However, the improvement of scale-based method is not very obvious and weaker then LSVMW-based one. The reason is that most detected local features are informative since no cluster background and noises are included in Fnt data. For the NS data which include more noises, scale-based feature selection strategy overruns both original data and LSVMW-based feature selection. The results show that the scale-based feature selection brings more benefits when a rare word number is used and the efforts of LSVMW is close to our method when the number of words increases. The reason is that when rare word number is used, the influence of noises is more obvious in that error code words will reduce the accuracy, and the proposed method is more effective for filtering non-informative features and avoiding the generation of error code words.

The accuracy for CH data is calculated under the same method. We can discover that the proposed scale-based feature selection method obviously overruns original and LSVMW-based method. This encouraging result further proves that we can filter non-informative local features by considering both feature scales and stroke width. To examine the improvement of the proposed method in greater detail, the recognition accuracy of original data and filtered data is calculated. Moreover, the improvement brought by stroke width information is evaluated as follows: noting the recognition accuracy on original feature set as C_{ori} and C_{sel} as accuracy on selected feature set, the improvement C_{imp} can be evaluated as $C_{imp} = (C_{sel} - C_{ori}) / C_{ori}$. The results are shown in Table I. From Table I, we can see that supervised feature selection algorithms such as LSVMW are more effective for clean data and the proposed method is more effective when samples contain more noises and degradation such as NS data and CHdata.

V. CONCLUSION

In this paper, we proposed a new approach for filtering text-independent local features by considering both stroke width information and feature scales. The proposed approach is tested on representative benchmarks and the encouraging experimental results (a maximum improvement of 25.56% for CH data and 19.34% for natural data) prove the existence of relevancy between stroke width and feature scales. Different from traditional methods which need a group of training data, the proposed approach can effectively filter on-informative local features when only a few samples are used. Moreover, it is notable that the proposed approach is evidently effective for degraded images and small sample size situations. These two advantages ensure the proposed method could be widely applied in the fields such as historical document analysis and text-associate image retrieval. At the same time, we can find that there is much room for improvement in recognition rate for local feature based algorithms. Therefore, our future work include developing probability model which aims at increasing the accuracy of local feature based STR and building end-to-end scene text analysis system.

ACKNOWLEDGMENT

This work was supported by National Natural Science Foundation of China (grant No.61073128).

REFERENCES

[1] K. Jung, K. In Kim, and A. K Jain, "Text information extraction in images and video: a survey," Pattern recognition, vol. 37, no. 5, pp. 977-997, 2004.

[2] X. Chen and A. L. Yuille, "Detecting and reading text in natural scenes," in Computer Vision and Pattern Recognition, 2004. CVPR 2004. Proceedings of the 2004 IEEE Computer Society Conference on, vol. 2. IEEE, 2004, pp. II-366.

[3] A. Coates, B. Carpenter, C. Case, S. Satheesh, B. Suresh, T. Wang, D. J. Wu, and A. Y. Ng, "Text detection and character recognition in scene images with unsupervised feature learning," in Document Analysis and Recognition (ICDAR), 2011 International Conference on. IEEE, 2011, pp. 440-445.

[4] K. Wang, B. Babenko, and S. Belongie, "End-to-end scene text recognition," in Computer Vision (ICCV), 2011 IEEE International Conference on. IEEE, 2011, pp. 1457-1464.

[5] A. Bosch, A. Zisserman, and X. Muoz, "Image classification using random forests and ferns," in Computer Vision, 2007. ICCV 2007. IEEE 11th International Conference on. IEEE, 2007, pp. 1–8.

[6] T. de Campos, B. R. Babu, and M. Varma, "Character recognition innatural images," 2009.

[7] Q. Zheng, K. Chen, Y. Zhou, C. Gu, and H. Guan, "Text localization and recognition in complex scenes using local features," in Computer Vision–ACCV 2010. Springer, 2011, pp. 121-132.

[8] D. G. Lowe, "Distinctive image features from scale-invariant keypoints," International journal of computer vision, vol. 60, no. 2, pp. 91-110, 2004.

[9] Q. Lv, W. Josephson, Z. Wang, M. Charikar, and K. Li, "Multiprobe lsh: efficient indexing for high-dimensional similarity search," in Proceedings of the 33rd international conference on Very large databases. VLDB Endowment, 2007, pp. 950–961.

[10] M. Diem and R. Sablatnig, "Are characters objects?" in Frontiers inHandwriting Recognition (ICFHR), 2010 International Conference on. IEEE, 2010, pp. 565-570.

[11] K. Das and Z. Nenadic, "An efficient discriminant-based solution for small sample size problem," Pattern Recognition, vol. 42, no. 5, pp. 857-866, 2009.

[12] T. Tuytelaars and C. Schmid, "Vector quantizing feature space with a regular lattice," in Computer Vision, 2007. ICCV 2007. IEEE 11th International Conference on. IEEE, 2007, pp. 1-8.

[13] D. Lee, S. Baek, and K. Sung, "Modified k-means algorithm for vectorquantizer design," Signal Processing Letters, IEEE, vol. 4, no. 1, pp.2-4, 1997.

[14] F. Jurie and B. Triggs, "Creating efficient codebooks for visual recognition,"in Computer Vision, 2005. ICCV 2005. Tenth IEEE InternationalConference on, vol. 1. IEEE, 2005, pp. 604-610.

[15] M. Vidal-Naquet and S. Ullman, "Object recognition with informativefeatures and linear classification." in ICCV, vol. 3, 2003, p. 281.

[16] J. Brank, M. Grobelnik, N. Milic-Frayling, and D. Mladenic, "Interactionof feature selection methods and linear classification models," inWorkshop on Text Learning held at ICML, 2002.

[17] Y. W. Chang and C. J. Lin, "Feature ranking using linear svm," Causationand Prediction Challenge Challenges in Machine Learning, Volume2, p. 47, 2008.

[18] F. Moosmann, B. Triggs, F. Jurie et al., "Fast discriminative visualcodebooks using randomized clustering forests," Advances in NeuralInformation Processing Systems 19, pp. 985-992, 2007.

[19] S. Zhang, Q. Tian, G. Hua, Q. Huang, and W. Gao, "Generatingdescriptive visual words and visual phrases for large-scale imageapplications," Image Processing, IEEE Transactions on, vol. 20, no. 9,pp. 2664-2677, 2011.

[20] L. Zhang, C. Chen, J. Bu, Z. Chen, S. Tan, and X. He, "Discriminativecodeword selection for image representation," in Proceedings of theinternational conference on Multimedia. ACM, 2010, pp. 173-182.

[21] X. He, D. Cai, and P. Niyogi, "Laplacian score for feature selection," inAdvances in neural information processing systems, 2005, pp. 507-514.

[22] L. Wolf and A. Shashua, "Feature selection for unsupervised andsupervised inference: The emergence of sparsity in a weight-basedapproach," The Journal of Machine Learning Research, vol. 6, pp.1855-1887, 2005.

[23] B. Epshtein, E. Ofek, and Y. Wexler, "Detecting text in natural sceneswith stroke width transform," in Computer Vision and Pattern Recognition(CVPR), 2010 IEEE Conference on. IEEE, 2010, pp. 2963-2970.

7

Compressed Sensing Based Encryption Approach for Tax Forms Data

Adrian Brezulianu
"Gheorghe Asachi" Technical University of Iasi
Iasi, Romania

Monica Fira
Romanian Academy
Institute of Computer Science
Iasi, Romania

Marius Daniel Peştină
"Gheorghe Asachi" Technical University of Iasi
Iasi, Romania

Abstract—**In this work we investigate the possibility to use the measurement matrices from compressed sensing as secret key to encrypt / decrypt signals. Practical results and a comparison between BP (basis pursuit) and OMP (orthogonal matching pursuit) decryption algorithms are presented. To test our method, we used 10 text messages (10 different tax forms) and we generated 10 random matrices and for distortion validate we used the PRD (the percentage root-mean-square difference), its normalized version (PRDN) measures and NMSE (normalized mean square error). From the practical results we found that the time for BP algorithm is much higher than for OMP algorithm and the errors are smaller and should be noted that the OMP does not guarantee the convergence of the algorithm. We found that it is more advantageous, for tax forms (or other templates that show no interest for encryption) to encrypt only the recorded data. The time required for decoding is significantly lower than the decryption for the entire form**

Keywords—compressed sensing; encryption; security; greedy algorithms

I. INTRODUCTION

The theory of compressed sensing, perfected in the past few years by prestigious researchers such as D. Donoho [1], E. Candès [2], M. Elad [3], demonstrates the feasibility of recovering sparse signals from a number of linear measurements, dependent with the signal sparsity. Compressed sensing (CS) is a new method which draws the attention of many researchers and it is considered to have an enormous potential, with multiple implications and applications, in all fields of exact sciences [1-4]. Specifically, CS is a new technique for finding sparse solutions to underdetermined linear systems. In the signal processing domain, the compressed sensing technic is the process of acquiring and reconstructing a signal that is supposed to be sparse or compressible.

The perfect secrecy together with the secret communication is a well-defined field of research, being a difficult problem in the domain of information theory. One of the requirements for the information theoretic secrecy is to assure that a spy who listens a transmission containing messages will collect only small number of information bits from message. Additionally, it should provide protection against of an computationally unlimited adversary based on the statistical properties of a system. Shannon introduced the idea of perfect secrecy, in his fundamental paper [5].

An encryption idea by utilizing CS has been mentioned for the first time in [7], but not been addressed in detail [6]. In paper [8], the secrecy of CS is researched, and whose result is that CS can provide a computational guarantee of secrecy. In [9] examine the security and robustness of the CS-based encryption method. In paper [10], the authors describe a new coding scheme for secure image using the principles of compressed sensing (CS) and they analyze the secrecy of the scheme.

II. BACKGROUND

A. Compressed Sensing

Compressed sensing studies the possibility of reconstructing a signal x from a few linear projections, also called measurements, given the a priori information that the signal is sparse or compressible in some known basis Ψ.

To define sparsity precisely, we introduce the following notation: for Ψ - a matrix whose columns form an orthonormal basis, we define a K-sparse vector $x \in R^n$ as $x = \Psi\theta$, where $\theta \in R^N$ has K non-zero entries (i.e., is K-sparse) and Ω_K as the set of K indices over which the vector θ is non-zero.

The vectors on which x is projected onto are arranged as the rows of a nxN projection matrix Φ, n < N, where N is the size of x and n is the number of measurements. Denoting the measurement vector as y, the acquisition process can be described as:

$$y = \Phi x = \Phi\Psi\gamma \quad (1)$$

$$\hat{\gamma} = \arg\min_{\gamma} \|\gamma\|_{l_0} \quad \textit{subject to} \quad y = \Phi\Psi\gamma \quad (2)$$

$$\hat{x} = \Psi\hat{\gamma} \quad (3)$$

The equations system (1) is obviously undetermined. Under certain assumptions on Φ and Ψ, however, the original expansion vector γ can be reconstructed as the unique solution to the optimization problem (2); the signal is then reconstructed with (3). Note that (2) amounts to finding the sparsest decomposition of the measurement vector y in the dictionary $\Phi\Psi$. Unfortunately, (2) is combinatorial and unstable when considering noise or approximately sparse signals.

For a K-sparse signal, only “K+1 projections of the signal onto the incoherent basis are required to reconstruct the signal with high probability”[5]. In this case, is necessary to use combinatorial search with huge complexity. In [1] and [2] is proposed tractable recovery procedures based on linear programming. In these papers is demonstrated that the tractable recovery procedures obtain the same results toward combinatorial search when for signal reconstruction are used aprox. 3 or 4 cK projections.

Two directions have emerged to circumvent these problems:

- Pursuit and thresholding algorithms seek a sub-optimal solution of (2)
- The Basis Pursuit algorithm [1] relaxes the l_0 minimization to, solving the convex optimization problem (4) instead of the original.

$$\hat{\gamma} = \arg\min_{\gamma} \|\gamma\|_{l_1} \quad \textit{subject to} \quad y = \Phi\Psi\gamma \quad (4)$$

The matrix Φ satisfies a restricted isometry property of order K whether there is a constant $\delta_K \in (0,1)$ such that the inequation (5),

$$(1-\delta_K)\|x\|_2^2 \le \|\Phi x\|_2^2 \le (1+\delta_K)\|x\|_2^2 \quad (5)$$

holds for all x with sparsity K.

A. Notions of secrecy and Model

In cryptography, “a secret key system is an encryption system where both sender and receiver use the same key to encrypt and respectively, decrypt the message” [11-12].

A conventional encryption scheme consists of five elements [13-14]:

• *Plain text:* This is the original message or input information for the encryption algorithm.

• *Encryption algorithm*: This algorithm performs various substitutions and modifications to the clear text.

• *Secret Key:* This key is an input to the encryption algorithm.

• *Ciphertext:* The text resulting from encryption algorithm and it is depends on the plaintext and the secret key. Thus, for a given message, two different secret keys produce two different ciphertexts.

• *Decryption algorithm:* This algorithm is the inverse of the encryption algorithm. The decryption algorithm is applied with the same secret key to the ciphertext in order to get the original clear text.

Following two elements must be taken into account in order to achieve a secure encryption [15]:

1) The encryption algorithm should be very strong. If an attacker knows the encryption algorithm (encryption) and has access to one or more ciphertext, he cannot decrypt the ciphertext or find the secret key.

2) Both the transmitter and the receiver must obtain the secret key in a safe manner (on a secure communication channel) and to keep it secret.

Based on previous remarks, in Figure 1 (in the upper half) is shown the basic model for CS and it includes two major aspects: measurements taking and signal recovery. The measurements taking involve an encryption algorithm and signal recovery is associated with a decryption algorithm from the perspective of symmetric-key cipher. The relationship between CS and symmetric cryptography indicates that some possible cryptographic features can be embedded in CS.

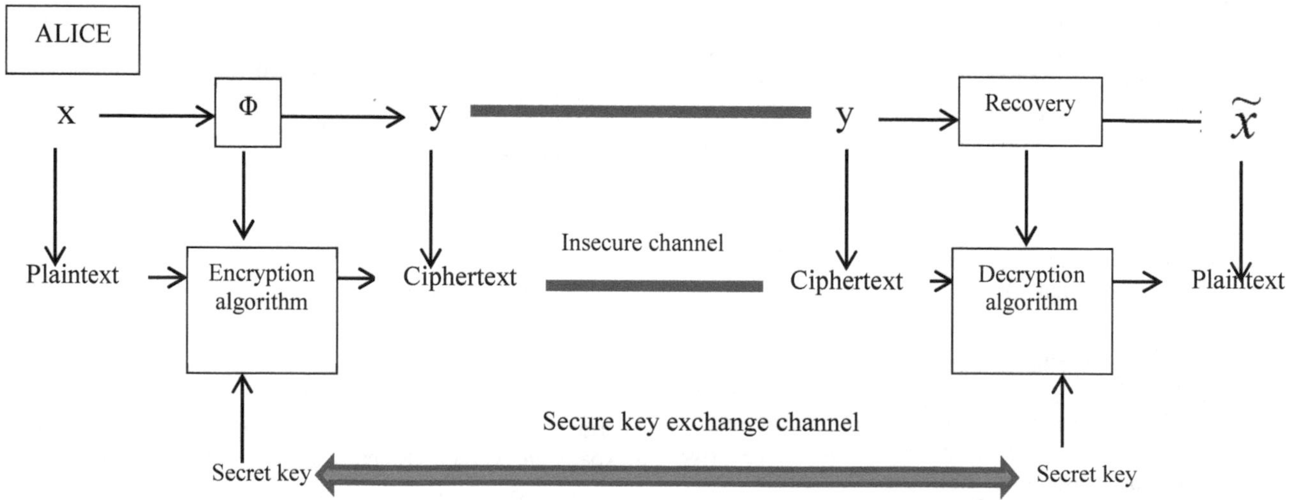

Fig. 1. The relationship between CS and symmetric-key cipher

The classical example of communication of a secret message from Alice to Bob assumes that Alice must use key from the set of keys. In this paper, let be i a key chosen by Alice with equal probability, and used to encrypt the message x with help of Φ_i matrix (via matrix multiplication operation). The result of multiplication is the cryptogram y which is

Identify applicable sponsor/s here. If no sponsors, delete this text box (*sponsors*).

transmitted to Bob. The recipient knows the key used for encryption of the message. Knowing Φ_i and y, the compressed sensing literature provides conditions for x and Φ_i to allow the recovery of the original message x. The classical example of secret message communication assumes that the Alice's encrypted message y is being intercepted by an eavesdropper named Eve. For the third person, the used key the message encryption is unknown.

In our case, the measurement matrix Φ can be selected from a set of keys that is known for the transmitter (Alice) and the permitted receiver (Bob). Each random measurement matrix Φ is generated with a seed which can be exchanged through a secure approach between two desired sides [16-17].

A computational encryption scheme is secure if the ciphertext has one or two properties:

- The cost of breaking ciphertext is much higher than the encrypted information.
- The time needed for breaking ciphertext is longer than the lifetime of the information.

A brute force attack on the compressive sampling based encryption scheme would be guessing the linear measurement matrix Φ_i. Thus, an eavesdropper, e.g. Eve, could directly try to do this by performing an exhaustive search over a "grid" of values for Φ_i. But, the step size of this grid is critical because a too large step size may cause the search to miss the correct value and a too small grid size will increase the computational task unnecessarily.

The computational cost of signal reconstruction is high. For the best optimization algorithm (BP), the computational cost is in the order of $O(N^3)$ and a random search will make the search too expensive.

III. Simulations and Discussions

To test our method, we used 10 text messages (10 different tax forms) and we generated 10 random matrices.

To validate the decoding results, we evaluate the distortion between the original plaintext and the reconstructed plaintext by means of the PRD (the percentage root-mean-square difference), its normalized version (PRDN) measures and NMSE (normalized mean square error).

The percentage root-mean-square difference (PRD) measure defined as (6):

$$PRD\% = 100\sqrt{\frac{\sum_{n=1}^{N}(x(n)-\tilde{x}(n))^2}{\sum_{n=1}^{N}x^2(n)}} \quad (6)$$

is employed, where $x(n)$ is the original signal, $\tilde{x}(n)$ is the reconstructed signal, and N is the length of the window over which the PRD is calculated. The normalized version of PRD, PRDN, which does not depend on the signal mean value, $\bar{x}$, is defined as (7):

$$PRDN\% = 100\sqrt{\frac{\sum_{n=1}^{N}(x(n)-\tilde{x}(n))^2}{\sum_{n=1}^{N}(x(n)-\bar{x})^2}} \quad (7)$$

The normalized mean square error (NMSE) measure defined as (8):

$$NMSE = \frac{1}{\sigma^2} * \frac{\sum_{n=1}^{N}(x(n)-\tilde{x}(n))^2}{N} = \frac{MSE}{\sigma^2} \quad (8)$$

Where σ are the variance and MSE are mean square error measure.

Because our messages are text type, ie contain characters and numbers, we chosen to transform the messages in numerical signals based on the ASCII codes.

To use the identity matrix as decoding dictionary, the plaintext is necessary to be a sparse signal [18]. Because our messages had not this property, we have modified them by artificial insertion of zeros, thus obtaining sparse signals.

We used random matrix for encryption and for reconstruction we used two different algorithms, and namely,

- Basis pursuit algorithm (BP), known in the CS domain as the optimal algorithm in terms of errors [19-20] and
- Orthogonal matching pursuit algorithm (OMP) known in CS domain for its speed far superior to BP [21].

The orthogonal matching pursuit algorithm (OMP) is an iterative greedy algorithm. In this algorithm, at each step, the dictionary element which has the maximum correlation with the residual part of the signal is selected. The Basis Pursuit algorithm (BP) is a more sophisticated approach comparatively with OMP. In case of the BP algorithm, the initial sparse approximation problem is reduced to a linear programming problem.

Generically, the greedy algorithms (such OMP) have the disadvantage that there are not general guarantees of optimality. The basis pursuit algorithm, namely the convex relaxation algorithms, has the disadvantage of high computational complexity, translated into large computing time [22-26].

To synthesize ideas, we present the encryption and decryption necessary steps, namely:

- The message transformation into digital signal using extended ASCII code. This achieves a 1D digital signal.
- The segmentation of message or digital signal into segments of length 100.

- Transforming of the signals (signals with length 100) in sparse signals by inserting a predefined number of zeros. The position of the zeros is random from one segment to another.
- Encryption of sparse segments using a random matrix. Encryption is done by multiplying the signal sparse with a random matrix (Φ), resulting a lower dimension signal than initially sparse signal. The signal thus obtained is not sparse.
- Transmission of the message text is achieved by transmitting the encrypted signals (ciphertext) on an insecure line. It is important that random matrix (encryption matrix representing the secret key) is not sent with the ciphertext; it should be sent on a secure line. Another variant is use case when there is an agreement between the transmitter and receiver to generate random matrices in the same way, for example, using the same random number generator which is started from the same initial conditions.
- Decryption of the message will be achieved using a greedy algorithm (either orthogonal matching pursuit (OMP), or matching pursuit (MP), or greedy LS etc.) or convex relaxation algorithm (basis pursuit (BP)). For decryption, it is necessary to know the following: random matrix encryption Φ, the encrypted message (the ciphertext) Y, and the base for sparsity Ψ (in case of this paper, it is the identity matrix, due the fact that the message that was encrypted was a sparse signal).
- Because there is a decryption error which is very small, to return to the decrypted text, a decryption correction will be necessary. This correction consists in rounding of decrypted values to the nearest integer because the ASCII code is built from integers.

Figure 2 shows an example of plaintext and figure 3 presents a plot of the plaintext in ASCII format.

Anexa nr.1
DECLARATIE
privind veniturile realizate
Agentia Nationala de Administrare
200
Fiscala
din România
Anul Se completeaza cu X în cazul declaratiilor rectificative
A. DATE PRIVIND ACTIVITATEA DESFASURATA Cod CAEN cote forfetare de cheltuieli norma de venit Nr. Data 7. Data începerii activitatii 4. Obiectul principal de activitate 5. Sediul/Datele de identificare a bunului pentru care se cedeaza folosinta 8. Data încetarii activitatii asociere fara personalitate juridica entitati supuse regimului transparentei fiscale individual 6. Documentul de autorizare/Contractul de asociere/Închiriere/Arendare 3. Forma de organizare: 2. Determinarea venitului net: comerciale profesii libere drepturi de proprietate intelectuala cedarea folosintei bunurilor operatiuni de vânzare-cumparare de valuta la termen, pe baza de contract transferul titlurilor de valoare, altele decât partile sociale si valorile mobiliare în cazul societatilor închise activitati agricole 1. Categoria de venit Venituri: cedarea folosintei bunurilor calificata în categoria venituri din activitati independente sistem real modificarea modalitatii/formei de exercitare a activitatii

Fig. 2. The plaintext

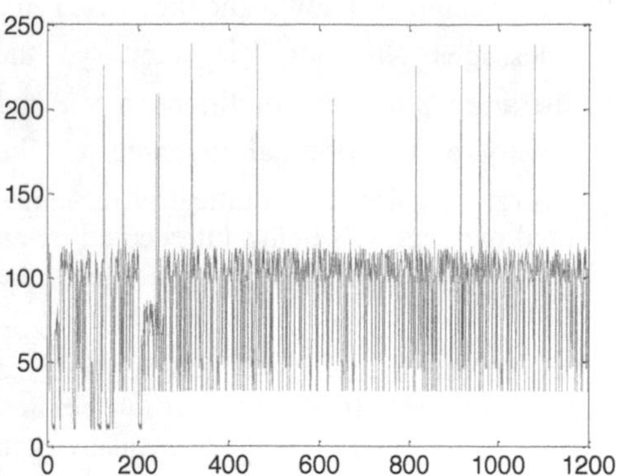

Fig. 3. The plaintext in ASCII format

We have chosen to split the signal into segments of length 100 and to insert a number of 800 by zeros for each ASCII codes segment. This means that each plaintext sequence with length 100 was transformed into a sequence with length 900. Figure 4 shows the plot of sparse plaintext.

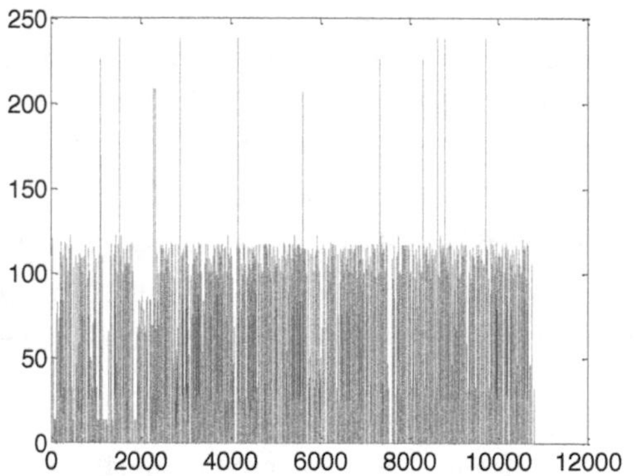

Fig. 4. The sparse plaintext

We used for encryption a random matrix of size 500x900. This random matrix represents the secret key. Figure 5 show the ciphertext obtained a random matrix for encryption. Note that the ciphertext contains positive and negative numbers and it has a different length than the plaintext.

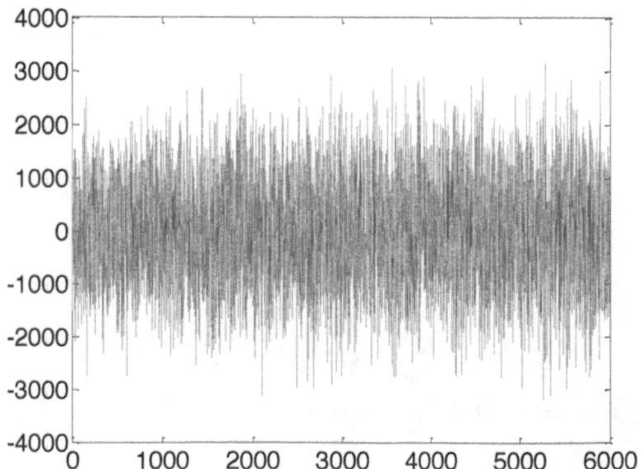

Fig. 5. The ciphertext

To decode the ciphertext we tested two known algorithms from compressed sensing domain, namely, orthogonal matching pursuit algorithm (OMP) and basis pursuit algorithm (BP).

OMP is an iterative greedy algorithm and selects at each step the column of Φ matrix which has the maximum correlation with the current residuals. A set of iteratively selected columns is built. The residuals are iteratively updated by projecting the observation y onto the subspace spanned by the previously selected columns. This algorithm has simpler and faster implementation toward similar methods.

The Basis Pursuit (BP) algorithm consists in finding a least L1 norm solution of the underdetermined linear system $y = \Phi x = \Phi \Psi \gamma$.

The both methods can be guaranteed to have bounded approximation solution of sparse coefficients estimation for the condition that the L0 norm of sparse coefficients is smaller than a constant decided by the dictionary [1].

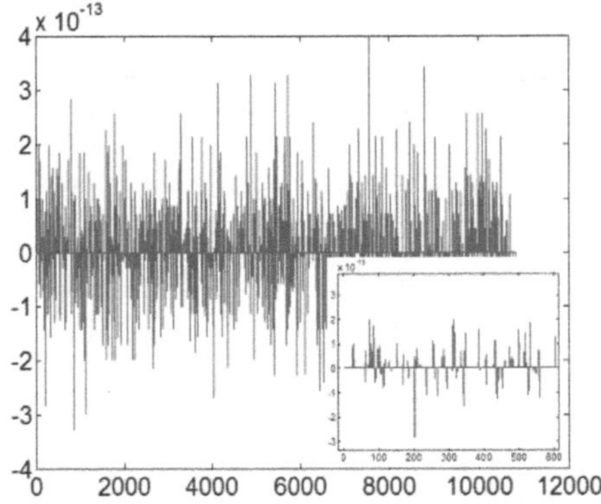

Fig. 6. Error for decoding with OMP, before decoding correction. In the bottom right corner there is a zoom for the first 600 samples

Figure 6 and figure 7 show errors for decoding with OMP, respectively BP algorithms.

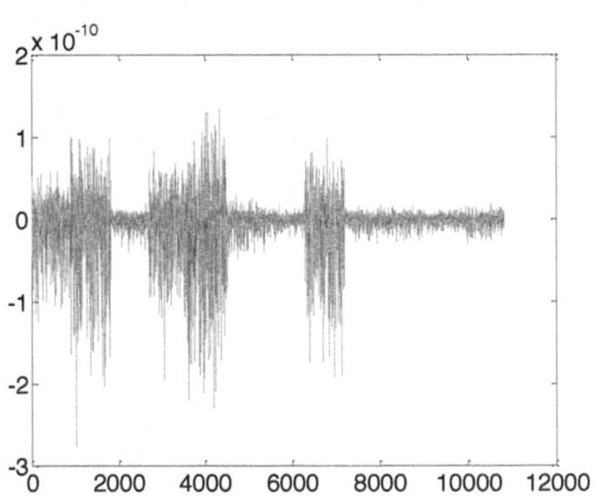

Fig. 7. Error for decoding with BP, before decoding correction

For the OMP based decoding, where the original signal (plaintext) was sparse (had null values), null values were obtained after decoding. In case of BP decoding, the algorithm approximates all values and it failed to return null values for the null values from plaintext, but it returned values very close to zero.

Because in the case of typical tax forms often it is required to encrypt only registration data and because the decryption time is higher for the completed form (data + template), we tested the proposed algorithm for encrypting data alone. In Figure 8 is an example of data belonging to the form shown in Figure 2.

12656
03.08.2010
Minerit
Str. Minei, Nr. 23, Hunedoara
02.08.2000
contract nr. 05/09.10.1999
SRL

Fig. 8. The plaintext with registration data from tax form

For a signal of dimension m with assumed sparsity s<<m, and a dictionary of N>>m atoms, computational costs for pursuits using general and fast dictionaries are:

$$complexity \quad for \quad OMP = smN(sN\log N + s^3))$$

where m stands for measurements, s stands for sparsity.

The popular basis pursuit algorithm (BP) has computational complexity

$$complexity \quad for \quad BP = O(N^3)$$

Alternatives to BP (e.g., greedy matching pursuit) also have computational complexities that depend on N.

Table 1 presents average results for 10 text messages and for 10 datasets from tax forms. The time for BP algorithm is much higher than for OMP algorithm and the errors are smaller.

TABLE I. AVERAGE RESULTS

Decoding algorithm	Time (seconds)	Error (PRD, PRDN, NMSE)
average results for 10 text messages, each with 1200 char		
Basis pursuit algorithm (BP)	867.40	PRD = 7.7521e-011 PRDN = 8.1579e-011 NMSE = 7.1476e-028
Orthogonal matching pursuit algorithm (OMP)	2.61	PRD = 1.0139e-013 PRDN = 1.0670e-013 NMSE = 1.2227e-033
average results for 10 registration data text messages, each with 103 char		
Basis pursuit algorithm (BP)	42.27	PRD = 3.9552e-011 PRDN = 4.1392e-011 NMSE = 3.2769e-028
Orthogonal matching pursuit algorithm (OMP)	0.09	PRD = 1.0979e-013 PRDN = 1.1489e-013 NMSE = 2.5248e-033

It should be noted that the OMP does not guarantee the convergence of the algorithm and for a smaller number of measurements; the results can be much worse for OMP comparatively with BP. Results depend on the number of measurements and on used decoding algorithm [24-26].

IV. CONCLUSIONS

In this paper, the perfect secrecy via compressed sensing was studied and discussed. We presented an analysis with practical results for tax forms as plaintexts. For decoding we used BP and OMP algorithms, and we presented a comparative analysis. The time for BP algorithm is much higher than for

OMP algorithm and the errors are smaller and should be noted that the OMP does not guarantee the convergence of the algorithm. According to average results from Table 1, it is more advantageous, for tax forms (or other templates that show no interest for encryption) to encrypt only the recorded data. The time required for decoding is significantly lower than the decryption for the entire form.

ACKNOWLEDGMENT

This work was supported by a grant of the Romanian National Authority for Scientific Research and Innovation, CNCS – UEFISCDI, project number PN-II-RU-TE-2014-4-0832 "Medical signal processing methods based on compressed sensing; applications and their implementation."

REFERENCES

[1] D. Donoho, "Compressed sensing", IEEE Transactions on Information Theory, vol. 52(4), pp. 1289–1306, 2006

[2] E. Candes, J. Romberg, and T. Tao, "Robust uncertainty principles: Exact signal reconstruction from highly incomplete frequency information" IEEE Transactions on Information Theory, vol. 52(2), pp. 489–509, 2006.

[3] M. Elad, "Optimized Projections for Compressed Sensing", IEEE Transactions on Signal Processing, Vol. 52, 2007

[4] Shuhui Bu, Zhenbao Liu, Tsuyoshi Shiina, Kazuhiko Fukutani, *Matrix Compression and Compressed Sensing Reconstruction for Photoacoustic Tomography*, Elektronika ir elektrotechnika, Vol 18, No 9 (2012)

[5] C. E. Shannon, "Communication theory of secrecy systems" Bell System Technical Journal, vol. 28(4), pp. 656–715, October 1949.

[6] E. Candes, J. Romberg, "Sparsity and incoherence in compressive sampling" Inverse Problems, Vol. 23, pp. 969–985, 2007.

[7] D. Takhar, J. N. Laska, M. B. Wakin, M. F. Duarte, D. B. S. Sarvotham, K. F.Kelly, and R. G. Baraniuk, "A new camera architecture based on optical-domain compression" in Proc. IST/SPIE Symposium on Electronic Imaging: Computational Imaging, vol. 6065, 2006, pp. 129-132

[8] Y. Rachlin, D. Baron, "The Secrecy of Compressed Sensing Measurements", 46th Annual Allerton Conference on Communication, Control, and Computing, 2008

[9] A. Orsdemir, H. Oktay Altun, G. Sharma, Mark F. Bocko,"On the Security and Robustness of Encryption via Compressed Sensing" Military Communications Conference, 2008, Milcom 2008, IEEE pp.1-7

[10] G. Zhang, S. Jiao, X. Xu, "Application of Compressed Sensing for Secure Image Coding", Lecture Notes in Computer Science Volume 6221, 2010, pp 220-224.

[11] Gary C. Kessler, An Overview of Cryptography, 2015, http://www.garykessler.net/library/crypto.html

[12] Sattar B. Sadkhan Al Maliky and Nidaa A. Abbas, Multidisciplinary Perspectives in Cryptology and Information Security, IGI Global, 2014

[13] W. Stallings, Cryptography and Network Security (4th Edition), pp. 30, Prentice Hall, 2005

[14] V. Preoteasa, Cryptography and Network Security, Lecture 2: Classical Encryption Techniques, Spring 2008, Abo Akademi University

[15] D.R. Stinson, Cryptography: Theory and Practice, 2nd edition, Chapman & Hall/CRC, 2002

[16] W. Diffie, M. E. Hellman, "New directions in cryptography" IEEE Transactions on Inform. Theory, vol. IT-22, no. 6, pp. 644–654, 1976.

[17] U. Maurer, S. Wolf, "Information-theoretic key agreement: From weak to strong secrecy for free" Advances in Cryptology—EUROCRYPT, Lecture Notes in Computer Science, 2000.

[18] J. Bowley, L. Rebollo - Neira, "Sparsity and "something else": an approach to encrypted image folding", IEEE signal processing letters, 18 (3), pp. 189-192., 2011

[19] D. Donoho, Y. Tsaig, "Fast solution of L1-norm minimization problems when the solution may be sparse," Stanford University Department of Statistics Technical Report, 2006.

[20] D. Donoho, "For most large underdetermined systems of linear equations, the minimal L1 norm solution is also the sparsest solution" Communications on Pure and Applied Mathematics, Vol. 59, pp. 797–829, June 2006.

[21] J. Tropp, A. Gilbert, "Signal recovery from random measurements via orthogonal matching pursuit" IEEE Trans. on Information Theory, Vol. 53, No. 12, pp. 4655–4666, December 2007.

[22] E. Candes, T. Tao, "Near optimal signal recovery from random projections: Universal encoding strategies?", IEEE Transactions on Information Theory, Vol. 52, No. 12, pp. 5406–5425, December 2006.

[23] M. J. Wainwright, "Sharp thresholds for noisy and high-dimensional recovery of sparsity using L1-constrained quadratic programming (Lasso)", IEEE Transactions on Information Theory, 2009.

[24] T.T. Cai, L. Wang, "Orthogonal Matching Pursuit for Sparse Signal Recovery with Noise", IEEE Transactions on Information Theory, vol. 57, 7, 4680–4688, 2011.

[25] G. Davis, S. Mallat, M. Avellaneda, "Greedy adaptive approximation", J. Constr. Approx., 13:57-98, 1997.

[26] J.A. Tropp, "Greed is good: Algorithmic results for sparse approximation", IEEE Transactions on Information Theory, 50, pp. 2231–2242, 2004.

[27] M. N. Chavhan, S.O.Rajankar, "Study the Effects of Encryption on Compressive Sensed Data", International Journal of Engineering and Advanced Technology, Volume 2, Issue 5, pp. 179 – 182, 2013

8

Changes in Known Statements After New Data is Added

Sylvia Encheva
Stord/Haugesund University College
Bjørnsonsg. 45,
5528 Haugesund,
Norway

***Abstract*—Learning spaces are broadly defined as spaces with a noteworthy bearing on learning. They can be physical or virtual, as well as formal and informal. The formal ones are customary understood to be traditional classrooms or technologically enhanced active learning classrooms while the informal learning spaces can be libraries, lounges, cafés, etc.. Students' as well as lecturers' preferences to learning spaces along with the effects of these preferences on teaching and learning have been broadly discussed by many researchers. Yet, little is done to employ mathematical methods for drawing conclusions from available data as well as investigating changes in known statements after new data is added. To do this we suggest use of ordering rules and ordered sets theories.**

Keywords*—*Ordering rules; Ordered sets; Implications

I. INTRODUCTION

Interest in building, teaching in, and researching the impact of technologically enhanced learning spaces appears to have grown exponentially, [3]. Learning spaces are usually divided into formal and informal, [7]. The first ones are customary understood to be classrooms while the latter can be libraries, lounges or cafés. At the same time, today's teaching and learning processes are heavily effected by an increasing use of laptop computers, smart phones and tablets. All these various opportunities can be viewed with respect to students' preferences and learning effectiveness. To do this we suggest support taken from ordering rules and ordered sets theories, [5].

Modern information technology tools allow good opportunities for collecting, storing and even classifying data. A number of scientific fields like for example automation and decision support systems, [4], require ordering of elements and rules.

The theory of ordering rules is often applied for ordering elements with respect to attributes' values. Ordering rules like "if the value of an object x on an attribute a is ordered ahead of the value of another object y on the same attribute, then x is ordered ahead of y", are presented in [10] and [11]. This rises a natural question on whether it is possible to combine several ordering rules and draw reasonable conclusions afterwards.

The rest of the paper goes as follows. Some terms from the fields of ordered sets and ordered relations are presented in Section II. Their application is discussed in Section III followed by a conclusion in Section IV.

II. ORDERED SETS AND RELATIONS

Two very interesting problems are considered in [2], namely the problem of determining a consensus from a group of orderings and the problem of making statistically significant statements about ordering.

A relation I is an *indifference* relation when given AIB neither $A > B$ nor $A < B$ has place in the componentwise ordering. A partial ordering whose indifference relation is transitive is called a *weak ordering*. A *total ordering* is a binary relation which is transitive, antisymmetric, and total ($p \leq q$ or $q \leq p$).

If given two alternatives, a person is finally choosing only one. the natural extension of to more than two elements is known as the 'majority rule' or the 'Condorcet Principle'. A relation $R(L_1, L_2, ..., L_k)$ is constructed by saying that the pair $(a, b) \in R$ if (a, b) belong to the majority of relations L_i.

The following three linear orderings

$$a \quad b \quad c$$

$$b \quad c \quad a$$

$$c \quad a \quad b$$

leading to

$$R = \{(a, b), (b, c), (c, a)\}$$

(three-way tie), illustrate the 'paradox of voting'. A 'social welfare function' maps k-tuples of the set of linear orderings of any $b \subset A$ to single linear orderings of B, where A is a set of at least three alternatives, [1].

Two elements a and b where $a \neq b$ and $a, b \in P$ are comparable if $a \leq b$ or $b \leq a$, and incomparable otherwise. If $\forall a, b$ where $a, b \in P$ are comparable, then P is chain. If $\forall a, b$ where $a, b \in P$ are incomparable, then P is antichain.

Interesting set-relational approach for computer administration of psychological investigations has been employed in [12].

Below we list some definitions and formulas as in [10]. The authors also introduce order relations into attribute values.

An information function I_a is a total function mapping an object of U to an exact value in V_a.

Definition 1: [9] An information table is a quadruple:

$$IT = (U, At, \{V_a | a \in At\}, \{I_a | a \in At\}),$$

where U is a finite nonempty set of objects, At is a finite nonempty set of attributes, V_a is a nonempty set of values for $a \in At$, $I_a : U \to V_a$ is an information function.

Definition 2: Let U be a nonempty set and $\succ$ be a binary relation on U. The relation $\succ$ is a weak order if it satisfies the two properties:

Asymmetry:

$$x \succ y \Longrightarrow \neg(y \succ x),$$

Negative transitivity:

$$(\neg(x \succ y), \neg(y \succ z)) \Longrightarrow \neg(x \succ z).$$

An important implication of a weak order is that the following relation,

$$x \sim y \Longleftrightarrow (\neg(x \succ y), \neg(y \succ x))$$

is an equivalence relation. For two elements, if $x \sim y$, we say x and y are indiscernible by $\succ$. The equivalence relation $\sim$ induces a partition $U/\sim$ on U, and an order relation $\succ^\star$ on $U/\sim$ can be defined by

$$[x]_\sim \succ^\star [y]_\sim \Longleftrightarrow x \succ y$$

where $[x]_\sim$ is the equivalence class containing x. Any two distinct equivalence classes of $U/\sim$, can be compared since $\succ^\star$ is a linear order.

Definition 3: An ordered information table is a pair

$$OIT = (IT, \{\succ_a | a \in At\}),$$

where IT is a standard information table and $\succ_a$ is a weak order on V_a.

An ordering of values of a particular attribute a naturally induces an ordering of objects, namely, for $x, y \in U$:

$$x \succ_{\{a\}} y \Longleftrightarrow I_a(x) \succ_a I_a(y),$$

where $\succ_{\{a\}}$ denotes an order relation on U induced by the attribute a. An object x is ranked ahead of another object y if and only if the value of x on the attribute a is ranked ahead of the value of y on a. The relation $\succ_{\{a\}}$ has exactly the same properties as that of $\succ_a$. That is, x is ranked ahead of y if and only if x is ranked ahead of y according to all attributes in A.

Data mining in an ordered information table may be formulated as finding association between orderings induced by attributes.

Definition 4: Consider two subsets of attributes $A, B \subseteq At$. For two expressions $\phi \in E(A)$ and $\psi \in E(B)$, an ordering rule is read "if ϕ then ψ" and denoted by $\phi \Rightarrow \psi$. The expression ϕ is called the rule's antecedent, while the expression ψ is called the rule's consequent.

III. Graphical Representation of Correlations Between Ordering Rules

Recent enthusiasm for shifting the manner in which institutions of higher education approach and conceptualize classroom space has been fueled by a host of articles extolling the potential transformative power of formal learning spaces on teaching practices and learning outcomes, [3]. Handling such large amount of data resulting from both physical and mental processes requires solid well formalized techniques like f. ex. those included in the field of artificial intelligence.

It is worth mentioning that ordinary majority voting is unable to handle situations with incomplete or changing data. Ordered sets and ordering rules can accommodate both very nicely. They are particularly applicable in cases when some of the elements are compared while others are not or different conclusions are drawn by different groups and non of them should be ignored in a decision making process.

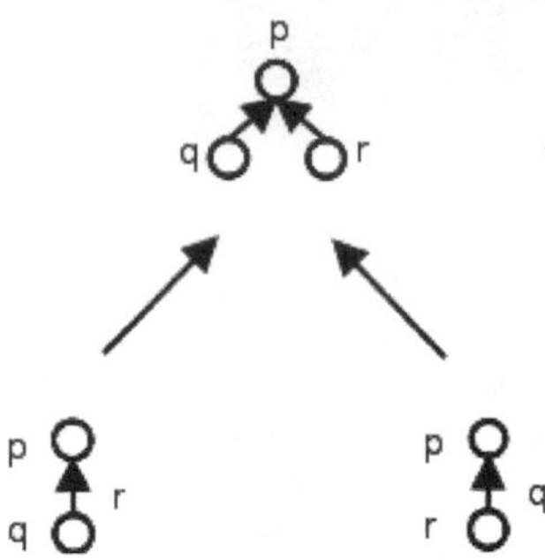

Fig. 1. q and r imply p

Suppose students are suggested to express their preferences with respect to two different learning spaces x, y based on three attributes a, b, c. Their preferences related to learning spaces are first collected and conclusions can be drawn afterwards supported by ordered sets theory.

An ordering rule states how orderings of objects by attributes in A determines orderings of objects by attributes in B. For example, an ordering rule,

$$(a, \succ) \wedge (b, \preceq) \Rightarrow (c, \succ),$$

can be re-expressed as

$$x \succ_{\{a\}} y \wedge x \preceq_{\{b\}} y \Rightarrow x \succ_{\{c\}} y.$$

That is, for two arbitrary objects x and y, if x is ranked ahead of y by attribute a, and at the same time, x is not ranked ahead of y by attribute b, then x is ranked ahead of y by attribute c.

In order to facilitate readability we introduce three new notations p, q, r, where:

p stands for $x \succ_{\{a\}} y$,

q stands for $x \preceq_{\{b\}} y$, and

r stands for $x \succ_{\{c\}} y$.

Obviously, p, q, r generate six implications where one of them implies any of the other two. When two of them imply the third we obtain a rule similar to

$$x \succ_{\{a\}} y \wedge x \preceq_{\{b\}} y \Rightarrow x \succ_{\{c\}} y.$$

The three possible rules are illustrated in Fig. 1, Fig. 2, and Fig. 3. They can be used to draw conclusions when only

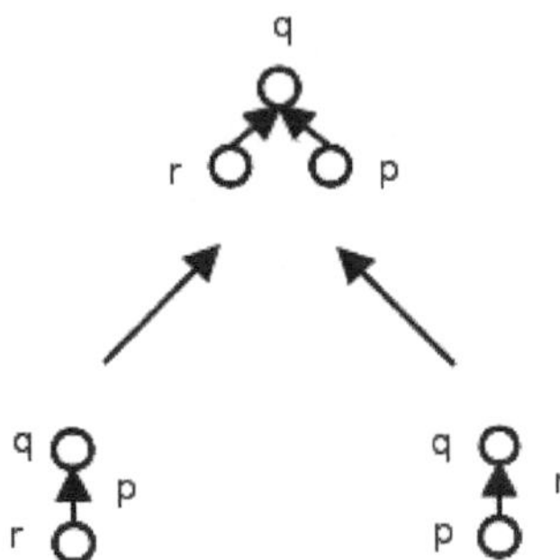

Fig. 2. p and r imply q

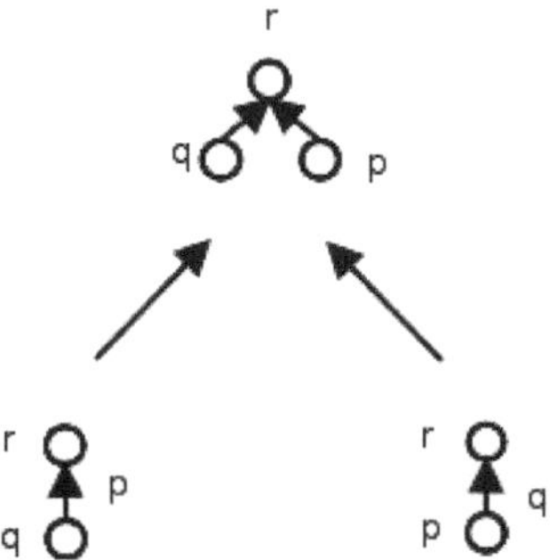

Fig. 3. p and q imply r

couples of spaces are evaluated and also when a conclusion involving all of them is required.

We can extract much more information if we connect all these implications as in Fig. 4. The applied implication rules are based on the theory presented in Section II.

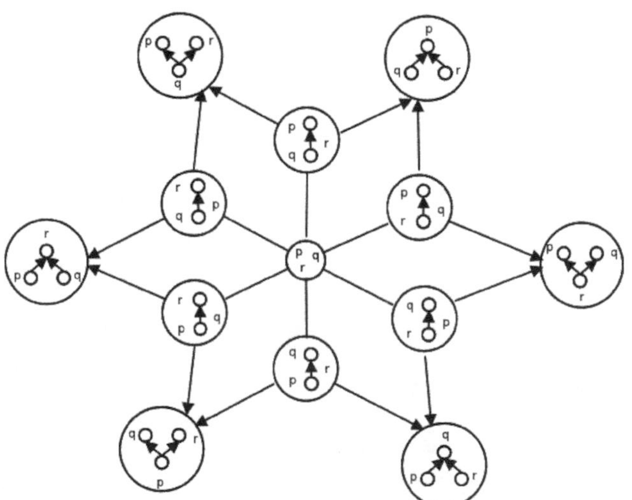

Fig. 4. Illustration of ordering rules

The graph in Fig. 5 shows that comparing two cases where an element is a rule's antecedent in one case or a rule's consequent in the other one is operating with two disjoint sets of implications.

Any two non disjoint sets of implications, i.e. pares of implications sharing one implication, generate a rule where

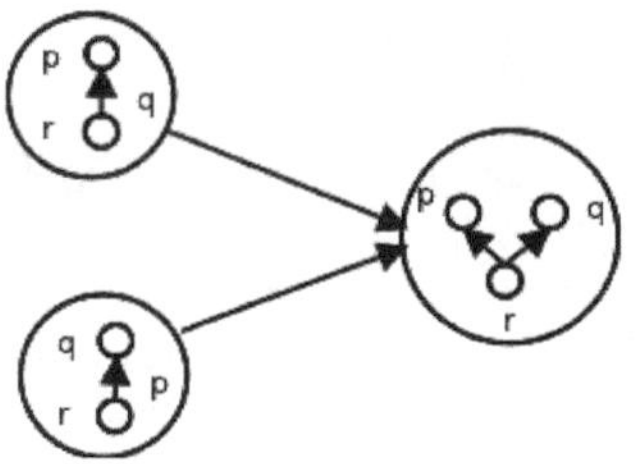

Fig. 5. p and q are implied by r

one element is an antecedent and another rule where another element is a consequent, see f. ex. Fig. 6.

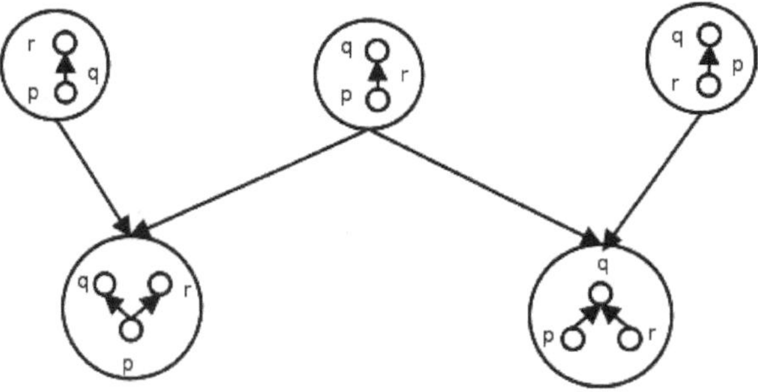

Fig. 6. Two sets of implications

Formal classroom space shapes the behavior of lecturers and students, [3]. A ranking system can be used for drawing automated conclusions when different initial rules are assumed. For ranking systems the rule illustrated in Fig. 6 can be used to show how one ordering combined with two different orderings can lead to two different conclusions.

Remark: Learning spaces should not be considered as a part of knowledge space theory, [6] and [8]. Knowledge spaces refer to states of knowledge of a person.

IV. Conclusion

Appropriate combination of ordering rules sharing common elements leads to new conclusions. Closed sets coupled with ordering rules can be applied in decision support processes and development of machine learning techniques.

In future work we intend to investigate opportunities to involve more than three elements and rules in an ordering rules deductive system.

References

[1] K. J. Arrow, *Social Choice and Individual Values*, Wiley, New York, 2nd ed. 1963.

[2] K. P. Bogart, *Some social sciences applications of ordered sets*, In: I. Rival, Editor, Ordered Sets, Reidel, Dordrecht, pp. 759–787, 1982.

[3] D. C. Brooks, *Space and Consequences: The Impact of Different Formal Learning Spaces on Instructor and Student Behaviour*, Journal of Learning Spaces, Vol.1, No 2, 2012.

[4] C. Carpineto and G. Romano, *Concept Data Analysis: Theory and Applications*, John Wiley and Sons, Ltd., 2004.

[5] B. A. Davey and H. A. Priestley, *Introduction to lattices and order*, Cambridge University Press, Cambridge, 2005.

[6] J.-P. Doignon and J.-C. Falmagne, *Knowledge Spaces*, Springer-Verlag, 1999.

[7] J.-P. Doignon and J.-C. Falmagne, *Learning Spaces*, Springer-Verlag, 2011.

[8] J. -C. Falmagne, D. Albert, C. Doble, D. Eppstein, and X. Hu, *Knowledge Spaces: Applications in Education*, Springer, 2013.

[9] Z. Pawlak, *Rough Sets, Theoretical Aspects of Reasoning about Data*, Kluwer Academic Publishers, Dordrecht, 1991.

[10] Y. Sai and Y. Y. Yao, *Analyzing and Mining Ordered Information Tables*, Journal of Computer Science and Technology, vol. 18, No.6, 771–779, 2003.

[11] Y. Sai and Y. Y. Yao, *On Mining Ordering Rules*, New Frontiers in Artificial Intelligence, Lecture Notes in Computer Science, vol. 2253, pp. 316–321, 2001.

[12] K. Yordzhev and I. Peneva, *Computer Administering of the Psychological Investigations: Set-Relational Representation*, Open Journal of Applied Sciences, 2, pp. 110–114, 2012. http://www.scirp.org/journal/PaperInformation.aspx?PaperID=20331

An Evaluation of the Implementation of Practice Teaching Program for Prospective Teachers at Ganesha University of Education Based on *CIPP-Forward Chaining*

I Putu Wisna Ariawan[1]
Lecturer of Mathematics Education,
Ganesha University of Education
Bali, Indonesia

Dewa Bagus Sanjaya[2]
Lecturer of Civics Education,
Ganesha University of Education
Bali, Indonesia

Dewa Gede Hendra Divayana[3]
Lecturer of IT Education
Ganesha University of Education
Bali, Indonesia

***Abstract*—The recognition of teacher status is very high and this is followed by a requirement of a high competence level that a teacher has to have that the existence of teachers has to gain a serious attention, including also the beginning of prospective teachers preparation. Ganesha University of Education (Undiksha) as one of public universities in Indonesia is given authority by the government to train prospective teachers. To produce quality prospective teachers that meet the requirement Undiksha requires all education students who will become prospective teachers to take Practice Teaching Program (PPL Real). However, in its implementation it can be said that it has not been effective yet. Thus there is a need to evaluate the implementation of Practice Teaching Program for prospective teachers at Undiksha. One of the techniques used is the CIPP model that is combined with Forward Chaining Method that is one of inferring strategies of Expert System. From the components of context, input, process, and product, the implementation of Practice Teaching Program (PPL-Real) of the education students of Undiksha in 2015 falls into an effective classification.**

Keywords—Evaluation; Practice Teaching Program; CIPP; Forward Chaining; Expert System

I. INTRODUCTION

The existence of teachers in Indonesia is strictly regulated in [1] that states that a teacher has a status as professional profession, and a teacher also has four major competences, namely pedagogical, personal, social and professional competencies. In [2] it is stated that the four competencies that can be used as the most important characteristics of teachers as profession.

Pedagogical competence is the ability of the teacher in managing instruction for the students that consists of the understanding of the students, the development of potentialities of the students, the planning and implementation of instruction, and the evaluation of learning achievement. Personal competence is the personal ability of the teacher that is reflected in a healthy, stable, matured, wise, charismatic, creative, polite, disciplined, honest, neat and becomes the model for the students. Professional competence is the ability to master the instructional materials well and has an expertise in the field of education that consists of: mastery of the instructional materials, understanding of curriculum and its development, classroom management, use of strategies, media, and learning resources, has an insight into educational innovations, gives help and guidance to the students, etc. Social competence is the teacher's ability to communicate and interact well with the students, students' parents and the community, all of his or her fellow teachers/colleagues and can work together with the education council/school committee, is able to participate actively in preservation and development of the community's culture, and participates actively in social activities.

The so high recognition of the teacher's status and the high requirement for the competences that the teacher has to meet, makes the existence of the teacher something that has to gain a serious attention, including the beginning of the prospective teachers' preparation.

Ganesha University of Education (Undiksha) as one of public universities in Indonesia is given authority by the government to train prospective teachers. The graduates are expected to be of good quality that have skill, knowledge and disposition to be effective educators and ready to teach and able to have impact on the students' learning at the local schools [3].

To be able to produce good quality prospective teachers, as to meet the above requirement Undiksha requires all education students who will become prospective teachers to take Practice Teaching Program (PPL Real). In [4] it is stated that PPL is a program that forms a forum of training to apply various knowledges, skills, and attitudes in the framework of developing teacher professionalism in keeping with the demand of developments in science and technology, and art as stipulated in the National Education Art. As a program, success of PPL Real that is run by Undiksha needs to be seen carefully to be of use as a consideration about its usefulness, process and results.

II. Literature Review

A. *Practice Teaching Program for Prospective Teachers at Ganesha University*

In [5], [6] it is stated that practice teaching has a key position in the teachers education program. PPL Real will give an opportunity to the student teachers to connect theories and concepts that they studied with experience in the real classroom [7]. Considering the essential nature of practice teaching program for the teacher's education program, PPL Real, which is the practice of real teaching at school, is established as a compulsory program that has to be taken by all education students at Undiksha.

In [4] some conditions have been decided that the education students at Undiksha have to meet to be eligible to take PPL Real, namely having been registered formally as participants, having taken PPL Awal, having passed major instructional program development (micro-teaching), having passed at least 120 semester credit units, and taken the whole of the preparation activity. The preparation activity is held before the students do practice teaching at school.

The implementation of PPL Real calls for the application and integration of all of the students' previous learning experiences into the training program, in the form of performances in all that are related to the teaching position, both the teaching activity and other teaching tasks. The activities are done gradually and in integrated way in the form of field orientation, limited practice, guided practice and independent practice that are systematically scheduled that are facilitated by a supervising lecturer and guiding teacher in collaboration.

The supervising lecturer and the guiding teacher assigned to guide the student during the teaching practice at school have to meet the conditions that are regulated in [4]. The *conditions* for the guiding teacher are : (a) having a status as a full time teacher and preferably with at least S1 educational background and D2 educational background for Early Child Education Teacher Program and Primary School Education Program, (b) having worked at least 5 years or having at least "guru dewasa" position, (c) experienced in teaching the subject that he or she guides at least 3 years, and, (4)having taken the practice teaching pattern training. While for the supervising lecturer, the conditions are : (a) for the lecturers who are the S1 certificate holders, having worked at least 5 years or possessing at least "lektor" position, (b) for those who are the S2 or S3 holders, having worked at least 3 years, and (c) having taken the practice teaching pattern training.

In [4] it is mentioned that the responsibility of the supervising lecturer and guiding teacher is to guide the student teachers in doing practice teaching, to discuss various problems faced at the school, and to direct the students towards the pursuit of various experiences obtained that can be used as the preparation as professional prospective teachers. In everyday activities the guiding process is largely done by the guiding teacher. While the supervising lecturer due to his or her tight schedule on campus, is required to supervise/ be present at the partner school at least 4 times, that is, at the first meeting, in the middle, prior to the student takes the practice teaching examination and at the practice teaching examination.

In doing PPL Real the student is also required to observe an experienced teacher. In [8] it is stated that observing an experienced teacher is the important part of the many kinds of practice teaching programs in the world.

Success in or passing PPL Real is determined based on the score obtained by the student from the guiding teacher and the supervising lecturer, the score in the preparation program and the score in the monitoring with the minimum score 70 (in the 100 scale).

The score given by the guiding teacher and the supervising lecturer refers to four components using Instrument for Evaluating Prospective Teacher (APKCG) that is Lesson Plan (N1), Instructional Procedures (N2), Non-instructional Task (N3), Practice Teaching Final Report (N4).

B. *Evaluation*

In [9], evaluation is an activity for collecting information about the effectiveness of something; the information is then used to determine an appropriate alternative in making a decision.

In [10], evaluation is a systematic collection of facts to determine whether in fact there is a change in students and establish the extent of change in the individual student.

In [11], the evaluation is an activity for data collecting, data analysing and data presenting into information about a particular object under study so that the results can be used to take a decision.

In [12], evaluation is a process starting from describing, obtaining and explaining various pieces of information that can be used for determining a decision.

From various points of view above, it can be concluded that evaluation is an activity that consists of the process of gathering, describing, and explaining various pieces of information about the effectiveness of something that can be used later as the consideration for making a decision and a recommendation.

C. *CIPP (Context, Input, Process, Product)*

In [13], CIPP model is a model that essentially has four stages of evaluation are: evaluation of the component context, component input, component process and component product.

In [14], the CIPP evaluation there are four components that must be passed is the evaluation of the component context, the evaluation of input component, the evaluation of process components, and the evaluation of product components.

In [15], one of the major strengths of CIPP model is it provides a useful and simple instrument to help the evaluator to generate important questions to be raised in the evaluation process.

From various points of view above it can be concluded that CIPP model is a model that consists of four major components, that is, context, input, process, and product that are often used by an evaluator to evaluate an object.

D. *Forward Chaining*

In [16], The inference engine contains the methodology used to perform reasoning on the information in the knowledge base and used to formulate conclusions. Inference engine is the part that contains the mechanism and function of thought patterns of reasoning systems that are used by an expert. The mechanism will analyse a specific problem and will seek answers, conclusions or decisions are best. Because the inference engine is the most important part of an expert system that plays a role in determining the effectiveness and efficiency of the system. There are several ways that can be done in performing inference, including the Forward Chaining.

In [17], If there are many competing hypotheses, and there is no reason to start with one rather than another, it may be better to chain forward. In particular, forward chaining is more natural in monitoring tasks where the data are acquired continuously and the system has to detect whether an anomalous situation has risen, a change in the input data can be propagated in the forward chaining fashion to see whether this change indicates some fault in the monitored process or a change in the performance level. If there are a few data nodes and many goal nodes then forward chaining looks more appropriate.

In [18], forward chaining is a top down method which takes facts from satisfied conditions in rules which lead to actions being executed.

From various points of view above, it can be concluded that forward chaining is a strategy of making an inference that is used in Expert System to obtain a conclusion / decision that starts by tracing facts and premises.

III. METHODOLOGY

A. *Object dan Research Site*

1) Research Object is Practice Teaching Program (PPL Real).

2) Research Site is at Ganesha University of Education, Bulleng, Bali.

B. *Subjects of Research*

The subjects were 250 Undiksha education students that took PPL Real in the Odd Semester in 2015 drawn at random out of 1640 students dispersed in 145 schools where the practice teaching program was implemented (kindergartens, primary schools, junior high schools, senior high schools, vocational high schools) in Buleleng and Denpasar.

C. *Data Type*

In this research, the authors use primary and secondary data, qualitative and also quantitative data.

D. *Data Collection Techniques*

In this research, the authors use data collection techniques such as observation, interviews, questionnaire, and documentations.

E. *Evaluation Model*

Evaluation model used in this research is CIPP model.

F. *Aspect of Evaluation*

The aspects evaluated in practice teaching program for prospective teachers at Ganesha University of Education can be seen in the table of evaluation criteria below.

TABLE I. EVALUATION CRITERIA FOR PRACTICE TEACHING PROGRAM FOR PROSPECTIVE TEACHERS AT GANESHA UNIVERSITY OF EDUCATION BULELENG

No	Component	Aspects	
1.	Context	A_1	Position of PPL Real in higher education curriculum
		A_2	Vision and Missions of PPL Real
		A_3	PPL Real Implementation Regulation
2.	Input	A_4	Syllabus, lesson plans
		A_5	Human resources
		A_6	Infrastructure and facilities
3.	Process	A_7	PPL Real implementation plan
		A_8	PPL Real implementation
		A_9	PPL Real evaluation
4.	Product	A_{10}	Impacts of PPL Real implementation
		A_{11}	Results expected from PPL Real implementation

IV. RESULT AND DISCUSSION

A. *Result*

1) Result of Model CIPP Analysis

The result of analysis using CIPP model obtained from the evaluation of the practice teaching program implementation of the prospective teachers at Ganesha University of Education in Buleleng can be described as follows.

a) In the Context component, the effectiveness level of 78. 52% is obtained with T-Score = 52. 235. Thus it can be concluded that in the context variable, the implementation of the practice teaching program (PPL-Real) of the education students of Undiksha in 2015 falls into effective classification.

b) In the Input component, the effectiveness level of 72. 47% is obtained with t-score = 51. 084. Thus it can be concluded that the implementation of the practice teaching program (PPL-Real) of the education students of Undiksha in 2015 falls into effective classification.

c) In the Process component, the effectiveness level of 70. 26% with t-score = 50. 003. Thus it can be concluded that in the Process component, the implementation of the practice teaching program (PPL-Real) of the education students of Undiksha in 2015 falls into effective classification.

d) In the Product component, the effectiveness level of 71. 94% is obtained with t-score = 50. 072. Thus it can be concluded that in the Product component, the implementation of the practice teaching program (PPL-Real) of the education students of Undiksha in 2015 falls into effective classification.

2) The result using Forward Chaining method

Before we look at the result of the analysis using forward method to evaluate the practice teaching program of the prospective teachers at Ganesha University of Education, first let us make it clear that the effectiveness level in all of the components of CIPP using *Glickman's pattern* is as follows[19].

TABLE II. EFFECTIVENESS LEVEL IN ALL OF THE COMPONENTS IN CIPP USING GLICKMAN'S PATTERN

Component	Effectiveness Level			
	Very Less Effective	Less Effective	Effective	Very Effective
Context, Input, Process, Product	- - - - - - - - - - - - - - - -	+ - - - - + - - - - + - - - - + + + - - + - - + + - + - - + - + - + + - - - + +	+ + + - + + - + + - + + - + + +	+ + + + + + + + + + + + + + + +

Notes:

T-Score > 50 → + (Positive)

T-Score ≤ 50 → - (Negative)

The result of analysis using Forward Chaining obtained from the evaluation of the practice teaching program of the prospective teachers at Ganesha University of Education in Buleleng is as follows.

TABLE III. RESULT OF ANALYSIS USING FORWARD CHAINING METHOD OBTAINED FROM THE EVALUATION OF THE PRACTICE TEACHING PROGRAM OF THE PROSPECTIVE TEACHERS AT GANESHA UNIVERSITY OF EDUCATION

No	Component	Effectiveness Level			
		%	T-Score	+ (Positive)	- (Negative)
1.	Context	78.52	52.235	√	X
2.	Input	72.47	51.084	√	X
3.	Process	70.26	50.003	√	X
4.	Product	71.94	50.072	√	X
Result				Very Effective	

From the above table a detailed description was made through a tracing chart with Forward Chaining Method as shown below.

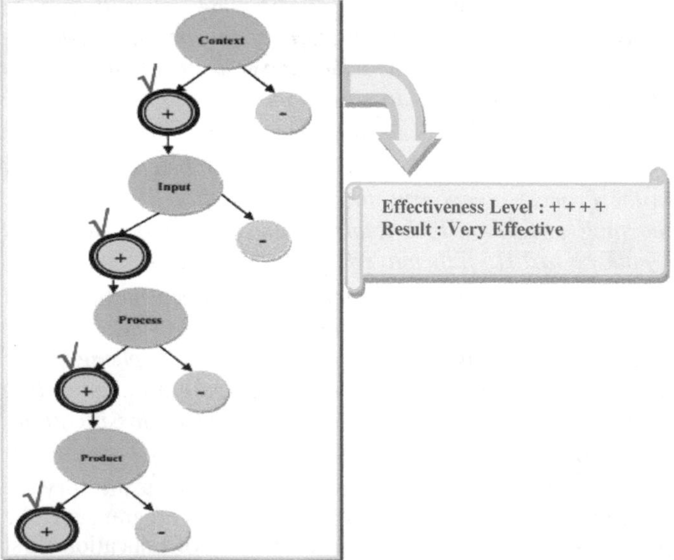

Fig. 1. Tracing with Forward Chaining to Evaluate the Practice Teaching Program Implementation for the Prospective Teachers at Undiksha

From figure 1 above we can see green circles marked with red check marks indicating that the Context component obtains an effective level of + a (positive) value, the Input component + a +(positive) value, the Process component a + (positive) value, and the Product component a + (positive) value. Thus if this is matched with the table of effectiveness level in all CIPP components pada semua komponen CIPP using Glickman's pattern, then the result of the evaluation of the practice teaching program implementation for the prospective teachers at Ganesha University of Education is very effective.

B. Discussion

In the Input component, of the 276 teachers involved as guiding teachers for the students taking the practice teaching (Real PPL) at schools, 118 teachers (42. 754%) had a status as full time teachers of foundations or "contract teachers". Especially for "contract teachers" they did not meet the requirement for becoming guiding teachers who have to be full time teachers.tentu tidak sesuai dengan persyaratan. However, considering the assignment of them by the school principals, the limitation in human resources and the teachers' competences, they were also included as guiding teachers. This policy turns out not to be wrong, as proven by the score of the Input component that remains high. The students guided by the "contract teachers" stated that the guiding teachers are those who match their fields, can guide and evaluate well. From this fact, it can be said that the requirement concerning the status of guiding teachers does not hinder the implementation of the practice teaching. Thus, a revision needs to be made to the requirement concerning the status of guiding teachers.

Although in the Context and the Process components of the implementation of the practice teaching program (PPL-Real) of the education students of Undiksha in 2015 falls into very effective classification, based on the result of monitoring there were still some students who left the schools to take courses. Although they did not ignore the main duties of practice teaching, this should be avoided since the essence of PPL-Real is not only having the teaching practice only. Thus a better regulation concerning the implementation of PPL Real needs to be made.

V. CONCLUSIONS

Based on the analysis that has been made and the results of the discussion in the previous section, then some conclusions can be drawn as follows:

a) From the components of contex, input, process, and product of the implementation of Practice Teaching (PPL-Real) of the education students in 2015 is effective.

b) Although in the components of contex, input, process, and product of the implementation of Practice Teaching (PPL-Real) of the education students of Undiksha in 2015 is effective, but there are some problems that need to be improved, including the requirement that has to be met by the guiding teachers concerning their status, and improvent of the regulation for the implementation of the practice teaching (PPL Real).

REFERENCES

[1] Law of the Republic of Indonesia No. 14, 2005 about Teachers and Lecturers.

[2] Suparlan, "Teachers as a Profession and Standards of Competence," 2005. Available on http://www.suparlan.com/pages/posts/guru-sebagai-profesi-dan-standar-kompetensinya44.php. Access at 03-01-2016.

[3] Lisa Barron, "Preparing Pre-Service Teachers for Performance Assessments," in Journal of Interdisciplinary Studies in Education, Vol. 3, Issue 2, 2015, pp. 70-76.

[4] LPPL, Guidebook of Undiksha Practice Teaching Program. Singaraja: Ganesha University of Education, 2014.

[5] M. A. Al-Malki and K. Weir, "A comparative Analysis between the Assessment Criteria Used to Assess Graduating Teachers at Rustaq College (Oman) and Griffith University (Australia) During The Teaching Practicum," in Australian Journal of Teacher Education, Vol. 39, Issue 12, 2014, pp. 28-42.

[6] Aijaz Ahmed Gujjar, Muhammad Ramzan, and Muhammad Jamil Bajwa, "An Evaluation of Teaching Practice: Practicum," in Pak. J. Commer. Soc. Sci., Vol. 5 (2), 2011, pp. 302-318.

[7] Darryl Roy T. Montebon, "A Needs Assessment Survey on Teacher Readiness of Science Pre-Service Teachers: Towards a Contextualized Student Teaching Enhancement Program (STEP)," in International Journal of Learning, Teaching and Educational Research, Vol. 10, No. 3, 2015, pp. 17-26.

[8] Bilal Genc, and Kagan Buyukkarci, "An Assessment Of Pre-Service Language Teachers' Practicum Observation Forms: Descriptive Observation vs. Critical Observation," in Educational Research, Vol. 2, No. 2, 2013, pp. 83-91.

[9] S. Arikunto, and Cepi Safruddin Abdul Jabar, Guidelines of Theoretical-Practical Education Program Evaluation for Students and Education Practitioners (Second Edition). Jakarta: Bumi Aksara, 2009.

[10] Bloom, Evaluation to Improve Learning. San Fransisco: McGraw Hill Book Company,1981.

[11] D. Bagus Sanjaya, and D. G. Hendra Divayana, "An Expert System-Based Evaluation of Civics Education as a Means of Character Education Based on Local Culture in the Universities in Buleleng," in International Journal of Advanced Research in Artificial Intelligence, Vol. 4, No. 12, 2015, pp. 17-21.

[12] D. Rosyanda, Democratic Education Paradigm, A Model Community Involvement in Education Implementation. Jakarta: Kencana Pronada Media, 2004.

[13] D. Bagus Sanjaya, and D. G. Hendra Divayana, "An Expert System-Based Evaluation of Civics Education as a Means of Character Education Based on Local Culture in the Universities in Buleleng," in International Journal of Advanced Research in Artificial Intelligence, Vol. 4, No. 12, 2015, pp. 17-21.

[14] Wirawan, Evaluation Theory, Model, Standards, Applications, and Profession (1st Edition). Jakarta: Rajawali Pers,2011.

[15] K. Hakan, dan F. Seval, "CIPP Evaluation Model Scale: Development, Reliability and Validity," in Procedia Social and Behavioral Sciences, Vol.1 No.1, 2011, pp. : 1-8.

[16] D.G. Hendra Divayana, "Development of Duck Diseases Expert System with Applying Alliance Method at Bali Provincial Livestock Office," International Journal of Advanced Computer Science and Applications, Vol. 5, No. 8, 2014, pp. 48-54.

[17] S.T. Deepa, and S. G. Packiavathy. "Expert System for Car Troubleshooting," International Journal for Research in Science & Advanced Technologies, Vol. 1, No. 1, 2012, pp. 46-49.

[18] N.L. Griffin, and F.D. Lewis, A Rule Base Inference Engine Optimal and VLSI Implementable. Lexington: Computer Science Dept. University of Kentucky, www.citeseerx.ist.psu.edu., 1993.

[19] I.W. Koyan, Assesment in Education. Singaraja: Undiksha Press, 2007.

10

A Heuristic Approach for Minimum Set Cover Problem

Fatema Akhter
IEEE student member
Department of Computer Science and Engineering
Jatiya Kabi Kazi Nazrul Islam University Trishal,
Mymensingh-2220, Bangladesh

***Abstract*—The Minimum Set Cover Problem has many practical applications in various research areas. This problem belongs to the class of NP-hard theoretical problems. Several approximation algorithms have been proposed to find approximate solutions to this problem and research is still going on to optimize the solution. This paper studies the existing algorithms of minimum set cover problem and proposes a heuristic approach to solve the problem using modified hill climbing algorithm. The effectiveness of the approach is tested on set cover problem instances from OR-Library. The experimental results show the effectiveness of our proposed approach.**

***Keywords*—Set Cover; Greedy Algorithm; LP Rounding Algorithm; Hill Climbing Method**

I. Introduction

For a given set system on a universe of items and a collection of a set of items, Minimum Set Cover Problem (MSCP) [1] finds the minimum number of sets that covers the whole universe. This is a NP hard problem proven by Karp in [2]. The optimization has numerous applications in different areas of studies and industrial applications [3]. The applications include multiple sequence alignments for computational biochemistry, manufacturing, network security, service planning and location problems [4]–[7].

Several heuristics and approximation algorithms have been proposed in solving the MSCP [8]. Guanghui Lan et al. proposed a Meta-RaPS (Meta-heuristic for Randomized Priority Search) [9]. Fabrizio Grandoni et al. proposed an algorithm based on the interleaving of standard greedy algorithm that selects the min-cost set which covers at least one uncovered element [10]. Amol Deshpande et al. [11] proposed an Adaptive Dual Greedy which is a generalization of Hochbaums [12] primal-dual algorithm for the classical Set Cover Problem.

This paper studies some popular existing algorithms of MSCP and proposes a heuristic approach to solve MSCP using modified hill climbing method. Within our knowledge, the same approach for MSCP of this paper has not been yet reported. Although this work implements two popular algorithms, Greedy Minimum Set Cover [14] and Linear Polynomial Rounding (LP) algorithm [15] to find solutions to MSCP, this work does not focus on the strength and weakness of the algorithms. The proposed approach starts with an initial solution from Greedy approach and LP rounding and then the result is optimized using modified hill climbing technique. The computational results shows the effectiveness of the proposed approach.

The rest of the paper is organized as follows: Section II describes the preliminary studies for proposed approach. Section III describes the proposed algorithm for MSCP. Section IV presents the experimental results. Section V provides the conclusion and future work.

II. Background Theory and Study

This section briefly describes MSCP and presents some preliminary studies. This includes Greedy Algorithm, LP Rounding Algorithm, Hill Climbing Algorithm and OR Library of SCP instances.

A. Minimum Set Cover Problem

Given a set of n elements $U = [e_1, e_2, ..., e_n]$ and a collection $S = \{S_1, S_2, ..., S_m\}$ of m nonempty subsets of U where $\bigcup_{i=1}^{m} S_i = U$. Every S_i is associated with a positive cost $c(S_i) \geq 0$. The objective is to find a subset $X \subseteq S$ such that $\sum_{S_i \in X} c(S_i)$ is minimized with respect to $\bigcup_{S \in X} S = U$.

B. Minimum k-Set Cover Problem

An MSCP (U, S, c) is a k-set cover problem [13] if, for some constant k, it holds that $|S_i| \leq k$, $\forall S_i \in S$ represented as (U, S, c, k). For an optimization problem, x^{OPT} presents an optimal solution of the problem where $OPT = f(x^{OPT})$. For a feasible solution x, the ratio $\frac{f(x)}{OPT}$ is regarded as its approximation ratio. If the approximation ratio of a feasible solution is upper-bounded by some value k, that is $1 \leq \frac{f(x)}{OPT} \leq k$, the solution is called an k-approximate solution.

C. Greedy Minimum Set Cover Algorithm

Data: Set system $(U, S), c : S \to Z+$

Input: Element set $U = [e1, e2, ..., en]$, subset set $S = \{S1, S2, ..., Sm\}$ and cost function $c : S \to Z+$

Output: Set cover X with minimum cost

Algorithm 1 Greedy MSCP

procedure GREEDY(U, S, c) ▷ Set system {U,S} and cost function, c (S)
$X \leftarrow \varphi$
while $\sum_{i \in X} X_i \neq U$ **do** ▷ Continue until X = U
Calculate cost effectiveness, $\alpha = \frac{c(S)}{|S-X|}$ for every unpicked set $\{S_1, S_2, ..., S_m\}$
Pick a set S, with minimum cost effectiveness
$X \leftarrow X \cup S$
end while
return X ▷ Output X, minimum number of subsets
end procedure

D. LP Rounding Algorithm

The LP formulation [15] of MSCP can be represented as

$$\begin{aligned} &\textit{Minimize:} \\ &\quad \sum_{i=1}^{m} c_i \times X_i \\ &\textit{Subject To:} \\ &\quad \sum_{i:e \in S_i} X_i \geq 1 \qquad \forall e \in U \\ &\quad X_i \leq 1 \qquad \forall e \in \{1, 2, ..., m\} \\ &\quad X_i \geq 0 \qquad \forall e \in \{1, 2, ..., m\} \end{aligned} \tag{1}$$

Algorithm 2 LP Rounding MSCP

procedure LPROUND(U, S, c)
Get an optimal solution x^* by solving the linear program for MSCP defined in Equation 1.
$X \leftarrow \varphi$
for each S_j **do** ▷ Continue for all members of S
if $x_j^* \geq \frac{1}{f}$ **then**
$X \leftarrow X \cup S_j$
end if
end for
return X ▷ The minimum number of sets
end procedure

E. Hill Climbing Algorithm

Hill climbing [16] is a mathematical optimization technique which belongs to the family of local search. It is an iterative algorithm that starts with an arbitrary solution to a problem, then attempts to find a better solution by incrementally changing a single element of the solution. If the change produces a better solution, an incremental change is made to the new solution, repeating until no further improvements can be found.

F. OR Library

OR-Library [17] is a collection of test data sets for a variety of Operations Research (OR) problems. OR-Library was originally described in [17]. There are currently 87 data files for SCP. The format is

Algorithm 3 Hill Climbing Algorithm

1: Pick a random point in the search space.
2: Consider all the neighbors of the current state.
3: Choose the neighbor with the best quality and move to that state.
4: Repeat 2 through 4 until all the neighboring states are of lower quality.
5: Return the current state as the solution state.

a) number of rows (m), number of columns (n)
b) the cost of each column $c(j), j = 1, 2, ..., n$

For each row $i(i = 1, ..., m)$: the number of columns which cover row i followed by a list of the columns which cover row i.

III. Proposed Algorithm

This work modified the conventional hill climbing algorithm for set cover problem. To avoid the local maxima problem, this work introduced random re-initialization. For comparisons, greedy algorithm and LP rounding algorithm are used to find the initial state for the modified hill climbing algorithm. The evaluation function for the modified hill climbing algorithm is described below.

A. Problem Formulation

- **Input:** $N = |U|$, $U = [e1, e2, ..., en]$, $M = |S|$, $S = \{S1, S2, ..., Sm\}$, $c = \{c_1, c_2, ..., c_m\}$
- **Output:**
  1) Minimum number of sets, $n(X) = |X|$.
  2) List of minimum number of Sets, $X = \{X_1, X_2, ..., X_{n(X)}\}$.
- **Constraint:** Universality of X must hold, that is $\sum_{i \in X} X_i = U$.
- **Objective:**
  1) Minimize the number of sets, X.
  2) Minimize the total cost, $c(X)$.

B. OR Library MSCP Formulation

The formulation of MSCP for OR Library is given below.

1) Let $M^{m \times n}$ be a 0/1 matrix, $\forall a_{ij} \in M_{ij}, a_{ij} = 1$ if element i is covered by set j and $a_{ij} = 0$ otherwise.
2) Let $X = \{x_1, x_2, ..., x_n\}$ where $x_i = 1$ if set i with cost $c_i \geq 0$ is part of the solution and $x_i = 0$ otherwise.

$$\begin{aligned} &\textit{Minimize:} \\ &\quad \sum_{i=1}^{n} x_i \times c(x_i) \\ &\textit{Subject To:} \\ &\quad 1 \leq \sum_{i=1}^{n} x_i \times a_{ij} \qquad j \in \{1, 2, ..., m\} \\ &\quad x_i \geq 0 \qquad \forall x_i \in \{0, 1\} \end{aligned} \tag{2}$$

C. Proposed Algorithm

This section describes our proposed algorithm for MSCP. The algorithm finds an initial solution and then optimizes the result using modified hill climbing algorithm.

IV. Experimental Results and Discussions

This section presents the computational results of the proposed approach. The effectiveness of the proposed approach is tested on 20 SCP test instances obtained from Beasley's OR Library. These instances are divided into 11 sets as in Table I, in which Density is the percentage of nonzero entries in the SCP matrix. All of these test instances are publicly available via electronic mail from OR Library.

The approach presented in this paper is coded using C on an Intel laptop with speed of 2.13 GHz and 2GB of RAM under *Windows 7* using the codeblock,version-13.12 compiler. Note here that this study presented here did not apply any kind of preprocessing on the instance sets received from OR-Library. This paper did not report the CPU times or running time of the algorithm as they vary machine to machine and compiler to compiler.

TABLE I: Test instance details

Problem Set	Number of instances	Number of rows(m)	Number of columns(n)	Range of cost	Density %
4	10	200	1000	1-100	2%
5	10	200	2000	1-100	2%
6	5	200	1000	1-100	5%
A	5	300	3000	1-100	2%
B	5	300	3000	1-100	5%
C	5	400	4000	1-100	2%
D	5	400	4000	1-100	5%
NRE	5	500	5000	1-100	10%
NRF	5	500	5000	1-100	20%
NRG	5	1000	10000	1-100	2%
NRH	5	1000	10000	1-100	5%

A. Experimental Results of Weighted SCP

Table II presents the experimental results for the proposed approach for weighted SCP instances. The first column represents the name of each instance. The optimal or best-known solution of each instance is given in the 2nd column. The 3rd and 4th column represent the solution found using greedy and LP rounding approach. The 5th and 6th column represent the solutions found in [5] and [7]. The last two columns contain the result found using proposed approach, started from greedy approach and LP rounding approach respectively.

B. Experimental Results of Unweighted SCP

Table III presents the experimental results of the proposed approach for unweighted SCP instances. This paper used the same 20 instances of weighted SCP and made them unweighted by replacing the weights to 1 on these instances. The first column represents the name of each instance. The optimal or best-known solution of each instance is given in the 2nd column. The 3rd and 4th column represent the solution found using greedy and LP rounding approach. The 5th and 6th column represent the solutions found in [18] and [19]. The last two columns contain the result found using proposed approach, started from greedy approach and LP rounding algorithm respectively.

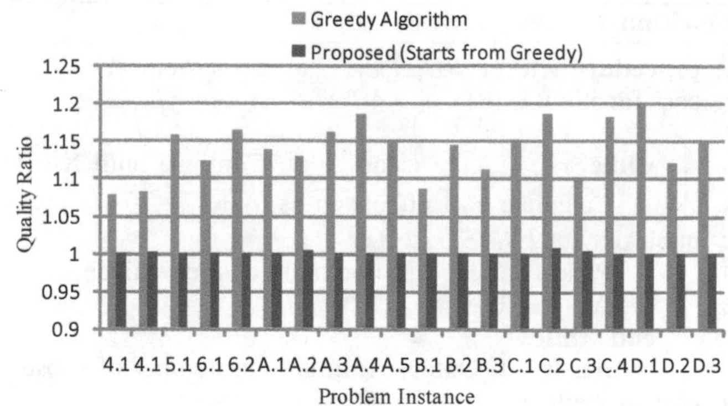

Fig. 1: Quality ratio of weighted problem instances for Greedy and Proposed Algorithm.

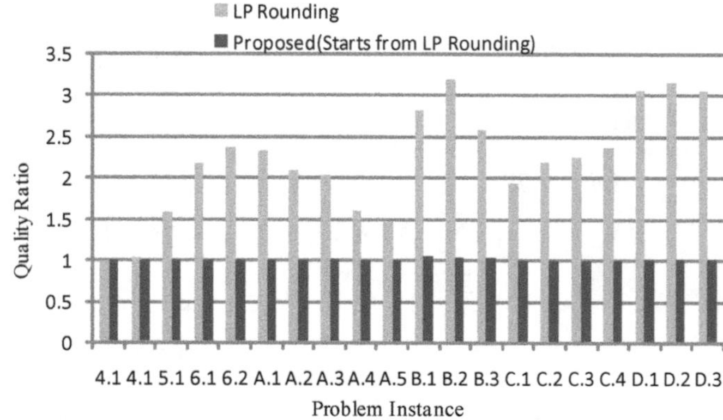

Fig. 2: Quality ratio of weighted problem instances for LP Rounding and Proposed Algorithm.

C. Result Summary

Summary: The optimal solution presented in Table II and Table III are taken from [7]. The quality of a solution derived by an algorithm is measured by *Quality Ratio* which is defined as a ratio of the *derived solution* to the *optimal solution*. The *quality ratio* for each instance for conventional greedy algorithm, LP rounding and Proposed algorithms, presented in this work are shown in Fig. 1, 2, 3 and 4. The figures show the ratio values, plotted as histogram for every instance, presented in this work.

$$Quality\ Measure\ Ratio = \frac{Derived\ Solution}{Optimal\ Solution} \quad (3)$$

Another popular quality measurement reported in literature is called *GAP* which is defined as the percentage of the *deviation of a solution* from the *optimal solution* or *best known solution*. The summarized results, in terms of average quality and average GAP, for weighted set covering instances are presented in Table IV. For unweighted set covering instances it is represneted in Table V.

Algorithm 4 Proposed Algorithm

1: Preparation: In this step, elements of Universal set U, subsets of sets S and cost c of each set are taken as inputs.
2: Initial Solution: This step finds a solution X using *Greedy method* and *LP Rounding algorithm* of MSCP. X is considered the initial state for hill climbing optimization step. This study uses both the solutions and further optimizes for comparisons.
3: Hill Climbing Optimization: This Phase uses modified hill climbing algorithm and optimizes the cost of set cover problem.
4: Find the cost $c(X)$ from X. ▷ Initial best found cost, $c(X)$
5: Keep this (X) as the best found sets. ▷ Initial best found set, X
6: Calculate $R = n(S) \times n(U)$ ▷ number of elements, $n(U) = |U|$, number of subsets, $n(S) = |S|$
7: **for** M times of R **do** ▷ Here M is the *Set Minimization Repetition Factor*
8:   Randomly select a set X^* from the selected sets. ▷ Random selection of a candidate redundant set
9:   Mark this set X^* as *Unselected Set*.
10:   **if** $X - X^* = U$ **then** ▷ Check whether the universality constraint holds
11:     Stay with this state and find the cost, C_{new}.
12:     Replace the best found cost C, with the current cost, C_{new}.
13:     Remove set X^* from the selected sets, X.
14:     Go back to step 8
15:   **end if**
16:   **for** K times **do** ▷ Here K is the Hill Climbing Repetition Factor
17:     Randomly select a set Y from the unselected sets, $S - X$
18:     Mark this set as *Selected*.
19:     **if** $(X - X^*) \cup Y \neq U$ **then** ▷ Check whether the universality constraint holds
20:       Go back to step 17
21:       Find cost C_{new} *of* $c((X - X^*) \cup Y)$
22:       **if** $C_{new} \leq C$ **then**
23:         Replace the best found cost C, with the current cost, C_{new}.
24:         Enlist Y in the *Selected Sets*.
25:         Go back to step 17
26:       **end if**
27:     **end if**
28:   **end for**
29: **end for**
30: Return best found list of sets X and minimum number of sets $n(X)$.

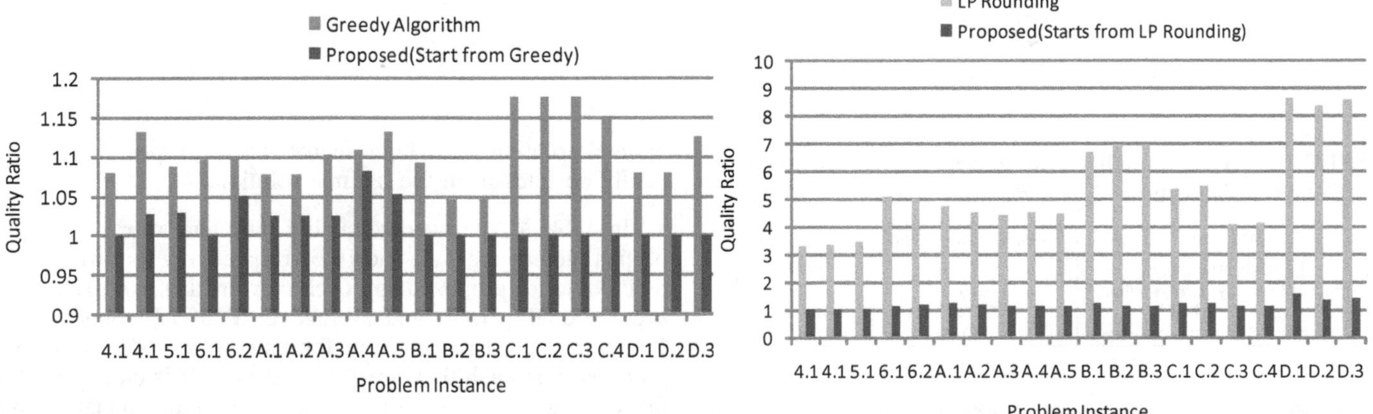

Fig. 3: Quality ratio of unweighted problem instances for Greedy and Proposed Algorithm.

Fig. 4: Quality ratio of unweighted problem instances for LP Rounding and Proposed Algorithm.

$$GAP = \frac{\textit{Derived Solution} - \textit{Optimal Solution}}{\textit{Optimal Solution}} \times 100\% \quad (4)$$

The proposed algorithm presented in this paper used conventional greedy algorithm and LP-Rounding Algorithm as initial solution. Then with the modified hill climbing method, these results are further optimized. Table IV and Table V compare the proposed heuristic approach to the original greedy approach and LP Rounding algorithm.

In Table IV, the average quality ratio and average GAP of original greedy are 1.14 and 14.10 respectively for weighted SCP while for proposed approach they are 1.00 and 0.09.

TABLE II: Experimental Results for Weighted SCP

Instance number	Optimal Solution	Greedy Algorithm	LP Rounding	[5] (Meta-RaPS)	[7] (Descent Heuristic)	Proposed Algorithm	
						Start from Greedy	Start from LP Rounding %
4.1	429	463	429	429	433	429	429
4.10	514	556	539	514	519	515	514
5.1	253	293	405	253	265	253	255
6.1	138	155	301	138	149	138	138
6.2	146	170	347	146	156	146	147
A.1	253	288	592	253	258	253	255
A.2	252	285	531	252	262	253	253
A.3	232	270	473	232	243	232	235
A.4	234	278	375	234	240	234	234
A.5	236	272	349	236	240	236	236
B.1	69	75	196	69	72	69	73
B.2	76	87	243	76	79	76	79
B.3	80	89	207	80	84	80	84
C.1	227	261	442	227	237	227	229
C.2	219	260	484	219	230	221	221
C.3	243	268	551	243	249	244	245
C.4	219	259	523	219	229	219	221
D.1	60	72	184	60	64	60	61
D.2	66	74	209	66	68	66	68
D.3	72	83	221	72	77	72	74

TABLE III: Experimental Results for Unweighted SCP

Instance number	Optimal Solution	Greedy Algorithm	LP Rounding	[18] (Tabu Search)	[19] Local Search for SCP	Proposed Algorithm	
						Start from Greedy	Start from LP Rounding %
4.1	38	41	125	38	38	38	38
4.10	38	43	127	38	38	39	39
5.1	34	37	117	35	34	35	34
6.1	21	23	107	21	21	21	23
6.2	20	22	101	21	20	21	23
A.1	39	42	186	39	39	40	47
A.2	39	42	176	39	39	40	46
A.3	39	43	172	39	39	40	44
A.4	37	41	167	38	37	40	41
A.5	38	43	170	38	38	40	43
B.1	22	24	147	22	22	22	27
B.2	22	23	154	22	22	22	25
B.3	22	23	154	22	22	22	25
C.1	40	47	214	43	43	40	49
C.2	40	47	220	44	43	40	49
C.3	40	47	163	43	43	40	45
C.4	40	46	165	43	43	40	45
D.1	25	27	216	25	25	25	39
D.2	25	27	209	25	25	25	34
D.3	24	27	206	25	24	24	33

TABLE IV: Average quality ratio and GAP for the Weighted Set Covering Problem

Algorithm	Average Quality Ratio	Average GAP
Greedy Algorithm	1.14	14.10
Proposed (greedy initial solution)	1.00	0.09
LP Rounding	2.22	122.57
Proposed (LP initial solution)	1.01	1.48

TABLE V: Average quality ratio and GAP for the Unweighted Set Covering Problem

Algorithm	Average Quality Ratio	Average GAP
Greedy Algorithm	1.11	10.66
Proposed (greedy initial solution)	1.02	1.58
LP Rounding	5.41	441.06
Proposed (LP initial solution)	1.18	17.6

The average quality ratio and average GAP of LP rounding are 2.22 and 122.57 respectively for weighted SCP while for proposed approach they are 1.01 and 1.48. It is clearly visible that original greedy and LP Rounding are deviated from the optimal solution by a high degree where proposed approach hardly deviates from the optimal solution.

In Table V, the average quality ratio and average GAP of original greedy are 1.11 and 10.66 respectively for unweighted SCP while for proposed approach they are 1.02 and 1.58. The average quality ratio and average GAP of LP rounding are 5.41 and 441.06 respectively for unweighted SCP while for proposed approach they are 1.18 and 17.6. It is clearly visible that original greedy and LP Rounding are highly deviated from the optimal solution where proposed approach hardly deviates from the optimal solution.

V. Conclusion and Future Work

This paper studies the existing approaches of MSCP and proposes a new heuristic approach for solving it. Appropriate theorems and algorithms are presented to clarify the proposed approach. The experimental results are compared with the existing results available in literature which shows the effectiveness of the proposed approach. This approach is tested only on OR-Libray in this work. In future this approach will be tested on some other libraries of SCP like *Airline and bus scheduling problems* and *Railway scheduling problems*. The proposed algorithm can also be tested in another popular NP hard problem called *Vertex Cover Problem*.

ACKNOWLEDGEMENT

The author would like to express her greatest gratitude to the anonymous reviewers for their constructive feedback and critical suggestions that helped significantly to elicit the utmost technical attribute of this research work.

REFERENCES

[1] G. Gens and E. Levner, *"Complexity of approximation algorithms for combinatorial problems: a survey,"* ACM SIGACT News, vol. 12, no. 3, pp. 52-65, Fall 1980.

[2] R. M. Karp, *"Reducibility among combinatorial problems,"* Springer US, pp. 85-103, March 1972.

[3] T. M. Chan, E. Grant, J. Knemann and M. Sharpe, *"Weighted capacitated, priority, and geometric set cover via improved quasi-uniform sampling,"* Proceedings of the twenty-third annual ACM-SIAM symposium on Discrete Algorithms, pp. 1576-1585, SIAM, January 2012.

[4] F. C. Gomes, C. N. Meneses, P. M. Pardalos and G. V. R. Viana, *"Experimental analysis of approximation algorithms for the vertex cover and set covering problems,"* Computers and Operations Research, vol. 33, no. 12, pp. 3520-3534, December 2006.

[5] G. Lan, G. W. DePuy and G. E. Whitehouse, *"An effctive and simple heuristic for the set covering problem,"* European Journal of Operational Research, vol. 176, no. 3, pp. 1387-1403, February 2007.

[6] L. Ruan, H. Du, X. Jia, W. Wu, Y. Li and K.-I Ko *"A greedy approximation for minimum connected dominating sets,"* Theoretical Computer Science, vol. 329, no. 1, pp. 325-330, December 2004.

[7] N. Bilal, P. Galinier and F. Guibault, *"A New Formulation of the Set Covering Problem for Metaheuristic Approaches,"* International Scholarly Research Notices, pp.1-10, April 2013.

[8] Y. Emek and A. Rosen, *"Semi-streaming set cover,"* Automata, Languages, and Programming, pp. 453-464, Springer Berlin Heidelberg, July 2014.

[9] G. Lan, G. W. DePuy and G. E. Whitehouse, *"An effective and simple heuristic for the set covering problem,"* European journal of operational research, vol. 176, no. 3, pp. 1387-1403, February 2007.

[10] F. Grandoni, A.Gupta, , S. Leonardi, P. Miettinen, P. Sankowski and M. Singh, *"Set covering with our eyes closed,"* SIAM Journal on Computing, vol. 42, no. 3, pp. 808-830, May 2013.

[11] A. Deshpande, L. Hellerstein and D. Kletenik, *"Approximation algorithms for stochastic boolean function evaluation and stochastic submodular set cover,"* Proceedings of the Twenty-Fifth Annual ACM-SIAM Symposium on Discrete Algorithms, pp. 1453-1467, SIAM, January 2014.

[12] F. A. Chudak, M. X. Goemans, D. S. Hochbaum and D. P. Williamson, *"A primaldual interpretation of two 2-approximation algorithms for the feedback vertex set problem in undirected graphs,"* Operations Research Letters, vol. 22, no. 4, pp. 111-118, May 1998.

[13] F. Colombo, R. Cordone and G. Lulli, *"A variable neighborhood search algorithm for the multimode set covering problem,"* Journal of Global Optimization, pp. 1-20, August 2013.

[14] V. Chvatal, *"A greedy heuristic for the set-covering problem,"* Mathematics of operations research, vol. 4, no. 3, pp. 233-235, August 1979.

[15] B. Saha and S. Khuller, *"Set cover revisited: Hypergraph cover with hard capacities,"* Automata, Languages, and Programming, pp. 762-773, Springer Berlin Heidelberg, July 2012.

[16] K. J. Lang, *"Hill climbing beats genetic search on a boolean circuit synthesis problem of koza's,"* Proceedings of the Twelfth International Conference on Machine Learning, pp. 340-343, June 2014.

[17] J. E. Beasley, *"OR-library: distributing test problems by electronic mail,"* Journal of the Operational Research Society, vol. 41, no. 11, pp. 1069-1072, November 1990.

[18] G. Kinney, J. W. Barnes and B. Colleti, *"A group theoretic tabu search algorithm for set covering problems,"* Working paper, online available at http://www.me.utexas.edu/barnes/research/, 2004.

[19] N. Musliu, *"Local search algorithm for unicost set covering problem,"* Springer Berlin Heidelberg, pp. 302-311, June 2006.

[20] B. Yelbay, S. I. Birbil and K. Bulbul, *"The set covering problem revisited: an empirical study of the value of dual information,"* European Journal of Operational Research, September 2012.

11

Framework for Knowledge–Based Intelligent Clinical Decision support to Predict Comorbidity

Ernest E. Onuiri, OludeleAwodele and Sunday A. Idowu
Department of Computer Science, Babcock University,
Ilishan-Remo, Ogun State, Nigeria

***Abstract*—Research in medicine has shown that comorbidity is prevalent among chronic diseases. In ophthalmology, it is used to refer to the overlap of two or more ophthalmic disorders. The comorbidity of cataract and glaucoma has continued to gain increasing prominence in ophthalmology within the past few decades and poses a major concern to practitioners. The situation is made worse by the dearth in number of ophthalmologists in Nigeria vis-à-vis Sub-Saharan Africa, making it most inevitable that patients will find themselves more at the mercies of General Practitioners (GPs) who are not experts in this domain of interest. To stem the tide, we designed a framework that adopts a knowledge-based Clinical Decision Support System (CDSS) approach to deal with predicting ophthalmic comorbidity as well as the generation of patient-specific care plans at the point of care. This research which is within the domain of medical/healthcare informatics was carried out through an in-depth understanding of the intricacies associated with knowledge representation/preprocessing of relevant domain knowledge. Furthermore, we present the Comorbidity Ontological Framework for Intelligent Prediction (COFIP) in which Artificial Neural Network and Decision Trees, both being mechanisms of Artificial Intelligence (AI) was embedded into the framework to give it an intelligent (predictive and adaptive) capability. This framework provides the platform for a CDSS that is diagnostic, predictive and preventive. This is because the framework was designed to predict with satisfactory accuracy, the tendency of a patient with either of cataract or glaucoma to degenerate into a state comorbidity. Furthermore, because this framework is generic in outlook, it can be adapted for other chronic diseases of interest within the medical informatics research community.**

Keywords—Framework; Knowledge-based; Comorbidity; Clinical Decision Support System (CDSS)

I. Introduction

In today's contemporary times, a trend that seems to be gaining a lot of ground is the integration of intelligent mechanisms in the development of applications to enable them make decisions and attempt to behave like humans. This is widespread in Expert Systems, a branch of AI which Professor Edward Feigenbaum of Stanford University defines as the use of knowledge and inference procedures to solve problems that are difficult enough to require significant human expertise for their solution [1]. Other areas gaining popularity include data mining, machine learning, neural networks, natural language processing, semantic web and so on. These different sub-fields of AI to a large extent make use of knowledge representation techniques. One of such techniques is ontology. "Ontology from the perspective of AI is a model that represents a set of concepts within a specific domain as well as the relationships between those concepts" for the purpose of communicating knowledge between entities and how they inter-relate [2]. In addition, ontology describes a formal specification of a certain domain. It is constituted by a specific vocabulary used to describe a certain phenomenon as well as a set of explicit assumptions as to the intended meaning of the vocabulary [3]. It is the implementation of concepts like ontology in the development of applications such as Decision Support Systems (DSSs) that make them act intelligently.

A DSS is a computer application designed to aid decision makers with the task of decision making. Hence, it can be said that CDSS associates health observations with health knowledge in order to influence the health choices made by clinicians to improve healthcare. A CDSS can be said to be an active knowledge system, where two or more items of patient data are used to generate case-specific recommendation(s) [4]. This means that a CDSS is a DSS that uses knowledge management to achieve clinical advice for patient care based on some number of items of patient data. This goes a long way in easing the job of healthcare practitioners, especially in areas where the number of patients is overwhelming. In medical domains like ophthalmology, there is a dire need of CDSSs, given the few specialists in this area of medical practice in regions of the world like Africa.

Co morbidity is defined as "any distinct clinical entity that has co-existed or may occur during the clinical course of a patient who has the primary disease under study" [5]. Within ophthalmology, co morbidity is commonly used to refer to the co-existence of two or more ophthalmic disorders. Comorbidity between cataract and glaucoma disorders have gained increasing prominence in ophthalmology within the past few decades. In a survey carried out in Sweden, it was substantiated that as much as 36% of patients with cataract also had ocular comorbidities [6].

More so, it has been observed that chronic diseases are often associated with comorbidities. In view of the foregoing National Centre of Health Statistics reported that some of the reasons that explain this trend range from inadequate hospital resources, long waits in hospitals and inadequate medical practitioners [7]. Therefore, one can conclude that comorbidity can be referred to as a condition caused by the debilitating effects of a prevailing ailment. Cataract (Catt) is one of such ailment that can trigger comorbidities especially glaucoma. It is worthy of note that in some instances, glaucoma can also be the prevailing ailment that triggers cataract. Hence, the comorbid condition of choice is that of cataract and Glaucoma (Gla) because both disorders are prevalent in sub-Saharan Africa.

Furthermore, in keeping with this established trend is the need for CDSSs of this nature to help GPs solve problems that are outside their knowledge-base/expertise (in this case, ophthalmology). Consequently, knowledge transition tools such as Evidence-Based Clinical Algorithms (EBCAs), which includes Clinical Practice Guidelines (CPGs) and Clinical Pathways (CPs), significantly go a long way in trying to reduce this care gap caused by the absence of an up-to-date knowledge [8]. CPGs are a function of a detailed and in-depth evaluation of scientific evidence about a specific medical condition/disease/procedure, designed for informed recommendations to aid clinicians in making decisions based on adequate evidence [9]. CPs on the other hand is used to implement the recommendations generated by the CPGs in actual clinical practice [10]. CPs also specifies the clinical processes as well as their workflow to implement the CPGs in a specific clinical setting. Consequently, a CPG entails *medical knowledge* whereas a CP entails *operational knowledge* about how to implement the CPG—i.e. the domain-specific protocols specifying the actual sequencing, decisions and scheduling of clinical tasks, as per the CPG, for the entire clinical course [11].

In view of the foregoing, a major challenge that arises is that which pertains to the alignment of multiple CPs of the comorbid diseases while conserving the integrity of medical knowledge and task pragmatics, and also ensuring patient safety. Hence, in this work, we modelled an intelligent/predictive CDSS for ophthalmic co-morbidity at the point of care. In addition, part of the emphasis was the examination and development of methods that use EBCAs to formalize, model, align and predict patient-specific clinical recommendations with a feedback mechanism, care coordination and decision support for ophthalmic comorbiditybased on substantial clinical evidence.

II. Objectives and Methodologies

The main aim of this research was to design a predictive framework that adopts a knowledge-based CDSS approach to deal with ophthalmic comorbidity at the point of care. Therefore, preliminary methods for the formalization, modelling, alignment and prediction of EBCAs in comorbidities were determined and outlined. Focus was on the prevalent comorbidity of Cataract (Catt) and Glaucoma (Gla), because both are very rampant and common in sub-Saharan Africa and blacks in general. Accordingly, the objectives are to:

1) *Outline a stepwise description/exposition of how to build useful and valid knowledge-based CPs (guidelines/framework) through knowledge identification, acquisition, synthesis, formalization and alignment of relevant datasets especially those related to ophthalmic diseases of interest (cataract and glaucoma). This was achieved through an extensive review of literature leading to an in-depth understanding of the intricacies associated with knowledge representation/preprocessing of relevant domain knowledge. In doing this, the background natures of the ophthalmic diseases of interest are revealed.*

2) *identify relevant classification/predictive algorithms to provide intelligence for the proposed intelligent framework that makes the system adaptive and consequently improves on the existing framework. This was carried out by investigating the appropriate/suitable AI mechanisms aimed at giving adaptability to the proposed framework. This was done by looking at results obtained when such mechanisms were deployed in existing frameworks.*

3) *Design a framework that is diagnostic, predictive and preventive. Hence, the system when operational zed, will predict the chances of a patient with either of cataract or glaucoma to degenerate into a state comorbidity. This led to the adoption and embedding of Artificial Neural Network (ANN – feed-forward back multi-layer propagation) and Decision Trees (DTs – C5.0 algorithm) which are tools of AI into the framework to give it an intelligent (predictive and adaptive) capability.*

III. Analysis of Existing and Related Frameworks, Their Features and Findings

This section introduces a discourse into the workings of existing and related frameworks especially COMET (Comorbidity Ontological Modeling &Execution) and PEDSS (Perinatal DSS).

A. COMET Framework

The system in Fig 1 is an ontology-based decision support framework called COMET for handling comorbidities by the alignment of ontologically modelled CPGs. It is built to formalize, model, align and operationalize the evidence-based clinical algorithms of co-morbid Chronic Heart Failure (CHF) and A trial Fibrillation (AF) in order to provide evidence-based clinical recommendations, care coordination and decision support to GPs for effective management of CHF and AF.

Consequently, the framework addressed the following healthcare knowledge modelling issues:

1) *Modelling of healthcare knowledge, especially in terms of CPGs and CPs, to develop an ontology-based healthcare knowledge model for handling comorbid diseases.*

2) *Computerization of CPs to offer point-of-care decision support*

3) *Alignment of ontologically-modelled disease-specific CPs to handle comorbid diseases; and*

4) *The provision of computerized decision support for GPs, based on modelled CPGs and CPs, to assist them in handling chronic and comorbid diseases [12]*

Also an elaborate OWL-CP (Web Ontology Language – Clinical Pathway) ontology for comorbid CHF and AF was developed that can be executed to support the diagnosis and management of comorbid CHF and AF in a general practice setting. Hence, the COMET framework was implemented to handle three patient care scenarios:

1) *Patient that has CHF*
2) *Patient that has AF and*
3) *Patient with comorbidity of both AF and CHF*

COMET is accessible by web and is designed for GPs. It has been evaluated, both by simulated cases and by health professionals (GP and specialist), for its ability to handle single disease and comorbid care scenarios based on patient data and related constraints. The output at every phase was compared with the expected output as per single disease or comorbid management. Their results showed that the resultant sequence of plans and their outcomes are comparable to the CP knowledge [12].

B. Web-Based PEDSS Framework Using a Knowledge-Based Approach to Estimate Clinical Outcomes: Neonatal Mortality and Preterm Birth in Twins Pregnancies

This system whose architecture is shown in Fig 2 and a description of its components in Table 1, adopts an improved classification method that was derived to improve the prediction of two distinct medical problems non-invasively:

1) *Neonatal mortality with information available at 10 minutes from birth and*
2) *Preterm birth in twin pregnancies before 22 weeks.*

The framework was developed by assessing various data mining methods with the aim of improving the classification of neonatal mortality and preterm birth in twin pregnancies. The major analyzed models were DTs and hybrid ANN to see which produced better outcomes. Positive findings related to the DT mechanism showed that same method can be applied to many other multi-factorial medical problems to improve its classification. This is given the fact that most published risk estimation models attempt to meet clinically acceptable sensitivity and specificity, in which case successful identification of positive cases have been met with much difficulty. Also, with unnecessary variables adding noise and complexity to the problem, it reduces the likelihood of identifying positive cases. A major aim of this framework was to incorporate the advantages of DTs to create a system able to predict the two perinatal problems already mentioned at an earlier stage while maintaining high sensitivity, specificity, Positive Predictive Value (PPV) and Negative Predictive Value (NPV)[13].

Thus, the new approach provides several improvements to better predict medical problems as outlined below:

1) *Pre-processed datasets run against C5.0 algorithm produced DTs superior to the DT-ANN hybrid method.*

2) *Two novel prediction models using DTs and hybrid ANN were evaluated. The DT prediction model had the highest performance outcome for predicting neonatal mortality (sensitivity = 62.24%, specificity = 99.95%, PPV = 72.34%, NPV = 99.92%) using information available within 10 minutes from birth, and preterm birth in twin pregnancies (sensitivity = 80.00%, specificity = 91.55%, PPV = 67.35%, NPV = 95.79%) before 22 weeks gestation. This was achieved using 5-by-2 cross validation. This indicates that the system is not over trained and provides good generalization. [NPV = Negative Predictive Value PPV = Positive Predictive Value].*

3) *Creation of a neonatal mortality prediction system for newborn to be assessed with data available from the first 10 minutes from birth non-invasively with excellent discrimination, exceeding the results of current standard predictions.*

4) *Creation of a preterm birth prediction system for a high risk population (for women pregnant with twins) non-invasively before 22 weeks gestation with excellent discrimination, exceeding the results of current standard predictions*

5) *The previous neonatal prediction method only focused on newborns after admission to NICU. This is the first attempt at predicting neonatal mortality in a heterogeneous population with data available at 10 minutes from birth.*

6) *Several improvements were made compared to past models: For the neonatal mortality case, the prediction of neonatal mortality non-invasively was reduced to data available at 10 minutes from birth using only 13 attributes, whereas the previous models required up to 12 hours from birth using 3 variables derived from invasive methods.*

7) *A conceptual framework for a secure web-based Perinatal Decision Support System (PEDSS) was consequently developed (with components as seen in Table 1) to provide audience targeted information and risk prediction of neonatal mortality and preterm birth in twin pregnancies [13].*

IV. The Proposed Framework [Comorbidity Ontological Framework For Intelligent Prediction (COFIP)]

Having analyzed the PEDSS and COMET frameworks, a description of COFIP is given in this section. The framework diagram is as represented in Fig 3.

A. Knowledge Representation/Preprocessing

This section is comprised of the knowledge identification, acquisition, synthesis, formalization and alignment layers. They are discussed below;

a) Knowledge Identification/Acquisition Layer: The cost and performance of an application depends directly on the quality of the knowledge acquired [14]. The purpose of this phase is to identify valid sources of relevant patient management knowledge as it pertains to two chronic disease conditions namely cataract and glaucoma. This is derivable from existing CPGs – a documentation that is predicated on evidence-based research and is thus a repository of knowledge aimed at providing guidance for decisions and criteria regarding diagnosis, management and treatment of specific disease conditions. The knowledge sources considered, not only entailed evidence-based recommendations but also specific tasks and procedures and their scheduling information. A number of knowledge sources are identified during this phase, including CPGs, institution specific drug management protocols, journal publications, and most importantly domain experts (in this case a consultant ophthalmologist and an optometrist at the Babcock University Teaching Hospital, Ilishan-Remo, Ogun State, Nigeria).

b) Knowledge Synthesis Layer: The knowledge synthesis phase involves the acquisition of the clinically useful task-specific heuristics from the identified knowledge sources (such as the CPGs) through the processes of selection, interpretation and augmentation of the guideline statements, tacit knowledge and logic. Where necessary, the heuristics will be further decomposed into atomic tasks and then organized in such a way as to develop two (cataract and glaucoma) CPs packages containing clear and relevant evidence-based diagnostic and therapeutic plans for patient care management, especially by GPs.Knowledge synthesis is a process in which one builds concepts in cooperation with others. It provides the opportunity for one's hypothesis or assumption to be tested. Social intercourse is one of the most powerful media for verifying one's own ideas. As such, participants in the dialogue can engage in the mutual co-development of ideas [15].

c) Knowledge Formalization Layer: Written sources such as textbooks and technical treatises are often not precise enough for transformation into descriptive logic: there may be competing accounts of the same phenomena, overlapping taxonomies and standards, or outright contradictions [16]. Hence in the knowledge formalization layer, the fused knowledge from the previous layer is modelled and formalized in terms of a dedicated CP ontology to be developed using the Web Ontology Language (OWL). Ontology is the standard knowledge representation mechanism for the Semantic Web framework. The choice of OWL is predicated on the fact that it offers declarative semantics that allows us to associate natural language descriptions with formal statements, thereby allowing human and machine readability of knowledge and subsequent execution of the knowledge. In this phase, the comorbid clinical processes in the CP ontology is hierarchically decomposed into component tasks that are based on the available evidence for specific single disease and comorbid scheduling constraints. This will ensure the conceptualization of the domain into an unambiguous model, thereby determining all implicit constraints on the relationships between the domain concepts, particularly to assist the alignment of concepts in handling comorbidities.

d) Knowledge Alignment Layer: The knowledge alignment layer involves ontology alignment—i.e. alignment of discrete and ontologically defined care plans in response to single disease or comorbid preconditions. The alignment of comorbid CPs is achieved at knowledge modelling level by developing a unified ontological model that encompasses the combined knowledge of aligned CPs. Also, knowledge alignment is tackled at the ontology level, implying that all ontological constraints about knowledge consistency will be observed in the ontologically-modelled Cataract-Glaucoma CP that entails a network of specific classes and the relationship between them. This is indeed a complex activity given the fact that the alignment of comorbid plans needed to take into account the medical correctness and clinical pragmatics of the resultant Cataract-Glaucoma CP.

B. Knowledge-Based Warehouse

The knowledge-based warehouse is the repository for all the relevant domain knowledge gotten from the knowledge stratified into precondition sets A, B and C by the knowledge representation/preprocessing section of the framework. This knowledge-base is also updated through the workings of the results classification/prediction algorithms section where the rules and prediction modules are housed. It is structured such that knowledge is represented in such a way as to promote an efficient system that gives results that tend towards what is obtainable in reality. Therefore explicit/domain knowledge is synergized with tacit knowledge leading to an optimized outcome that helps to inform a patient-specific care plan/recommendation.

The optimized outcome from the prediction algorithms prior to generating the CPs is also used to update the knowledge-base. When this happens, a similar problem can be taken care of without subjecting it to the ANN and DT algorithms, an instance of learning having taken place. This implies a smooth transition from the knowledge-based warehouse to the CP formulation. However, where the patient already exhibits a full-blown comorbid condition it is not subjected to the results classification/prediction algorithms since the essence of that section is to predict the percentage tendency for the emergence of comorbidity where one of both diseases has developed. The section is composed of sub-modules, namely:

- Pre-Condition Set A – Cataract Disease: this houses the knowledge set akin to the cataract condition and sets up a need to predict whether the patient is likely to develop glaucoma vis-à-vis the comorbid condition.
- Pre-Condition Set B – Glaucoma Disease: this contains the knowledge set akin to the glaucoma condition and sets up a need to predict whether the patient is likely to develop cataract vis-à-vis the comorbid condition.
- Pre-Condition Set C – Cataract-Glaucoma Comorbid Condition: this houses the knowledge set akin to the cataract-glaucoma comorbid condition that sets up a need to initiate patient treatment plan and management.

C. Results Classification/Prediction Algorithm

This section entails the mechanisms that make the framework adaptive. They include DTs and ANN which are both techniques of AI.

a) Artificial Neural Network: ANNs are powerful non-linear mapping structures and are especially useful for modelling relationships which are unknown. ANNs function similar to the human brain and can solve problems involving data that is complex, non-linear, imprecise and/or noisy [17]. The human brain is a collection of more than 10 billion interconnected neurons that is able to receive process and transmit data. The human brain also consists of a highly parallel computing structure to support computationally demanding perceptual acts and control activities [18]. ANNs were developed as generalized mathematical models to represent the biological nervous system [18]. The ANN is trained to detect a pattern between the inputted data and the related output value from a dataset. After training the set, the ANN can be used to predict the result of a newly inputted data [17].There is various types of ANNs including feed-forward, recurrent neural network and probabilistic network. The ANN structure used in this thesis is referred to as feedback oriented propagation multi-layer perception.

b) Decision Trees: Decision trees are favoured in the data mining community due to its highly interpretable structure, allowing business end users and analysts to understand the models, whereas neural networks are difficult to understand and interpret [19]. A decision tree consists of a root node, branch nodes and leaf nodes. The tree starts with a root node, is further split into branch nodes (each of the nodes represent a choice of various alternatives), and terminates with a leaf node which are un-split nodes (represents a decision) [20]. Classification of decision trees are conducted in two phases, including the tree building (top down) and tree pruning (bottom-up). Tree building is computationally intensive, and requires the tree to be recursively partitioned until all data items belong to the same class. Tree pruning is conducted to improve the prediction and classification of the algorithm and to minimize the effects of over-fitting, which may lead to misclassification errors [21]. There are a number of decision tree algorithms that exist including Classification and Regression Trees (CART), Iterative Dichotomiser 3 (ID3), C4.5 and C5.0. This thesis work uses C5.0 based decision tree algorithm which is an improvement over C4.5, which itself is an improvements over the earlier ID3 method.

c) Result Comparison and Optimization: The result comparison and optimization module is responsible for synergizing the outputs generated by the different classification/prediction algorithms i.e. the ANN and DT mechanisms so as to settle for an optimized output. The entire workings of the different modules in this unit are all geared towards finding a healthy association between the domain knowledge and tacit knowledge in other to make the overall system operation with some measure of expertise akin to a human expert. Tacit knowledge has to do with unwritten, unspoken, and hidden vast storehouse of knowledge held by an individual, based on the persons' emotions, experiences, insights, intuition, observations and internalized information. Tacit knowledge is integral to the entirety of a person's consciousness, is acquired largely through association with other people, and requires joint or shared activities to be imparted from one person to another. Like the submerged part of an iceberg it constitutes the bulk of what one knows, and forms the underlying framework that makes explicit knowledge possible. When the inputs from the knowledge-base is subjected to analysis by the ANN and DT mechanisms their outputs are compared and the outcome are optimized for the generation of rules with which predictions get carried out.

D. *Intelligent Clinical Pathway Generator System*

The intelligent clinical generator system unit is composed of rules that are a function of the results comparison/optimization module which is used to make predictions that gets to inform the generation of patient-specific care plans. It also contains a feedback mechanism.

a) Rules and Prediction: The rules module receives an optimized output upon which rules are generated and subjected to further coordinated analysis that becomes the yardstick for the prediction of the tendency for comorbidity. Once this is established the predicted values become the benchmark upon which generation of patient-specific care plans and recommendations take place.

b) Patient-Specific Care Plans/Recommendations: This system adopts patient-specific CPs/recommendations as against case-based CPs/recommendations because the modern patient wants to be treated as an individual person and not just as a statistic [22]. Patients want to know their own risk, not just a parameter regarding a class of people similar to them. This feature is highly enhanced through the deployment of ANNs which are able to reproduce the dynamical interaction of multiple factors simultaneously, allowing the study of complexity which is very important for a researcher interested in in-depth knowledge of a specific disease or to better understand the possible implications relative to strange associations among variables. This has to do with what is called "intelligent data mining". But on the other hand ANNs can also help medical doctors in making decisions under extreme uncertainty and to draw conclusions on individual basis and not as average trends.

c) Feedback Mechanism: In view of the fact that this framework serves as a platform for implementation and translation into a real system, a feedback mechanism is sacrosanct and must be included. More so, a typical developer may find it difficult to adapt a framework that is without a feedback system during implementation [23]. Hence, the proposed framework holds an extension to one. The schematic diagram in Fig 4 shows a typical feedback mechanism which is a sub-set of the patient-specific care plan generator system. The feedback mechanism is composed of functionalities that:

1) *Describe diagnosis:*
- Inform patient about disease (you're diagnosed as suffering from...)
- present supporting evidence for disease

2) *Alleviate patient fears:*
- describe improvements

3) *Describe future prospects.*

4) *Describe disease triggers:*
- present background information
- List triggers mentioned by the patient.
- List triggers mentioned by doctor.
- Suggests methods of avoidance.

5) *Describe drug prescription.*

6) *Describe need for long term effort.*

d) Discussion of the Underlying Processes:

1) *The patient information is fed into the system (See Fig 5). The patient information includes age, race, gender etc. medical history such as surgery, ocular disease, eye surgery, trauma etc. Questions about the patient's sight are also included. The patient's information is then subjected to the knowledge representation and preprocessing.*

2) *Knowledge representation/preprocessing has four layers: identification/acquisition, synthesis, formalization and alignment layers.*

- Knowledge identification/acquisition layer identifies valid sources of relevant patient management knowledge as it pertains to two chronic disease conditions namely cataract and glaucoma. The knowledge is derived from existing CPGs.
- Knowledge synthesis layer involves the acquisition of the clinically useful task-specific heuristics from CPGs through the processes of selection, interpretation and augmentation of the guideline statements, tacit knowledge and logic.
- In the knowledge formalization layer (semantic layer) the synthesized knowledge will be modelled and formalized in terms of a dedicated CP ontology to be developed using the Web Ontology Language (OWL).
- Knowledge Alignment layer involves ontology alignment of discrete and ontologically defined care plans in response to single disease or comorbid preconditions.

3) *The preprocessed results from the knowledge representation unit are passed into knowledge base warehouse for dynamic storage and updating. The knowledge-based warehouse is updated through the rules and prediction unit. Here, the preconditions are segmented and classified into one of cataract, glaucoma or comorbid precondition sets. If it is the comorbid condition the patient-specific care plan is generated straight away. Else, the disease identified (cataract or glaucoma) is subjected to the classification/prediction algorithm. The two algorithm used are ANN and DT. The two are used to complement each other's strengths and weaknesses as the case may be. The result from ANN and DT are compared for optimization and the best values are chosen and used to generate the rules.*

4) *The rules generated are made from the result of the optimization and they are used to update the precondition sets A and B (cataract and glaucoma) contents in the knowledge base warehouse. Once the condition is ascertained, the patient-specific care plan is generated.*

V. Conclusions and Implications

This research paper entails the design of an adaptive framework (COFIP) that is diagnostic, preventive and predictive. This is because COFIP was designed to predict with satisfactory accuracy, the tendency of a patient with either cataract or glaucoma to degenerate into a state of comorbidity.

Furthermore, because this framework is generic in outlook, it can be adapted for other chronic diseases of interest within the medical informatics research community.

A. Recommendations and Future Work

Having built the predictive framework, full implementation will be carried out as follows:

1) *Build useful and valid knowledge-based CPs (guidelines/framework) through the acquisition of relevant data sets especially those related to ophthalmological diseases (Catt-Gla).*

2) *Model the selected ophthalmological diseases such that the diagnostic and treatment concepts show inter-relationships in formal language in ways that curb every form of possible ambiguity. This is carried out in such a way as to ensure that the encoded knowledge and the underlying decision logic can be executed through computerized clinical decision support systems to provide patient-specific CPG-based recommendations.*

3) *Systematically align the designed model to handle ophthalmologic diseases without compromising the integrity of medical knowledge, care coordination and safety of patients.*

4) *Fully operationalize the designed framework that is diagnostic, predictive and preventive. Thus, the system when implemented will predict the chances of a patient with either of Catt or Gla to degenerate into a state comorbidity.*

B. Summary

The coexistence of cataract and glaucoma accounts for alarming levels of visual impairment in our society vis-à-vis impairment of quality of life and hence increases the burden of illness, care plan, patient management and related concerns. A viable method to reduce the menace of blindness and other related conditions is to engage GPs in the management of cataract and glaucoma as well as its co-morbidities, because GPs are the first point of care for most patients. There are challenges in the diagnosis of glaucoma for instance given that many of its clinical features are time-specific and sometimes not obvious though present.

To make matters worse, concurrent presence of cataract complicates the management of either condition as the choice of treatment depends on individual factors of each disease as it manifests in the patient. EBCAs such as CPGs and CPs have the propensity to narrow this gap [8]. They can by assisting GPs to undertake complex diagnostic and management scenarios resulting from the comorbidity of cataract and glaucoma. In view of the foregoing we have designed an improved framework that enhances the classification/prediction of comorbidities that bring with it adaptability that helps to guide patients with one chronic disease to manage their condition so as to prevent the degeneration of their condition to full-blown comorbidities. COFIP was designed to improve on the existing COMET framework that is not adaptive.

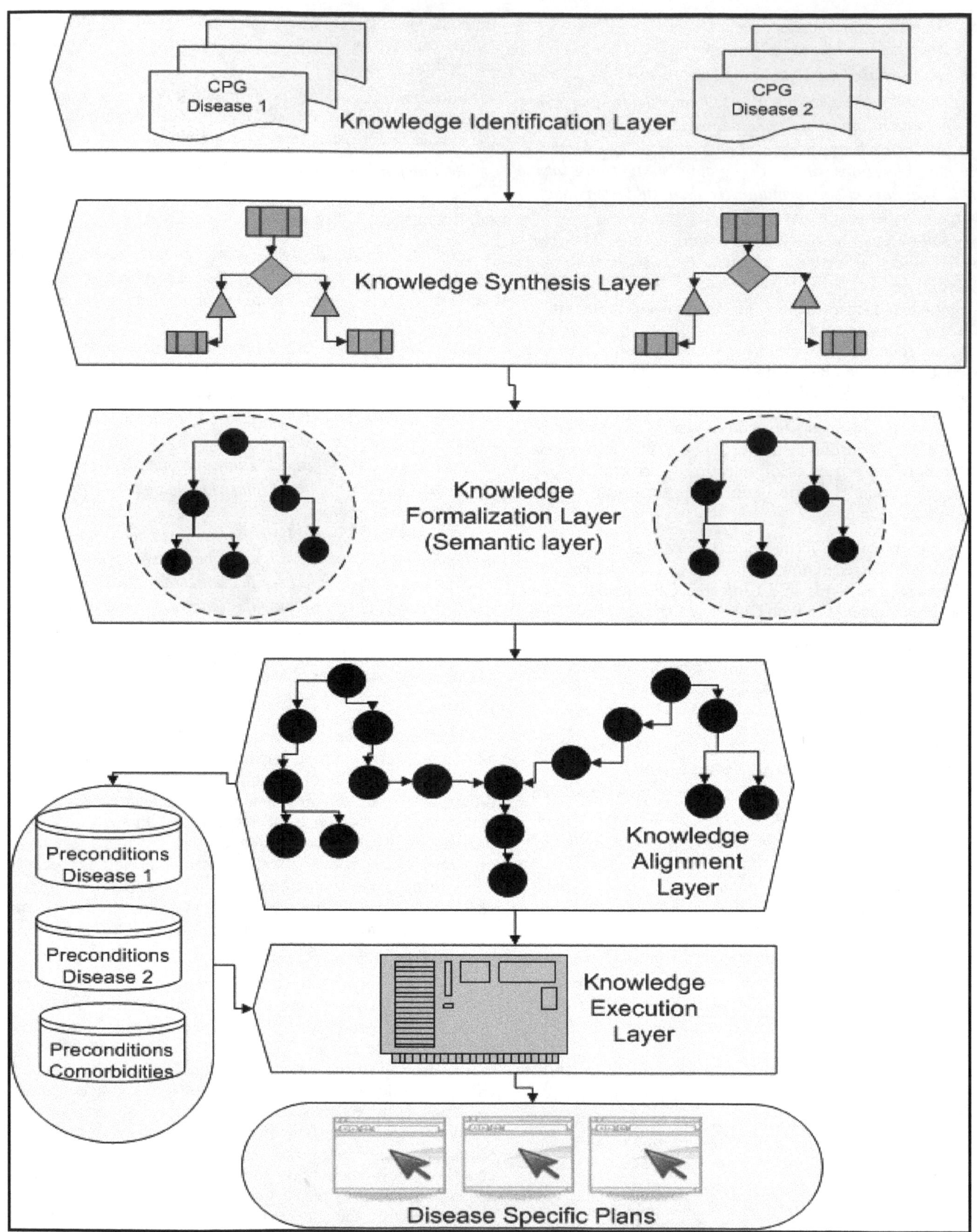

Fig. 1. COMET Framework [12]

TABLE I. COMPONENTS FOR A WEB-BASED PEDSS [13]

	Components	Description
1	Authentication Server	Authenticates users into the system
2	Content Management System (CMS)	The heart of the system used to display, search, and process the data, based upon the user request
3	Workflow Engine	Required to automate alerts, warning and actions
4	External Data Source	A repository of the patient, or user information
5	Directories	A database of user information, etc.
6	Other Web Servers	Other servers required to operate the PEDSS
7	ASP.Net, XML, HTML	The interface presented to the user

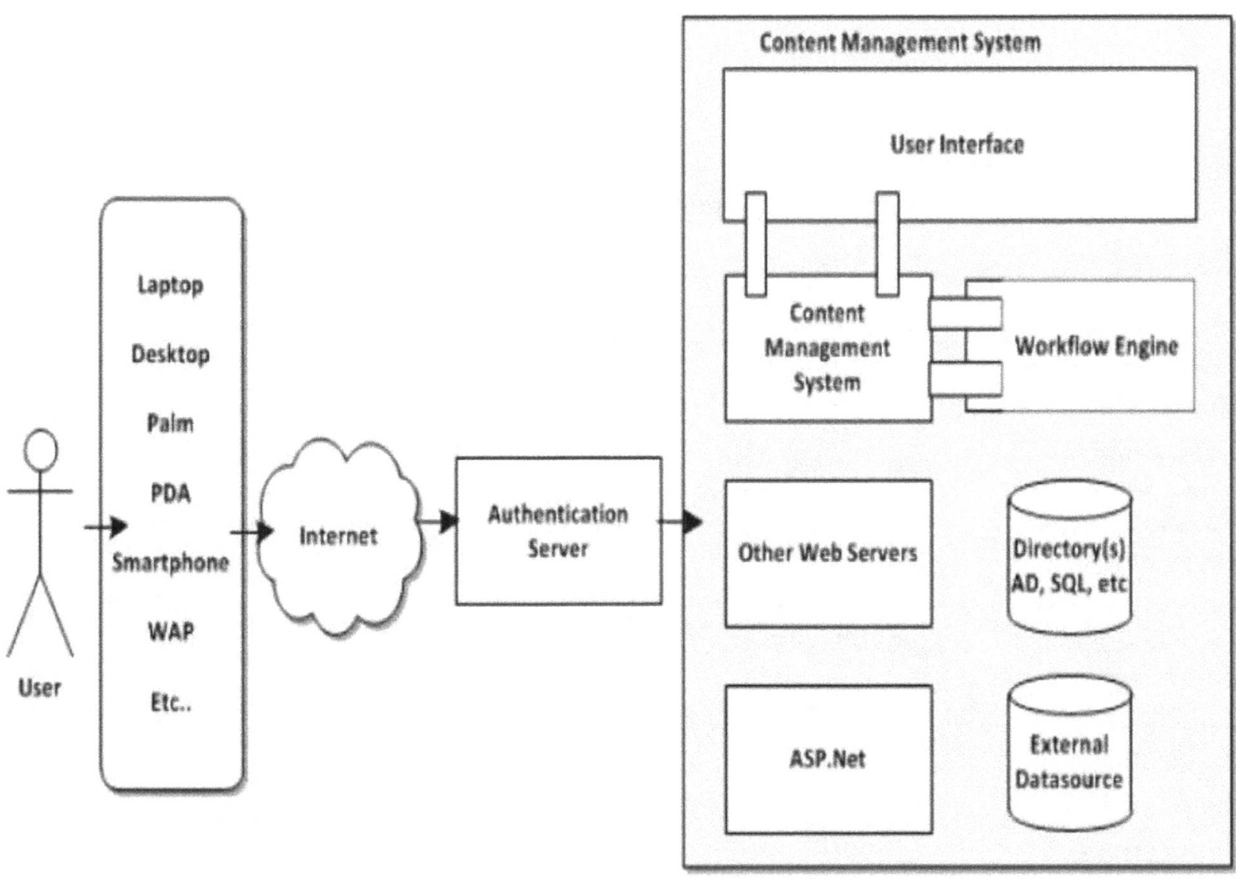

Fig. 2. System Architecture of Web-based PEDSS [13]

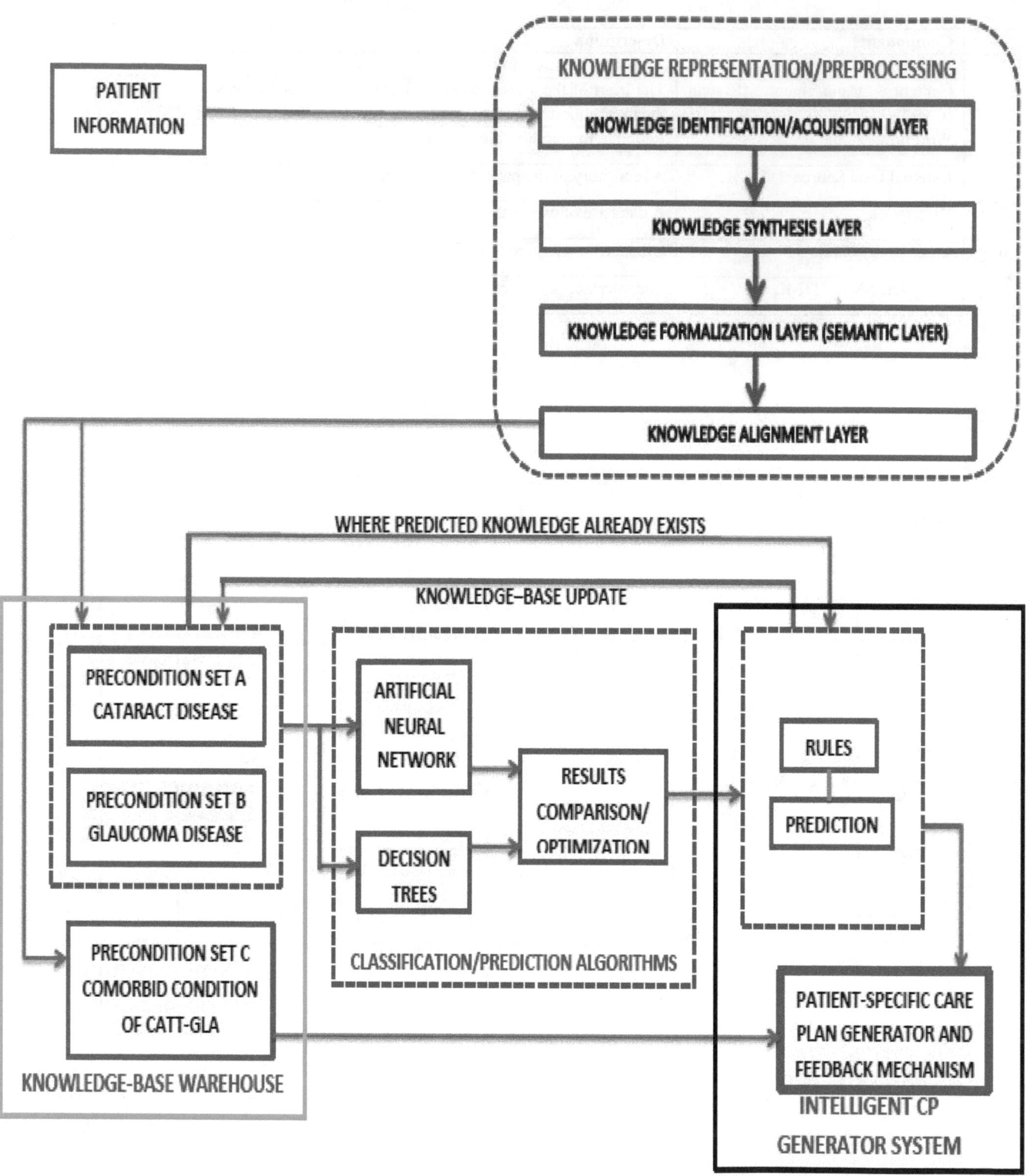

Fig. 3. The Proposed (COFIP) Framework

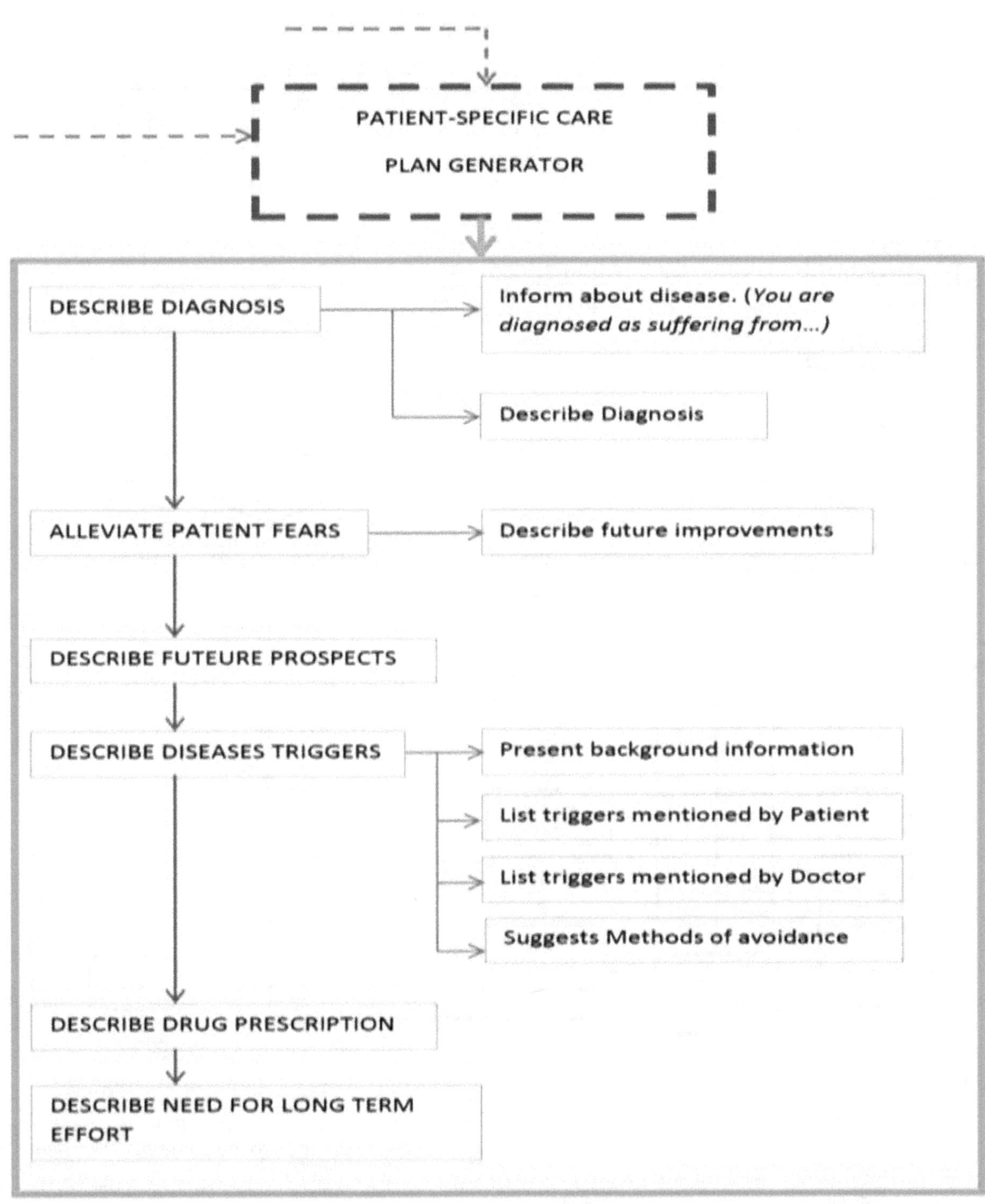

Fig. 4. Feedback Mechanism

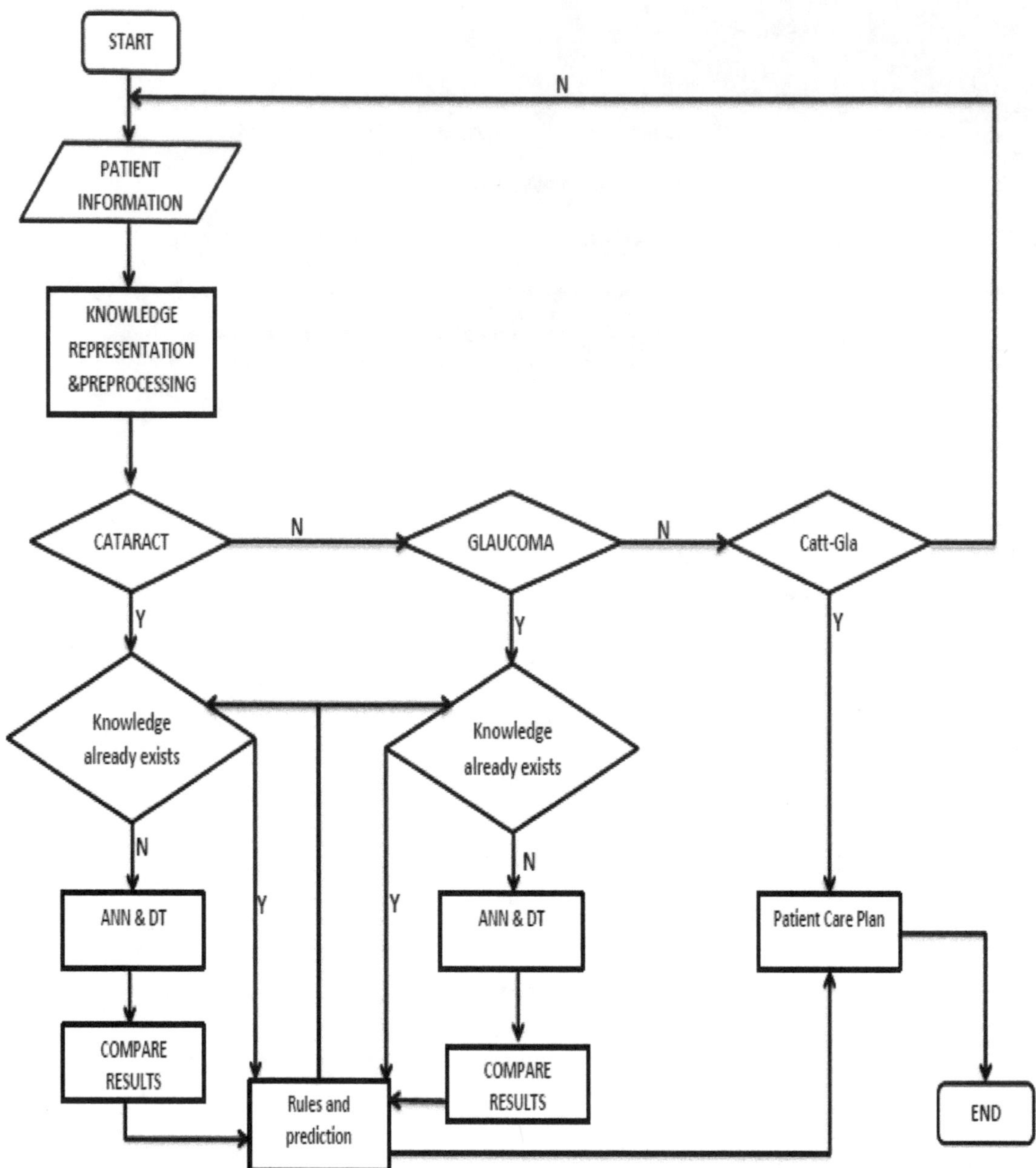

Fig. 5. Flowchart of COFIP

REFERENCES

[1] Angeli C. (2010). *Diagnostic Expert Systems: From Expert's Knowledge to Real-Time Systems.* Advanced Knowledge Based Systems: Model, Applications & Research. (Eds. Sajja&Akerkar), Vol. 1, pp 50 – 73, 2010

[2] Jones T. M. (2008). *Artificial Intelligence, a System Approach.*

[3] Fonseca, F. (2007). *The Double Role of Ontologies in Information Science Research.* Journal of the American Society for Information Science and Technology, 58(6), pp. 786-793.

[4] Chen J.Q & Lee S.M. (2002). *An Exploratory Cognitive DSS for Strategic Decision Making.* 2002 Elsevier Science B.V. All rights reserved. doi:10.1016/S0167-9236(02)00139-2

[5] Feinstein, A. R. (1970). *The Pre-therapeutic Classification of Comorbidity in Chronic Disease.* Journal of Chronic Diseases, 23, pp456–7.

[6] O'hEineachain R. (2005). *Glaucoma and Cataract surgery. Comorbidities: Patients With Cataracts Often Have Other Ocular Conditions Affecting Outcomes.*

[7] National Center of Health Statistics (2007). *N.C.H.S. Definitions.* http://www.cdc.gov/nchs/datawh/nchsdefs/comorbidities.htm. Retrieved November 12, 2008

[8] Brush, J.E, Radford, M.J. &Krumholz, H.M. (2005). *Integrating Clinical Practice Guidelines into Routine of Everyday Practice.*Crit Pathway Cardio, 4:161-167

[9] Woolf, S.H., Grol, R., Hutchinson, A., Eccles, M., &Grimshaw, J., (1999). *Clinical Guidelines: Potential Benefits, Limitations, and Harms of Clinical Guidelines.* BMJ, 318: 527-530

[10] Kitchiner, D.J. &Bundred, P.E. (1996). *Integrated Care Pathways.* Arch DisChild, 75: 166-168

[11] Pearson, S. D., Goulart-Fisher, D., & Lee, T.H. (1995). *Critical Pathways as a Strategy for Improving Care: Problems and Potential.* Annals of Internal Medicine, 123(12), 941-948

[12] Abidi S.R. (2010). *A Knowledge Management Framework to Develop, Model, Align and Operationalize Clinical Pathways to Provide Decision Support for Comorbid Diseases.* http://dalspace.library.dal.ca/handle/10222/13009. Retrieved February 5, 2013.

[13] Gunaratnam M. (2013*). A Web-Based Perinatal Decision Support System Framework Using a Knowledge-Based-Approach to Estimate Clinical Outcomes: Neonatal Mortality and Preterm Birth in Twins Pregnancies.* http://www.sce.carleton.ca/faculty/schwartz/RCTI/Seminar%20Day/Autonomous%20vehicles/theses/Thesis%20-%20Marry%20Gunaratnam%20_%20100550026.docx. Retrieved March 28, 2013.

[14] Rhem A.J & Associates. (2002). *A framework for Knowledge Acquisition.* http://www.ajrhem.com/framework.pdf. Retrieved March 11, 2013.

[15] Markova, I. &Foppa, K. (1990). *The Dynamic of Dialogue*. New York: Harvester Wheatsheaf.

[16] Bowker, G. C., (2000). *"Biodiversity Datadiversity."* Social Studies of Science 30(5): 643-683.

[17] Jha, G.K. (2011). *Aritificial Neural Networks.* http://www.iasri.res.in/ebook/EB_SMAR/e-book_pdf%20files/Manual%20IV/3-ANN.pdf

[18] Lisbboa, P.J.G. (2011). *A Review of Evident of Health Benefit from Artificial Neural Networks in Medical Intervention.*Neural Networks.2,11-39.

[19] Apte, C., et al. (2002). Data Intensive Analytics for Predictive Modelling.http://www.research.ibm.com/dar/papers/pdf/dar_overview_ibmjrd.pdf. Retrieved March 12, 2013.

[20] Peng, W., Chen J. & Zhou H. (2006). *An Implementation of ID3 – Decision Tree Learning Algorithm*. University of New South Wales. Last accessed: December17, 2011 from http://web.arch.usyd.edu.au/~wpeng/DecisionTree2.pdf.

[21] Anyanwu, M.N., and Shiva, A.G., (2009). *Comparative Analysis of Serial Decision Tree Classiication Algorithms*. International Journal of Computer Science and Security. 3, 230-237

[22] Grossi E. (2011).*Artificial Neural Networks and Predictive Medicine: a Revolutionary Paradigm* Shift. http://cdn.intechopen.com/pdfs/14887/InTech-Artificial_neural_networks_and_predictive_medicine_a_revolutionary_paradigm_shift.pdf. Retrieved April 7, 2013.

[23] Buchanan B.G, Carenini G, Mittal V.O. & Moore J.D. (1998). *Designing Computer-based frameworks that facilitate doctor-parient collaboration*. Elsevier, Journal of Artificial Intelligence in Medicine. http://goo.gl/uGgxV. Retrieved March 5, 2013.

12

Optimal Network Reconfiguration with Distributed Generation Using NSGA II Algorithm

Jasna Hivziefendić
Faculty of Engineering and Information Technologies
International Burch University
Sarajevo, Bosnia and Herzegovina

Amir Hadžimehmedović
University of Tuzla
Tuzla, Bosnia and Herzegovina

Majda Tešanović
Faculty of Electrical Engineering
University of Tuzla
Tuzla, Bosnia and Herzegovina

***Abstract*—This paper presents a method to solve electrical network reconfiguration problem in the presence of distributed generation (DG) with an objective of minimizing real power loss and energy not supplied function in distribution system. A method based on NSGA II multi-objective algorithm is used to simultaneously minimize two objective functions and to identify the optimal distribution network topology. The constraints of voltage and branch current carrying capacity are included in the evaluation of the objective function. The method has been tested on radial electrical distribution network with 213 nodes, 248 lines and 72 switches. Numerical results are presented to demonstrate the performance and effectiveness of the proposed methodology.**

Keywords—radial distribution network; distributed generation; genetic algorithms; NSGA II; loss reduction

I. Introduction

Newly formed market conditions articulate the need for adjusted approach in managing distribution network in order to meet not only the requirements imposed by technical conditions of the system but also the requirements imposed by consumers and network regulators. Significant changes in distribution system have been caused by installing distribution generation units which have considerable impact on system voltage profile and power losses, both being important quantities in the process of planning and reconfiguration of electrical network.

DGs are grid-connected or stand-alone electric generation units located within the electric distribution system at or near the end user. The integration of DGs in distribution system would lead to improving the voltage profile, reliability improvement such as service restoration and uninterruptible power supply and increase in energy efficiency. The distribution feeder reconfiguration (DFR) is one of the most significant control schemes in the distribution networks which can be affected by the interconnection of DGs [1].

Generally, the DFR is defined as altering the topological structure of distribution feeders by changing the open/closed status of automatic and tie switches or protective devices located in strategic places in distribution system. By changing the statuses of the sectionalizing and tie switches, the configuration of distribution system is varied, and loads are transferred among the feeders while the radial configuration format of electrical supply is still maintained.

Network reconfiguration is a very effective and efficient way to ensure more even load distribution of network's elements, improve system reliability and voltage profile, and to reduce power losses. All modes are subject to reconfiguration: normal, critical and failure. Provided that all variables are within acceptable limits, network reconfiguration will achieve optimal working conditions in normal mode.

Taking into consideration a large number of switches in distribution network, whose on/off switching affects the network topology, reconfiguration problem can be defined as a complex combinatorial, non-differentiable, and constrained multi-objective optimization problem. Radial network conditions, explicit voltage constraints in all node, line capacities, etc. are viewed as some of the constraints that have to be taken into consideration.

In recent years, different methods and approaches have been used to solve the problem of distribution system reconfiguration with distribution generators installed. The literature related to this problem mainly refers to application of heuristic algorithms and artificial intelligence-based algorithms such as Genetic algorithms, Fuzzy logic, Particle swarm optimization, Tabu search, etc. [2-7]. Most cases address reducing power losses and load balancing, taking into account the effect of generators distributed in the network, while very little attention is paid to system reliability. However, special attention should been paid to the issue of reliability of power supply, in order to increase economic efficiency of distribution companies [14]. Network reconfiguration process can be used as one possibility to improve network reliability indicators. Furthermore, reliability improvement by DGs is possible when intended islanding operation is allowed [8].

In paper [15] NSGA II (Non-Dominated Genetic Algorithm II) is applied to the planning of distribution electrical network problem. This paper is focused on the application of NSGA II on resolving the problem of reconfiguration of distribution network with distributed generation. The effect of distributed generation on voltage in the network, taking into account two objective functions: power losses and reliability function is presented. Depending on characteristics of the power distribution networks (network parameters, characteristics of power lines, failure rates, types of consumers, etc) simultaneous optimization of these functions can be disagreeing, that is the optimum topology for one objective can be very different by the optimum obtained with the other function. Since the proposed method optimizes two objective functions simultaneously, the problem is defined as multi-objective problem taking into account the defined system constraints. The effectiveness of the methodology is demonstrated on real distribution network consisting of 213-buses showing its potential of applicability to the large distribution systems.

The problem formulation is discussed in detail in section II. The network reconfiguration algorithm using a multi-objective NSGA II is described in section III. The simulation results in terms of power loss and energy not supplied are discussed in Section IV and finally the last section presents the conclusion of the study.

II. Problem Formulation

The objective of the proposed solution model for reconfiguration of distribution network problem is to minimize two functions as follows: power losses function and reliability function presented by Energy not supplied index (ENS). The optimal results of the defined functions do not lead to the same optimal network topology what creates trade-off between reliability and power losses function.

Electrical power losses are one of the most important factors which point to business cost-effectiveness and quality of distribution. Energy losses in electrical distribution network in the amount of 1% cause the increase in company's business costs of up to 2% to supply energy to cover the losses [9]. Therefore, the reduction of power losses is one of the most important issues in distribution system operation.

In addition to the power losses function, reliability function is also defined in the paper, with the objective to increase reliability in consumers supply by minimizing expected energy not supplied due to power interruptions. The essential attributes of interruptions in the power supply of the customers are the frequency and duration. While duration is predominantly influenced by the distribution system structure (radial, meshed, weak meshed) and the existing automations, the frequency is mainly influenced by the adopted operational configuration; it can be minimized by the suitable choice of the effective configuration [7]. Since there is no 100% reliable system, it is in the best interest of both suppliers and consumers to minimize power supply interruptions. This function will be presented through Energy not Supplied function (ENS).

These two objectives can be met by identifying optimal network topology. Efficient solution of the described problem requires the choice of optimal topology of radial network within the set of possible solutions.

A. MathematicalFormulation of Problem

The purpose of distribution network reconfiguration is to find optimal radial operating structure that minimizes two functions: the system power losses and ENS function within the operating constraints. Accorindg to the literature [2], [14] thus the problem can be formulated as follows:

$$\min P_{loss} = \sum_{i=1}^{n} (I_i + D_i I_{DI})^2 R_i \tag{1}$$

where $D_i = 1$ if line $i \in \alpha$, otherwise is equal to 0.

$$\min f_{ens} = \sum_{i=1}^{N_V} P_i \lambda_i r_i \tag{2}$$

where P_{loss} is total real power losses function, f_{ens} is the ENS function, I_i current in the branch i; R_i resistance in the branch i; α is set of branches connected to the distribution generators node m, P_i real power flow through branch i; λ_i failure rate of branch i (number of failure per year and per kilometer of branch i); r_i failure duration of branch i; N_v number of branches. The DG produces active current I_{DI}, and for a radial network it changes only the active component of current of branch set α.

Subject to the system constraints:

- $I_i \le I_{\max_i}$ - the current in each branch cannot exceed the branch capacity,
- $V_{\min} \le V_j \le V_{\max}$ - voltage constraint.

The voltage in each load buses in the system has to be within the defined limits. The minimum voltage is 0.95 and maximum voltage is 1.05 (±5%).

- $P_S + \sum_{i=1}^{n} P_{g,i} = \sum_{i=1}^{n} P_i + P_{loss}$ - power balance constraint
- $\sum i = n-1$ - topological constraint.

Electrical distribution systems are operated in radial configuration.

Generator operation constraints:

DG units are only allowed to operate within the acceptable limit where $P_i^{\min}$ and $P_i^{\max}$ are the lower and upper bound of DG output and $P_i^{\min} \le P_g \le P_i^{\max}$.

The search space for this problem is the set of all possible network configurations.

In order to check the defined constraints, voltage magnitude and angle at each bus in the system have to be known at any time. When the voltage magnitude at any bus in the system is not within the defined limits, network configuration cannot be considered as a possible solution.

Since this information is included in the state variable *x*, it can be presented as follows:

$$x = [q_2, q_3, \ldots, q_n, |V_2|, |V_3|, \ldots, |V_n|]^T \tag{3}$$

and state space as is $IR^{6(n-1)}$, where

$|V_i| = \left[|V_i^a|, |V_i^b|, |V_i^c|\right]^T$ and $\theta_i = \left[\theta_i^a, \theta_i^b, \theta_i^c\right]^T$ are

voltage magnitude and angels respectively for load buses *i*, and the bus 1 is substation.

For three-phase distribution network with n buses, bus 1 presents substation and buses 2, 3, ...n, are load buses. Equation (3) can be solved by power flow calculation solving the system of (6n-6) non-linear algebraic equations.

To calculate the second objective function, ENS index due to interruption in supply, it is necessary to consider two elements: failure rate and the length of interruption in power supply for each load point. The latter is consisted of two components: time necessary to locate failure and the time necessary to repair it. Automatic sectionalizers and switches separate the part of the network where the failure occurred, reducing the risk for other consumers in the network. The time needed to repair the failure is usually the time needed to isolate the failure, to connect the affected consumers to reserve power supply (if possible) and to repair the fault itself [10]. In order to calculate this function, load flow studies should be performed to calculate not-distributed energy in all consumer nodes without supply, which are located "under" the fault in the network.

III. Optimization Method for Multi-Objective Reconfiguration Network

The development of heuristic algorithms and computer performances have contributed towards solving the problem of multi-objective optimization. While solving multi-objective optimization problems, it is necessary to pay attention to convergence to optimal set of solutions (Pareto set) and maintain diversity of solutions within the set of current solutions [11]. Suggested methodology for solving the defined multi-objective optimization problem is based on multi-objective Non-Dominated Sorting Genetic Algorithm II. (NSGA II).

Genetic algorithms use population of solutions in every optimization path within optimization process. The objective is to come as close as possible to the true Pareto-front and simultaneously gain as many solutions as possible. This ensures that the decision-maker will have a wider choice of quality solutions with a better overview of all possible optimal topologies of a distribution network [11].

A. NSGA II Algorithm

Multi-objective evolutionary algorithms are suitable for multi-objective optimization due to their ability to handle complex problems, involving features such as discontinuities, multimodality, disjoint feasible spaces and noise functions evaluation [12]. NSGA II is a multi-objective genetic algorithm developed by Deb, 2003 [13]. Basic advantage of NSGA II over other multi-objective genetic algorithms is reflected in possibilities for diversity preservation of population, which further enables uniform distribution of solutions within Pareto front. The crowding distance approach is introduced into NSGA II as the fitness measure to make comparison of solutions in the same Front. This approach estimates the density of solutions surrounding a particular solution by calculating the average distance of two points on either side of the observed solution for all objective functions defined for particular problem. The fast non-dominated sort strategy is used to evaluate solution dominance and classify the solution into Pareto fronts that corresponds to the cluster with the same solution dominance. Furthermore, NSGA II uses elite strategy that significantly helps in speeding up the performance of the genetic algorithm [13].

B. Proposed Methodology

Algorithm starts with randomly selected radial functional solution that is typical for electrical distribution network, as a basis for a first generation of trade-offs in the part of a genetic algorithm code. By applying NSGA II algorithm new potential solutions of the network are generated. The binary alphabet has been used to implement the optimization model, in which every bite of chromosome represents the status of switches (open/closed). Every bite can have value of 0 or 1, which identifies the status of every electric line, 0-open, 1-closed. In the reconfiguration network problems, only a certain number of lines have a changeable bit in chromosome, and therefore only those lines are subject to genetic operators, crossover and mutation, while other lines have a fixed value in chromosome (always have the value of 1 in operation).

If newly created solutions meet topological constraints (radial conditions), evaluation of the objective functions is carried out, i.e. power flow and calculation of objective functions are performed. Power flow calculation is done in MATLAB. For that purpose, a part of the code for power flow calculation based on Newton-Raphson method is modified for the need of objective functions evaluation, transfer of variables, storage of diverging solutions and visualization. Based on the power flow results, convergence of specific network configuration is verified, as well as other constraints which refer to the capacities of lines, power stations and distributed generations. Solutions which do not satisfy defined constraints are eliminated or penalized, depending on convergence of power flow calculation. For other solutions, which meet defined limits, evaluation of objective function is done.

The procedure is repeated until stopping criteria are met. The criteria for stopping calculation can be based on a maximum number of generations, minimum of evaluated solutions, time limits to simulation, average change in solution distribution, etc.

Suggested model uses the concept of Pareto domination in the evaluation of the objective functions. Input data to describe multi-objective optimization problem are system parameters and constraints, lines, the loads, reliability parameters, failure rates, and the repair times.

MATLAB functions for genetic algorithms which are used for calculation are modified for specific discrete function for calculating power flow, power losses and testing system's constraints for network solutions.

CPU time spent for calculating and identifying the set of possible solutions depends on the time necessary for the objective functions evaluation, time for verification of defined constraints by power flow calculation and active losses. To speed up the calculation, parallel processing of genetic algorithm, in the part of evaluation of the objective function and constraints, is done.

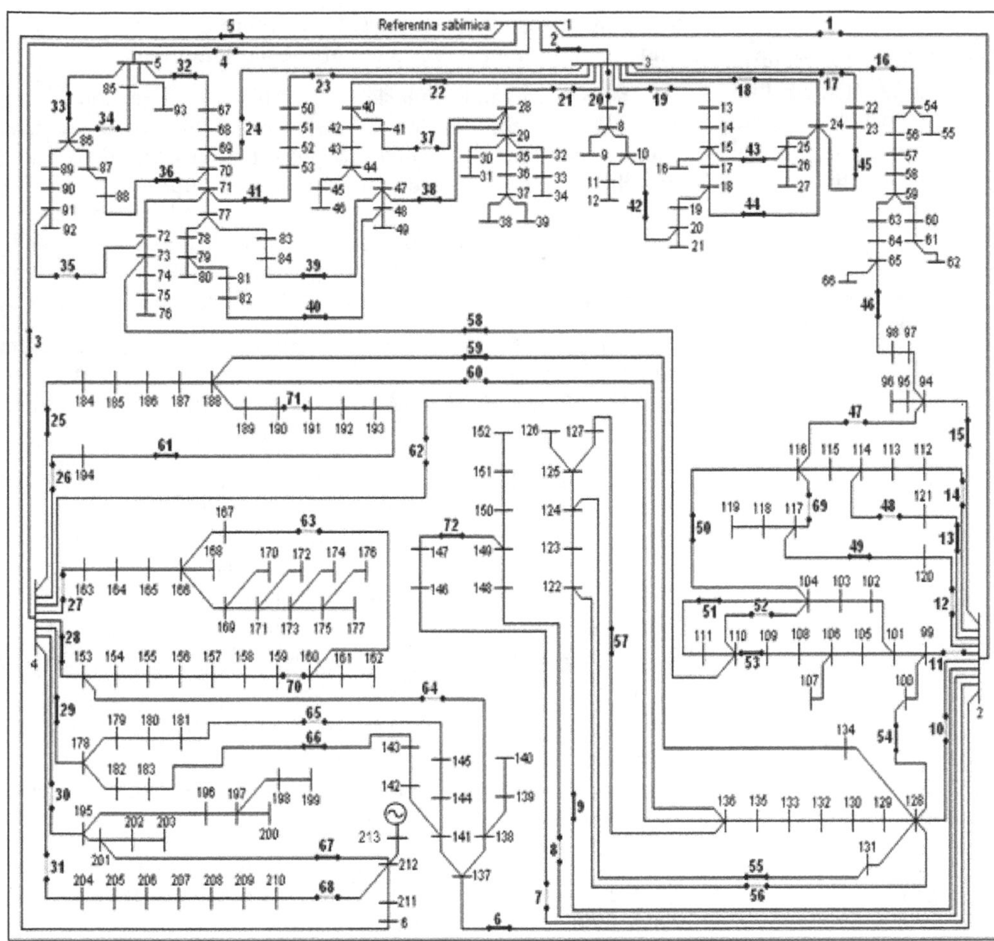

Fig. 1. Part of a distribution network with distributed generation

Taking into consideration that evaluation of one individual (solution) is completely independent of some other individual, independent of some other individual, evaluation of the entire population is distributed to 8 available processors, which significantly speeds up the calculation process.

The results of the described algorithm is a Pareto front of possible optimal topological solutions for the electrical distribution network

IV. Case Study and Numerical Results

The test system for the case study is 10 KV radial distribution network with distributed generation (Fig. 1.) with 213 buses, 248 lines and 72 switches. Distributed generation is connected to the node 213 and generates the active power of 5 MW. It is assumed that DG does not generate reactive power to the network.

All simulations are done on Intel Xeon E5-2699V3 with 32 GB RAM, that enables parallel data processing on 18 processor units. In initial network topology, presented in Fig. 1., switches 2–3, 5–6, 9, 13, 15, 22, 25, 28–29, 32–33, 36, 38–46, 49–51, 53–55, 57–59, 61, 66–67 and 72 are closed, while 1, 4, 7-8, 10–12, 14, 16–21, 23–24, 26–27, 30–31, 34–35, 37, 47–48, 52, 56, 60, 62–65 i 68–71 switches are open. For initial solutions, total power losses are 51.630 KW and ENS due to interruption in supply is 24.1387 KWh.

NSGA II parameters

In the consideration of optimal parameters 60 simulations were run with different values. Only the best performing parameters are presented in this paper and they are: initial population size is 150, crossover probability pc is 0.8, maximum number of generation is 200 and Pareto fraction is 0.45. Tournament selection is used, as well as two-point crossover. The stopping criterion of the algorithm is an imposed maximum number of generations or limit of the average change in distribution of solutions within the Pareto set (less than 10). Certain number of simulations is done by using adaptive mutation which search the larger solution space in the smaller number of generation (in this case the number of generation was 114), but evaluation time is longer. After the multiple independent runs of algorithm, it was concluded that the best results are achieved with fixed mutation probability of 0.01.

Furthermore, it was observed that fitness values significantly improve in early generations, when the solutions are farther from optimal values. The best fitness values slowly improve in later generations, whose population is closer to the optimal solutions. Number of generations for achieved solutions was in the range from 110-140 generations, for all tests with fixed crossover and mutation factor.

A. Result Analysis

Application of the described methodology to identify optimal network configuration based on NSGA II algorithm resulted in a set of possible solutions, out of which 9 Pareto optimal solutions are presented in Fig. 2. Pareto set of optimal solutions is achieved for 138 generations, while total number of achieved possible solutions is 49.

Total algorithm execution time on 18 processor units was 15 minutes and 36 seconds.

Table 1. shows values of objective functions for solutions from Pareto optimal set, as well as the on/off change of switches.

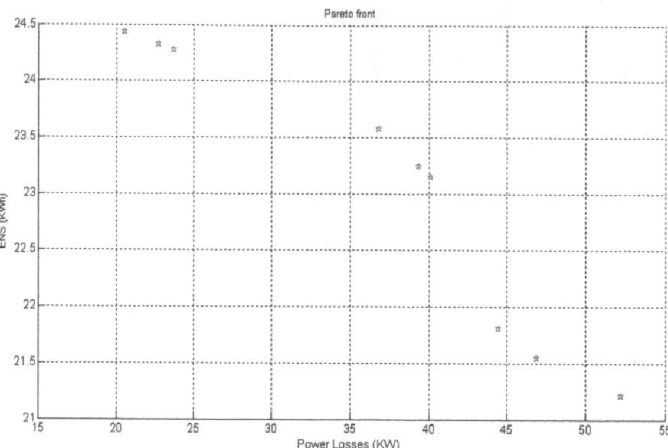

Fig. 2. Pareto optimal solutions for network reconfiguration with distributed generation

TABLE I. VALUES OF OBJECTIVE FUNCTION FOR PARETO OPTIMAL SOLUTIONS FROM FIG. 2

No.	Switch on	Switch off	Losses KW	IS Vh
1.	62	3	20.6344	24.4875
2.	7	72	23.7643	24.2683
3.	63	6	25.8799	24.2139
4.	68	67	37.1474	23.5844
5.	60	59	38.7454	23.2314
6.	31	5	40.1339	23.1442
7.	62	5	44.4429	21.8045
8.	71	61	46.8351	21.6585
9.	63	8	52.2096	21.2066

Considering the results shown in Table 1, the best solution for losses function is solution 1. However, this is the worst solution for ENS function. It is evident that changes in the value for ENS function are smaller than for losses function. Therefore, based on practical network topology implementation, functional and economical benefit the best compromise solution can be solution 3, given the evident small changes in ENS function between solutions 3, 4 and 5, while the difference in losses function is somewhat bigger. Solution 3 has power losses of 25.8799 KW, while ENS is 24.2139 KWh.

The achieved near-optimal solutions show traits of each solution from the Pareto front (the fact that not a single individual solution from Pareto front can be improved for one function without affecting the other in the opposite manner). This trait does not apply for all permissible searched solutions. The character of the achieved searched and near-optimal solutions depends on all set values, where different intensities and length of fault on lines are of special importance.

If intensity and fault length at all lines are equal, the variability of solution would be considerably lower, with a unique optimum for both functions. It is obvious that many searched solutions can be simultaneously improved from the aspect of both functions which are optimized; subsequently, the considered objectives are not necessarily in conflict with each other. This does not provide values which are approximately optimal for either of the objective functions.

Objective function assessing reliability, interruptions in supply, is incidental. Therefore, when descending sort order strategy is applied, the probability for local minimum is higher for reliability criterion that for power losses function.

Optimal minimization of losses will be achieved when the voltage in lines is closer to the maximum allowed value U_{max}. Since calculations were done with assumed constant load values (values of peak loads), maintaining voltage at lines as closer as possible to U_{max} ensures considerably less values for losses in the network. If voltage limitation is $U_{max} = 1.20$, losses values for solution 3 would be 11.256 KW.

Power losses for line 211-212 for initial solution are 0.73 KW, and for the identified optimal solution it is 0.49 KW (solution 3). Values for losses at the same line without distributed generation connected is 0.58 KW with the same switch state.

Since distributed generation is of small capacity in relation to the strength of distributive network into which it is connected, there is a reduction in the losses in the line onto which it is connected.

Impact on voltage

On Fig. 3. is shown voltage profile of bus system for solution 3 from Table 1. It is clear that the voltage is within the allowed limits. On Fig. 4. is shown the change in voltage in network nodes when distributed generation is not connected into the network (voltage shown in blue) and when the distributed generation is connected (voltage shown in red). The shapes of voltage profiles are almost the same in both cases, except for minor changes in the voltage strength at end lines, which is a consequence of connected distributed generation. Lines with distributive generation connected have an increase in voltage from 0.9723 p.u. to 0.9813 p.u. after installing the distributed generation into the network.

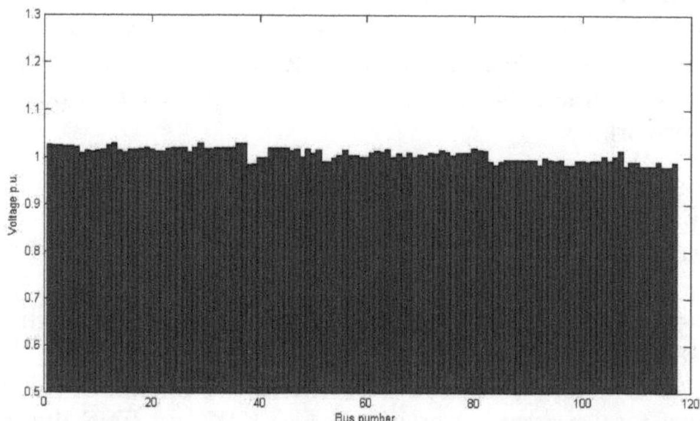

Fig. 3. Voltage profile for solution 3

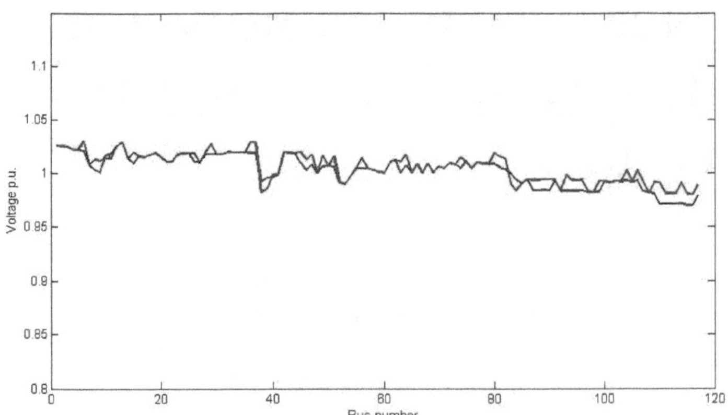

Fig. 4. Voltage profile with distributed generation connected and disconnected

V. CONCLUSION

Reconfiguration represents one of the most important measures which can improve performance in the operation of a distribution system. The paper shows application of multi-objective genetic algorithm NSGA II on resolving the problem of reconfiguration of distributive network with distributed generation, in order to identify optimal topological solution taking into account the set limitations. Algorithm is tested on a part of a network with 213 nodes, 248 lines and 72 switches. Multi-objective problem is formulated in order to decrease overall network losses and improve system reliability through minimizing ENS. The achieved results show the efficiency of the proposed methodology. Identification of near-optimal network configuration is presented. It is evident in decreasing overall network losses and ENS index compared to network initial state. The paper also shows the effect of distributed generation on voltage profile in distributive network. The results show that network reconfiguration in the presence of DG improve the voltage profile in the network.

It is obvious that application of the proposed methodology enables a more complex approach to improving the operational conditions in distributive networks, compared to traditional methods. The methodology proposed is also a useful tool for quick identification of optimal network configuration in case of faults, and it can also be beneficial for planning and upgrading existing network.

So, both the NSGA II efficiency in finding solutions and the increased efficiency of the the distribution network after using NSGA II are presented in the paper.

The achieved results for both objective functions can be represented in financial terms as well, and these are economic indicators to improve efficiency in managing a distributive network.

REFERENCES

[1] W. M. Dahalan, H. Mokhlis, "Network Reconfiguration for Loss Reduction with Distributed Generations Using PSO", IEEE International Conference on Power and Energy (PECon), Malaysia, pp. 823-828G, December 2012.

[2] N. Rugthaicharoencheep, S. Sirisumranukul, Optimal Feeder Reconfiguration with Distributed Generation in Three-Phase Distribution System by Fuzzy Multiobjective and Tabu Search, Energy Technology and Management, 1st ed, Tauseef Aized, ISBN 978-953-307-742-0, October , 2011 under CC BY-NC-SA 3.0 license

[3] M. N. M. Nasir, N. M. Shahrin, M. F. Sulaima, M. H. Jali, M. F. Baharoma, "Optimum Network Reconfiguration and DGs Sizing With Allocation Simultaneously by Using Particle Swarm Optimization (PSO)", International Journal of Engineering and Technology (IJET), Vol 6, No 2, pp. 773-780, Apr-May 2014.

[4] T.P. Sai Sree, N.P. Chandra Rao, "Optimal Network Reconfiguration and Loss Minimization Using Harmony Search Algorithm In The Presence Of Distributed Generation", International Journal of Electrical and Electronics Research, Vol. 2, Issue 3, pp. 251-265,July - September 2014.

[5] Y. Shu-jun, Y.Zhi, W.Yan, Y. Yu-xin, S.Xiao-yan, "Distribution network reconfiguration with distributed power based on genetic algorithm", 4th International Conference on Electric Utility Deregulation and Restructuring and Power Technologies (DRPT), ,pp. 811-815, 6-9 July 2011.

[6] A.Hajizadeh, E. Hajizadeh, "PSO-Based Planning of Distribution Systems with Distributed Generators", World Academy of Science, Engineering and Technology 45, pp.598-603, 2008.

[7] B. Tomoiagă , M. Chindriş, A. Sumper, A. Sudria-Andreu, R. Villafafila-Robles, "Pareto Optimal Reconfiguration of Power Distribution Systems Using a Genetic Algorithm Based on NSGA-II", Energies 2013, 6, pp.1439-1455, 2013., www.mdpi.com/journal/energies

[8] S.M. Cho, H.S. Shin, J.H. Park, J.C. Kim, "Distribution System Reconfiguration Considering Customer and DG Reliability Cost", Journal of Electrical Engineering & Technology Vol. 7, No. 4, pp. 486~492, 2012.

[9] K.Šimleša, Z.Vrančić, A.Matković, F.Lulić, G.Juretić, "Gubitak električne energije u distribucijskoj elektroenergetskoj mreži", CIRED hrvatski ogranak, 1. savjetovanje, S01-2, Šibenik, Croatia 2008.

[10] A.Hajizadeh, E.Hajizadeh : PSO-based Planning of Distribution Systems with Distributed Generators, World Academy of Science, Engineering and Technology, No. 45, pp.598–603, 2008.

[11] F.Rivas-Davalos, E.Moreno-Goytia, G.Gutierrez-Alacaraz, J.Tovar-Hernandez, "Evolutionary Multi-Objective Optimization in Power Systems: State-of-the-Art", Proceedings on IEEE PowerTech Conference, pp.2093-2098, Lausanne, Switzerland, 2007..

[12] C.M.Fonseca, P.J.Fleming, "An overview of evolutionary algorithms in multiobjective optimization", Evolutionary Computation, pp.1-16, 3(1), Spring 1995.

[13] K.Deb, A.Pratpat, S.Agarwal, T.Meyarivan, "A Fast and Elitist Multi-Objective Genetic Algorithm: NSGA-II", IEEE Transaction on Evolutionary Computation, Vol. 6, No2, pp.182-197, 2002.

[14] I.J.Ramirez-Rosado, J.L.Bernal-Agustin, "Reliability and Costs Optimization for Distribution Networks Expansion Using an Evolutionary Algorithm", IEEE Power Engineering Review, Vol. 21, pp. 70-76, 2001

[15] Muhammad Ahmadi, Ashkan Yousefi, Alireza Soroudi, Mehdi Ehsan, "Multi Objective Distributed Generation Planning Using NSGA-II", 13th International Power Electronics and Motion Control Conference EPE-PEMC, pp.1847-1851, 2008.

A Minimal Spiking Neural Network to Rapidly Train and Classify Handwritten Digits in Binary and 10-Digit Tasks

Amirhossein Tavanaei
The Center for Advanced Computer Studies
University of Louisiana at Lafayette
Lafayette, LA, USA

Anthony S. Maida
The Center for Advanced Computer Studies
University of Louisiana at Lafayette
Lafayette, LA, USA

***Abstract*—This paper reports the results of experiments to develop a minimal neural network for pattern classification. The network uses biologically plausible neural and learning mechanisms and is applied to a subset of the MNIST dataset of handwritten digits. The research goal is to assess the classification power of a very simple biologically motivated mechanism. The network architecture is primarily a feedforward spiking neural network (SNN) composed of Izhikevich regular spiking (RS) neurons and conductance-based synapses. The weights are trained with the spike timing-dependent plasticity (STDP) learning rule. The proposed SNN architecture contains three neuron layers which are connected by both static and adaptive synapses. Visual input signals are processed by the first layer to generate input spike trains. The second and third layers contribute to spike train segmentation and STDP learning, respectively. The network is evaluated by classification accuracy on the handwritten digit images from the MNIST dataset. The simulation results show that although the proposed SNN is trained quickly without error-feedbacks in a few number of iterations, it results in desirable performance (97.6%) in the binary classification (0 and 1). In addition, the proposed SNN gives acceptable recognition accuracy in 10-digit (0-9) classification in comparison with statistical methods such as support vector machine (SVM) and multi-perceptron neural network.**

Keywords*—Spiking neural networks; STDP learning; digit recognition; adaptive synapse; classification

I. INTRODUCTION

Neural networks that use biologically plausible neurons and learning mechanisms have become the focus of a number of recent pattern recognition studies [1, 2, 3]. Spiking neurons and adaptive synapses between neurons contribute to a new approach in cognition, decision making, and learning [4-8].

Recent examples include the combination of rank order coding (ROC) and spike timing-dependent plasticity (STDP) learning [9], the calculation of temporal radial basis functions (RBFs) in the hidden layer of spiking neural network [10], and linear and non-linear pattern recognition by spiking neurons and firing rate distributions [11]. The studies mentioned utilize spiking neurons, adaptive synapses, and biologically plausible learning for classification.

Learning in the present paper combines STDP with competitive learning. STDP is a learning rule which modifies the synaptic strength (weight) between two neurons as a function of the relative pre- and postsynaptic spike occurrence times [12]. Competitive learning takes the form of a winner-take-all (WTA) policy. This is a computational principle in neural networks which specifies the competition between the neurons in a layer for activation [13]. Learning and competition can be viewed as two building blocks for solving classification problems such as handwritten digit recognition. Nessler et al. (2009) utilized the STDP learning rule in conjunction with a stochastic soft WTA circuit to generate internal models for subclasses of spike patterns [14]. Also, Masquelier and Thorpe (2007) developed a 5-layer spiking neural network (SNN) consisting of edge detectors, subsample mapping, intermediate-complexity visual feature extraction, object scaling and position adjustment, and categorization layers using STDP and WTA for image classification [15].

Auditory and visual signals have special authentication processes in the human brain. Thus, one or more neuron layers are required to model the signal sequences in one and two-dimensional feature vectors in addition to the learning phase. Wysoski et al. (2008 and 2010) proposed a multilayer SNN architecture to classify audiovisual input data using an adaptive online learning procedure [16, 17]. The combination of Izhikevich's neuron firing model, the use of conductance-based synaptic dynamics, and the STDP learning rule can be used for a convenient SNN for pattern recognition. As an example, Beyeler et al. (2013) developed a decision making system using this combination in a large-scale model of a hierarchical SNN to categorize handwritten digits [18]. Their SNN architecture consists of 3136 plastic synapses which are trained and simulated in 500 (ms). They trained the system by 10/100/1000/2000 samples of the MNIST dataset in 100 iterations and achieved a 92% average accuracy rate. In another study, Nessler et al. (2013) showed that Bayesian computation is induced in their proposed neural network through STDP learning [2]. They evaluated the method, which is an unsupervised method for learning a generative model, by MNIST digit classification and achieved an error rate of 19.86% (80.14% correctness). Their proposed neural network for this experiment includes 708 input spike trains and 100 output neurons in a complete-connected feedforward network.

Some previous studies (c.f. [18], [2]) have attempted to develop an autonomous and strong artificial intelligence based on human brain anatomy in a large network of neurons and synapses. However, two inevitable and important aspects of the brain simulation are 1) the size of the network that is, number of the neurons and synapses, and 2) rapid learning and decision making. In some cases, a concise network is needed to be tuned and make a decision quickly in a special environment such as binary classification in the real time robot vision. Although large networks provide convenient circumstances for handling the details and consequently desirable performance, they are resource intensive. Our goal is to develop a fast and small neural network to extract useful features, learn their statistical structure, and make accurate classification decisions quickly.

This paper presents an efficient 3-layer SNN with a small number of neurons and synapses. It learns to classify handwritten MNIST digits. The training and testing algorithms perform weight adaptation and pattern recognition in a time and memory efficient manner while achieving good performance. The proposed SNN provides a robust solution for the mentioned challenge in three steps. First, the digit image is converted to spike trains so that each spike is a discriminative candidate of a row pixel in the image. Second, to reduce the network size and mimic human perception of the image, the spike trains are integrated to a few sections. In this part, each output spike train specifies a special part of the image in the row order. Third, training layer which involves STDP learning, output spike firing, and WTA competition by inhibitory neuron modifies a fast pattern detection strategy. The remarkably simple SNN is implemented for binary ("0, 1" c.f. Fig. 1) and 10-digit task (0-9) handwritten digit recognition problem to illustrate efficiency of the proposed strategy in primitive classifications. Furthermore, the obtained results are compared with statistical machine learning models in the same circumstances (same training/testing data without feature mapping) to depict the trustworthy of our model in similar situations.

II. Spiking Neural Network Architecture

The proposed SNN architecture is shown in Fig. 2. It includes three components: 1) a neural spike generator, 2) image segmentation, and 3) learning session and output pattern creator. Theory and implementation of each component will be explained.

A. First layer: Presynaptic spike train generator

Each row in the 28×28 binary image (c.f. Fig. 3) is transcribed into a spike train in a left-to-right fashion. Fig. 3 shows an example digit "0" with $N \times M$ binary pixels. Rows are converted to spike trains where a pixel value of "1" represents a spike occurrence. To apply the discriminative features of the image in a small network architecture, the digit image is recoded to N presynaptic spike trains with $A \times M$ discrete time points. A controls the interspike spacing and is interpreted as the refractory period of the neuron. In summary, the first layer converts the binary digit image into N rows of spike trains according to the white pixels of the digit foreground.

Fig. 1. Sample of handwritten digit images "0" and "1"

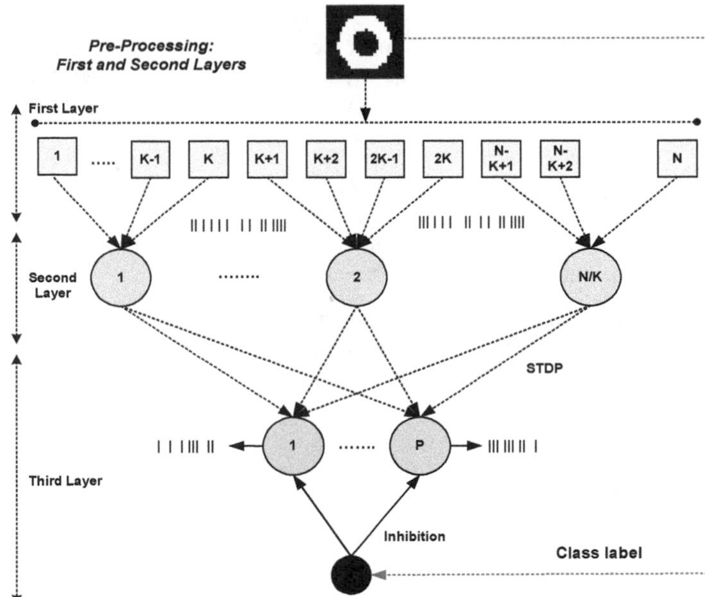

Fig. 2. Supervised SNN architecture containing spike transcription layer, spike train segmentation, STDP learning, output pattern firing, and inhibitory neuron. N: number of rows. K: number of adjacent rows connected to one neuron. P: number of classes. Black circle inhibits all output neurons except the one designated by the class label

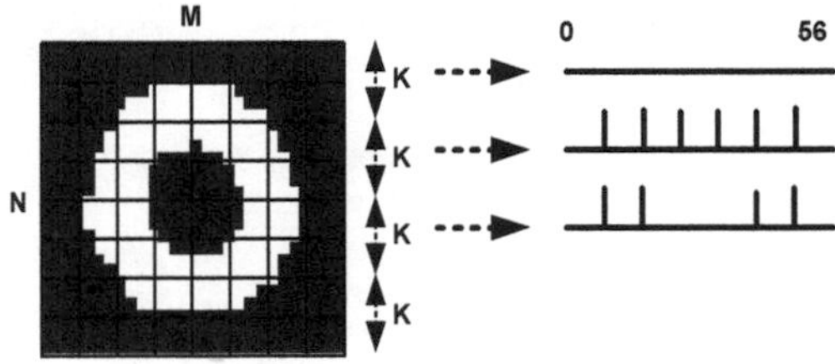

Fig. 3. A digit image with $N \times M$ pixels divided into N/K segments, consisting of k rows per segment

B. Second Layer: Image segmentation

The first layer generates N spike trains, where N is the number of the rows, encoding the image features row by row. However, it does not consider the slight change in orientation and thickness of the digit foreground in comparison with its background. To address this, the second layer illustrated in Fig. 2 merges every K spike trains (rows) onto one neuron. Then the digit image is segmented into N/K parts while preserving the spike train order. This preprocessing layer reduces the number of trainable parameters. Fig. 4 shows three instances of digit "1" (from the MNIST). The second layer converts these different shapes to similar N/K rows of spike trains. In addition, combining the sequential rows increases the network flexibility in pattern classification by decreasing its size. In summary, without the second layer, spike trains are sensitive to noise, outlier data, and diverse writing styles.

Fig. 4. Three instances of the handwritten digit "1"

To simulate the conductance of the postsynaptic membrane of the second layer units upon receipt of a presynaptic spike, the α-function is used. Equation (1) and Fig. 5 show the formula and graph of the postsynaptic conductance based on the α-function.

$$G_{\text{syn}}(t) = K_{\text{syn}}\, t\, e^{-t/\tau} \tag{1}$$

K_{syn} controls the peak conductance value and τ controls the time at which the peak conductance is reached.

Additionally, the total conductance of N input synapses with $N_{\text{rec},k\,(k=1:N)}$ spikes is calculated by (2)

$$G^{\text{tot}}(t) = \sum_{k=1}^{N}\sum_{j=1}^{N_{\text{rec},k}} K_{\text{syn},k}\,(t - t^{\text{f}}_{k,j})\, e^{-(t-t^{\text{f}}_{k,j})/\tau} \tag{2}$$

where t^f is spike firing time. This formula performs linear spatio-temporal summation across the received spike train. The total postsynaptic current is obtained by (3)

$$I_{\text{syn}}(t) = \sum_{k=1}^{N} E_{\text{syn},k} G^{\text{tot}}_{\text{syn},k}(t) - V(t)\sum_{k=1}^{N} G^{\text{tot}}_{\text{syn},k}(t) \tag{3}$$

In this investigation, spike generation in the second and third layer is controlled by Izhikevich's model [19] (4) specified by two coupled differential equations.

$$\begin{aligned} C\frac{dV}{dt} &= k(V - V_{\text{rest}})(V - V_{\text{th}}) - U + I_{\text{inj}} \\ \frac{dU}{dt} &= a\left[b(V - V_{\text{rest}}) - U\right] \end{aligned} \tag{4}$$

There is also a reset condition after a spike is generated, given in (5).

$$\begin{aligned} &if\ V \geq V_{\text{peak}} : V = c,\ U = U + d \\ &an\ AP\ is\ emitted \end{aligned} \tag{5}$$

Where, V denotes membrane potential and U specifies the recovery factor preventing the action potential (AP) and keeping the membrane potential close to the resting point. *"a"*, *"b"*, *"k"*, *"c"*, *"d"*, and "V_{peak}" are predefined constants controlling the spike shapes. The time of spike events is taken to occur at reset.

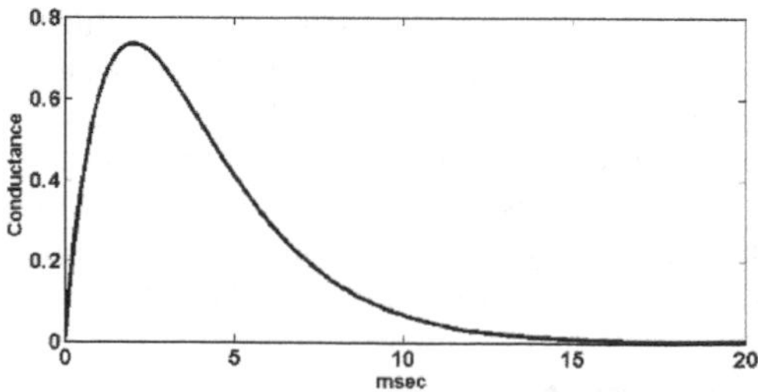

Fig. 5. Conductance graph with K_{syn}=1 (α-function) and τ=2 (msec)

C. *Third layer: Learning and output neurons*

Third layer of the SNN shown in Fig. 2 learns the input spike patterns and generates output spikes based on the evolving synaptic weights. STDP is controlled by relative pre- and postsynaptic spike times. Equation (6) specifies that postsynaptic spikes which follow presynaptic spikes cause the synaptic weight to be increased (LTP) and in contrast, synapses are weakened when presynaptic spike occurs after postsynaptic spike generation time (LTD).

$$\Delta K_{ji} = \begin{cases} 0.01A_{\text{ltp}} \exp(-(|\,t_j^f - t_i^f\,|)/_{\tau+})\ ,\ t_j^f - t_i^f > 0 \\ 0.01A_{\text{ltd}} \exp(-(|\,t_j^f - t_i^f\,|)/_{\tau-})\ ,\ t_j^f - t_i^f < 0 \end{cases} \tag{6}$$

In (6), A_{ltp} and τ+ (A_{ltd} and τ-) are maximum and time constant strengthening (weakening) constants respectively. In addition, the change in synaptic weights contributes to change in conductance amplitude, K_{syn}, in α-function derivation. The learning strategy used in this investigation is basically derived from the STDP concept. The proposed network in the first layer emits spike trains with maximum M spikes, where M is the number of columns in the image matrix. The second layer presents new information of spike trains at which spikes depict explicit foreground pixel information. In addition, the membrane potential is accumulated based on the received action potentials. Therefore, in the proposed minimal network architecture which models the patterns by exact object coordinates, a modified STDP learning is defined in (7).

$$\Delta K_{ji} = \begin{cases} \dfrac{A_{\text{ltp}}}{K_{ji}} \exp(-(|\,t_j^f - t_i^f\,|)/_{\tau+})\ , & \begin{pmatrix} P_j = 1, \\ t_i^f \in [\min(t_j^{f-1}, t_j^f - \sigma), t_j^f] \end{pmatrix} \\ -A_{\text{ltd}} \exp(-\beta) & ,\ \begin{pmatrix} P_j = 1, \\ t_i^f \notin [\min(t_j^{f-1}, t_j^f - \sigma), t_j^f] \end{pmatrix} \end{cases} \tag{7}$$

where A_{ltp}, A_{ltd}, β>1, and σ are constant parameters. In (7), if output neuron P_j fires, the synaptic weights can be either increased or decreased. Presence of the presynaptic spikes in the σ time interval before current time strengthens the synaptic conductance. In contrast, absence of the presynaptic spikes reduces the synaptic conductance. To prevent aliasing between σ time interval and previous output spike, presynaptic spikes after the last emitted postsynaptic spike are counted. Also, the inverse value of the conductance amplitude (K_{ji}) controls rate of the LTP in the high conductance conditions.

In addition, output neurons in the third layer receive N/K spike trains and generate P (as number of the output patterns) output spike trains based on the current synaptic weights, presynaptic spike trains, and Izhikevich's model for the spike generation mechanism. Furthermore, each output neuron specifies one class. The learning strategy in the output layer is supervised. This is implemented by using an inhibitory neuron that imposes a WTA discipline across the output units. Specifically, the inhibitory unit uses the category label for the current training stimulus to inhibit all the output neurons that do not match the label. The net learning effect is that the nonmatching units undergo LTD, while the single matching unit undergoes LTP. Equation (8) specifies the LTD rule for inactive neurons. The synaptic conductance reduction in this formula depends on the presynaptic spikes "y_i" conveyed to the neuron in the time interval [$\min(t_j^{f-1}, t_j^f - \sigma)$, t_j^f].

$$C = \frac{-\gamma}{\sum_{i=\min(t_j^{f-1}, t_j^f - \sigma)}^{t_j^f} y_i} \quad , \quad \Delta K'_{ji} = -0.01e^C \tag{8}$$

Where, γ is rate of the inhibition. In the last step, conductance magnitude of the synapses (which can be interpreted as synaptic weights) are updated by (9)

$$K_{ji}^t = K_{ji}^{t-1} + \mu \Delta K_{ji}^t \tag{9}$$

where, μ is learning rate. Finally, the result will be an array of synaptic weights and output spike patterns. Fig. 6 shows pseudocode for the SNN architecture and learning strategy.

D. Justification

Digits belonging to the same categories are not entirely similar due to different handwriting styles, variations in orientation, and variations in line thickness. The second layer converts the various images of one digit into a small number of similar patterns. It combines K spike trains to adjust the thickness and presents the image in N/K row segments. The slight diversity of the images in a digit category can be manipulated by foreground adjustment in height and width which is implemented by row segmentation and regular spiking (RS) neurons respectively. In addition, N input spike trains are mapped to N/K spike trains to minimize the network size.

To explain the learning procedure and justify its function in classification, an example consisting of the digits 2, 4, 1, and 9 is described step by step. Fig. 7 shows the digits. They are divided into 4 horizontal segments which are mapped into 4 adaptive synapses. If an output spike occurs, the synapses carrying more frequent and closer presynaptic spikes (white pixels) before the output spike have more casual effects. Thus, their weights are increased based on the LTP rule. For example, the synapses {1,4}, {3}, and {1,2} in digits 2, 4, 9 respectively carry frequent presynaptic spikes, so their weights are increased more than the other synapses in each digit. In digit "1", all of the synapses have analogous influences onto the output neuron firing. So, the synaptic weights should be almost unbiased. After the first training period including weight augmentation and reduction, in the next iteration, the synaptic weights are tuned better according to the input digit patterns. Additionally, synaptic weights, which are connected to the same neurons in the second layer and different output neurons, are adapted in a competition due to the inhibitory neuron. Therefore, the synaptic weights demonstrate discriminative weight vectors for different digit patterns. In Fig. 7, some nominal synaptic weights (Ex. {0.20, 0.20, 0.45, 0.15} for digit 4) have been shown.

In the test session, if the digit spike trains are matched to the synaptic weights, the target output neuron releases a spike train close to the target pattern. Otherwise, due to discriminative synaptic weights, if the input spike trains are not compatible with the synaptic weights and target pattern, the output neuron might release a spike train either with 0 frequency or arbitrary pattern. Finally, the digit having maximum correlation with training data will be recognized in a small and fast neural network.

III. Experiments and Results

A subset of the MNIST machine learning data set consisting of handwritten digit images was used for evaluation of the proposed method [20]. Digital images in the dataset are 28 pixels in height and 28 pixels in width for a total of 784 pixels composing each grayscale image.

A. Binary classification

In the first experiment, 750 images of the digits "0" and "1" were sampled. Each grayscale image was converted to a binary image by applying the middle point threshold (threshold pixel=128). The 750 digit samples were divided into training and testing sets by 3-fold cross validation to guarantee the generality of the method.

The first layer scans the rows pixel by pixel and generates spikes where the digit points occur. Pixel values equal to 1 denote spike occurrences. In addition, a refractory period, A, is assumed to be 2 (ms). Therefore, the spike trains represent a row fall into a 28×2=56 (ms) temporal window. Fig. 3 gives an example of spike train generation for a sample digit "0". Finally, 28 spike trains with 56 (ms) discrete time points are obtained as presynaptic spikes conveyed to the second layer.

To segment the image into groups of rows, presynaptic spikes are collected to the N/K layer-2 neurons where N=28 and K=4. That is, every 4 consecutive sequential spike trains are connected to one neuron in the second layer. Spike generation of the neuron in this layer is computed by Izhikevich's RS model with parameters given in Table 1. Spike trains from seven layer-2 neurons submit information to the output neurons in the layer 3.

```
Function OneDataPassTraining(image, &
weights):OutputSpike
{
   [N,M]=size(image);
   r=2; % refractory period
   % Layer 1
   for each row of the B/W image
      spikes=generate spikes in r*M time points (1: spike
occurrence)
   %Layer 2
   for i=1:N/K
   {
     for j=1:K
        preSpikes{i}.append(spikes{(i-1)*K+j})
     middleSpikes{i}=Izhikevich's model (preSpikes{i});
   }
   %Layer 3
   for p=1:#classes
   {
      OutputSpike=Izhikevich's model (middleSpikes);
      STDP learning for target output
      Inhibition for non-target output based on STDP
      weights=update Synaptic weights
   }
```

Fig. 6. SNN pseudocode for handwritten digit classification

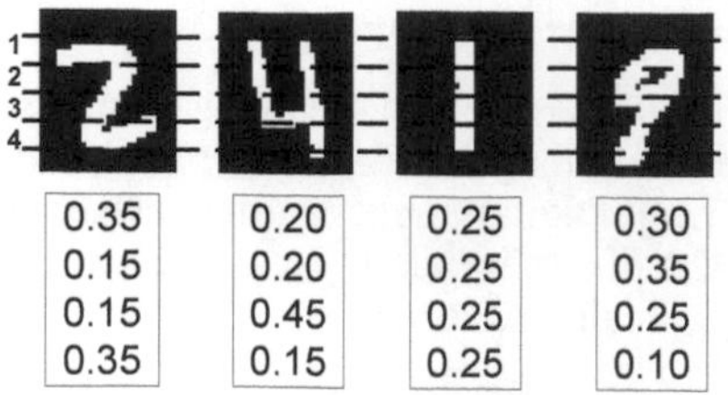

Fig. 7. Synaptic weights for digits 2, 4, 1, and 9. The intervals between dashed lines specify synapses. The numbers below each digit denote the synaptic weights

The output neurons use the same parameters in Izhikevich's model of the second layer to generate the spikes. In the third layer, synaptic weights projecting to the output neurons are initialized uniformly and updated by the STDP rule with parameters of LTP and LTD in Table 2. Furthermore, the inhibition neuron prevents the non-target (0 or 1) neuron to fire while receiving the presynaptic spikes. Hence, synaptic weights are changed according to the relative pre- and postsynaptic spike times.

After one batch of training (500 training samples), 14 synaptic weights (7 synapses for output "0" and 7 synapses for output "1") and a set of output spike patterns for "0" and "1" are obtained. Fig. 8 shows the simulation results (with ΔT=0.1 (msec)) of output spike trains of some handwritten digit images in "0" and "1" categories (each row shows a spike train). The illustrated spike trains in Fig. 8 show 1) specific first spike times and 2) discriminative spike time patterns for class "1" and class "0". Therefore, extracted target patterns are appropriate sources for pattern recognition. The output patterns of the testing samples are compared with the average target patterns for each class. Finally, the similarity measure denotes the objective function for the classification that is shown in (10).

$$Similarity^{\text{pattern}} = \frac{1}{\sqrt{\frac{1}{T}\sum_{i\in \text{spike train}} (Target_i^{\text{pattern}} - Output_i)^2}} \quad (10)$$

Where T is size of the target pattern.

Five different simulation step sizes (ΔT=0.05, 0.1, 0.2, 0.5, 1 (msec)) were studied. Table 3 specifies scaled synaptic weights connected to the output neurons "0" and "1" in the five temporal resolutions. Synaptic weights in Table 3 claim that ΔT in range of 0.05 to 0.5 (msec) give discriminative weight vectors for different classes. On the other hand, in ΔT=1 (msec), the training is biased to "0" because the simulation step size is so large and the learning procedure is not converged.

A subset of disjoint training and testing data was applied to the trained SNN to evaluate the accuracy rate of the proposed method. The results are shown in Table 4. The average accuracy rate is 97.6 for the testing sets. In addition, values of ΔT in the range of 0.05 to 0.5 (msec) give acceptable performance. ΔT=1 (msec), as explained, is not applicable. According to the results in Table 4, ΔT=0.2 (msec) has the best performance. It is also a fast neuron simulation for training and testing sessions.

TABLE I. REGULAR SPIKING NEURON PARAMETERS

Parameter	Value	Parameter	Value
V_{rest}	-60 (mv)	a	0.03
$V_{threshold}$	-40 (mv)	b	-2
V_{peak}	35 (mv)	c	-50
C	100	d	100
K	0.7	U_0	0
ΔT	0.05, 0.1, 0.2, 0.5, 1	I_{inj}	0

TABLE II. STDP PARAMETERS

Parameter	value
A	103
B	-40
$\tau+$	14
$\tau-$	34
K Default	10

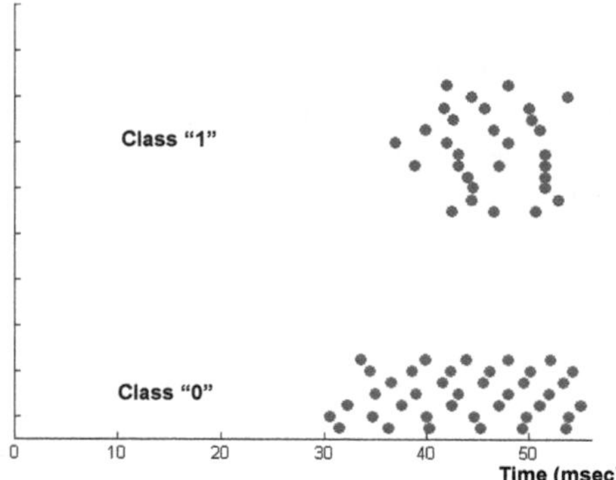

Fig. 8. Example output spike trains for the digits "0" (Blue) and "1" (Red) after learning. ΔT=0.1 (msec)

TABLE III. SYNAPTIC WEIGHTS (AFTER TRAINING) PROJECTING TO THE OUTPUT NEURONS REPRESENTING CATEGORIES "0" AND "1" IN DIFFERENT SIMULATION STEP SIZES (ΔT). EACH COLUMN INDICATES THE IMPORTANCE OF ONE OF THE N/K=7 IMAGE SEGMENTS TO EACH CATEGORY. THE BOLD WEIGHTS SHOW ACCEPTABLE LEARNING. ΔT IMPACTS ON MEMBRANE POTENTIAL COMPUTATION AND SYNAPTIC WEIGHT ALTERNATIONS. THEREFORE, EACH COLUMN SHOWS SOME SLIGHT VARIATIONS IN THE SYNAPTIC WEIGHTS. HOWEVER, THE SYNAPTIC WEIGHTS FOR EACH SIMULATION SHOW SEPARATE CATEGORIES

ΔT	Digit	Syn1	Syn2	Syn3	Syn4	Syn5	Syn6	Syn7
0.05	0	**0.00**	**1.30**	**6.60**	**5.24**	**0.26**	**6.60**	**0.00**
	1	**0.00**	**0.00**	**0.00**	**10.19**	**4.29**	**5.52**	**0.00**
0.1	0	**0.00**	**2.98**	**5.66**	**4.17**	**1.55**	**5.66**	**0.00**
	1	**0.00**	**0.00**	**0.00**	**10.43**	**4.20**	**5.37**	**0.00**
0.2	0	**0.00**	**3.69**	**5.94**	**4.16**	**0.26**	**5.94**	**0.00**
	1	**0.00**	**0.00**	**0.00**	**9.83**	**7.36**	**2.81**	**0.00**
0.5	0	**0.00**	**2.42**	**7.24**	**1.05**	**1.27**	**8.03**	**0.00**
	1	**0.00**	**0.00**	**0.00**	**17.40**	**2.25**	**0.36**	**0.00**
1	0	0.00	5.06	5.06	4.59	0.24	5.06	0.00
	1	0.03	0.00	0.00	0.00	0.00	0.00	19.97

TABLE IV. ACCURACY RATE OF THE PROPOSED SNN FOR HANDWRITTEN DIGIT CLASSIFICATION OF HANDWRITTEN DIGITS "0" AND "1"

ΔT	Accuracy Rate (Testing Data) %	Accuracy Rate (Training Data) %
0.05	**97.20**	---
0.1	**97.60**	---
0.2	**98.00**	**99.00**
0.5	**97.60**	---
1	56.00	---

B. 10-digit classification task

In the second experiment, 320 image samples of the MNIST handwritten digits were randomly selected and converted to binary images. The first and second layers of the SNN are the same as in the binary classification experiment except the segmentation factor, *K*, is set to 2. Therefore, the learning component consists of 28/2=14 adaptive synapses connected to 10 output neurons representing the digits 0 to 9 (140 adaptive synapses total and 24 layer-2 neurons). According to the mentioned theory, the second layer should generate candidate spike trains for a large variety of the input patterns. Fig. 9b illustrates 14 spike trains in the second layer which show a schematic of the input digits in Fig. 9a.

These discriminative spike trains invoke STDP learning in the next layer to adapt the synaptic weights and generate distinguishable spike patterns for digit categories (0-9). Fig. 10 shows the convergence scenario of the training process in 1000 iterations. This chart determines total distance between synaptic weights in sequential trials that is calculated by (11). It is concluded that, the training algorithm converges in 84 iterations and more training trials will not change the synaptic weights considerably. The synaptic weight matrix after 84 training iterations is shown in Table 5.

$$\Delta W^t_{Total} = \sum_{i=0}^{9} \left(\sum_{j=1}^{N/K=14} (w^t_{ij} - w^{t-1}_{ij})^2 \right)^{\frac{1}{2}} \tag{11}$$

The adjusted synaptic weights and input digits provide 10 patterns of output spike trains shown in Fig. 11. The membrane potential and spike times in Fig. 11 illustrates discriminative patterns for different digits. For example, spikes in time stream of the digit "1" are close together in the center of the time window because all of the presynaptic spikes are gathered in a small range of simulation time (30-40 (ms)).

However, the proposed method as a minimal SNN architecture with 10 × 14 adaptive synapses (14-D weight vector) is designed for small classification problems such as binary categorization, not optimized for 10 categories, the performance of 10-digit task is 75.93% in average.

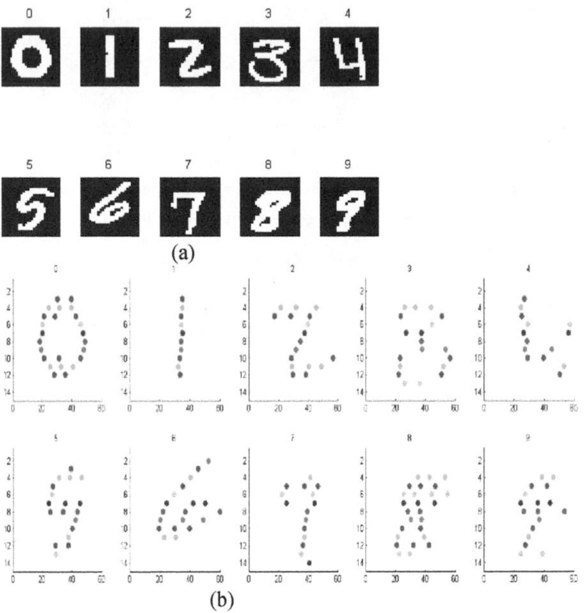

Fig. 9. a) Handwritten digits 0-9. b) 14 spike patterns emitted from RS neurons of the second layer

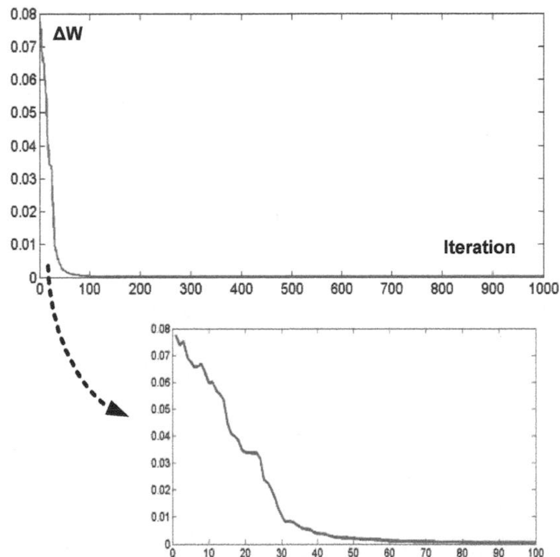

Fig. 10. Convergence plot obtained from 1000 iterations

TABLE V. SYNAPTIC WEIGHTS OF 14 SYNAPSES OF 10 DIGITS 0-9. THE FIRST TWO SYNAPTIC WEIGHTS AND THE LAST ONE ARE SMALLER THAN OTHER SYNAPSES BECAUSE THEY MOSTLY CONVEY BACKGROUND INFORMATION. THUS, RELATIVELY MUCH WEAK SYNAPSES PROVIDE A FAST METHOD OF BACKGROUND ELIMINATION IN THE ROW ORDER

Syn / Digit	1	2	3	4	5	6	7	8	9	10	11	12	13	14
1	0.008	0.008	0.041	0.028	0.033	0.070	0.109	0.137	0.158	0.155	0.132	0.105	0.008	0.008
2	0.009	0.010	0.062	0.147	0.100	0.034	0.026	0.051	0.147	0.190	0.142	0.063	0.009	0.009
3	0.009	0.009	0.130	0.190	0.083	0.045	0.076	0.062	0.018	0.026	0.122	0.186	0.033	0.010
4	0.010	0.010	0.010	0.028	0.082	0.138	0.187	0.208	0.159	0.055	0.036	0.038	0.029	0.010
5	0.008	0.008	0.020	0.065	0.092	0.117	0.137	0.152	0.051	0.105	0.144	0.088	0.008	0.008
6	0.009	0.033	0.051	0.043	0.051	0.071	0.132	0.165	0.171	0.151	0.097	0.011	0.009	0.009
7	0.010	0.010	0.010	0.079	0.199	0.114	0.069	0.045	0.051	0.044	0.066	0.091	0.151	0.063
8	0.006	0.006	0.032	0.084	0.114	0.127	0.113	0.055	0.089	0.113	0.113	0.112	0.029	0.006
9	0.008	0.008	0.008	0.059	0.133	0.154	0.152	0.155	0.052	0.027	0.041	0.072	0.124	0.008
0	0.008	0.008	0.021	0.047	0.083	0.092	0.122	0.126	0.114	0.146	0.156	0.061	0.008	0.008

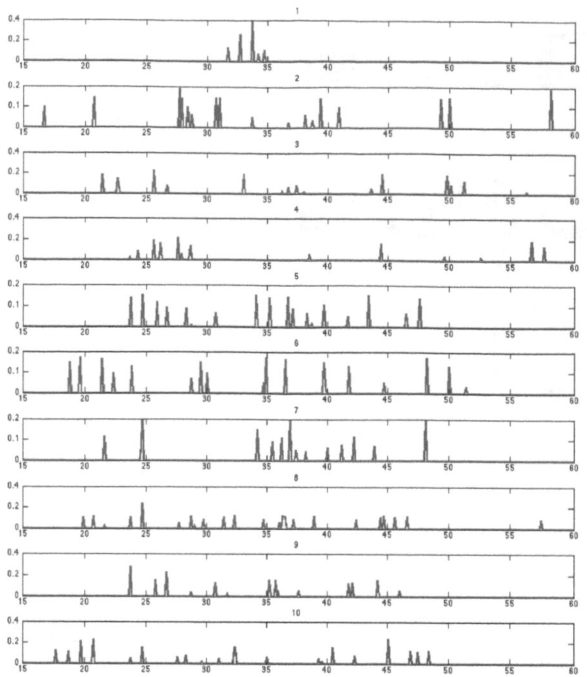

Fig. 11. Membrane potential and spike patterns for digits 0-9 after training

C. Comparison with other models

To compare our model with statistical machine learning strategies, the same training and testing datasets were applied to 1) a support vector machine (SVM) which maximizes the border distances between the classes [21]; and 2) a back propagation multi perceptron artificial neural network (BP-ANN) which learns the synaptic weights using error-feedback adjustment [22]. The obtained results are shown in Table 6. They have been implemented by the R software package [23, 24]. If more data are used in statistical models (with modified parameters) and some preprocessing algorithms such as principle components analysis (PCA) are applied, the performance should be higher than the rates reported in Table 6. However, based on the same situations at which the SNN performs, the SVM and ANN methods show accuracy rates that are slightly lower than the proposed SNN. We claim that the minimal SNN in this investigation has sufficient capacity to be improved more by the required preprocessing and experiments while using the biologically plausible principles.

IV. Conclusion and Discussion

A minimal time and memory efficient SNN architecture for classification was presented. This research shows that phenomenological STDP in a minimal model can support pattern recognition learning. The algorithms and neuron models were chosen to be biologically plausible. The proposed method represented an architecture which specifies a remarkably simple and fast network in comparison with previous investigations.

TABLE VI. Accuracy of 10-Digit Recognition Using Statistical Machine Learning Models. 200 Training Data Without Preprocessing

Method	Acc. %	Description
SVM	73.44	Polynomial kernel n=3.
ANN	70.87	40 hidden neurons, Decay=0.0001, 550 iterations.

Our SNN was applied to handwritten digit classification for a subset of images in the MNIST dataset. First, the initial layer interpreted the image logically based on the exact foreground pixel locations. Therefore, digit image was scanned row by row to generate the spikes as impulse reaction to the object perception. Also, this layer extracted the feature spikes directly from the image and represented a quick and natural image perception without complex computations. Second, every K (Ex. 2 or 4) spike trains were accumulated in a sequential order to provide the segmentation aspect of object detection in order to reduce the working space using Izhikevich's neuron model. This part of the network guaranteed to reduce number of the computational neurons and kept the order of the image segments from top to bottom. This layer provided a structure to produce set of spike trains invariant to diverse handwriting styles, outlier points, and slight changes in foreground orientation and thickness. The extracted sections mimicked digital scanning methods in a fast and implementable manner. Third, STDP learning and inhibitory neuron prepared the required environment for training the network and competition among dissimilar categories. The third layer's algorithm focused on supervised learning of the summarized input patterns. Additionally, the STDP rule was applied in two different sets of the synapses (connected to the target and non-target neurons) simultaneously and the training process converged after a small number of iterations. Thus, The SNN was tuned to categorize the input spike patterns quickly and it did not need many feature spike trains.

In summary, the introduced strategy was implemented in a simple and fast way due to the small number of the neurons and adaptive synapses (totally, {10 and 25} computational neurons and {14 and 140} adaptive synapses for binary and 10-digit classifications respectively). Finally, evaluation of the presented model demonstrated admirable performance of 98.0% maximum and 97.6% average accuracy rates for binary ("0" and "1") handwritten digit recognition. Furthermore, in spite of the minimal architecture of the presented SNN, acceptable performance of 75.93% was obtained in 10-digit recognition. The comparison between accuracy rate of the proposed method and statistical machine learning approaches (basic models without preprocessing and a small number of training data) determined slightly better performance of our SNN in the same and basic situations as well as incremental learning ability of the SNN. The minimal SNN worked much better for binary classification than 10-digit task. However, the reported results showed the potential capability of the SNN to be shrunk and work fast in training and prediction phases.

Therefore, the proposed SNN architecture and learning procedure can be a trustworthy model for classification due to its simple structure, quick feature extraction and learning, robust synaptic adaptation, and feasible implementation on VLSI chips in the future experiments.

REFERENCES

[1] Maass, Wolfgang. "Networks of spiking neurons: the third generation of neural network models." Neural networks 10.9 (1997): 1659-1671.

[2] Nessler, Bernhard, et al. "Bayesian computation emerges in generic cortical microcircuits through spike-timing-dependent plasticity." PLoS computational biology 9.4 (2013): e1003037.

[3] Kappel, David, Bernhard Nessler, and Wolfgang Maass. "STDP installs in winner-take-all circuits an online approximation to hidden markov model learning." PLoS computational biology 10.3 (2014): e1003511.

[4] Gerstner, Wulfram, and Werner M. Kistler. "Spiking neuron models: Single neurons, populations, plasticity". Cambridge University Press, 2002.

[5] Rozenberg, Grzegorz, Thomas Back, and Joost N. Kok. "Handbook of natural computing". Springer Publishing Company, Incorporated, 2011.

[6] Wade, John J., et al. "SWAT: a spiking neural network training algorithm for classification problems." Neural Networks, IEEE Transactions on 21.11 (2010): 1817-1830.

[7] Yousefi, Ali, Alireza A. Dibazar, and Theodore W. Berger. "Synaptic dynamics: Linear model and adaptation algorithm." Neural Networks 56 (2014): 49-68.

[8] Chrol-Cannon, Joseph, and Yaochu Jin. "Computational modeling of neural plasticity for self-organization of neural networks" BioSystems 125 (2014): 43-54.

[9] Kasabov, Nikola, et al. "Dynamic evolving spiking neural networks for on-line spatio-and spectro-temporal pattern recognition." Neural Networks 41 (2013): 188-201.

[10] Wang, Jinling, et al. "An Online Supervised Learning Method for Spiking Neural Networks with Adaptive Structure." Neurocomputing (2014).

[11] Vázquez, Roberto A. "Pattern recognition using spiking neurons and firing rates." Advances in Artificial Intelligence–IBERAMIA 2010. Springer Berlin Heidelberg, 2010. 423-432.

[12] Caporale, Natalia, and Yang Dan. "Spike timing-dependent plasticity: a Hebbian learning rule." Annu. Rev. Neurosci. 31 (2008): 25-46.

[13] Maass, Wolfgang. "On the computational power of winner-take-all." Neural computation 12.11 (2000): 2519-2535.

[14] Nessler Bernhard, Michael Pfeiffer, and Wolfgang Maass. "STDP enables spiking neurons to detect hidden causes of their inputs." Advances in neural information processing systems. 2009.

[15] Masquelier, Timothée, and Simon J. Thorpe. "Unsupervised learning of visual features through spike timing dependent plasticity." PLoS computational biology 3.2 (2007): e31.

[16] Wysoski, Simei Gomes, Lubica Benuskova, and Nikola Kasabov. "Fast and adaptive network of spiking neurons for multi-view visual pattern recognition." Neurocomputing 71.13 (2008): 2563-2575.

[17] Wysoski, Simei Gomes, Lubica Benuskova, and Nikola Kasabov. "Evolving spiking neural networks for audiovisual information processing." Neural Networks 23.7 (2010): 819-835.

[18] Beyeler, Michael, Nikil D. Dutt, and Jeffrey L. Krichmar. "Categorization and decision-making in a neurobiologically plausible spiking network using a STDP-like learning rule." Neural Networks 48 (2013): 109-124.

[19] Izhikevich, Eugene M. "Simple model of spiking neurons." IEEE Transactions on neural networks 14.6 (2003): 1569-1572.

[20] LeCun, Y., Bottou, L., Bengio, Y., and Haffner, P. (1998). "Gradient-based learning applied to document recognition". Proceedings of the IEEE, 86, 2278–2324.

[21] Bishop, Christopher M. "Pattern recognition and machine learning". New York: Springer, 2006.

[22] Cross, Simon S., Robert F. Harrison, and R. Lee Kennedy. "Introduction to neural networks." The Lancet 346.8982, pp. 1075-1079, 1995.

[23] Karatzoglou, Alexandros, David Meyer, and Kurt Hornik. "Support vector machines in R." 2005.

[24] Günther, Frauke, and Stefan Fritsch. "Neuralnet: Training of neural networks." The R Journal 2.1, pp.30-38, 2010.

14

A Multi_Agent Advisor System for Maximizing E-Learning of an E-Course

Khaled Nasser ElSayed
Computer Science Department, Umm Al-Qura University

***Abstract*—Web-based learning environments have become popular in e-teaching throw WWW as distance learning. There is an urgent need to enhance e-learning to be suitable to the level of learner knowledge. The presented paper uses intelligent multi-agent technology to advise and help learners to maximize their learning of an offered e-course. It will build its advices on the performance and level of education of learners including past and current learning. Most of advices are to guide learner to make exercises as quizzes or passing tests in different level of difficulties.**

Keywords—AI; Agent; Multi_Agents; distant learning; e-Learning; e-Teaching; Education; e-Course

I. Introduction

In the time being, Distance learning is the hot issue in computer science. Online learning through the web has become popular in the decade [1]. E-learning is nowadays recognized as one of the efficient methods to respond to the requirements of open and distance learning. In the e-learning system, several traditional learning styles should be combined with the learner-centered approach. It needs a good notation to represent the requirements of the e-learning system [2].

In the dynamic changes information environment without prior modeling, it can independently plan complex operation steps to solve practical problems, can independently discover and obtain the available resources the learners needed and then provide the corresponding services under the circumstance that the learners do not take part in [3].

An agent is something that perceives and acts in an environment. The agent function for an agent specifies the action taken by the agent in response to any percept sequence [4]. Intelligent agents are task-oriented software components that have the ability to act intelligently. They may contain more knowledge about the needs, preferences and pattern of the behaviors of a person or a process as in [5].

The agent has to collect users' personal interests and give fast response according to the pre-specified demands of users. The personal agent can discover users' personal interests voluntarily without bothering the users. It is very suitable for personalized e-learning by voluntarily recommending learning materials [6].

Intelligent agents should have the ability of adaptive reasoning. They must have the capability to access information from other sources or agents and perform actions leading to the completion of some task. Also, they must control over their internal state and behavior and work together to perform useful and complex tasks. Thus, they should be able to examine the external environment and the success of previous actions taken under similar conditions and adapt their actions [7].

Educators, using Web-based learning environments, are in desperate need for non-instructive and automatic ways to get objective feedback from learner in order to better follow the learning process and appraise the online course structure effectiveness. On the learner side, it would be very useful if the system could automatically guide the learner's activities and intelligently recommend online activities and resources that would favour and improve the learning. The automatic recommendation could be based on the instructor's intended sequence of navigation in the course material, or, more interestingly, based on navigation patterns, of other successful learners [8].

Currently, the state of intelligent is focused on one-to-one learning instruction. Some examples include ACT systems [8], DEBUGGY [9], and PIXIE [10]. Specifically, the kind of learning modality used is centered on learning by being told [11].

There are too much work done in the field of e-learning and e-teaching based on agent. Gascuena and Fernadez-Caballeroe [12] introduced an Agent-based Intelligent Tutoring System for enhancing E-Learning/E-Teaching, where agents monitor the progress of the students and propose new tasks. De Antonio presented architecture of intelligent virtual environment based on agent technology [13]. Also, a similar one for nurse training is offered in [14]. Tang offered the implementation of a multi-agent intelligent tutoring system for learning the programming languages [15]. According to Java Agent for distance education (JADE) frame work, Silveira and Vicari carried out their system Electrotutor which is Electrodynamics distance teaching environment [16].

Since the students and teachers are on different time and spare in an e-learning environment, the learning status of a student is difficult to be controlled by teachers. In current learning platforms, they neither analyze the causes of learning inefficiency of users, nor generate new learning material and testing. The former keeps the learners from not using these learning systems anymore because they are confusing; the latter leads to out-of-date materials and the learners could not get any new knowledge [17].

In the proposed work, there is multi-agent system that could get learner profile knowledge at his logging to the e-course. Then system can help users and advises them in their on line learning. It will enhance e-learning of e-courses through advising learners for better navigation through e-course contents by offering some links or jumping over course resources, or by guiding learner to make exercises in a quiz or passing through an exam.

II. E-COURSE DELIVERY

One of the main goals of e-teaching is that the learner learns more and better to enhance teaching as well as learning. E-teaching should be able to facilitate the learning facilities, and to take into account in learning to introduce concepts to each learner.

The presented system incorporates multi agents, collaborated together to help in maximizing the learning process of an e-course. The course tested in this system is the Programming Language Concepts, as taught in Computer Science Department in Umm Al-Qura University, in Saudi Arabia. The task of the system is to enhance e-course navigation, which, by the way, improves e-learning process.

The main goal of the proposed system is to maximize the course learning. It will acquire knowledge directly and indirectly about learners. Direct knowledge includes preferences and level of education of learners (current knowledge). While indirect knowledge includes learner's ability and efficiency of learning (new knowledge), which is gathered from results of any assessment (exercise, quiz or test). All of these knowledge are stored in the learning KDB.

Before using the system to navigate course materials (domain), the learner should open an account, and get a password, to be able to log in. The learner should feed the system with some personal knowledge, to be stored in the Learning KDB. This knowledge includes historical education level.

At logging in the system to navigate the material of the e-course, the learner will see the menu which include main topics of the course, which represent the main part which is the theory pages. Each of these pages could posses with any media: text, graphic, image, audio, video, or even links to an external page.

Also, the e-course material includes two important parts: quizzes and tests which are all considered assessment for each topic or the whole course topics. Quizzes are created from exercises, in a way to complete the understanding of the theory material pages. Delivering of any part of the e-course (material, quiz, or test) and relative advices is done inconsequence manner as will be described in section IV.

III. STRUCTURE OF MULTI-AGENT E-LEARNING SYSYTEM

The proposed e-learning advisor system is structured basically, as shown in Fig. 1, from three modules; each of them represents a knowledge level. The domain module includes the material and assessments (exercises and questions) of topics of the e-course to be taught to the learner. While the learning module represents the knowledge that already known by the learner (personal knowledge, historical learning level and the newly acquired knowledge from the coming e-courses) . Finally, pedagogical module holds rules and strategies of teaching the course materials (fundamentals of teaching).

Strategies of pedagogy specify how the sequence of materials, what kind of feedback has to be given during education, when and how the course contents (problems, definition, example, and so on) have to be shown or explained [18].

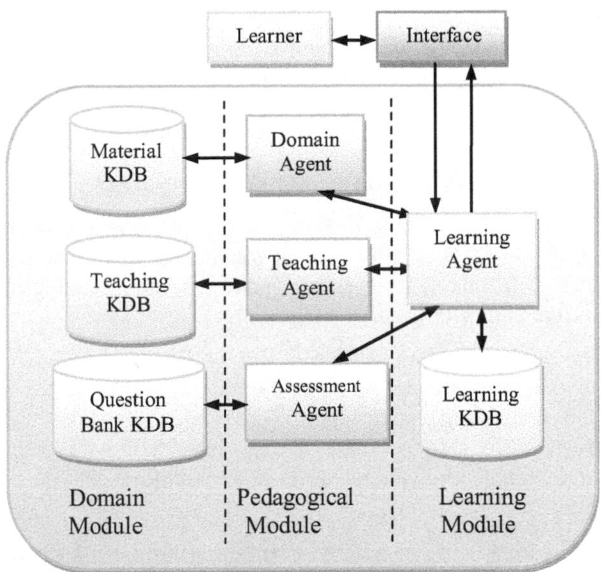

Fig. 1. Strucure of the muti-agent advisor system.

The presented system includes the following knowledge Databases (KDB) and agents:

- Material KDB holds the e-course material or pages. Each page could include text, graphics, audio, video, or links to external pages
- Question Bank KDB holds question and exercises in two level of difficulties for each topic of the e-course.
- Teaching KDB holds the perquisite and the sequence of presenting each topic. It also holds guidance and advices.
- Learning KDB holds account and personal information of and learning performance level of learners. It includes historical and new learning knowledge of learners.
- *Learning Agent* [Lagent} is the main agent in the system. It is responsible of many tasks including managing the learning process, controlling all other agent in the system.. Also, it interacts with the learner to acquires his account personal information and stores it in the learning KDB and consult all materials, assessments and advices to him. It receives assessment results from Aagent and evaluate learning efficiency of learner and update the learning KDB.
- Domain Agent [Dagent] receives a request to consult pages of certain topic from Lagent.
- Assessment Agent [Aagent] which is an external agent system for creating an assessment (quiz or test) automatically [19]. It receives a request from Lagent to build an assessment to be conducted to the learner, under some conditions. This agent selects exercise or questions randomly to creates quizzes or tests with two level of difficulties for each topic(s) from the course

material. It also grades the assessment and gives correct answers for each question.

- Teaching Agent [Tagent] retrieves the prerequisite of each topic or page in the course material page. It also retrieves learning level and performance of each learner from learning KDB through Lagent. Then, it passes its advice and guidance as a message to learner.

IV. THE LEARNING PROCESS

The learning process is done in the presented system as shown in Fig. 2. Sometimes, the system offer its advice for all learners, while navigating certain page, as help, or suggesting alternative pages, or guidance page

The main target of the proposed system is to advise the learners of an e-course to read certain pages or to navigate through some suggested links. Those pages or inks, which represent actions to be done by learners, will help them to improve their knowledge and understanding of e-course materials.

Suggestion of those actions is triggered by events (learning activities) done by learners such as starting or finishing certain part of e-course material, jumping to advanced part of e-course material, attempting to perform a simulation, passing through an assessment (quiz or test), and accessing certain part of e-course material or even external link.

According to learner's level and performance in the Learning KDB, Lagent will decide if the learner needs an advice. This decision is done by accessing the learning KDB, to enable Lagent to evaluate the performance of the learner,. This KDB includes his level of education (already known knowledge), prerequisites (knowledge should be known before learning) for each page-material accessed by learners, assessments (quizzes, exercises, tests) should be passed by the learner in what minimum correctness percentage and maximum time.

In this case, Lagent will constitute its advice to the learner accordingly. If the learner has to be advised, the agent will look up in the e-course for the materials or media that should be taught to that learner to maximize his learning for that e-course.

This advice will guide the learner to access a link(s) which will include a media. This media could be one of t following classes: theory page(s), an assessment(s), and/or event voice or video files. All of these classes may be included in the same course, other courses, or Web sites.

Finally, the agent offers its advice to the learner. If the learner followed the advisor agent, his learning and performances will be improved.

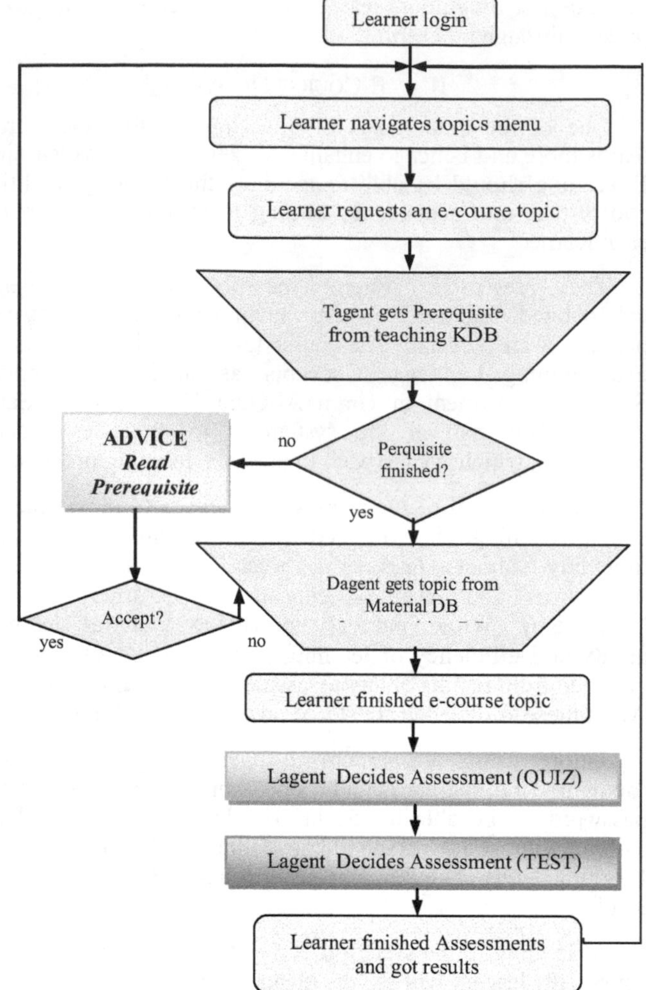

Fig. 2. The Learning Process.

V. ADVISING THROUGH ASSESSMENTS

The system will improve the learning process by an intelligent agent for advice and assistance. It causes learners to be as have an e-course suitable to their level of education, learning ability and assessment results. The intelligent agent will guide learners to their needed course materials to decrease any learning confusion. Fig. 3 and Fig. 4 demonstrate the advising process

A. Advising Actions

Testing scores of learners is always used to estimate their efficiency, and is divided into different levels in the traditional learning. During the learning activity, the behaviors of learners can be recorded in a database. This information can find out learners adaption to the teaching material and modify the level of learners.

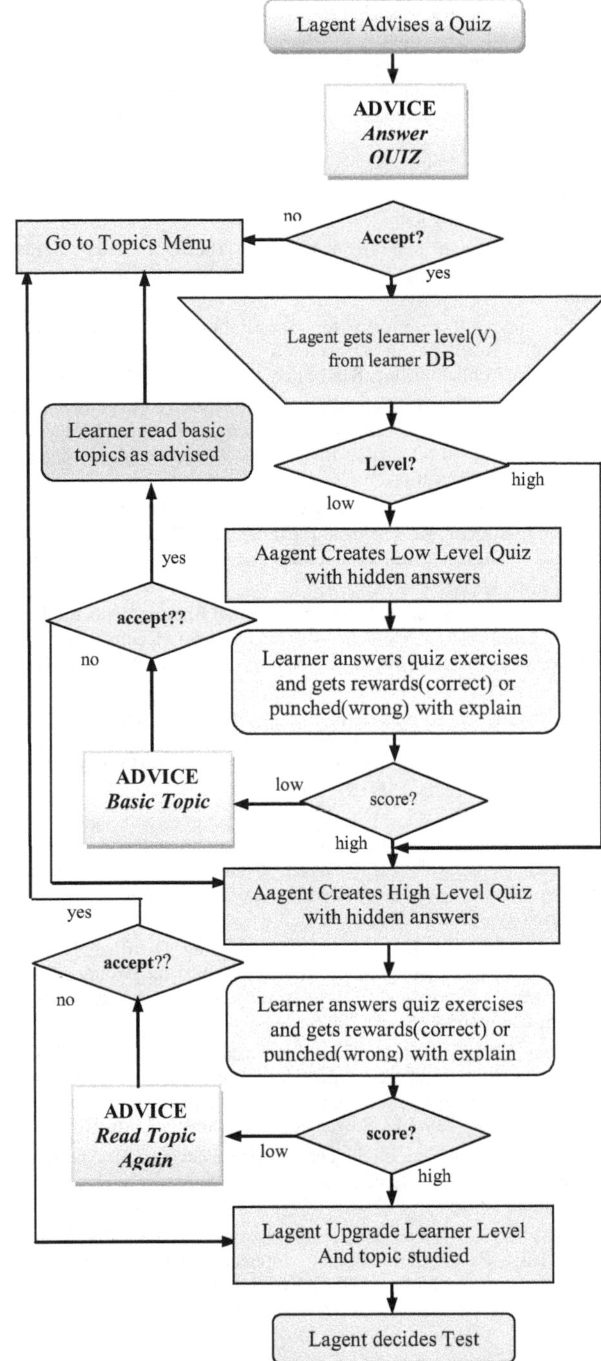

Fig. 3. The Quiz Process.

Sometimes, certain learner gets bad results after passing through an assessment (exam or quiz). This event will trigger an advising action. That event will activate Tagent, which will try to locate one sequence of action from previous sequences of actions taken with the event.

Then the agent will advise the current learner to visit and read some pages, which include important course material.

As example of advice, if the event is passing through an assessment, the advice is navigating a sequence of pages or media. It is not applicable to advise a learner by certain page or media that could directly be reached form the current page or media or by shortcut. So, the system takes care at offering its advice, not to include that case.

Tagent checks test results of assessments (as passed from Aagent) for each learner. According to these results, the *agent* will find the appropriate learning sequence for each learner and advises him through the learning process. The *agent* will advise the learner to get efficient learning time with useful e-course material.

B. Passing Through a Quiz

After navigating all pages of certain topic, Lagent will decide to enforce learner to pass through assessments like quizzes and test. Each quiz consists of 2-4 exercises. There is two level of exercise. The quiz level is specified, as shown in Fig. 3, according to the performance level of the learner.

Lagent gets the level of performance of the current learner from the learning KDB and decides the level of quiz. It asks Aagent to create the suitable quiz accordingly. Exercises will not be not too difficult. In the low level quiz, exercises will be small and in similar words as taught in topic. It will be accompanied with helpful figure or images. While in the high level quiz, question will be more difficulty.

While the learner is making exercises of a quiz, he will input his answer for each exercise to the system. The system reaction will be accompanied by reward or punishing for a correct or wrong answer for any assessment. This is always done through blinking text or image to show certain media such as a text, table, video, audio, picture or even graph. This is done as in the most of the learning systems. Then, Lagent gives its advice to the learner according to his results, as shown in Fig. 3.

C. Passing Through a Test

When the learner finishes the high level quiz, Lagent will update the learning KDB according to quiz results. Then Lagent will advise learner to pass through a Test, as shown in Fig. 4. It will ask Aagent to create a test consists of multiple type of questions.

Tagent is able to create exams from bank of questions randomly for certain topic(s). There are four types of questions: True/False questions, Multiple Choice questions, Fill in the Blanks questions, and Non-standard questions.After finishing the exam, Tagent will evaluate answers and pass scores to Lagent. Then Lagent will update learner level and performance in the learning KDB. Also, it stores that the topic is navigated and tested by that learner.

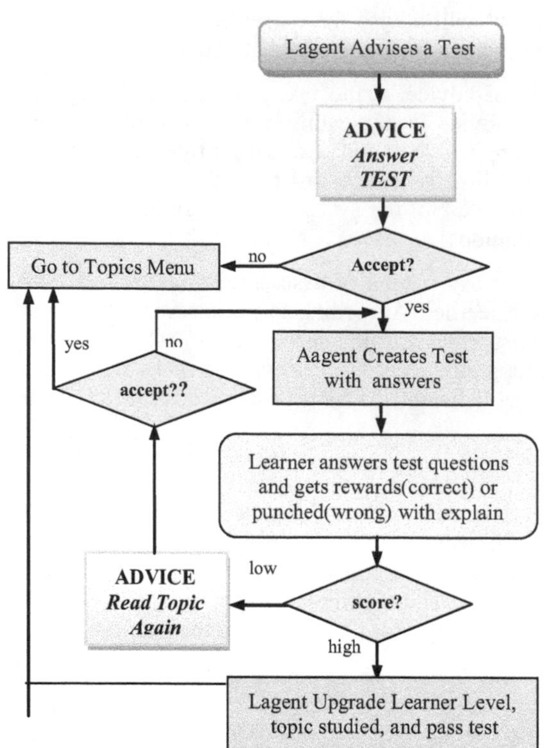

Fig. 4. The Test Process.

VI. Conclusion

The presented paper provided a multi-agent based advisor system to guide and advise learners of e-courses. It is suggested to advise learners in the Concepts of Programming Languages course. It is based on multi-agent technology. Agent built its advice on the past and current knowledge learnt by learners. It calls another agent was built to create quizzes and exams automatically in different level of difficulties and grades answers. Future work will extend applying that system in Computer Sciences courses. Also, it will be upgrading it to offer more adapted e-course.

References

[1] H.W. Tsang, L.M. Hung, and S.C. Ng, "A multimedia distance learning system on the Internet", proceedings of IEEE International Conference on Systems, Man, and Cybernetics (SMC), Vol. 2, 243 –246, 1999.

[2] Zhi Liu & Bo Chen, "Model and Implement an Agent Oriented E-Learning System", Proceedings of the 2005 International Conference on Computational Intelligence for Modelling, Control and Automation, and International Conference Intelligent Agents, Web Technologies and Internet Commerce (CIMCA-IAWTIC'05) 0-7695-2504-0/05.

[3] Ying-Han Fangl and Lei Shao2, "Agent-Based E-Learning System Research and Implementation", proceedings of the 7th International Conference on Machine Learning and Cybernetics, Kunming, 12-15 July 2008, 4080-4084.

[4] S. J. Russell, and P. Norvig, "Artificial Intelligence – A Modern Approach", 3rd edition, Prentice Hall, 2013.

[5] H. K. Mohammed, "An Intelligent Agent to Teach C-Language", proceedings of ICECS'97, Cairo, Egypt, December 15-18, 1997, 483-488.

[6] Jin-Ling Lin and Ming-Hung Chen, "An Intelligent Agent for Personalized E-Learning", 8th International Conference on Intelligent Systems Design and Applications - Volume 01,27-31,2008.

[7] M. Nissen, "Intelligent Agents: A Technology and Business Application Analysis", Telecommunications and Distributed Processing, November 1994.

[8] O. R. Zarane, "Building a Recommender Agent for e-Learning Systems", proceedings of the international Conference on Computers in Education(ICCE'02), pp 55-59, Dec. 2002.

[9] R. Burton, "Diagnosing Bugs in a Simple Procedural Skills", Xerox Palo Alto Research Center, Cognitive and Instructional Sciences Group, Palo, Alto, CA 94304, 1986.

[10] D. Sleeman, "Frameworks for Cooperation in Distributed Problem Solving", Readings in Distributed Artificial Intelligence. Morgan Kaufmann Publishers, Inc. San Mateo, California, 1982, 61-69.

[11] A. M. Florea, "An Agent-Based Collaborative Learning System", Advanced Research in Computers and Communications in Education, G. Cuming ming et al (eds), IOS Press, International Conference on Computers in Education, 1999, 161-164.

[12] J.M. Gascuena and A. Fernandez-Caballero, "An Agent-based Intelligent Tutoring System for Enhancing E-Leaning/E-Teaching", International Journal of instructional Technology and Distance Education, itdl.org, vol. 2, No. 11, pp 15-26, Nov. 2005.

[13] A. de Antonio, and et al., "Intelligent Virtual Environments for Training: An Agent-based Approach. 4th International Central and Eastern European Conference on Multi-Agent Systems (CEEMAS'05). Hungary, 15-17, 2005.

[14] M. Hospers, E. Kroezen, A. Nijholt, H.J.A. op den Akker, and D. Heylen, "An agent-based intelligent tutoring system for nurse education", Application of Intelligent Agents in Health Care, J. Nealon and A. Moreno (eds), 143-159, 2003.

[15] T. Y. Tang and A. Wu, "The implementation of a multi-agent intelligent tutoring system for the learning of a computer programming", Proceedings of 16th IFIP World Computer Congress-International Conference on Educational Uses of Communication and Information Technology, ICEUT, 2000.

[16] R.A. Silveira and R.M. Vicari, "Developing Distributed Intelligent Learning Environment with JADE –Java Agents for Distance Education Framework", International Conference on Intelligent Tutoring Systems, LNCS, 2363, 105-118, ITS 2002.

[17] HL.Tsai, CJ. Lee, WH. HSU, and YH. Chang, "An Adaptive E-learning System based on Intelligent Agents", Proceeding of the 11th International Conference on Applied Computer and Applied Computer Science, WSEAS.US, pp 139-142, Steven point, Wisconsigin, April 2012.

[18] T, Murray, "Authoring Intelligent Tutoring Systems : An Analysis of the state of the art", International Journal of Artificial Intelligence in Eductaion, 10, 98-129, 1999.

[19] Kh.N. ElSayed," A Tool for Creating Exams Automatically From an OO Knowledge Base Question Bank", International Journal of Information and Education Technology, IJIET Vol. 3, No.1, 27-31, Feb. 2013.

15

A Comparison between Regression, Artificial Neural Networks and Support Vector Machines for Predicting Stock Market Index

Alaa F. Sheta
Computers and Systems Department
Electronics Research Institute Giza, Egypt

Sara Elsir M. Ahmed
Computer Science Department
Sudan University of Science and Technology
Khartoum, Sudan

Hossam Faris
Business Information Tech. Dept.
The University of Jordan
Amman, Jordan

Abstract—**Obtaining accurate prediction of stock index significantly helps decision maker to take correct actions to develop a better economy. The inability to predict fluctuation of the stock market might cause serious profit loss. The challenge is that we always deal with dynamic market which is influenced by many factors. They include political, financial and reserve occasions. Thus, stable, robust and adaptive approaches which can provide models have the capability to accurately predict stock index are urgently needed. In this paper, we explore the use of Artificial Neural Networks (ANNs) and Support Vector Machines (SVM) to build prediction models for the S&P 500 stock index. We will also show how traditional models such as multiple linear regression (MLR) behave in this case. The developed models will be evaluated and compared based on a number of evaluation criteria.**

Index Terms—**Stock Market Prediction; S&P 500; Regression; Artificial Neural Networks; Support Vector Machines.**

I. Introduction

Understanding the nature of the relationships between financial markets and the country economy is one of the major components for any financial decision making system [1]–[3]. In the past few decades, stock market prediction became one of the major fields of research due to its wide domain of financial applications. Stock market research field was developed to be dynamic, non-linear, complicated, nonparametric, and chaotic in nature [4]. Much research focuses on improving the quality of index prediction using many traditional and innovative techniques. It was found that significant profit can be achieved even with slight improvement in the prediction since the volume of trading in stock markets is always huge. Thus, financial time series forecasting was explored heavenly in the past. They have shown many characteristics which made them hard to forecast due to the need for traditional statistical method to solve the parameter estimation problems. According to the research developed in this field, we can classify the techniques used to solve the stock market prediction problems to two folds:

- **Econometric Models**: These are statistical based approaches such as linear regression, Auto-regression and Auto-regression Moving Average (ARMA) [5], [6]. There are number of assumptions need to be considered while using these models such as linearity and stationary of the the financial time-series data. Such non-realistic assumptions can degrade the quality of prediction results [7], [8].
- **Soft Computing based Models**: Soft computing is a term that covers artificial intelligence which mimic biological processes. These techniques includes Artificial Neural Networks (ANN) [9], [10], Fuzzy logic (FL) [11], Support Vector Machines (SVM) [12], particle swarm optimization (PSO) [13] and many others.

ANNs known to be one of the successfully developed methods which was widely used in solving many prediction problem in diversity of applications [14]–[18]. ANNs was used to solve variety of problems in financial time series forecasting. For example, prediction of stock price movement was explored in [19]. Authors provided two models for the daily Istanbul Stock Exchange (ISE) National 100 Index using ANN and SVM. Another type of ANN, the radial basis function (RBF) neural network was used to forecast the stock index of the Shanghai Stock Exchange [20]. In [21], ANNs were trained with stock data from NASDAQ, DJIA and STI index. The reported results indicated that augmented ANN models with trading volumes can improve forecasting performance in both medium-and long-term horizons. A comparison between SVM and Backpropagation (BP) ANN in forecasting six major Asian stock markets was reported in [22]. Other soft computing techniques such as Fuzzy Logic (FL) have been used to solve many stock market forecasting problems [23], [24].

Evolutionary computation was also explored to solve the prediction problem for the S&P 500 stock index. Genetic Algorithms (GAs) was used to simultaneously optimize all of a Radial Basis Function (RBF) network parameters such that an efficient time-series is designed and used for business forecasting applications [25]. In [26], author provided a new prediction model for the S&P 500 using Multigene Symbolic Regression Genetic Programming (GP). Multigene GP shows more robust results especially in the validation/testing case than ANN.

In this paper, we present a comparison between traditional regression model, the ANN model and the SVM model for predicting the S&P 500 stock index. This paper is structured as follows. Section II gives a brief idea about the S&P 500 Stock Index in the USA. In Section III, we provide an introduction to linear regression models. A short introduction to ANN and SVM is provided in Section IV and Section V, respectively. The adopted evaluation methods are presented in Section VI. In Section VII, we describe the characteristics of the data set used in this study. We also provide the experimental setup and results produced in this research.

II. S&P 500 Stock Index

The S&P 500, or the Standard & Poor's 500, is an American stock market index. The S&P 500 presented its first stock index in the year 1923. The S&P 500 index with its current form became active on March 4, 1957. The index can be estimated in real time. It is mainly used to measure the stock prices levels. It is computed according to the market capitalization of 500 large companies. These companies are having stock in the The New York Stock Exchange (NYSE) or NASDAQ. The S&P 500 index is computed by S&P Dow Jones Indices. In the past, there were a growing interest on measuring, analyzing and predicting the behavior of the S&P 500 stock index [27]–[29]. John Bogle, Vanguard's founder and former CEO, who started the first S&P index fund in 1975 stated that:

> The rise in the S&P 500 is a virtual twin to the rise in the total U.S. stock market, so of course investors, and especially index fund investors, who received their fair share of those returns, feel wealthier,"

In order to compute the price of the S&P 500 Index, we have to compute the sum of market capitalization of all the 500 stocks and divide it by a factor, which is defined as the Divisor (D). The formula to calculate the S&P 500 Index value is given as:

$$Index\ Level = \frac{\sum P_i \times S_i}{D}$$

P is the price of each stock in the index and S is the number of shares publicly available for each stock.

III. Regression Analysis

Regression analysis have been used effectively to answer many question in the way we handle system modeling and advance associations between problem variables. It is important to develop such a relationships between variables in many cases such as predicting stock market [13], [14], [30], [31]. It is important to understand how stock index move over time.

A. Single Linear Regression

In order to understand how linear regression works, assume we have n pairs of observations data set $\{x_i, y_j\}_{i=1,..,n}$ as given in Figure 1. Our objective is to develop a simple relationship between the two variables x (i.e. input variable) and y (i.e output variable) so that we can develop a line equation (see Equation 1).

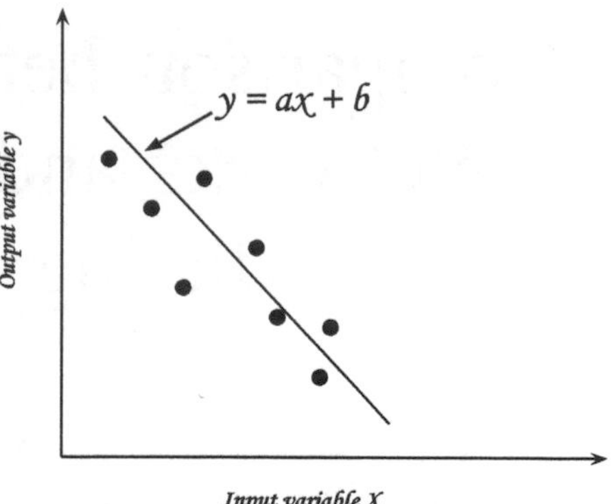

Fig. 1. Simple Linear Model

$$y = a + bx \tag{1}$$

where a is a constant (i.e. bias) and b is the slope of the line. It is more likely that the straight line will not pass by all the points in the graph. Thus, Equation 1 shall be re-written as follows:

$$y = a + bx + \epsilon \tag{2}$$

where ϵ represents the error difference between the values of x_i and y_i at any sample i. Thus, to find the best line that produce the most accurate relationship between x and y. We have to formulate the problem as an optimization problem such that we can search and find the best values of the parameters (i.e. $\hat{a}$ and $\hat{b}$). In this case, we need to solve an error minimization problem. To minimize the sum of the error over the whole data set. We need to minimize the function L given in Equation 3.

$$L = \sum_{i=1}^{n} \epsilon_i^2 = \sum_{i=1}^{n} (y_i - a - bx_i)^2 \tag{3}$$

To find the optimal values for $\hat{a}$ and $\hat{b}$ we have to differentiate L with respect to a and b.

$$\begin{aligned} \frac{\partial L}{\partial \hat{a}} &= -2\sum_{i=1}^{n}(y_i - \hat{a} - \hat{b}x_i) = 0 \\ \frac{\partial L}{\partial \hat{b}} &= -2\sum_{i=1}^{n}(y_i - \hat{a} - \hat{b}x_i)x_i = 0 \end{aligned} \tag{4}$$

By simplification of Equations 4, we get to the following two equations:

$$\begin{aligned} n\,\hat{a} + \sum_{i=1}^{n} x_i\,\hat{b} &= \sum_{i=1}^{n} y_i \\ \sum_{i=1}^{n} x_i\,\hat{a} + \sum_{i=1}^{n} x_i^2\,\hat{b} &= \sum_{i=1}^{n} x_i y_i \end{aligned} \tag{5}$$

Equations 5 is called least square (LS) normal equations. The solution of these normal equations produce the least square estimate for $\hat{a}$ and $\hat{b}$.

B. Multiple Linear Regression

The simple linear model Equation 2 can be expanded to a multivariate system of equations as follows:

$$y = a_1 x_1 + a_2 x_2 + \cdots + a_n x_j \tag{6}$$

where x_j is the jth independent variable. In this case, we need to use LS estimation to compute the optimal values for the parameters $a_1, \ldots, a_j$. Thus, we have to minimize the optimization function L, which in this case can be presented as:

$$L = \sum_{i=1}^{n} \epsilon_i^2 = \sum_{i=1}^{n} (y_i - \hat{a_1} x_1 - \hat{a_2} x_2 - \ldots \hat{a_n} x_n)^2 \tag{7}$$

To get the optimal values of the parameters $\hat{a_1}, \ldots, \hat{a_n}$, we have to compute the differentiation for the functions:

$$\frac{\partial L}{\partial \hat{a_1}} = \frac{\partial L}{\partial \hat{a_2}} = \cdots = \frac{\partial L}{\partial \hat{a_j}} = 0 \tag{8}$$

Solving the set of Equations 8, we can produce the optimal values of the model parameters and solve the multiple regression problem. This solution is more likely to be biased by the available measurements. If you we have large number of observations the computed estimate of the parameters shall be more robust. This technqiue provide poor results when the observations are small in number.

IV. Artificial Neural Networks

ANNs are mathematical models which were inspired from the understanding of some ideas and aspects of the biological neural systems such as the human brain. ANN may be considered as a data processing technique that maps, or relates, some type of input stream of information to an output stream of processing. Variations of ANNs can be used to perform classification, pattern recognition and predictive tasks [15], [19], [20], [22], [30].

Neural network have become very important method for stock market prediction because of their ability to deal with uncertainty and insufficient data sets which change rapidly in very short period of time. In Feedforward (FF) Multilayer Perceptron (MLP), which is one of the most common ANN systems, neurons are organized in layers. Each layer consists of a number of processing elements called neurons; each of which contains a summation function and an activation function. The summation function is given by Equation 9 and an activation function can be a type of sigmoid function as given in Equation 10.

Training examples are used as input the network via the input layer, which is connected to one, or more hidden layers. Information processing takes place in the hidden layer via the connection weights. The hidden layers are connected to an output layer with neurons most likely have linear sigmoid function. A learning algorithms such as the BP one might be used to adjust the ANN weights such that it minimize the error difference between the actual (i.e. desired) output and the ANN output [32]–[34].

$$S = \sum_{i=0}^{n} w_i x_i \tag{9}$$

$$\phi(S) = \frac{1}{1 + e^{-S}} \tag{10}$$

There are number of tuning parameters should be designated before we can use ANN to learn a problem. They include: the number of layers in the hidden layer, the type of sigmoid function for the neurons and the adopted learning algorithm.

V. Support Vector Machines

Support vector machine is a powerful supervised learning model for prediction and classification. SVM was first introduced by Vladimir Vapnik and his co-workers at AT&T Bell Laboratories [35]. The basic idea of SVM is to map the training data into higher dimensional space using a nonlinear mapping function and then perform linear regression in higher dimensional space in order to separate the data [36]. Data mapping is performed using a predetermined kernel function. Data separation is done by finding the optimal hyperplane (called the Support Vector with the maximum margin from the separated classes. Figure 2 illustrates the idea of the optimal hyperplane in SVM that separates two classes. In the left part of the figure, lines separated data but with small margins while on the right an optimal line separates the data with the maximum margins.

A. Learning Process in SVM

Training SVM can be described as follows; suppose we have a data set $\{x_i, y_j\}_{i=1,..,n}$ where the input vector $x_i \in \Re^d$ and the actual $y_i \in \Re$. The modeling objective of SVM is to find the linear decision function represented in the following equation:

$$f(x) \leq w, \quad \phi_i(x) > +b \tag{11}$$

where w and b are the weight vector and a constant respectively, which have to be estimated from the data set. ϕ is a nonlinear mapping function. This regression problem can be formulated as to minimize the following regularized risk function:

$$R(C) = \frac{C}{n} \sum_{i=1}^{n} L_\varepsilon(f(x_i), y_i) + \frac{1}{2} \|w\|^2 \tag{12}$$

where $L_\varepsilon(f(x_i), y_i)$ is known as $\varepsilon-$intensive loss function and given by the following equation:

$$L_\varepsilon(f(x), y) = \begin{cases} |f(x) - y| - \varepsilon & |f(x) - y| \geq \varepsilon \\ 0 & otherwise \end{cases} \tag{13}$$

To measure the degree of miss classification to achieve an acceptable degree of error, we use slack variables ξ_i and

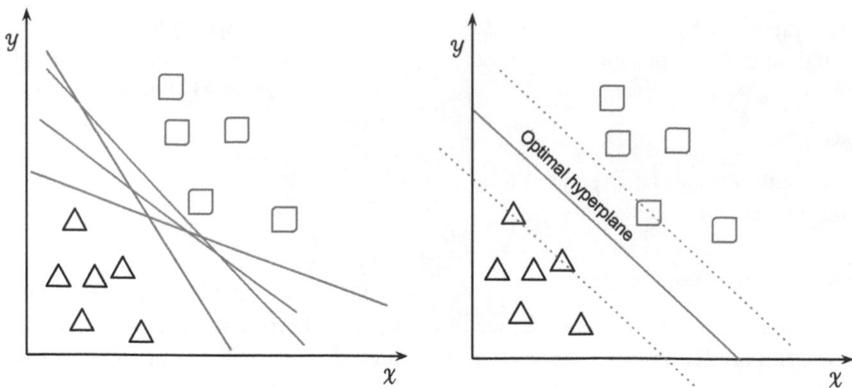

Fig. 2. Optimal hyperplane in Support Vector Machine

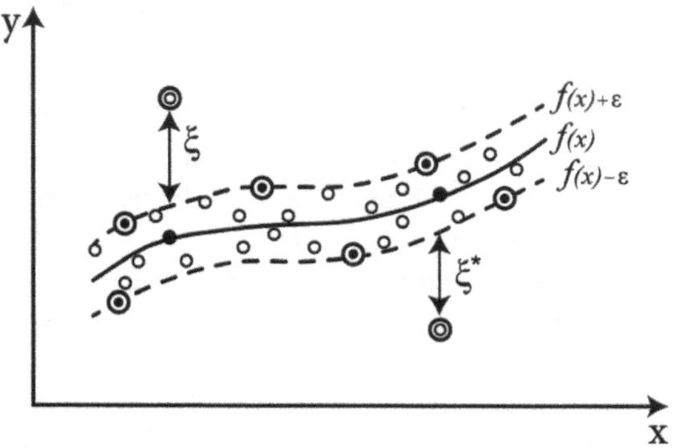

Fig. 3. Optimal hyperplane with slack variables

ξ_i^* as shown in Figure 3. This addition makes the problem presented as a constrained minimum optimization problem (See Equation 14).

$$Min.\ R(w, \xi_i^*) = \frac{1}{2}\|w\|^2 + C\sum_{i=1}^{n}(\xi_i + \xi_i^*) \quad (14)$$

Subject to:

$$\begin{cases} y_i - < w, x_i > -b & \leq \varepsilon + \xi_i \\ < w, x_i > +b - y_i & \leq \varepsilon + \xi_i^* \\ \xi_i, \xi_i^* & \geq 0 \end{cases} \quad (15)$$

where C is a regularized constant greater than zero. Thus it performs a balance between the training error and model flatness. C represents a penalty for prediction error that is greater than ε. ξ_i and ξ_i^* are slack variables that form the distance from actual values to the corresponding boundary values of ε. The objective of SVM is to minimize ξ_i, ξ_i^* and w^2.

The above optimization with constraint can be converted by means of Lagrangian multipliers to a quadratic programming problem. Therefore, the form of the solution can be given by the following equation:

$$f(x) = \sum_{i=1}^{n}(\alpha_i - \alpha_i^*)K(x_i, x) + b \quad (16)$$

TABLE I
COMMON SVM KERNEL FUNCTIONS

Polynomial Kernel	$K(x_i, x_j) = (x_i.x_j + 1)^d$
Hyperbolic Tangent Kernel	$K(x_i, x_j) = tanh(c_1(x_i.x_j) + c_2)$
Radial Basis Kernel	p: $K(x_i, x_j) = exp(\|x_j - x_i\|/2p^2)$

where α_i and α_i^* are Lagrange multipliers. Equation 16 is subject to the following constraints:

$$\sum_{i=1}^{n}(\alpha_i - \alpha_i^*) = 0 \quad (17)$$

$$0 \leq \alpha_i \leq C \quad i = 1, \ldots, n$$

$$0 \leq \alpha_i^* \leq C \quad i = 1, \ldots, n$$

$K(.)$ is the kernel function and its values is an inner product of two vectors x_i and x_j in the feature space $\phi(x_i)$ and $\phi(x_j)$ and satisfies the Mercer's condition. Therefore,

$$K(x_i, x_j) = \phi(x_i).\phi(x_j) \quad (18)$$

Some of the most common kernel functions used in the literature are shown in Table I. In general, SVMs have many advantages over classical classification approaches like artificial neural networks, decision trees and others. These advantages include: good performance in high dimensional spaces; and the support vectors rely on a small subset of the training data which gives SVM a great computational advantage.

VI. Evaluation Criterion

In order to assess the performance of the developed stock market predication models, a number of evaluation criteria will be used to evaluate these models. These criteria are applied to measure how close the real values to the values predicted using the developed models. They include Mean Absolute Error (MAE), Root Mean Square Error (RMSE) and correlation coefficient R. They are given in Equations 19, 20 and 21, respectively.

$$MAE = \frac{1}{n}\sum_{t=1}^{n}|(y_i - \hat{y}_i)| \tag{19}$$

$$RMSE = \sqrt{\frac{1}{n}\sum_{i=1}^{n}(y_i - \hat{y}_i)^2} \tag{20}$$

$$R = \frac{\sum_{i=1}^{n}(y_i - \overline{y})(\hat{y}_i - \overline{\hat{y}})}{\sqrt{\sum_{i=1}^{n}(y_i - \overline{y})^2 \sum_{i=1}^{n}(\hat{y}_i - \overline{\hat{y}})^2}} \tag{21}$$

where y is actual stock index values, $\hat{y}$ is the estimated values using the proposed technqiues. n is the total number of measurements.

VII. Experimental Results

A. S&P 500 Data Set

In this work, we use 27 potential financial and economic variables that impact the stock movement. The main consideration for selecting the potential variables is whether they have significant influence on the direction of (S&P 500) index in the next week. While some of these features were used in previous studies [30]. The list, the description, and the sources of the potential features are given in Table III show the 27 features of data set.

The categories of these features include: S&P 500 index return in three previous days SPY(t-1), SPY(t-2), SPY(t-3); Financial and economic indicators (Oil, Gold, CTB3M, AAA); The return of the five biggest companies in S&P 500 (XOM, GE, MSFT, PG, JNJ); Exchange rate between USD and three other currencies (USD-Y, USD-GBP, USD-CAD); The return of the four world major indices (HIS, FCHI, FTSE, GDAXI); and S&P 500 trading volume (V).

S&P 500 stock market data set used in our case consists of 27 features and 1192 days of data, which cover five-year period starting 7 December 2009 to 2 September 2014. We sampled the data on a weekly basis such that only 143 samples were used in our experiments. The S&P 500 data were split into 100 samples as training set and data for 43 samples as testing set.

B. Multiple Regression Model

The regression model shall have the following equation system.

$$y = a_0 + \sum_{i=1}^{27} a_i x_i \tag{22}$$

The values of the parameters $a's$ shall be estimated using LS estimation to produce the optimal values of the parameters $\hat{a}'s$. The produced linear regression model can be presented as given in Table II. The actual and Estimated S&P 500 index values based the MLR in both training and testing cases are shown in Figure 4 and Figure 5. The scattered plot of the actual and predicted responses is shown in Figure 6.

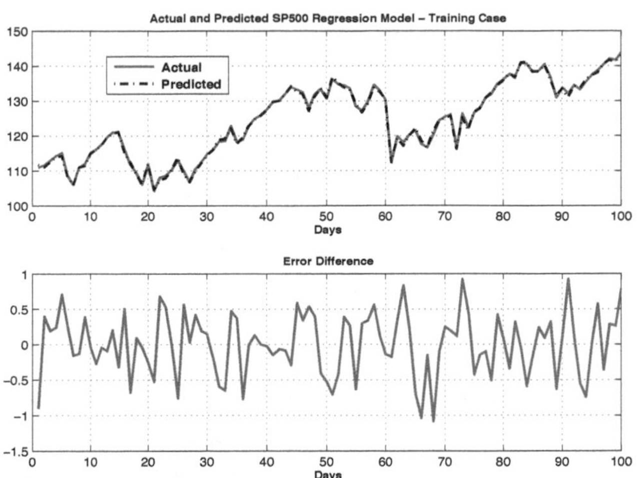

Fig. 4. Regression: Actual and Estimated S&P 500 Index values in Training Case

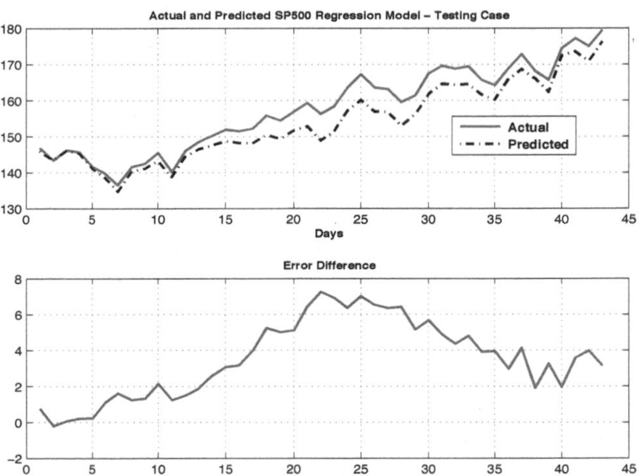

Fig. 5. Regression: Actual and Estimated S&P 500 Index values in Testing Case

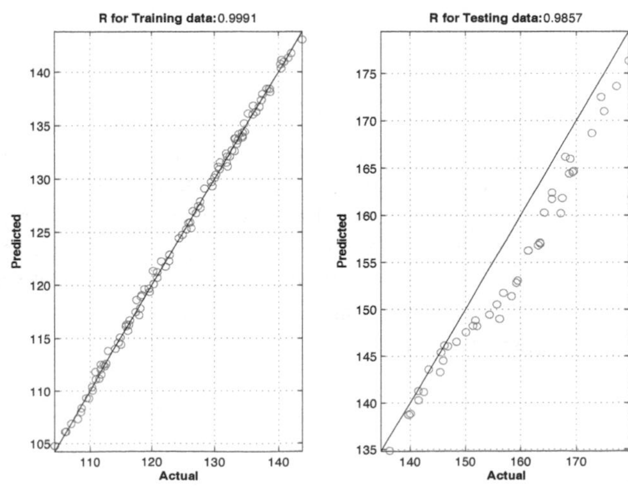

Fig. 6. Regression Scattered Plot

TABLE II
A REGRESSION MODEL WITH INPUTS: $x_1, \ldots, x_{27}$

$$
\begin{aligned}
\hat{y} &= -0.0234 * x_1 + 0.13 * x_2 + 0.021 * x_3 + 0.021 * x_4 - 0.021 * x_5 \\
&- 10.303 * x_6 + 6.0031 * x_7 + 0.7738 * x_8 + 0.2779 * x_9 - 0.43916 * x_{10} \\
&- 0.27754 * x_{11} + 0.12733 * x_{12} - 0.058638 * x_{13} + 13.646 * x_{14} + 9.5224 * x_{15} \\
&- 0.0003 * x_{16} + 0.24856 * x_{17} - 0.0016 * x_{18} + 0 * x_{19} - 2.334 \times 10^{-9} * x_{20} \\
&+ 0.16257 * x_{21} + 0.63767 * x_{22} - 0.14301 * x_{23} + 0.08 * x_{24} + 0.074 * x_{25} \\
&- 0.0002 * x_{26} + 0.026301 * x_{27} + 6.9312
\end{aligned} \tag{23}
$$

TABLE III
THE 27 POTENTIAL INFLUENTIAL FEATURES OF THE S&P 500 INDEX [30]

Variable	Feature	Description
x_1	SPY(t-1)	The return of the S&P 500 index in day $t-1$ Source data: finance.yahoo.com
x_2	SPY(t-2)	The return of the S&P 500 index in day $t-2$ Source data: finance.yahoo.com
x_3	SPY(t-3)	The return of the S&P 500 index in day $t-3$ Source data: finance.yahoo.com
x_4	Oil	Relative change in the price of the crude oil Source data: finance.yahoo.com
x_5	Gold	Relative change in the gold price Source data: www.usagold.com
x_6	CTB3M	Change in the market yield on US Treasury securities at 3-month constant maturity, quoted on investment basis Source data: H.15 Release - Federal Reserve Board of Governors
x_7	AAA	Change in the Moody's yield on seasoned corporate bonds - all industries, Aaa Source data: H.15 Release - Federal Reserve Board of Governors
x_8	XOM	Exxon Mobil stock return in day t-1 Source data: finance.yahoo.com
x_9	GE	General Electric stock return in day t-1 Source data: finance.yahoo.com
x_{10}	MSFT	Micro Soft stock return in day t-1 Source data: finance.yahoo.com
x_{11}	PG	Procter and Gamble stock return in day t-1 Source data: finance.yahoo.com
x_{12}	JNJ	Johnson and Johnson stock return in day t-1 Source data: finance.yahoo.com
x_{13}	USD-Y	Relative change in the exchange rate between US dollar and Japanese yen Source data: OANDA.com
x_{14}	USD-GBP	Relative change in the exchange rate between US dollar and British pound Source data: OANDA.com
x_{15}	USD-CAD	Relative change in the exchange rate between US dollar and Canadian dollar Source data: OANDA.com
x_{16}	HIS	Hang Seng index return in day t-1 Source data: finance.yahoo.com
x_{17}	FCHI	CAC 40 index return in day t-1 Source data: finance.yahoo.com
x_{18}	FTSE	FTSE 100 index return in day t-1 Source data: finance.yahoo.com
x_{19}	GDAXI	DAX index return in day t-1 Source data: finance.yahoo.com
x_{20}	V	Relative change in the trading volume of S&P 500 index Source data: finance.yahoo.com
x_{21}	CTB6M	Change in the market yield on US Treasury securities at 6-month constant maturity, quoted on investment basis Source data: H.15 Release - Federal Reserve Board of Governors
x_{22}	CTB1Y	Change in the market yield on US Treasury securities at 1-year constant maturity, quoted on investment basis Source data: H.15 Release - Federal Reserve Board of Governors
x_{23}	CTB5Y	Change in the market yield on US Treasury securities at 5-year constant maturity, quoted on investment basis Source data: H.15 Release - Federal Reserve Board of Governors
x_{24}	CTB10Y	Change in the market yield on US Treasury securities at 10-year constant maturity, quoted on investment basis Source data: H.15 Release - Federal Reserve Board of Governors
x_{25}	BBB	Change in the Moody's yield on seasoned corporate bonds - all industries, Baa Source data: H.15 Release - Federal Reserve Board of Governors
x_{26}	DJI	Dow Jones Industrial Average index return in day t-1 Source data: finance.yahoo.com
x_{27}	IXIC	NASDAQ composite index return in day t-1 Source data: finance.yahoo.com

C. Developed ANN Model

The proposed architecture of the MLP Network consists of three layers with single hidden layer. Thus input layer of our neural network model has 27 input nodes while the output layer consists of only one node that gives the predicted next week value. Empirically, we found that 20 neurons in the hidden layer achieved the best performance. The BP algorithm is used to train the MLP and update its weight. Table IV shows the settings used for MLP. Figure 7 and Figure 8 show the actual and predicted stock prices for training and testing cases of the developed ANN. The scattered plot for the developed ANN model is shown in Figure 9.

TABLE IV
THE SETTING OF MLP

Maximum number of epochs	500
Number of Hidden layer	1
Number of neurons in hidden layer	20
Learning rate	0.5
Momentum	0.2

D. Developed SVM Model

SVM with an RBF kernel is used to develop the S&P 500 index model. The RBF kernel has many advantages such as the ability to map non-linearly the training data and the ease of implementation [37]–[39]. The values of the parameters C and σ have high influence on the accuracy of the SVM model. Therefore, we used grid search to obtain these values. It was found that the best performance can be obtained with $C = 100$ and $\sigma = 0.01$. Figure 10 and Figure 11 show the actual and predicted stock prices for training and testing cases of the developed SVM model. The scattered plot for the developed SVM model is shown in Figure 12.

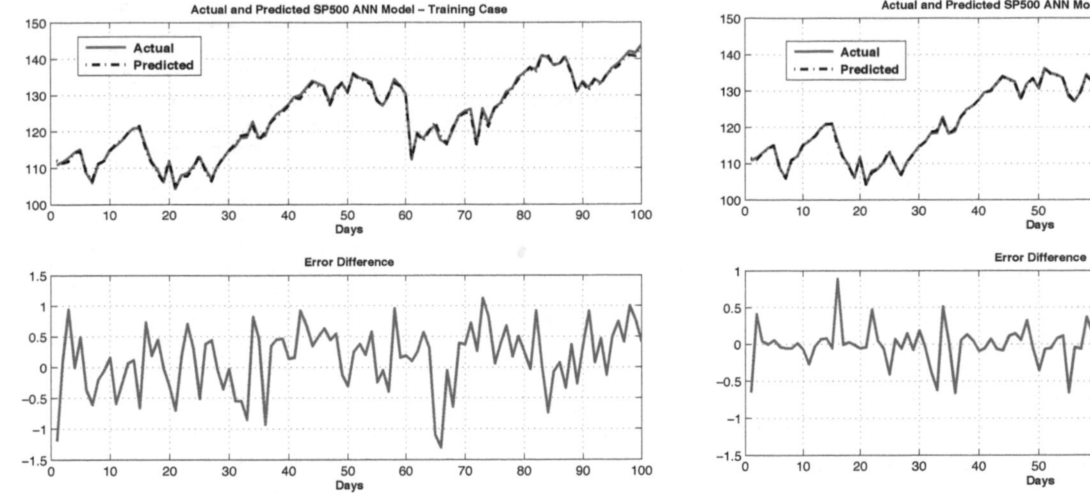

Fig. 7. ANN: Actual and Estimated S&P 500 Index values in Training Case

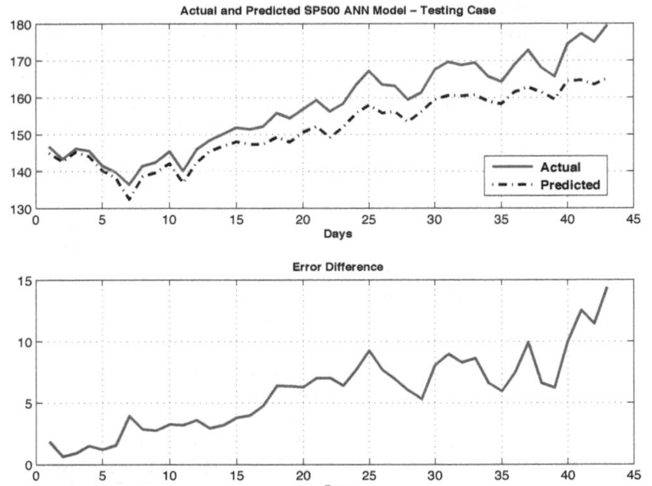

Fig. 8. ANN: Actual and Estimated S&P 500 Index values in Testing Case

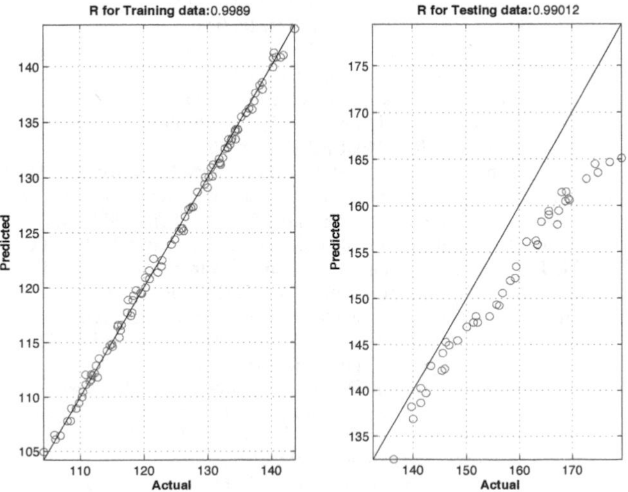

Fig. 9. ANN Scattered Plot

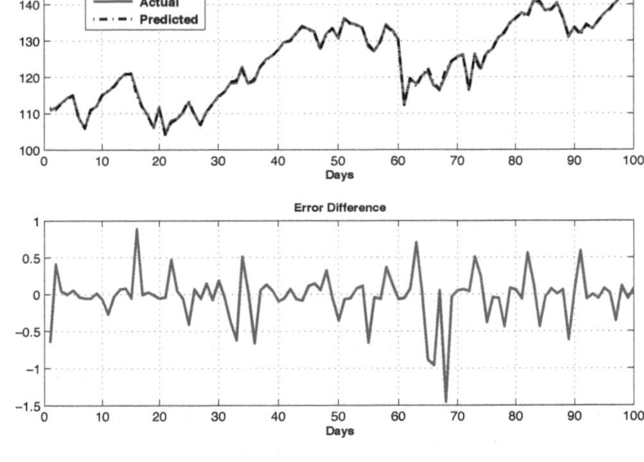

Fig. 10. SVM: Actual and Estimated S&P 500 Index values in Training Case

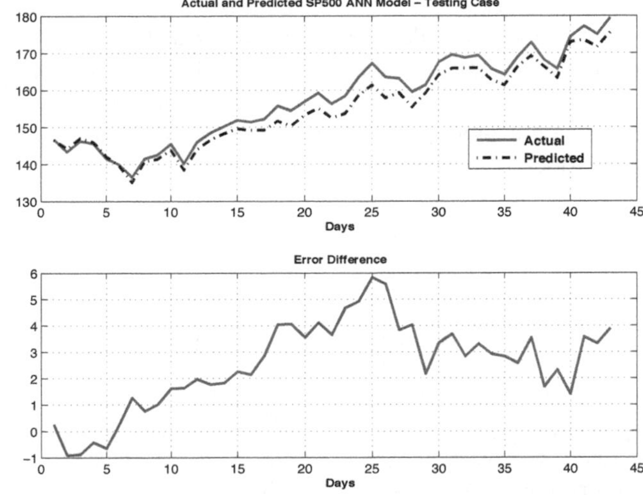

Fig. 11. SVM: Actual and Estimated S&P 500 Index values in Testing Case

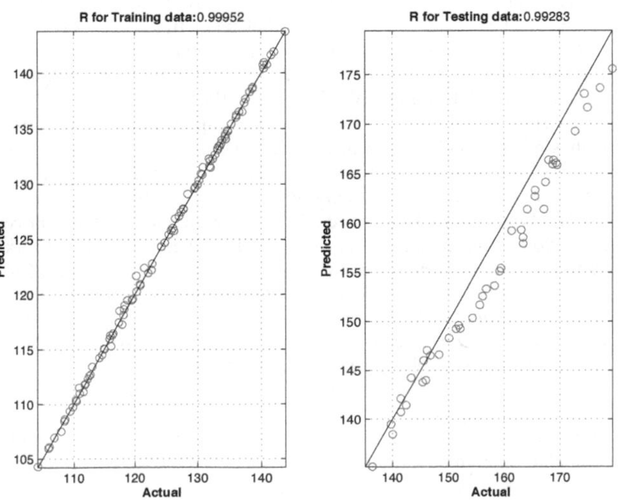

Fig. 12. SVM Scattered Plot

E. Comments on the Results

The calculated evaluation criterion of the regression, MLP and SVM models for training and testing cases are shown in Table V. Based on these results it can be noticed that SVM outperformed the MLP and MLR models in both training and testing cases. SVMs has many advantages such as using various kernels which allows the algorithm to suits many classification problems. SVM are more likely to avoid the problem of falling into local minimum.

VIII. Conclusions and Future Work

In this paper, we explored the use MLP and SVM to develop models for prediction the S&P 500 stock market index. A 27 potential financial and economic variables which impact the stock movement were adopted to build a relationship between the stock index and these variables. The basis for choosing these variables was based on their substantial impact on the course of S&P 500 index. The data set was sampled on a weekly bases. The developed SVM model with RBF kernel model provided good prediction capabilities with respect to the regression and ANN models. The results were validated using number of evaluation criteria. Future research shall focus on exploring other soft computing techniques to solve the stock market prediction problems.

References

[1] S. Hoti, M. McAleer, and L. L. Pauwels, "Multivariate volatility in environmental finance," *Math. Comput. Simul.*, vol. 78, no. 2-3, pp. 189–199, Jul. 2008. [Online]. Available: http://dx.doi.org/10.1016/j.matcom.2008.01.038

[2] Q. Wen, Z. Yang, Y. Song, and P. Jia, "Automatic stock decision support system based on box theory and svm algorithm," *Expert System Application*, vol. 37, no. 2, pp. 1015–1022, Mar. 2010. [Online]. Available: http://dx.doi.org/10.1016/j.eswa.2009.05.093

[3] M. Kampouridis, A. Alsheddy, and E. Tsang, "On the investigation of hyper-heuristics on a financial forecasting problem," *Annals of Mathematics and Artificial Intelligence*, vol. 68, no. 4, pp. 225–246, Aug. 2013. [Online]. Available: http://dx.doi.org/10.1007/s10472-012-9283-0

[4] T. Z. Tan, C. Quek, and G. S. Ng, "Brain-inspired genetic complementary learning for stock market prediction," in *Proceedings of the IEEE Congress on Evolutionary Computation, CEC 2005, 2-4 September 2005, Edinburgh, UK*, 2005, pp. 2653–2660. [Online]. Available: http://dx.doi.org/10.1109/CEC.2005.1555027

[5] A. C. Harvey and P. H. J. Todd, "Forecasting economic time series with structural and box-jenkins models: A case study," *Journal of Business & Economic Statistics*, vol. 1, no. 4, pp. 299–307, 1983. [Online]. Available: http://dx.doi.org/10.2307/1391661

[6] Y. B. Wijaya, S. Kom, and T. A. Napitupulu, "Stock price prediction: Comparison of ARIMA and artificial neural network methods - an indonesia stock's case," in *Proceedings of the 2010 Second International Conference on Advances in Computing, Control, and Telecommunication Technologies*, ser. ACT'10. Washington, DC, USA: IEEE Computer Society, 2010, pp. 176–179. [Online]. Available: http://dx.doi.org/10.1109/ACT.2010.45

[7] L. Yu, S. Wang, and K. K. Lai, "A neural-network-based nonlinear metamodeling approach to financial time series forecasting," *Appl. Soft Comput.*, vol. 9, no. 2, pp. 563–574, Mar. 2009. [Online]. Available: http://dx.doi.org/10.1016/j.asoc.2008.08.001

[8] S. Walczak, "An empirical analysis of data requirements for financial forecasting with neural networks," *J. Manage. Inf. Syst.*, vol. 17, no. 4, pp. 203–222, Mar. 2001. [Online]. Available: http://dl.acm.org/citation.cfm?id=1289668.1289677

[9] M.-D. Cubiles-de-la Vega, R. Pino-Mejías, A. Pascual-Acosta, and J. Muñoz García, "Building neural network forecasting models from time series ARIMA models: A procedure and a comparative analysis," *Intell. Data Anal.*, vol. 6, no. 1, pp. 53–65, Jan. 2002. [Online]. Available: http://dl.acm.org/citation.cfm?id=1293993.1293996

[10] R. Majhi, G. Panda, and G. Sahoo, "Efficient prediction of exchange rates with low complexity artificial neural network models," *Expert System Application*, vol. 36, no. 1, pp. 181–189, Jan. 2009. [Online]. Available: http://dx.doi.org/10.1016/j.eswa.2007.09.005

[11] M. R. Hassan, "A combination of hidden markov model and fuzzy model for stock market forecasting," *Neurocomputing*, vol. 72, no. 1618, pp. 3439 – 3446, 2009. [Online]. Available: http://www.sciencedirect.com/science/article/pii/S0925231209001805

[12] W. Huang, Y. Nakamori, and S. Wang, "Forecasting stock market movement direction with support vector machine," *Computers & Operations Research*, vol. 32, no. 10, pp. 2513–2522, 2005.

[13] R. Majhi, G. Panda, G. Sahoo, A. Panda, and A. Choubey, "Prediction of S&P 500 and DJIA stock indices using particle swarm optimization technique," in *The IEEE World Congress on Computational Intelligence on Evolutionary Computation (CEC2008)*. IEEE, 2008, pp. 1276–1282.

[14] Y. Zhang and L. Wu, "Stock market prediction of S&P 500 via combination of improved bco approach and bp neural network," *Expert Syst. Appl.*, vol. 36, no. 5, pp. 8849–8854, 2009.

[15] S. O. Olatunji, M. Al-Ahmadi, M. Elshafei, and Y. A. Fallatah, "Saudi arabia stock prices forecasting using artificial neural networks," pp. 81–86, 2011.

[16] T.-S. Chang, "A comparative study of artificial neural networks, and decision trees for digital game content stocks price prediction," *Expert Syst. Appl.*, vol. 38, no. 12, pp. 14846–14851, Nov. 2011. [Online]. Available: http://dx.doi.org/10.1016/j.eswa.2011.05.063

[17] E. Guresen, G. Kayakutlu, and T. U. Daim, "Using artificial neural network models in stock market index prediction," *Expert System Application*, vol. 38, no. 8, pp. 10389–10397, Aug. 2011. [Online]. Available: http://dx.doi.org/10.1016/j.eswa.2011.02.068

[18] P.-C. Chang, D.-D. Wang, and C.-L. Zhou, "A novel model by evolving partially connected neural network for stock price trend forecasting," *Expert Syst. Appl.*, vol. 39, no. 1, pp. 611–620, Jan. 2012. [Online]. Available: http://dx.doi.org/10.1016/j.eswa.2011.07.051

[19] Y. Kara, M. Acar Boyacioglu, and O. K. Baykan, "Predicting direction of stock price index movement using artificial neural networks and support vector machines: The sample of the istanbul stock exchange," *Expert Syst. Appl.*, vol. 38, no. 5, pp. 5311–5319, May 2011. [Online]. Available: http://dx.doi.org/10.1016/j.eswa.2010.10.027

[20] W. Shen, X. Guo, C. Wu, and D. Wu, "Forecasting stock indices using radial basis function neural networks optimized by artificial fish swarm algorithm," *Know.-Based Syst.*, vol. 24, no. 3, pp. 378–385, Apr. 2011. [Online]. Available: http://dx.doi.org/10.1016/j.knosys.2010.11.001

[21] X. Zhu, H. Wang, L. Xu, and H. Li, "Predicting stock index increments by neural networks: The role of trading volume under different horizons," *Expert System Application*, vol. 34, no. 4, pp. 3043–3054, May 2008. [Online]. Available: http://dx.doi.org/10.1016/j.eswa.2007.06.023

[22] W.-H. Chen, J.-Y. Shih, and S. Wu, "Comparison of support-vector machines and back propagation neural networks in forecasting the six major asian stock markets," *International Journal of Electron Finance*, vol. 1, no. 1, pp. 49–67, Jan. 2006. [Online]. Available: http://dx.doi.org/10.1504/IJEF.2006.008837

[23] P.-C. Chang, C.-Y. Fan, and J.-L. Lin, "Trend discovery in financial time series data using a case based fuzzy decision tree," *Expert System Application*, vol. 38, no. 5, pp. 6070–6080, May 2011. [Online]. Available: http://dx.doi.org/10.1016/j.eswa.2010.11.006

[24] M. R. Hassan, K. Ramamohanarao, J. Kamruzzaman, M. Rahman, and M. Maruf Hossain, "A HMM-based adaptive fuzzy inference system for stock market forecasting," *Neurocomput.*, vol. 104, pp. 10–25, Mar. 2013. [Online]. Available: http://dx.doi.org/10.1016/j.neucom.2012.09.017

[25] A. F. Sheta and K. De Jong, "Time-series forecasting using ga-tuned radial basis functions," *Inf. Sci. Inf. Comput. Sci.*, vol. 133, no. 3-4, pp. 221–228, Apr. 2001. [Online]. Available: http://dx.doi.org/10.1016/S0020-0255(01)00086-X

[26] A. Sheta, S. E. M. Ahmed, and H. Faris, "Evolving stock market prediction models using multi-gene symbolic regression genetic programming," *Artificial Intelligence and Machine Learning AIML*, vol. 15, pp. 11–20, 6 2015.

TABLE V
EVALUATION CRITERIA FOR THE DEVELOPED MODELS

	Regression		ANN		SVM-RBF	
	Training	Testing	Training	Testing	Training	Testing
Correlation coefficient	0.998	0.995	0.999	0.990	0.9995	0.9928
Mean absolute error	0.373	4.961	0.433	5.869	0.1976	2.6454
Root mean squared error	0.482	5.749	0.529	6.666	0.3263	3.0006
Relative absolute error	4.429%	11.838%	4.683%	58.335%	2.134%	7.993%
Root relative squared error	4.982%	12.313%	5.040%	57.569%	3.109%	8.5579%

[27] D. D. Thomakos, T. Wang, and J. Wu, "Market timing and cap rotation," *Math. Comput. Model.*, vol. 46, no. 1-2, pp. 278–291, Jul. 2007. [Online]. Available: http://dx.doi.org/10.1016/j.mcm.2006.12.036

[28] S. Lahmiri, "Multi-scaling analysis of the S&P500 under different regimes in wavelet domain," *Int. J. Strateg. Decis. Sci.*, vol. 5, no. 2, pp. 43–55, Apr. 2014. [Online]. Available: http://dx.doi.org/10.4018/ijsds.2014040104

[29] S. Lahmiri, M. Boukadoum, and S. Chartier, "Exploring information categories and artificial neural networks numerical algorithms in S&P500 trend prediction: A comparative study," *Int. J. Strateg. Decis. Sci.*, vol. 5, no. 1, pp. 76–94, Jan. 2014. [Online]. Available: http://dx.doi.org/10.4018/IJSDS.2014010105

[30] S. T. A. Niaki and S. Hoseinzade, "Forecasting s&p500 index using artificial neural networks and design of experiments," *Journal of Industrial Engineering International*, vol. 9, no. 1, pp. 1–9, 2013.

[31] A. Sheta, H. Faris, and M. Alkasassbeh, "A genetic programming model for S&P 500 stock market prediction," *International Journal of Control and Automation*, vol. 6, pp. 303–314, 2013.

[32] A. Nigrin, *Neural networks for pattern recognition*, ser. A Bradford book. Cambridge, Mass, London: The MIT Press, 1993. [Online]. Available: http://opac.inria.fr/record=b1126290

[33] J. Leonard and M. A. Kramer, "Improvement of the backpropagation algorithm for training neural networks," *Computer Chemical Engineering*, vol. 14, pp. 337–343, 1990.

[34] A. K. Jain, J. Mao, and K. Mohiuddin, "Artificial neural networks: A tutorial," *IEEE Computer*, vol. 29, pp. 31–44, 1996.

[35] V. Vapnik, *The Nature of Statistical Learning Theory*. Springer, New York, 1995.

[36] ——, "An overview of statistical learning theory," *IEEE Transactions on Neural Networks*, vol. 5, pp. 988–999, 1999.

[37] Y. B. Dibike, S. Velickov, D. Solomatine, and M. B. Abbott, "Model induction with support vector machines: introduction and applications," *Journal of Computing in Civil Engineering*, 2001.

[38] R. Noori, M. Abdoli, A. A. Ghasrodashti, and M. Jalili Ghazizade, "Prediction of municipal solid waste generation with combination of support vector machine and principal component analysis: A case study of mashhad," *Environmental Progress & Sustainable Energy*, vol. 28, no. 2, pp. 249–258, 2009.

[39] W. Wen-chuan, X. Dong-mei, C. Kwok-wing, and C. Shouyu, "Improved annual rainfall-runoff forecasting using pso–svm model based on eemd," 2013.

16

Iris Compression and Recognition using Spherical Geometry Image

Rabab M. Ramadan
College of Computers and Information Technology University of Tabuk Tabuk, KSA

***Abstract*—this research is considered to be a research to attract attention to the 3D iris compression to store the database of the iris. Actually, the 3D iris database cannot be found and in trying to solve this problem 2D iris database images are converted to 3D images just to implement the compression techniques used in 3D domain to test it and give an approximation results or to focus on this new direction in research. In this research a fully automated 3D iris compression and recognition system is presented. We use spherical based wavelet coefficients for efficient representation of the 3D iris. The spherical wavelet transformation is used to decompose the iris image into multi-resolution sub images. The representation of features based on spherical wavelet parameterization of the iris image was proposed for the 3D iris compression system. To evaluate the performance of the proposed approach, experiments were performed on the CASIA Iris database. Experimental results show that the spherical wavelet coefficients yield excellent compression capabilities with minimal set of features. Haar wavelet coefficients extracted from the iris image was found to generate good recognition results.**

Keywords—3D Iris Recognition; Iris Compression; Geometry coding; Spherical Wavelets

I. INTRODUCTION

Biometric identification is the process of associating an identity to the input biometric data by comparing it against the enrolled identities in a database [1]. To design and implement robust systems capable of mass deployment, one needs to address key issues, such as human factors, environmental conditions, system interoperability, and image standard [2].The iris, the colored portion of the eye surrounding the pupil, contains unique patterns which are prominent under near-infrared illumination. These patterns remain stable from a very young age, barring trauma or disease, allowing accurate identification with a very high level of confidence. Commercial iris systems are used as access to secure facilities or other resources, even Criminal/terrorist identification. The enrollment of an individual into a commercial iris system requires capturing one or more images from a video stream [3].

The question arises how to store and handle the acquired sensor data. Typically, the database for such system does not contain actual iris images, but rather it stores a binary file that represents each enrolled iris (the template). Most commercial iris systems today use the Daugman algorithm [4-6]. The recognition system used the template as the input to its process and the iris image is discarded to speed up the recognition process and decrease the storage requirements of the system. From the other point of view, if the data have to be transferred via a network link to the respective location, a minimization of the amount of data must be taken into account. The problem here is that a template alone cannot allow the recreation of the iris image from that it is derived, while the original iris imagery is still valuable for research.

A lot of researches concern on the recognition system which depends on the template of the iris image extracted from the original image. In this paper, the attention is transferred to the iris compression.

This research is considered to be a research to attract attention to the 3D iris compression to store the database of the iris. Actually, the 3D iris database cannot be found and in trying to solve this problem 2D iris database images are converted to 3D images just to implement the compression techniques used in 3D domain to test it and give an approximation results or to focus on this new direction in research. These results may encourage researchers to establish a new 3D iris database image to benefit from all techniques in 3D domain.

Geometry image is an image used to remesh an arbitrary surface onto a completely regular structure [7]. One important use for such a representation is shape compression, the concise encoding of surface geometry [8]. Geometry images can be encoded using traditional image compression and decompression algorithms, such as wavelet-based coders. The mesh-based spherical scheme more natural for coding geometry, and provide good reconstruction of shape detail at very low bit budgets [9].

In this paper, we detail how the geometry image is used to compress the 3D iris images. The 3D iris image is mapped to the spherical parameterization domain then the geometry image is obtained as a color image and a surface. Finally, the spherical based wavelet coefficient are computed for efficient representation and compression of the 3D iris image.

The rest of this paper is organized as follows: overview of related work in 3D image compression will be in section II. Section III contains the overview of image preprocessing stage including segmentation and normalization. Spherical geometry image is discussed in section IV. Section V reports the experimental results. Finally, section VI contains the conclusion of this paper.

II. RELATED WORK

A major advance in the field of iris recognition results from the expiration of two patents [10]. The first one is the pioneer patent dealing with the general idea of the iris recognition process. It was developed by the ophthalmologists Flom and

Safir (1987) and it expired in 2005. The second one, developed by the professor John Daugman (1994), was used to protect the iris-code approach and expired in 2011.

Flom and Safir first proposed the concept of automated iris recognition in 1987 [11]. Since then, some researchers worked on iris representation and matching and have achieved great progress [12], [13], [14], [15].

The iris recognition process starts with the segmentation of the iris ring. Further, data is transformed into a double dimensionless polar coordinate system, through the Daugman's Rubber Sheet process. Regarding the feature extraction stage, existing approaches can be roughly divided into three variants: phase-based [16], zero-crossing [17] and textureanalysis methods [18]. Dauman [16] used multi-scale quadrature wavelets to extract texture phase-based information and obtain an iris signature with 2048 binary components. Boles and Boashash [19] calculated a zero-crossing representation of one-dimensional (1-D) wavelet transform at various resolution levels of a concentric circle on an iris image to characterize the texture of the iris.. Wildes et al. [20] represented the iris texture with a Laplacian pyramid constructed with four different resolution levels and used the normalized correlation to determine whether the input image and the model image are from the same class. Tisse et al. [21] analyzed the iris characteristics using the analytic image constructed by the original image and its Hilbert transform. Emergent frequency functions for feature extraction were in essence samples of the phase gradient fields of the analytic image's dominant components [22], [23]. Similar to the matching scheme of Daugman, they sampled binary emergent frequency functions to form a feature vector and used Hamming distance for matching. Park et al. [24] used a directional filter bank to decompose an iris image into eight directional subband outputs and extracted the normalized directional energy as features. Iris matching was performed by computing Euclidean distance between the input and the template feature vectors.

Kumar et al. [25] utilized correlation filters to measure the consistency of iris images from the same eye. The correlation filter of each class was designed using the two-dimensional. In [26], Hong and Smith proposed the octave band directional filter banks which are capable of both directional decomposition and an octave band radial decomposition. Finally, in the feature comparison stage, a numeric dissimilarity value is produced, which determines the subject's identity. Here, it is usual to apply different distance metrics (Hamming [16], Euclidian [27] or weighted Euclidian [28]), or methods based on signal correlation [20]. Many image compression and representation methods depend on Gabor analysis or phase information, which are two important components in IrisCode. Daugman demonstrated that Gabor filters are effective for image compression [26]. Behar et al. showed that images can be reconstructed from localized phase [29].

This research is considered to be a research to attract attention to the 3D iris compression to store the database of the iris. Actually, the 3D iris database cannot be found and in trying to solve this problem 2D iris database images are converted to 3D images just to implement the compression techniques used in 3D domain to test it and give an approximation results or to focus on this new direction in research. These results may encourage researchers to establish a new 3D iris database image to benefit from all techniques in 3D domain. Geometry images and Spherical representations are used in the compression algorithm.

The construction of a geometry image involves parametrizing a given surface onto a planar domain, and resampling the surface geometry on a regular domain grid. The original work [30] heuristically cuts an arbitrary surface into a disk using a network of cut paths, with 2g loops for a genus g surface. The resulting cut surface is mapped onto a square using a stretch-minimizing parametrization to reduce under sampling.

For shapes with high genus or long extremities, forcing the whole surface to map into a square can introduce high distortion. To mitigate this, we can instead cut the surface into several pieces to produce a multi-chart geometry image [31]. The challenge is to join these piecewise regular charts into a watertight surface.

For genus-zero models, a geometry image may be constructed via spherical parametrization [32], which does not require any a priori surface cuts. The spherical domain is unfolded into a square using a simple cut with elegant boundary symmetries. These boundary symmetries permit the construction of a smooth (C1) polynomial surface, and the regular control grid structure lets the surface be evaluated entirely within the GPU rasterization pipeline [33]. In addition, a spherical geometry image can be compressed using traditional image wavelet Geometry images for static objects can be generalized to geometry videos for animated shapes [34]. Excellent survey of the various 3D mesh compression algorithms has been given by Alliez and C. Gotsman in [34, 30]. The recent development in the wavelet transforms theory has spurred new interest in multi-resolution methods, and has provided a more rigorous mathematical framework. Wavelets give the possibility of computing compact representations of functions or data. Additionally, wavelets are computationally attractive and allow variable degrees of resolution to be achieved. All these features make them appear as an interesting tool to be used for efficient representation of 3D objects.

3D Face recognition is one of the imperative applications calling for compact storage and rapid processing of 3D meshes. Face recognition based on 3D information is not a new topic. It has been extensively addressed in the related literature since the end of the last century [35-40]. Further surveys of the state-of-the-art in 3D face recognition can be found in [36, 37]. Spherical representations permit to efficiently represent facia surfaces and overcome the limitations of other methods towards occlusions and partial views. In our previous work [41], an innovative fully automated 3D face compression and recognition system is presented. We use spherical based wavelet coefficients for efficient representation of the 3D face. The spherical wavelet transformation is used to decompose the face image into multi-resolution sub images. To the best of our knowledge, the representation of 3D iris point clouds as spherical signals for iris recognition has however not been investigated yet. We therefore propose to take benefit of the spherical representations in order to build an effective and automatic 3D iris recognition system.

III. Iris Image Preprocessing

Image processing techniques can be employed to extract the unique iris pattern from a digitised image of the eye, and encode it into a biometric template, which can be stored in a database. This biometric template contains an objective mathematical representation of the unique information stored in the iris, and allows comparisons to be made between templates. When a subject wishes to be identified by an iris recognition system, their eye is first photographed, and then a template created for their iris region. This template is then compared with the other templates stored in a database until either a matching template is found and the subject is identified, or no match is found and the subject remains unidentified [42]. There are four main stages of an iris recognition and compression system. They are, image preprocessing, feature extraction and template matching [43], and compression algorithm.

A. Image preprocessing

The iris image is to be preprocessed to obtain useful iris region. Image preprocessing contains, iris localization that detects the inner and outer boundaries of iris [44], [45] and iris normalization, in this step, iris image is converted from Cartesian coordinates to Polar coordinates. In this paper, We are not focusing on the segmentation instead we are interested in iris compression hence we have used the existing algorithms [42]for image preprocessing normalization feature extraction and segmentation but focusing only on iris compression and matching algorithm. Figure 1 shows the output of the segmentation process using Masek algorithm.

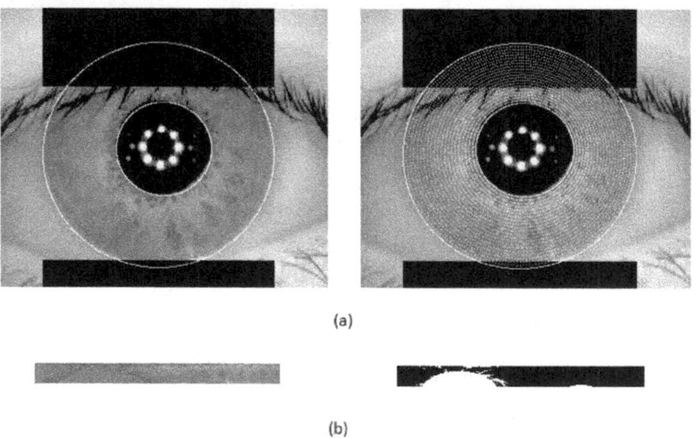

Fig. 1. Example the output of the segmentation process using Masek algorithm. (a) Automatic segmentation of an iris image from the CASIA database. Black regions denote detected eyelid and eyelash regions. (b) Illustration of the normalization process (polar array – noise array)

B. Feature Extraction

Feature extraction is the process of getting the iris features, Wavelet transform is used for this purpose.

C. Template Matching

Template matching compares the user template with templates from the database using a matching algorithm. The matching metric will give a measure of similarity between two iris templates. Finally, a decision with high confidence level is made through matching methods to identify whether the user is an authentic or imposter.

D. Compression Algorithm

Geometry image and spherical wavelet transform will be used for the compression algorithm as shown in the next section. Figure. 2 shows the stages of iris compression algorithm.

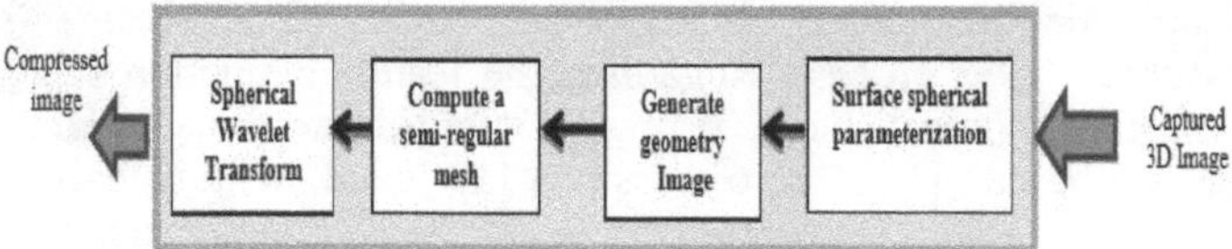

Fig. 2. Stages of iris compression algorithm

IV. Spherical Geometry Image

Surfaces in computer graphics are commonly represented using irregular meshes. While such meshes can approximate a given shape using few vertices, their irregularity comes at a price, since most mesh operations require random memory accesses through vertex indices and texture coordinates. Also, filter kernels must handle arbitrary mesh neighborhoods, and techniques like morphing, level-of-detail (LOD) control, and compression are complicated. As an alternative, we have introduced the geometry image representation, which captures shape using a completely regular sampling, i.e. a 2D grid of (x,y,z) values [46]. The benefits of uniform grids are often taken for granted. Grids allow efficient traversal, random access, convolution, composition, down-sampling, and compression.

A. Spherical Parameterization

Geometric models are often described by closed, genus-zero surfaces, i.e. deformed spheres. For such models, the sphere is the most natural parameterization domain, since it does not require cutting the surface into disk(s). Hence the parameterization process becomes unconstrained [47]. Even though we may subsequently resample the surface signal onto a piecewise continuous domain, these domain boundaries can be determined more conveniently and a posteriori on the sphere. Spherical parameterization proves to be challenging in practice, for two reasons. First, for the algorithm to be robust it must prevent parametric "foldovers" and thus guarantee a 1-to-1 spherical map. Second, while all genus-zero surfaces are in essence sphere-shaped, some can be highly deformed, and creating a parameterization that adequately samples all surface regions is difficult. Once a spherical parameterization is obtained, a number of applications can operate directly on the sphere domain, including shape analysis using spherical harmonics, compression using spherical wavelets [46, 48], and mesh morphing [49].

Given a triangle mesh M, the problem of spherical parameterization is to form a continuous invertible map φ: S→ M from the unit sphere to the mesh. The map is specified by assigning each mesh vertex v a parameterization φ-1(v) ϵ S. Each mesh edge is mapped to a great circle arc, and each mesh triangle is mapped to a spherical triangle bounded by these arcs. To form a continuous parameterization φ, we must define the map within each triangle interior. Let the points {A, B, C} on the sphere be the parameterization of the vertices of a mesh triangle {A'= φ (A), B'= φ (B), C'= φ (C)}. Given a point P'= αA'+βB'+γC' with barycentric coordinates α+β+γ=1 within the mesh triangle, we must define its parameterization P =φ-1(P'). Any such mapping must have distortion since the spherical triangle is not developable.

B. Geometry Image

A simple way to store a mesh is using a compact 2D geometry images. Geometry images was first introduced by Gu et al. [46, 50] where the geometry of a shape is resampled onto a completely regular structure that captures the geometry as a 2D grid of [x, y, z] values. The process involves heuristically cutting open the mesh along an appropriate set of cut paths. The vertices and edges along the cut paths are represented redundantly along the boundary of this disk. This allows the unfolding of the mesh onto a disk-like surface and then the cut surface is parameterized onto the square. Other surface attributes, such as normals and colors, are stored as additional 2D grids, sharing the same domain as the geometry, with grid samples in implicit correspondence, eliminating the need to store a parameterization. Also, the boundary parameterization makes both geometry and textures seamless. The simple 2D grid structure of geometry images is ideally suited for many processing operations. For instance, they can be rendered by traversing the grids sequentially, without expensive memory-gather operations (such as vertex index dereferencing or random-access texture filtering). Geometry images also facilitate compression and level-of-detail control. Figure 3 presents geometric representation of the iris image.

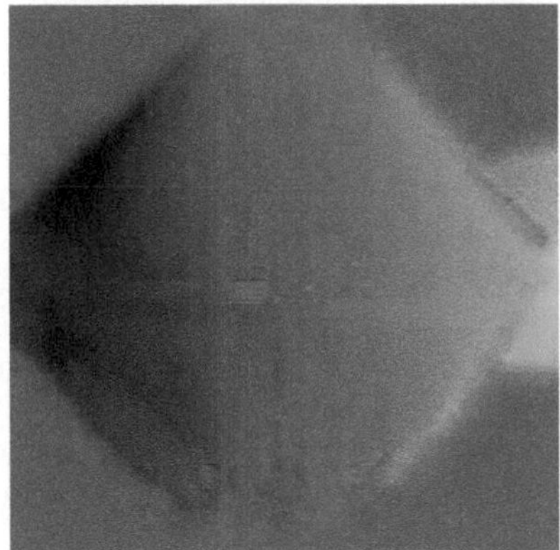

Fig. 3. Geometric representation of the iris image

C. Wavelet Transform

- Haar Transform:

Geometry images are regularly sampled 2D images that have three channels, encoding geometric information (x, y and z) components of a vertex in R3 [50]. Each channel of the geometry image is treated as a separate image for the wavelet analysis. The Haar wavelet transform has been proven effective for image analysis and feature extraction. It represents a signal by localizing it in both time and frequency domains. The Haar wavelet transform is applied separately on each channel creating four sub bands LL, LH, HL, and HH where each sub band has a size equal to 1/4 of the original image. The LL sub band captures the low frequency components in both vertical and horizontal directions of the original image and represents the local averages of the image. Whereas the LH, HL and HH sub bands capture horizontal, vertical and diagonal edges, respectively. In wavelet decomposition, only the LL sub band is used to recursively produce the next level of decomposition. The biometric signature is computed as the concatenation of the Haar wavelet coefficients that were extracted from the three channels of the geometry image.

- Spherical Wavelets:

To be able to construct spherical wavelets on an arbitrary mesh, this surface mesh should be represented as a multi-resolution mesh, which is obtained by regular 1:4 subdivision of a base mesh [51, 52, 53]. A multi-resolution mesh is created by recursive subdivision of an initial polyhedral mesh so that each triangle is split into four "child" triangles at each new subdivision

Denoting the set of all vertices on the mesh before the jth subdivision as K(j) a set of new vertices M(j) can be obtained by adding vertices at the midpoint of edges and connecting them with geodesics. Therefore, the complete set of vertices at the j+1th level is given by K(j+1) =K(j) ∪ M (j). Consequently, the number of vertices at level j is given by: 10*4j+2. This process is presented in Figure 4 (a)-(d) where the iris image is shown at 4 different subdivision levels.

In this research, we use the discrete bi-orthogonal spherical wavelets functions defined on a 3-D mesh constructed with the lifting scheme proposed by Schröder and Sweldens [51, 52, 53, 54]. Spherical wavelets belong to second generation wavelets adapted to manifolds with non-regular grids. The main difference with the classical wavelet is that the filter coefficients of second generation wavelets are not the same throughout, but can change locally to reflect the changing nature of the surface and its measure.

They maintain the notion that a basis function can be written as a linear combination of basis functions at a finer, more subdivided level. Spherical wavelet basis is composed of functions defined on the sphere that are localized in space and characteristic scales and therefore match a wide range of signal characteristics, from high frequency edges to slowly varying harmonics [52, 55].

The basis is constructed of scaling functions defined at the coarsest scale and wavelet functions defined at subsequent finer scales. If there exist N vertices on the mesh, a total of N basis functions are created, composed of scaling functions and where N0 is the initial number of vertices before the base mesh is subdivided. An interpolating subdivision scheme is used to construct the scaling functions on the standard unit sphere S denoted by φ j,k. The function is defined at level j and node k ϵ k(j) such that the scaling function at level j is a linear combination of the scaling function at level j and j+1. Index j specifies the scale of the function and k is a spatial index that specifies where on the surface the function is centered. Using these scaling functions, the wavelet ψ_(j,m)at level j and node m ϵ M(j) can be constructed by the lifting scheme.

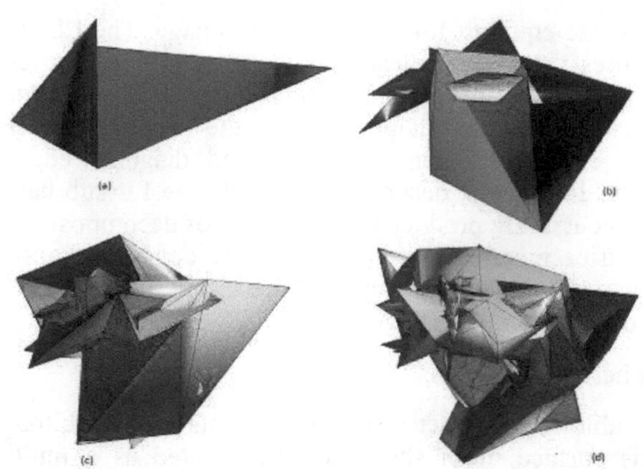

Fig. 4. Visualization of recursive partitioning of the iris mesh at different subdivision levels. (a) Initial icosahedron (scale 0). (b) Single partitioning of icosahedron (scale 1). (c) Two recursive partitioning of icosahedron (scale 2). (d) Three recursive partitioning of icosahedron (scale 3)

A usual shape for the scaling function is a hat function defined to be one at its center and to decay linearly to zero. As the j scale increases, the support of the scaling function decreases. A wavelet function is denoted by ψ_(j,k) : S → R. The support of the functions becomes smaller as the scale increases. Together, the coarsest level scaling function and all wavelet scaling functions construct a basis for the function space L2:

$$L^2 = \{\varphi_{0,k} \mid k \in N_o\} \cup \{\psi_{j,m} \mid j \geq 0, m \in N_{j+1}\} \quad (1)$$

A given function f: S →R can be expressed in the basis as a linear combination of the basis functions and coefficients

$$f(x) = \sum_k \lambda_{0,k}\, \varphi_{0,k}(x) + \sum_{0\leq j} \sum_m \gamma_{j,m}\, \psi_{j,m}(x) \quad (2)$$

Scaling coefficients λ_(0,k) represent the low pass content of the signal f, localized where the associated scaling function has support; whereas, wavelet coefficients γ_(j,m) represent localized band pass content of the signal, where the band pass frequency depends on the scale of the associated wavelet function and the localization depends on the support of the function. Figure 5 (a)-(g) presents the spherical wavelets of the iris image.

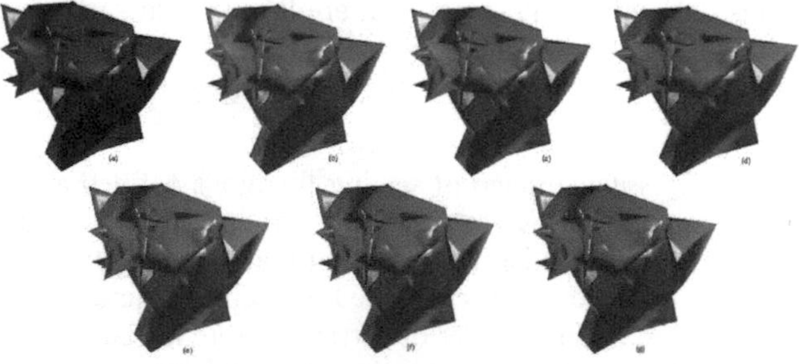

Fig. 5. Spherical wavelet transform of iris image. (a) using 5% of wavelet coefficients (b) using 10% of wavelet coefficients (c) using 20% of wavelet coefficients (d) using 40% of wavelet coefficients (e) using 60% of wavelet coefficients (f) using 80% of wavelet coefficients (g) Using all coefficients

V. Experimental Results

The CASIA-IrisV4 data base was used to evaluate the performance of the proposed system. CASIA- IrisV4 is said to be an extension of CASIA-IrisV3 and contains six subsets. It contains a total of 54,601 iris image from more than 1,800 genuine subject and 1,000 virtual subjects. All iris images are 8 bit gray-level JPEG files. In our experiment CASIA-Iris-Interval and CASIA-Iris-Lamp will be used.

A. Iris Recognition

In this experiment, CASIA-Iris-Interval contains 249 subjects. Only 99 subjects will be included in this experiment. For each one only 7 images for each eye is taken. The total number of classes is 198 which have 1386 images. 2-dimentional Haar wavelet transform is applied to the templates of the iris images which have dimension of (20 * 480) for each image. Each application of the Haar wavelet decomposition reduces the size of the image to ¼ of its original size so the input to the K-fold cross validation method is a pattern of (10 * 240) features.

K-fold cross-validation is a statistical method used to evaluate the performance of a learning algorithm [9]. In K-fold cross validation, the input data is divided into k nearly equal subsets. K iterations are performed. In each iteration, one of the k subsets is considered the test set while the other k-1 subsets are put together to form a training set. The output is the average error across all k trials.

The average recognition rate is 94% is achieved in this experiment. This result confirms that iris recognition is a reliable and accurate biometric technology. But as mentioned before, the iris recognition is not the objective of this research. We want to focus on 3D iris compression image which will be discussed in the next experiments.

B. Iris compression

In this experiment, the algorithm is applied to two sets from CASIA Iris-V4, CASIA-Iris-Interval and CASIA-Iris-Lamp. The input image is converted to 3D image then generate the geometry image. After this, a semi-regular mesh from a gim file is computed. Finally, the spherical wavelet transform on the mesh is computed. The total number of the features after computing wavelet transform is 774 features by keeping only the biggest coefficients. Different percentages of these coefficients were tested and each time the inverse wavelet transform was used to reconstruct the iris image. Figure 6 (a)-(g) shows the reconstructed images for the iris using different percentages of wavelet coefficient. These figure shows us that there is no visually difference between the original image and the corresponding reconstructed images.

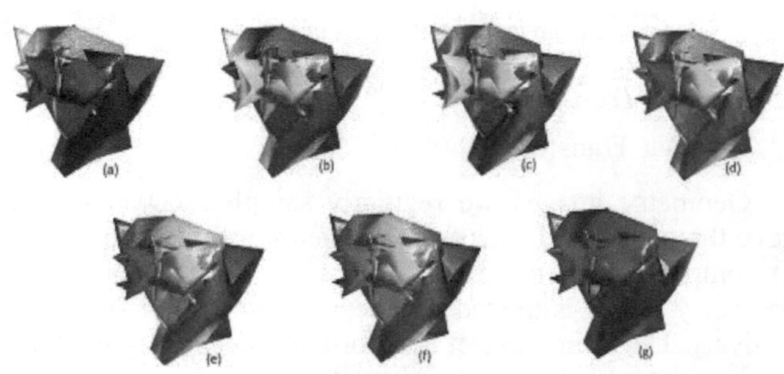

Fig. 6. Wavelet approximation of iris image. (a) using 2% of wavelet coefficients (b) using 5% of wavelet coefficients(c) using 10% of wavelet coefficients (d) using 20% of wavelet coefficients (e) Using all coefficient

To evaluate the quality of the reconstructed iris image, the Normalized Error (NE) and Normalized Correlation (NC) were used to. NE is given as follows:

$$NE = \frac{||x-y||}{||x||} \quad (3)$$

Where X is the original image and Y is the reconstructed image. i.e. NE is the norm of the difference between the original and reconstructed signals, divided by the norm of the original signal. The NC is given:

$$NC = \frac{\sum_{i=1}^{M}\sum_{j=1}^{N} X(i,j)Y(i,j)}{\sum_{i=1}^{M}\sum_{j=1}^{N} X(i,j)X(i,j)} \quad (4)$$

Where MxN is the size of the image. The NE and the NC values of the reconstructed images are presented in Table () for the different wavelet subsets.

TABLE I. NE AND NC FOR VARIOUS WAVELET SUBSETS FOR IRIS IMAGE FROM CASIA-IRIS-INTERVAL SUBSET

Percentage	5%	10%	20%	40%	60%	80%	100%
NE	0.477	0.1662	0.0912	0.0664	0.0517	0.0446	0.0311
Nc	0.6165	0.9278	0.9614	0.9671	0.977	0.9795	0.9979
No. of coefficient	39	77	155	310	464	619	774

TABLE II. NE AND NC FOR VARIOUS WAVELET SUBSETS FOR IRIS IMAGE FROM CASIA-IRIS-LAMP SUBSET

Percentage	5%	10%	20%	40%	60%	80%	100%
NE	0.3593	0.0411	0.0275	0.0210	0.0166	0.0130	0.0107
Nc	0.7318	0.9757	0.9959	1	0.9980	0.9973	0.9993
No. of coefficient	39	77	155	310	464	619	774

The NE and NC values indicate that the reconstructed images are the very similar to the original image. In the case of using only 20 % of the wavelets coefficients, the relative error of reconstruction is 9%. The reconstructed signal retains approximately 96.14% of the energy of the original signal while the number of coefficient is only 155.

VI. CONCLUSION

In this paper an innovative approach for 3D iris compression and recognition based on spherical wavelet parameterization was proposed. First. The iris image is to be preprocessed to obtain useful iris region. Image preprocessing contains, iris localization that detects the inner and outer boundaries of iris and iris normalization, in this step, iris image is converted from Cartesian coordinates to Polar coordinates. We are not focusing on the segmentation instead we are interested in iris compression hence we have used the existing algorithms for Masek. Next, the spherical wavelet features were extracted which provide a compact descriptive biometric signature. Spherical representation of iris permits effective dimensionality reduction through simultaneous approximations. The dimensionality reduction step preserves the geometry information, which leads to high performance matching in the reduced space. Multiple representation features based on spherical wavelet parameterization of the iris image were proposed for the 3D iris compression and recognition. The database CASIA was utilized to test the proposed system. Experimental results show that the spherical wavelet coefficients yield excellent compression capabilities with minimal set of features. Furthermore, it was found that Haar wavelet coefficients extracted from the templet of the iris yield good recognition results.

ACKNOWLEDGEMENTS

The authors would like to acknowledge financial support for this work from the Deanship of Scientific Research (DSR), University of Tabuk, Tabuk, Saudi Arabia, under grant no. 0024/1435.

REFERENCES

[1] A. Jain, A. Ross, and S. Prabhakar, “An Introduction to Biometric Recognition,” IEEE Transactions on Circuits and Systems for Video Technology, vol. 14, no. 1, pp. 4–20, January 2004.

[2] Rakshit, Soumyadip; Monro, Donald M., "An Evaluation of Image Sampling and Compression for Human Iris Recognition," Information Forensics and Security, IEEE Transactions on , vol.2, no.3, pp.605,612, Sept. 2007 doi: 10.1109/TIFS.2007.902401

[3] Ives, R.W.; Bishop, D.A.D.; Yingzi Du; Belcher, C., "Effects of image compression on iris recognition performance and image quality," Computational Intelligence in Biometrics: Theory, Algorithms, and Applications, 2009. CIB 2009. IEEE Workshop on , vol., no., pp.16,21, March 30 2009-April 2 2009

[4] J. Daugman "How iris recognition works," IEEE Trans. on Circuits and Systems for Video Technology., Vol. 14, No. 1, pp. 21-30.

[5] J. G. Daugman, “High confidence visual recognition of persons by a test of statistical independence,” IEEE Trans. Pattern Analysis and Machine. Intelligence, vol. 15, no. 11, pp. 1148 1161, Nov. 1993.

[6] J. G. Daugman, “The importance of being random: Statistical principles of iris recognition,” Pattern Recognition, vol. 36, no. 2, pp. 279–291, Feb. 2003.

[7] Gu, Xianfeng, Steven J. Gortler, and Hugues Hoppe. "Geometry images." ACM Transactions on Graphics (TOG) 21.3 (2002): 355-361.

[8] Hoppe, Hugues, and Emil Praun. "Shape compression using spherical geometry images." Advances in Multiresolution for Geometric Modelling. Springer Berlin Heidelberg, 2005. 27-46.

[9] Blum, Avrim, Adam Kalai, and John Langford. "Beating the hold-out: Bounds for k-fold and progressive cross-validation." Proceedings of the twelfth annual conference on Computational learning theory. ACM, 1999.

[10] J. Daugman, “Statistical richness of visual phase information: Update on recognizing persons by iris patterns,” Int. J. Comput. Vis., vol. 45, pp. 25–38, 2001.

[11] L. Flom and A. Safir, “Iris Recognition system,” U.S. Patent 4 641 394, 1987.

[12] R. Johnson, “Can iris patterns be used to identify people?,” Chemical and Laser Sciences Division LA-12 331-PR, Los Alamos Nat. Lab., Los Alamos, CA, 1991.

[13] K. Bae, S. Noh, and J. Kim, “Iris feature extraction using independent component analysis,” in Proc. 4th Int. Conf. Audio- and Video-Based Biometric Person Authentication, 2003, pp. 838–844.

[14] J. Daugman, “Biometric personal identification system based on iris analysis,” U.S. Patent 5 291 560, 1994.

[15] R.Wildes, J. Asmuth, S. Hsu, R. Kolczynski, J. Matey, and S. Mcbride, “Automated, noninvasive iris recognition system and method,” U.S. Patent 5 572 596, 1996.

[16] J. G. Daugman, “Phenotypic versus genotypic approaches to face recognition,” in Face Recognition: From Theory to Applications. Heidelberg: Springer-Verlag, 1998, pp. 108–123.

[17] W. Boles and B. Boashash, “A human identification technique using images of the iris and wavelet transform,” Signal Processing, IEEE Transactions on, vol. 46, no. 4, pp. 1185–1188, April 1998.

[18] R. P. Wildes, "Iris recognition: an emerging biometric technology," Proceedings of the IEEE, vol. 85, no. 9, pp. 1348–1363, September 1997.

[19] W. Boles and B. Boashash, "A human identification technique using images of the iris and wavelet transform," IEEE Trans. Signal Processing, vol. 46, pp. 1185–1188, Apr. 1998.

[20] R. Wildes, J. Asmuth, G. Green, S. Hsu, R. Kolczynski, J. Matey, and S. McBride, "A machine-vision system for iris recognition," Mach. Vis. Applic., vol. 9, pp. 1–8, 1996.

[21] C. Tisse, L. Martin, L. Torres, and M. Robert, "Person identification technique using human iris recognition," in Proc. Vision Interface, 2002, pp. 294–299.

[22] T. Tangsukson and J. Havlicek, "AM-FM image segmentation," in Proc. EEE Int. Conf. Image Processing, 2000, pp. 104–107.

[23] J. Havlicek, D. Harding, and A. Bovik, "The mutli-component AM-FM image representation," IEEE Trans. Image Processing, vol. 5, pp. 1094–1100, June 1996.

[24] C. Park, J. Lee, M. Smith, and K. Park, "Iris-based personal authentication using a normalized directional energy feature," in Proc. 4th Int. Conf. Audio- and Video-Based Biometric Person Authentication, 2003, pp. 224–232.

[25] B. Kumar, C. Xie, and J. Thornton, "Iris verification using correlation filters," in Proc. 4th Int. Conf. Audio- and Video-Based Biometric Person Authentication, 2003, pp. 697–705.

[26] P. Hong and M. 1. T. Smith, "An octave-band family of nonredundantdirectional filter banks", IEEE proceedings of ICASSP, vol. 2, pp.l165-1168, 2002.

[27] Y. Huang, S. Luo, and E. Chen, "An efficient iris recognition system," in Proceedings of the First International Conference on Machine Learning and Cybernetics, China, November 2002, pp. 450–454.

[28] L. Ma, T. Tan, Y. Wang, and D. Zhang, "Efficient iris recognition by characterizing key local variations," Image Processing, IEEE Transactions on, vol. 13, no. 6, pp. 739–750, June 2004

[29] J. Behar, M. Porat, and Y.Y. Zeevi, "Image Reconstruction from Localized Phase," IEEE Trans. Signal Processing, vol. 40, no. 4, pp. 736-743, Apr. 1992.

[30] GU, X., GORTLER, S., AND HOPPE, H. 2002. Geometry images. ACM SIGGRAPH 2002, pp. 355-361.

[31] SANDER, P,WOOD, Z., GORTLER, S., SNYDER, J., AND HOPPE, H. 2003. Multi-chart geometry images. Eurographics Symposium on Geometry Processing 2003, pp. 146-155.

[32] PRAUN, E., AND HOPPE, H. 2003. Spherical parametrization and remeshing. ACM SIGGRAPH 2003, pp. 340-349.

[33] LOSASSO, F., HOPPE, H., SCHAEFER, S., AND WARREN, J. 2003. Smooth geometry images. Eurographics Symposium on Geometry Processing 2003, pp. 138-145.

[34] BRICEÑO, H., SANDER, P, MCMILLAN, L, GORTLER, S., AND HOPPE, H. 2003. Geometry videos: A new representation for 3D animations. ACM Symposium on Computer Animation 2003, pp. 136-146.

[35] M. H. Mahoora and M. Abdel-Mottalebb, "Face recognition based on 3D ridge images obtained from range data," Pattern Recognition, Vol. 42, pp. 445 – 451, 2009.

[36] K.W. Bowyer, K.Chang, and P. Flynn, "A survey of approaches and challenges in 3D and multi-modal3D + 2D face recognition," Computer Vision and Image Understanding, Vol. 101, No.1, pp. 1-15, 2006.

[37] A.F. Abate, M. Nappi, D. Riccio and G. Sabatino, "2D and 3D face recognition: A survey," Pattern Recognition Letters, Vol. 28, pp. 1885-1906, 2007.

[38] K. I. Chang, K. W. Bowyer and P. J. Flynn, "Multiple nose region matching for 3D face recognition under varying facial expression," IEEE Transactions on Pattern Analysis and Machine Intelligence, Vol. 28, pp. 1695-1700, 2006.

[39] G. Günlü and H. S. Bilge, "Face recognition with discriminating 3D DCT coefficients," The Computer Journal, Vol. 53, No. 8, pp. 1324-1337, 2010.

[40] L. Akarun, B. Gokberk, and A. Salah, "3D face recognition for biometric applications," In Proceedings of the European Signal Processing Conference, Antalaya, 2005.

[41] Ramadan, Rabab M., and Rehab F. Abdel-Kader. "3D Face Compression and Recognition using Spherical Wavelet Parametrization." International Journal of Advanced Computer Science and Applications (IJACSA) 3.9 (2012).

[42] Masek, Libor. Recognition of human iris patterns for biometric identification. Diss. Master's thesis, University of Western Australia, 2003.

[43] Kulkarni, S. B., R. S. Hegadi, and U. P. Kulkarni. "Improvement to libor masek algorithm of template matching method for iris recognition." Proceedings of the International Conference & Workshop on Emerging Trends in Technology. ACM, 2011.

[44] Kevin W. Boyer, Karan Hollingsworth and Patrick J. Flynn. Image understanding for iris biometric: A survey Computer Vision and Image Understanding 110 (2), 281-307, May 2008.

[45] Richard Yew Fatt Ng, Yong Haur Tay and Kai Ming Mok. An Effective segmentation method for iris recognition system, The Institution of Engineering and Technology, 2008, PP-548 to 553

[46] P. Alliez and C. Gotsman, "Shape compression using spherical geometry images," In Proceedings of the Symp. Multi-resolution in Geometric Modeling, 2003.

[47] Praun and H. Hoppe, "Spherical parameterization and remeshing," in ACM SIGGRAPH 2003, pp. 340–349, 2003.

[48] L. Pastor, A. Rodriguez, J. M. Espadero, and L. Rincon, "3D wavelet-based multi-resolution object representation," Pattern Recognition, Vol. 34, pp. 2497-2513, 2001.

[49] M. Alexa, "Recent advances in mesh morphing," Computer Graphics Forum, Vol. 21, No. 2, pp. 173-196, 2002.

[50] X. Gu, S. J. Gortler and H. Hoppe, "Geometry images," ACM SIGGRAPH, pp. 355-361, 2002.

[51] P. Schröeder and W. Sweldens, "Spherical wavelets: Efficiently representing functions on a sphere," In Proceedings of Computer Graphics (SIGGRAPH 95), pp. 161-172, 1995.

[52] P. Schröder and W. Sweldens, "Spherical wavelets: Texture processing," in Rendering Techniques, New York, 1995, Springer Verlag.

[53] P. Schröeder and W. Sweldens, "Spherical wavelets: efficiently representing functions on a sphere," In Proceedings of Computer Graphics (SIGGRAPH 95), pp. 161-172, 1995.

[54] S. Campagna and H.-P. Seidel, Parameterizing meshes with arbitrary topology. In H.Niemann, H.-P. Seidel, and B. Girod, editors, Image and Multidimensional Signal Processing' 98, pp. 287-290, 1998.

[55] P. Yu, P. Ellen Grant, Y. Qi, X. Han, F. Ségonne and Rudolph Pienaar, " Cortical surface shape analysis based on spherical wavelets," IEEE Transactions on Medical Imaging, Vol. 26, No. 4, pp. 582-597, 2007.

17

External analysis of strategic market management can be realized based upon different human mindset – A debate in the light of statistical perspective

Prasun Chakrabarti
Head, Department of Computer Science and Engineering
Sir Padampat Singhania University
Udaipur-313601,Rajasthan, India
prasun9999@rediffmail.com

Prasant Kumar Sahoo
The Vice-Chancellor
Utkal University
Bhubaneswar - 751004,Orissa
prasantsahoo12@yahoo.co.in

***Abstract* - The paper entails the statistical correlation of the investigations carried out for the sales and profit prediction and analysis by persons of different mindsets in case of strategic uncertainty . The paper by virtue of statistical and fuzzy logic based justifications has pointed out certain discovered facts in this perspective. The normal , optimistic , pessimistic and fickle-minded based individual mindsets significantly contribute to varying external analysis of business statistics.**

***Keywords* -** *statistical correlation , fuzzy logic , optimistic , pessimistic , fickle-minded , business statistics*

I. Introduction

Strategic development or review[1] deals with an analysis of the factors external to a business that affect strategy. In strategic market management, estimation of sales and profit plays a significant role . Sometimes a separate statistical analyst team is solely recruited in certain business companies. A running business can be investigated on the basis of apriori events and statistical trend analysis[2,3]. However in certain cases due to some external stochastic events, statistical analysis has to be carried out based upon prediction and forecasting and in this perspective of strategic uncertainty , the business estimate varies from individual to individual depending on his nature viz. normal , optimistic , pessimistic and fickle-minded.

II. Variation of External Analysis of Strategic Market Management Based on Human Mindset

Certain discovered facts can be pointed out pertaining to the variation of external analysis of strategic market management depending on the human nature. We propose certain mathematically established axioms in this context. An opportunity or a threat results in a significant change in pattern of the sales and profit of a business. Marketing Myopia[4] also indicates the essence of investigation of sales and profit in case of strategic uncertainty. Furthermore, profit and loss are two mutually exclusive events at any specific timing instant of the observation period. Therefore, R , the Bernoulli random variable[5] for the external analysis of business strategy in this situation , can be viewed as –

R = 1 if profit occurs
else = 0 if loss occurs.

R is a statistical indicator of X or Y.

Claim 1 - If prediction of occurrence of gain in a strategic market management by a normal individual is based upon estimation of weight of single associated parameter and hypothesis of fairness by pessimistic individual is rejected , then for unit negative bias, the estimate of weight of the single parameter by either historical or predictive means by a normal person is represented as a complex variable.

Illustration of Claim 1 –

In case of business uncertainty, predictive decisions among various business analysts differ considerably. A normal person will efficiently judge the current status of the business and try to predict in a concise manner.

In many cases it can be observed that optimistic , pessimistic and fickle minded persons predict the sales and profit status defying the current status and hence the statistical hypothesis as per their predictions are likely to be biased.

In this claim, we propose the correlation of estimation of normal person with a pessimistic individual.
The proposed mathematical equation of neuro-fuzzy based

event (gain) estimation between a pessimistic and a normal individual in case of strategic uncertainty is as follows-

$$Tp + \beta = AWn = [(\sum_{i=1}^{x} AW_{x,i} * AW_{y,i}) / x] \quad \text{................(1)}$$

where Tp = average accuracy estimation of gain by pessimistic individual,

β = unit negative bias value

AWn = effective weight of the associated parameters per prediction by a normal individual

$AW_{x,i}$ = estimate of weight of ith parameter on the basis of sampled historical information ,

$AW_{y,i}$ = estimate of weight of ith parameter on the basis of present hypothesis,

x = total number of instances of the arrival of the event gain

As per our proposal , single incidence of gain takes place and hypothesis of fairness by pessimistic individual is rejected .

Therefore, $AW_{x,1} * AW_{y,1} = 0 + \beta$ or, $(AW_{y,1})^2 = \beta$

or, $(AW_{y,1})^2 = -1$ [since unit negative bias]

or, $AW_{y,1} = (-1)^{1/2}$ ....(2)

Similarly, we can show that $AW_{x,1} = (-1)^{1/2}$ ....(3)

Hence it is justified to state that "If prediction of occurrence of gain by a normal individual is based upon estimate of weight of single parameter and hypothesis of fairness by pessimistic individual is rejected , then for unit negative bias, the estimate of weight of the single parameter by either historical or predictive means by a normal person is represented as a complex variable".

Claim 2 - Accuracy estimate of future prediction of occurrence of an uncertain event (gain or loss) is governed by the principle of hypothesis of fairness rule in case of both optimistic and pessimistic individuals.

Illustration of Claim 2 –

Strategic uncertainties focus on specific unknown parameters that will affect the outcome of strategic decisions. In this claim we have proposed that the principle of hypothesis rule plays a pivotal role in strategic decisions. A statistical hypothesis[6] is an assertion about the distribution of one or more random variables which we want to verify on the basis of a sample.

In this claim we represent mathematically the relation among predictive gain estimates done by normal , optimistic and pessimistic individuals.

$$P (| \beta o - \alpha n | \geq \mu) = P (| \alpha n - \beta p | \geq \mu) = V \quad \text{..............(4)}$$

where βo = predicted value of percentage of gain by optimistic person in higher crisp form

αn = predicted value percentage of gain by normal person being 0.5

βp = predicted value percentage of gain by pessimistic person in higher crisp form

μ = estimate of deviation of both optimistic and pessimistic from actual outcome V ; V € { 0,1 }.

If V = 1, $P (| 1 - 0.5 | \geq (1\text{-}1))$ is valid and it reveals that prediction of optimistic individual is accurate and we reject hypothesis of fairness of pessimistic individual as $P (| 0.5 - 0 | \geq (1\text{-}0))$ is absurd.

Similarly, if V = 0, $P (| 1 - 0.5 | \geq (1\text{-}0))$ is absurd and it reveals that hypothesis of fairness of optimistic individual is rejected.

Hence it is justified to state that "Accuracy estimate of future prediction of occurrence of an uncertain event (gain or loss) is governed by the principle of hypothesis of fairness rule in case of both optimistic and pessimistic individuals".

Claim 3 - In case of sales and profit estimation of strategic market management done by a fickle-minded person , the predicted value (Tv) clearly acts as a reference parameter for identifying the output (To) trends towards both rare and frequent fuzzy domains.

Illustration of Claim 3 –

Fuzzy set theory was proposed in 1965 by Lotfi A. Zadeh. A fuzzy set[7] can be defined mathematically by assigning to each possible individual in the universe of discourse, a value representing its grade of membership in the fuzzy set.

In definite form , crisp value is coined and it is in bivalent or binary variable state {0,1}, while the fuzzy value is in probabilistic form and lower and higher crisps indicate the lower and upper boundary limits of a fuzzy range. The average (0.5) is a threshold that indicates rare range ($0 \leq L_R \leq 0.5$) and frequent range ($0.5 \leq F_R \leq 1$).

The following table illustrates that in case of the sales and profit estimation of strategic market management done by a fickle-minded person, the predicted value (Tv) clearly acts as a reference parameter for identifying the output (To) trends towards both rare fuzzy (L_R) and frequent fuzzy domains (F_R). C_L and C_H represent lower and higher crisp values respectively.

Nature	If (Tv < To)	If (Tv > To)
Optimistic	Tv = { F_R }	(i) Tv = { F_R, C_H

	To = { C_H }	} To = { C_L , L_R , A_{VG} } (ii) Tv = { C_H } To = { C_L , L_R , A_{VG} , F_R }
Pessimistic	(i) Tv = { C_L , F_R } To = { A_{VG} , F_R , C_H } (ii) Tv = { C_L } To = { L_R , A_{VG} , F_R , C_H }	Tv = { L_R } To = { C_L }
Fickle-minded	Tv = { A_{VG} } ***To = { F_R , C_H }***	Tv = { A_{VG} } ***To = { C_L L_R , }***

Table 1 : Gain estimation depending on different human mindset and

Hence it is justified to state that "In case of sales and profit estimation of strategic market management done by a fickle-minded person , the predicted value (Tv) clearly acts as a reference parameter for identifying the output (To) trends towards both rare and frequent fuzzy domains".

Claim 4 - The null hypothesis of validity of an unknown event (gain or loss) for a biased individual is identical to alternate hypothesis of the same for a normal person.

Illustration of Claim 4 –

In this claim we have proposed that in case of strategic uncertainty, the general statistical rules of null and alternated hypothesis can be significantly correlated with the external analysis of strategic market management in the view of a biased (either optimistic or pessimistic) and normal person.
Let p be unknown binary state of validity of an event (gain or loss) in case of biased individual
and q_s is specific fuzzy estimate.

Therefore, $H_0 : p = q_s$ ....(5)

and $H_A : p \neq q_s$..(6)

where H_0 and H_A are null and alternate hypothesis respectively of biased person. In this context biased person indicates optimistic and pessimistic nature of a person.

Now, (1-p) is unknown binary state of validity of an event (gain or loss) in case of normal person and q_n be specific fuzzy estimate.

Since the event is valid, hence $q_s = q_n = 1$.

Hence, $H_0 : (1\text{-}p) = q_n$ ....(7)

and $H_A : (1\text{-}p) \neq q_n$..(8)

where H_0 and H_A are null and alternate hypothesis respectively of normal person.

Biased property reflects false belief which means Eq(5) is invalid. In that case the validity of Eq(6) concludes that $p \neq q_s$ .Now q_s has to be in higher crisp whereby p = 0.

Let us examine whether Eq(7) is valid under this circumstance.

For Eq(7) to be valid , $(1\text{-}p) = q_n = 1$. Now q_n has to be 1 whereby p = 0. It indicates that alternate hypothesis of schizophrenic patient is identical to the null hypothesis of normal person, and vice-versa.

Hence it is justified to state that "The null hypothesis of validity of an unknown event (gain or loss) for a biased individual is identical to alternate hypothesis of the same for a normal person."

III. CONCLUSION

The paper points out the following discovered facts –

1. If prediction of occurrence of gain in a strategic market management by a normal individual is based upon estimation of weight of single associated parameter and hypothesis of fairness by pessimistic individual is rejected , then for unit negative bias, the estimate of weight of the single parameter by either historical or predictive means by a normal person is represented as a complex variable.

2. Accuracy estimate of future prediction of occurrence of an uncertain event (gain or loss) is governed by the principle of hypothesis of fairness rule in case of both optimistic and pessimistic individuals.

3. In case of sales and profit estimation of strategic market management done by a fickle-minded person , the predicted value (Tv) clearly acts as a reference parameter for identifying the output (To) trends towards both rare and frequent fuzzy domains.

4. The null hypothesis of validity of an unknown event (gain or loss) for a biased individual is identical to alternate hypothesis of the same for a normal person.

REFERENCES

[1] Aaker, D. A. (2005). *Strategic Market Management.* John Wiley & Sons ,Inc.

[2] Cobb ,Charles W.,and Paul H.Douglas , "A Theory of Production", *Amer. Econ. Rev., Supplement* ,pp.139-165, 1925

[3] Samuelson , Paul A. , *Economics* , (10th edition),chap.20.,New York : McGraw-Hill Book Company,1976

[4] Levitt, T. (1960) "Marketing Myopia" *Harvard Business Review.* July-August ,pp. 45-56

[5] Olofsson, P. (2005). *Probability, Statistics, and StochasticProcesses.* John Wiley & Sons ,Inc.

[6] Giri, P. K. & Banerjee, J. (1999). *Introduction to Statistics.* Academic Publishers

[7] Zadeh , L.A. (1965), *Fuzzy Sets* , *Information and Control,* Vol.8, pp. 338-353

18

One of the Possible Causes for Diatom Appearance in Ariake Bay Area in Japan In the Winter from 2010 to 2015 (Clarified with AQUA/MODIS)

Kohei Arai
Graduate School of Science and Engineering
Saga University
Saga City, Japan

Abstract—**One of the possible causes for diatom appearance in Ariake bay area I Japan in the winter seasons from 2010 to 2015 is clarified with AQUA/MODIS of remote sensing satellite. Two months (January and February) AQUA/MODIS derived chlorophyll-a concentration are used for analysis of diatom appearance. Match-up data of AQUA/MODIS with the evidence of the diatom appearance is extracted from the MODIS database. Through experiments, it is found that diatom appears after a long period time of relatively small size of red tide appearance. Also, it depends on the weather conditions and tidal effect as well as water current in the bay area in particular.**

Keywords—chlorophyl-a concentration; red tide; diatom; MODIS; satellite remote sensing

I. INTRODUCTION

The Ariake Sea is the largest productive area of Nori (Porphyra yezoensis1) in Japan. In winters in 2012, 2013, 2014 and 2015, a massive diatom bloom appeared in the Ariake Bay, Japan [1]. In case of above red tides, bloom causative was Eucampia zodiacus2. This bloom has being occurred several coastal areas in Japan and is well reported by Nishikawa et al. for Harimanada sea areas [2]-[10]. Diatom blooms have recurrently appeared from late autumn to early spring in the coastal waters of western Japan, such as the Ariake Bay [11] and the Seto Inland Sea [12], where large scale "Nori" aquaculture occurs. Diatom blooms have caused the exhaustion of nutrients in the water column during the "Nori" harvest season. The resultant lack of nutrients has suppressed the growth of "Nori" and lowered the quality of "Nori" products due to bleaching with the damage of the order of billions of yen [3].

In particular in winter since 2012, almost every year, relatively large size of diatoms of *Eucampia zodiacus* appears in Ariake Bay areas. That is one of the causes for damage of *Porphyra yezoensis*. There is, therefore, a strong demand to prevent the damage from Nori farmers. Since 2007, *Asteroplanus karianus* appears in the Ariake Bay almost every year. In addition, *Eucampia zodiacus* appears in Ariake Bay since 2012. Meanwhile, *Eucampia zodiacus* did not appeared in 2011, 2010. Therefore, there is a key for the diatom appearance. By comparing Ariake Bay situations in winter seasons in 2010 and 2011 and after 2012, it might be possible to find out possible causes for diatom appearance.

The chlorophyll-a concentration algorithm developed for MODIS[3] has been validated [13]. The algorithm is applied to MODIS data for a trend analysis of chlorophyll-a distribution in the Ariake Bay in the winter from 2010 to 2015 is made [14]. Also, locality of red tide appearance in Ariake Sea including Ariake Bay, Isahaya Bay and Kumamoto offshore is clarified by using MODIS data derived chlorophyll-a concentration [15]. On the other hand, red tide appearance (location, red tide species, the number of cells in unit water volume by using microscopy) are measured from the research vessel of the Saga Prefectural Fishery Promotion Center: SPFPC by once a 10 days. The location and size of the red tide appearance together with the red tide source would be clarified by using SPFPC data. Match-up data of MODIS derived chlorophyll-a concentration is used for investigation of relations between MODIS data and truth data of the red tide appearance. Through time series data analysis of MODIS derived chlorophyll-a concentration, one of the possible causes of diatom appearance is clarified with the evidence of research Bessel observations.

In the next section, the method and procedure of the experimental study is described followed by experimental data and estimated results. Then conclusion is described with some discussions.

II. EXPERIMENTAL METHOD AND RESULTS

A. *Intensive Study Areas*

Fig.1 shows the intensive study areas of Ariake Bay, Kyushu, Japan. Ariake Bay is a portion of Ariake Sea of which the width is around 20km (in direction of east to west) and the length is approximately 100km (in direction of north to south). It is almost closed sea area because the mouth of Ariake Sea is quite narrow. Sea water exchanges are, therefore, very small.

1 http://en.wikipedia.org/wiki/Porphyra

2 http://www.eos.ubc.ca/research/phytoplankton/diatoms/centric/eucampia/e_zodiacus.html

3 http://modis.gsfc.nasa.gov/

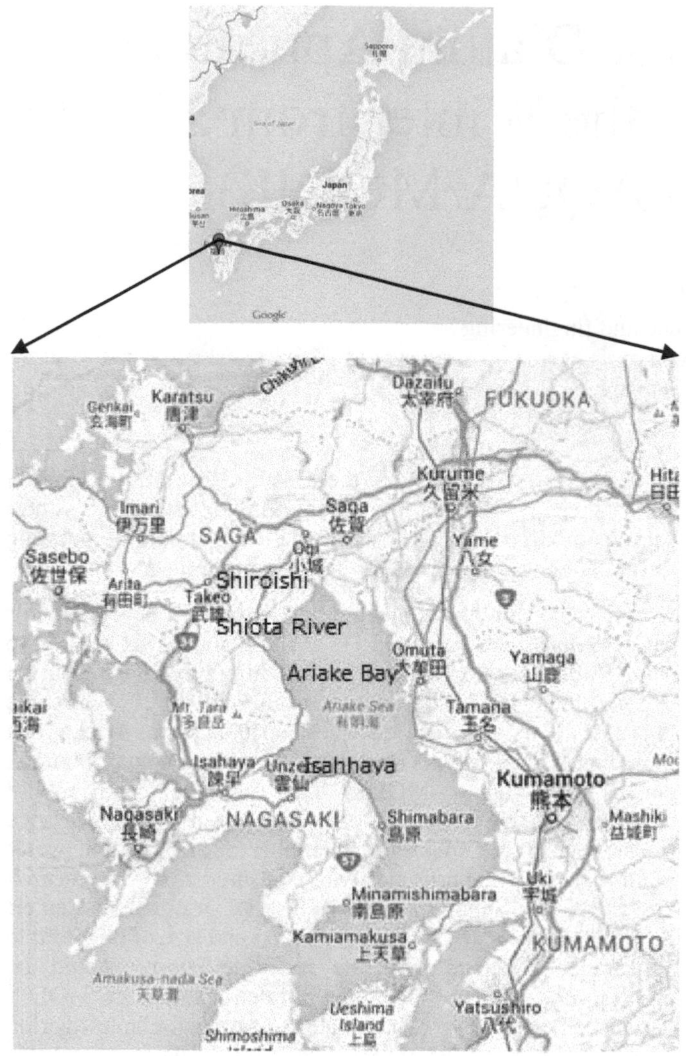

Fig. 1. Intensive study areas

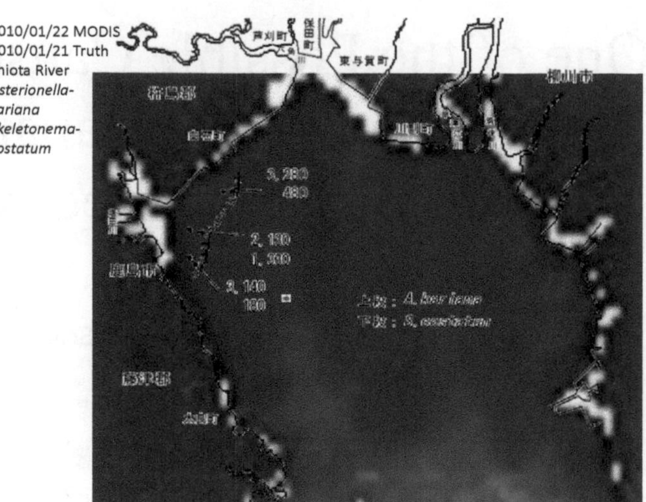

Fig. 2. Example of the superimposed image with MODIS data derived chlorophyll-a concentration and truth data which is provided by Saga Prefectural Fishery Promotion Center

B. MODIS Data Derived Chlorophyll-a Concentration and Truth Data and Truth Data of Red Tide in 2010 to 2015

MODIS derived chlorophyll-a concentration which area acquired for the observation period of two months (in January and February) in 2010 to 2015 is used for the experiments. On the other hand, Fig.2 shows the example of the superimposed image with MODIS data derived chlorophyll-a concentration and truth data which is provided by Saga Prefectural Fishery Promotion Center on 21 January 2010. The number in the figure denotes the number of red tide cells / ml. Such the number of red tide is reported every 10 days.

It is found the following red tide at around the Shiota river mouth and its surrounding areas on January 21 2010,

Asterionella kariana; 3280 cells/ml

Skeletonema costatum: 1330 cells/ml

On January11 2011, it is found the following red tide along with the Shiroishi town offshore to the Shiota river mouth and its surrounding areas,

Asterionella kariana; 10150 cells/ml

Although the truth data say that the red tide is distributed at around Shiota river mouth and its surrounding areas as well as Shiroishi offshore, it cannot be seen due to the fact that it is covered with cloud in the MODIS data derived chlorophyll-a concentration. It is found the following red tide at around the Kashima offshore on February 25 2011,

Asterionella kariana; 4950 cells/ml

It is found the following red tide at around the Shiota River Mouth and its surrounding areas on December 30 2011,

Asterionella kariana; 5150 cells/ml

On January 23 2012, it is found the following red tide at the Shiroishi offshore,

Skeletonema spp .: 5150 cells/ml

The red tide is distributed at around Shiota river mouth and its surrounding areas as well as Shiroishi offshore.

The following red tide is found widely along with the Kawazoe offshore to the Tara offshore on February 22 2012,

Eucampia zodiacus: 1,090 cells/ml

Also it is found the following red tide along with the Shiota river mouth and its surrounding areas to the Kashima offshore on December 31 2012,

Skeletonema spp .: 6110 cells/ml

On January 7 2013, the following red tide are observed along with the Shiota river mouth and its surrounding areas to the Shiroishi offshore,

Asterionella kariana; 5630 cells/ml

Skeletonema costatum: 3390 cells/ml

The red tide distribution derived from MODIS data is almost coincident to the truth data.

It is found the following red tide at the Shiroishi offshore on January 6 2014,

Asterionella kariana; 4830 cells/ml

The following red tide is observed at the Shiroishi offshore on January 16 2014,

Skeletonema spp .: 6110 cells/ml

Thalassiosira spp.: 1510 cells/ml

On February 6 2014, the following red tide is observed almost whole Ariake bay area except the Shiroishi offshore,

Eucampia zodiacus: 568 cells/ml

It is observed the following red tide along with the Shiroishi offshore to the Tara offshore on December 30 2014,

Asterionella kariana; 3890 cells/ml

Skeletonema costatum: 8750 cells/ml

On March 6 2015, the following red tide is observed along with the Kashima offshore to the Tara offshore,

Eucampia zodiacus: 1310 cells/ml

It is clear that the diatom of *Eucampia zodiac* appeared in the winter in 2012, 2013, 2014 and 2015. The differences between the situations in the time period of 2010, 2011 and the other time period from 2012 to 2015 are (1) relatively small size of red tides, *Asterionella kariana* and *Skeletonema costatum* appeared at around Shiota river mouth and its surrounding areas for a long time period, (2).

C. Chlorophyll-a Concentration Trends in the Different Areas in theWInter of the Different Year

Chlorophyll-a concentration trends in the different areas, Isahaya Bay, Around the Shiota river mouth and its surrounding areas, and the middle of the Ariake Bay are investigated with the MODIS derived chlorophyll-a concentrations acquired on

(2010) January 1, 3, 9, 14, 16, 17, 18, 22, 24, 26, 27, 29, February 3, 4, 5, 6, 20, 21, 23, and 28 in 2010

(2011) January 1, 2, 7, 8, 14, 17, 22, 26, 27, February 1, 3, 4, 15, 21, 22, 24, and 26 in 2011

(2012) January 2, 6, 7, 12, 17, 20, 21, 23, 26, 29, 30, 31, February 4, 11, 12, 20, 24, and 29 in 2012

(2013) January 4, 6, 10, 11, 12, 15, 18, 25, 28, 30, 31, February 2, 3, 10, 13, 16, 20, 22, 23, 24, and 29 in 2013

(2014) January 10, 13, 15, 16, 19, 23, 24, 26, 27, 29, 30, February 4, 8, 11, 12, 20, 21, 23, and 24 in 2014

(2015) January 4, 6, 7, 8, 9, 10, 12, 17, 18, 20, 23, February 1, 3, 6, 9, 13, 14, 20, and 27 in 2015

The results from the trend analysis are shown in Fig.3.

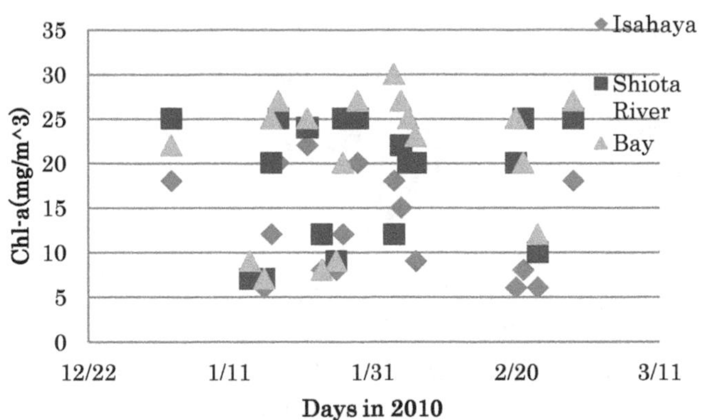

(a)2010

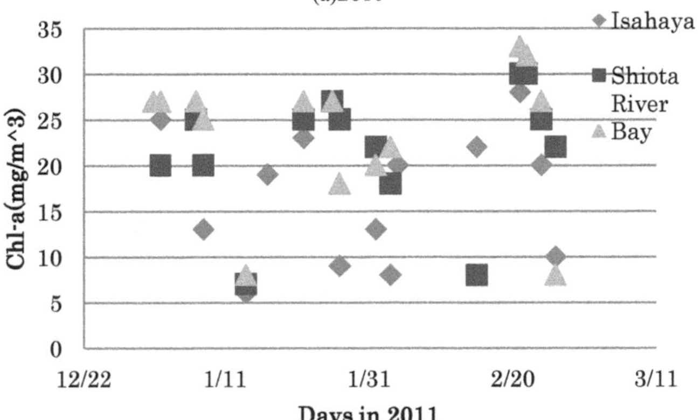

(b)2011

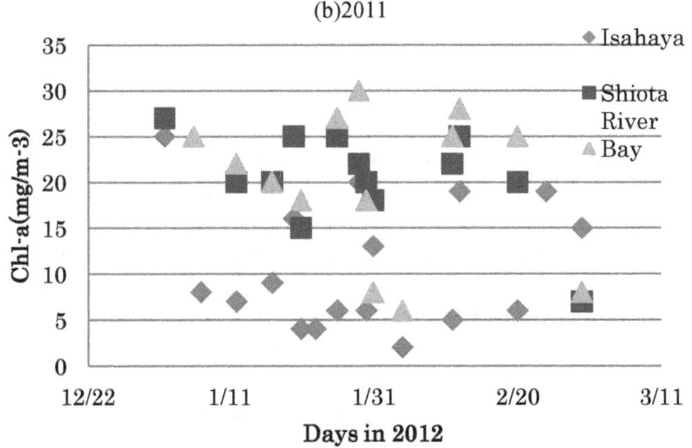

(c)2012

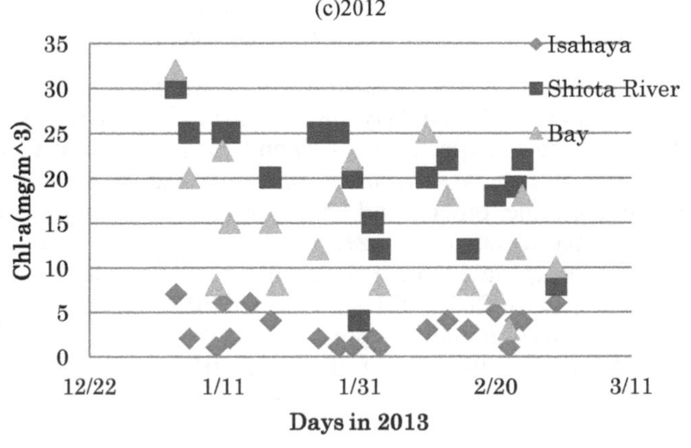

(d)2013

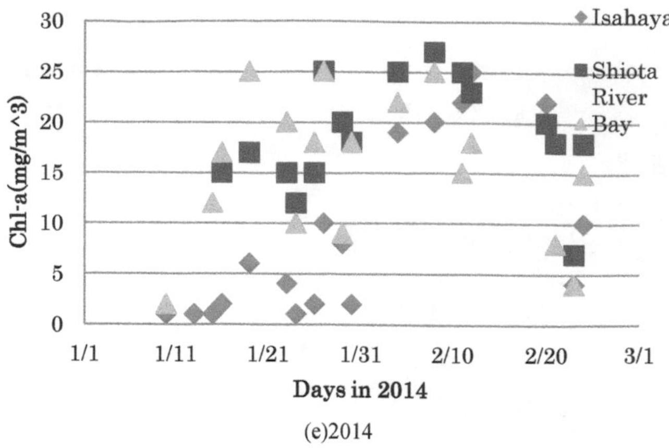

(e)2014

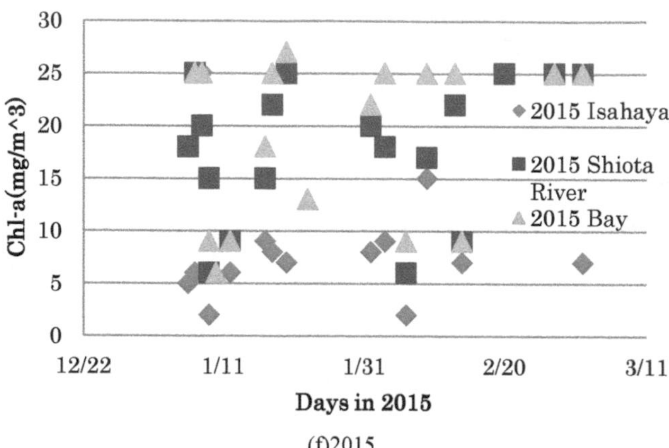

(f)2015

Fig. 3. Trends of chlorophyll-a concentration derived from MODIS data in the winter seasons in 2010-2015

The trends of chlorophyll-a concentrations of Shiota river mouth and its surrounding areas and the middle of Ariake Bay area are quite similar while that of Isahaya Bay area is not resemble to those of the trends of Shiota river mouth and its surrounding area and Ariake Bay area. Therefore, the origin of chlorophyll-a of Isahaya Bay area is different from those of Ariake Bay area as well as Shiota river mouth and its surrounding areas. Not only time series of trend of chlorophyll-a concentrations, but also spatial characteristics of chlorophyll-a concentration distributions between Isahaya Bay and the other Ariake Bay area as well as Shiota river mouth and its surrounding areas are different each other.

Relatively large sized diatom of *Eucampia zodiacus*: appeared in entire Ariake Bay areas in 2012, 2013, 2014, and 2015 and it did not appeare in 2010 and 2011. Comparatively small sized diatom of *Asterionella kariana* and *Skeletonema costatum*, on the other hand, appeared almost every year though. In particular, large sized diatom appeared after small sized diatom appeared for a long time period in the winter seasons in year of 2012, 2013, 2014, and 2015. On the other hand, small sized diatom disappeared in a short time period in the winter seasons in years of 2010 and 2011 results in large sized diatom did not appeared.

D. Relations Between Chlorophyll-a Concentration and the Meteorological Data

It may say that red tide appears when the following conditions are situated, nutrition rich water and rich solar illumination and less wind. Therefore, the relations between chlorophyll-a concentration and rainfall in a day, averaged air-temperature in a day, and the averaged wind speed in a day are investigated together with MODIS data derived 8 day composite of the Photosynthetically Available Radiance: PAR (Einstein/m^2/day). The meteorological data are gotten from the Japanese Meteorological Agency: JMA of Shiroishi Automated Meteorological Data Acquisition System: AMEDAS data[4] which meteorological station is situated at the longitude and latitude of 33:11.0' North and 130:8.9' East. The elevation of the station is 2 m. The results from these analyses are shown in Fig.4. Meanwhile, MODIS data derived PAR is obtained from the NASA/GSFC site [5]. Solar illumination condition can be replaced to PAR.

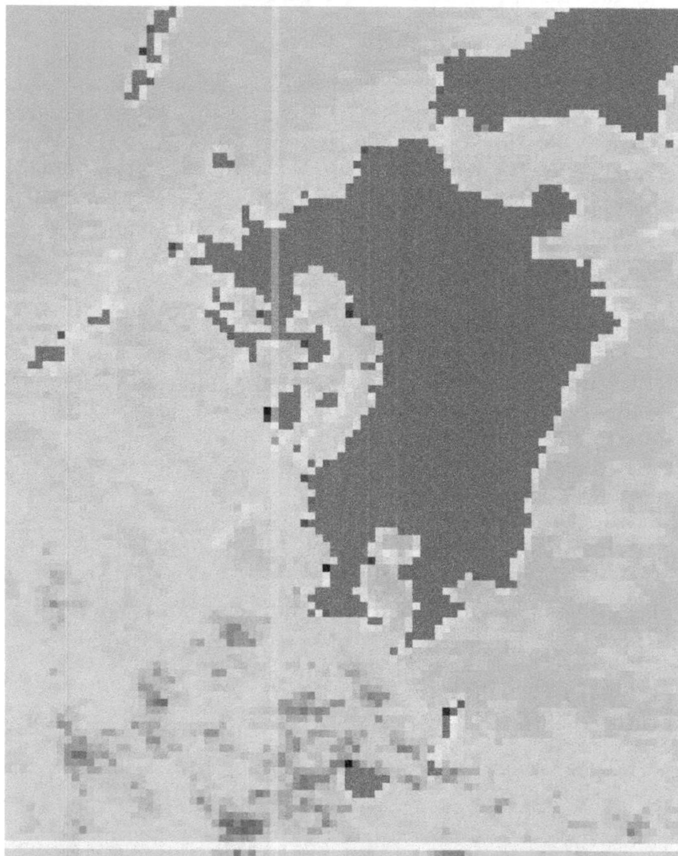

Photosynthetically Available Radiation (Einstein / m^2 / day)

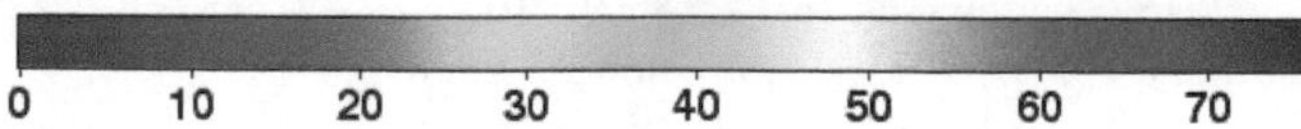

Fig. 5. Example of the 8 day composite of the PAR derived from MODIS data for the period from March 6 to March 14 in 2011

[4] http://www.data.jma.go.jp/obd/stats/etrn/index.php

[5] http://oceancolor.gsfc.nasa.gov/cgi/l3?per=8D&prd=PAR_par&sen=T&res=4km&num=24&ctg=Standard&date=1Jan2014

Eucampia zodiacus may grow a wide variety of water temperature and salinity conditions, the most preferable water temperature is around 25 °C tough. In order to grow, *Eucampia zodiacus* requires relatively strong solar illumination. It can maintain capability of consume nitrogen even if the water temperature is not so high. Therefore, the conditions of water temperature, solar illumination are key issues for *Eucampia zodiacus*

One the other hand, rainfall provide nutrient to the Bays from the rivers. Therefore, *Eucampia zodiacus* did not appear in 2010 and 2011 because of less of the rainfall while *Eucampia zodiacus* appeared in 2012 and after because of rich of the rainfall.

Meanwhile, PAR in the first half of January in both 2010 and 2011 is relatively small in comparison to the PAR in that period in 2012 and after. Therefore, diatom did not appear in both 2010 and 2011.

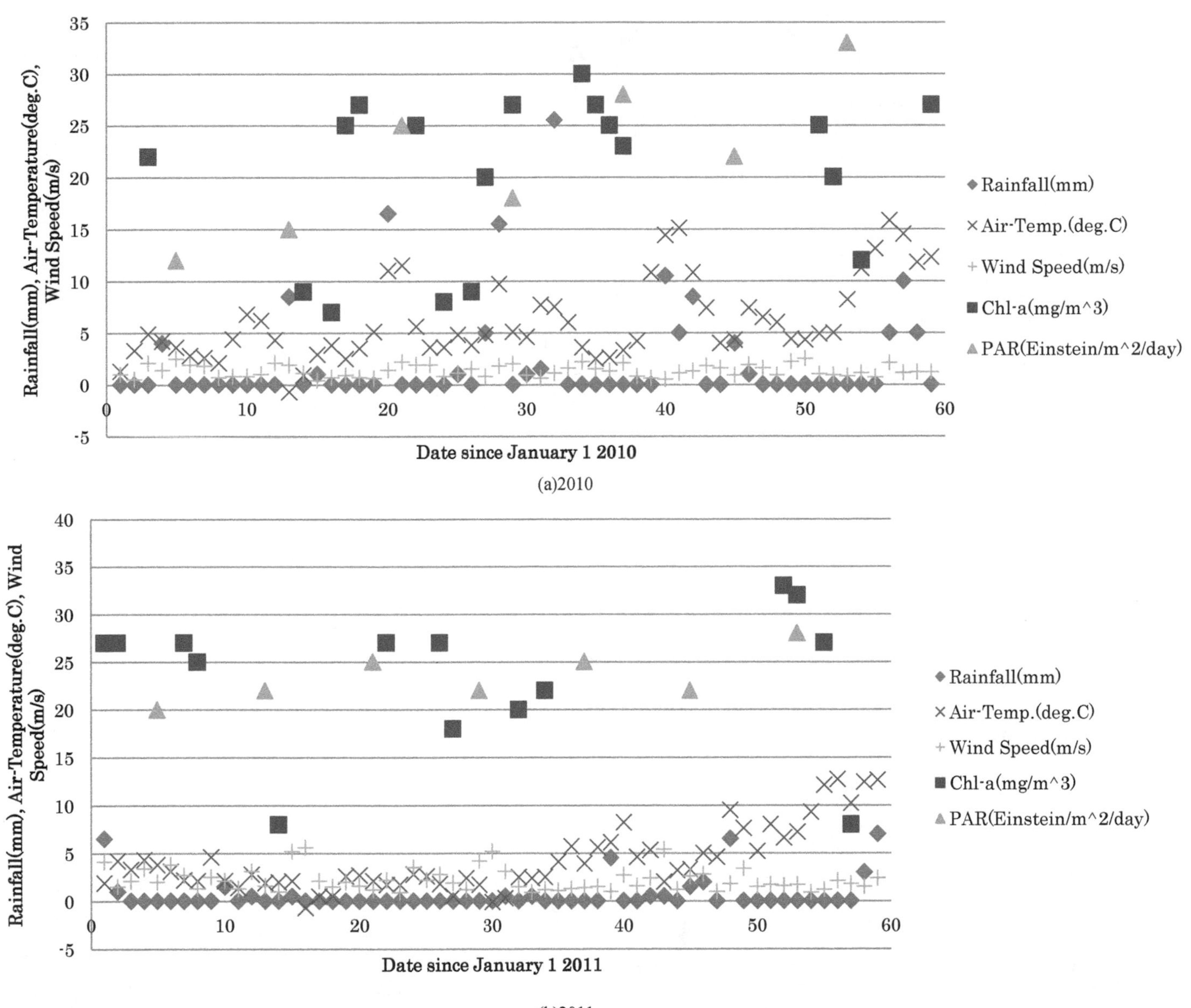

(a)2010

(b)2011

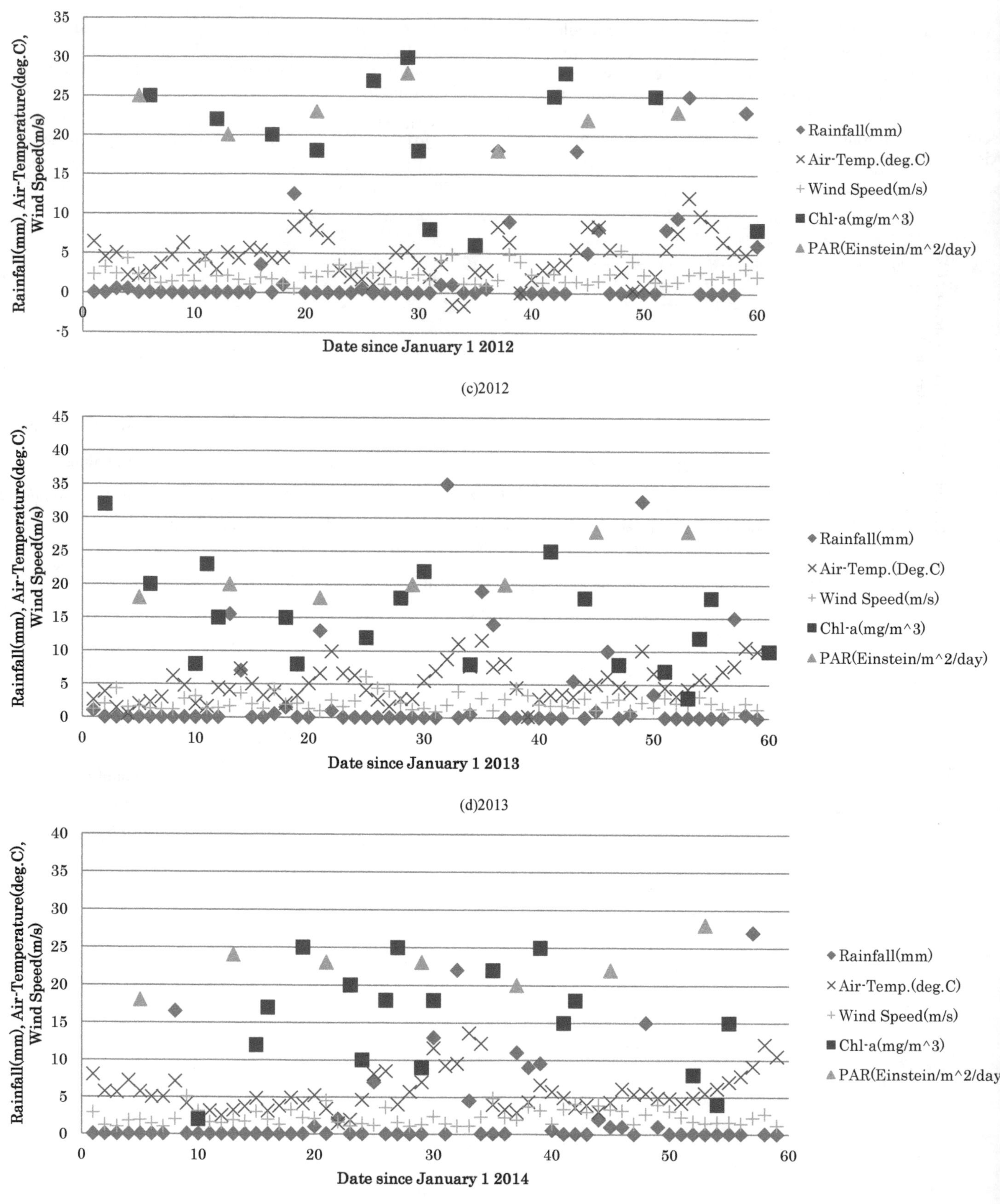

(c)2012

(d)2013

(e)2014

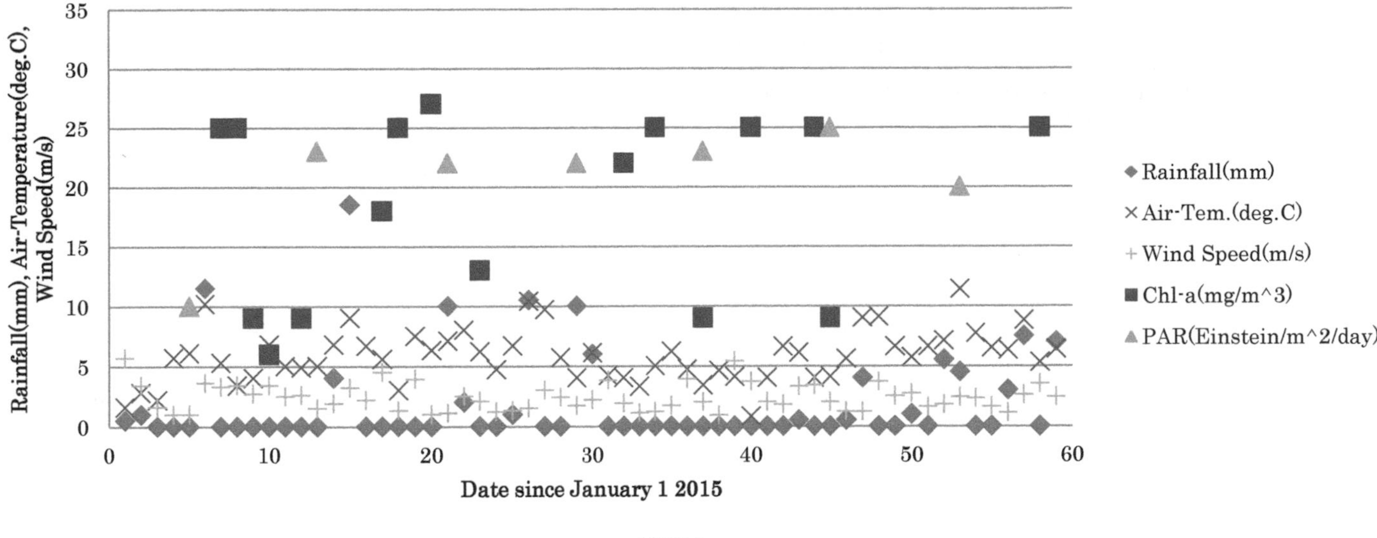

(f)2015

Fig. 4. Trends of meteorological data and chlorophyll-a concentration in the winter seasons of 2010 to 2015

III. Conclusion

Asterionella kariana and *Skeletonema costatum* are used to be appeared in the Ariake Bay area in the winter seasons followed by *Eucampia zodiacus* appearance in the early spring almost every year after 2012 in particular, on February 22 2012, February 26 2013, February 6 2014 and March 6 2015.

Through the trend analysis with the superimposed images of the truth data and the MODIS data derived chlorophyll-a concentration which are acquired in the period starting from February 27 to March 5 2015, it is found that chlorophyll-a is distributed densely in the Ariake bay area and Isahaya bay area on February 27. Then the densely distributed chlorophyll-a is flown to the south direction along with the sea water current in the Ariake bay while the densely distributed chlorophyll-a is flown from the Isahaya bay to the Tara-machi and far beyond the Shimabara offshore. Therefore, it may say that the sources of red tide are different between Ariake bay and Isahaya bay.

Further investigations are required to clarify the mechanism of red tide appearance with the consideration three dimensional of cross section analysis the red tide source movement.

Acknowledgment

The authors would like to thank Dr. Toshiya Katano of Tokyo University of Marine Science and Technology, Dr. Yuichi Hayami, Dr. Kei Kimura, Dr. Kenji Yoshino, Dr. Naoki Fujii and Dr. Takaharu Hamada of Institute of Lowland and Marine Research, Saga University for their great supports through the experiments.

References

[1] Yuji Ito, Toshiya Katano, Naoki Fujii, Masumi Koriyama, Kenji Yoshino, and Yuichi Hayami, Decreases in turbidity neap tides initiate late winter large diatom blooms in a macrotidal embayment, Journal of Oceanography,69: 467-479. 2013.

[2] Nishikawa T (2002) Effects of temperature, salinity and irradiance on the growth of the diatom *Eucampia zodiacus* caused bleaching seaweed *Porphyra* isolated from Harima-Nada, Seto Inland Sea, Japan. Nippon Suisan Gakk 68: 356–361. (in Japanese with English abstract)

[3] Nishikawa T (2007) Occurrence of diatom blooms and damage tocultured *Porphyra* thalli by bleaching. Aquabiology 172: 405–410. (in Japanese with English abstract)

[4] Nishikawa T, Hori Y (2004) Effects of nitrogen, phosphorus and silicon on the growth of the diatom *Eucampia zodiacus* caused bleaching of seaweed *Porphyra* isolated from Harima-Nada, Seto Inland Sea, Japan. Nippon Suisan Gakk 70: 31–38. (in Japanese with English abstract)

[5] Nishikawa T, Hori Y, Nagai S, Miyahara K, Nakamura Y, Harada K, Tanda M, Manabe T, Tada K (2010) Nutrient and phytoplankton dynamics in Harima-Nada, eastern Seto Inland Sea, Japan a 35-year period from 1973 to 2007. Estuaries Coasts 33: 417–427.

[6] Nishikawa T, Hori Y, Tanida K, Imai I (2007) Population dynamics of the harmful diatom *Eucampia zodiacus* Ehrenberg causing bleachings of *Porphyra* thalli in aquaculture in Harima- Nada, the Seto Inland Sea, Japan. Harmful algae 6: 763–773.

[7] Nishikawa T, Miyahara K, Nagai S (2000) Effects of temperature and salinity on the growth of the giant diatom *Coscinodiscus wailesii* isolated from Harima-Nada, Seto Inland Sea, Japan. Nippon Suisan Gakk 66: 993–998. (in Japanese with English abstract)

[8] Nishikawa T, Tarutani K, Yamamoto T (2009) Nitrate and phosphate uptake kinetics of the harmful diatom *Eucampia zodiacus* Ehrenberg, a causative organism in the bleaching of aquacultured *Porphyra* thalii. Harmful algae 8: 513–517.

[9] Nishikawa T, Yamaguchi M (2006) Effect of temperature on lightlimited growth of the harmful diatom *Eucampia zodiacus* Ehrenberg, a causative organism in the discoloration of *Porphyra* thalli. Harmful Algae 5: 141–147.

[10] Nishikawa T, Yamaguchi M (2008) Effect of temperature on lightlimited growth of the harmful diatom *Coscinodiscus wailesii*, a causative organism in the bleaching of aquacultured *Porphyra* thalli. Harmful Algae 7: 561–566.

[11] Syutou T, Matsubara T, Kuno K (2009) Nutrient state and nori aquaculture in Ariake Bay. Aquabiology 181: 168–170. (in Japanese with English abstract)

[12] Harada K, Hori Y, Nishikawa T, Fujiwara T (2009) Relationship between cultured *Porphyra* and nutrients in Harima-Nada, eastern part of the Seto Inland Sea. Aquabiology 181: 146–149. (in Japanese with English abstract)

[13] Arai K., T. Katano, Trend analysis of relatively large diatoms which appear in the intensive study area of the ARIAKE Sea, Japan, in winter (2011-2015) based on remote sensing satellite data, Internationa Journal

of Advanced Research in Artificial Intelligence (IJARAI), 4, 7, 15-20, 2015.

[14] Arai, K., Locality of Chlorophyll-a Concentration in the Intensive Study Area of the Ariake Sea, Japan in Winter Seasons Based on Remote Sensing Satellite Data, Internationa Journal of Advanced Research in Artificial Intelligence (IJARAI), 4, 8, 18-25, 2015.

Trend Analysis of Relatively Large Diatoms Which Appear in the Intensive Study Area of the Ariake Sea, Japan in Winter (2011-2015) based on Remote Sensing Satellite Data

Kohei Arai 1
1Graduate School of Science and Engineering
Saga University, Saga City, Japan

Toshiya Katano 2
2Tokyo University of Marine Science and Technology
Tokyo Japan

***Abstract*—Behavior of relatively large size of diatoms which appear in the Ariake Sea areas, Japan in winter based on remote sensing satellite data is clarified. Through experiments with Terra and AQUA MODIS data derived chlorophyll-a concentration and truth data of chlorophyll-a concentration together with meteorological data and tidal data which are acquired for 5 years (winter 2011 to winter 2015), it is found that strong correlation between the chlorophyll-a concentration and tidal height changes. Also it is found that the relations between ocean wind speed and chlorophyll-a concentration. Meanwhile, there is a relatively high correlation between sunshine duration a day and chlorophyll-a concentration.**

***Keywords*—chlorophyl-a concentration; red tide; diatom; sunshine duration; ocean winds; tidal effect**

I. INTRODUCTION

The Ariake Sea is the largest productive area of Nori (Porphyra yezoensis1) in Japan. In winters of 2012 and 2013, a massive diatom bloom occurred in the Ariake Sea, Japan [1]. In case of above red tides, bloom causative was Eucampia zodiacus2. This bloom has being occurred several coastal areas in Japan and is well reported by Nishikawa et al. for Harimanada sea areas [2]-[10]. Diatom blooms have recurrently occurred from late autumn to early spring in the coastal waters of western Japan, such as the Ariake Sea [11] and the Seto Inland Sea [12], where large scale "Nori" aquaculture occurs. Diatom blooms have caused the exhaustion of nutrients in the water column during the "Nori" harvest season. The resultant lack of nutrients has suppressed the growth of "Nori" and lowered the quality of "Nori" products due to bleaching with the damage of the order of billions of yen [3].

This bloom had been firstly developed at the eastern part of the Ariake Sea. However, as the field observation is time-consuming, information on the developing process of the red tide, and horizontal distribution of the red tide has not yet been clarified in detail. To clarify the horizontal distribution of red tide, and its temporal change, remote sensing using satellite data is quite useful.

In particular in winter, almost every year, relatively large size of diatoms of *Eucampia zodiacus* appears in Ariake Sea areas. That is one of the causes for damage of *Porphyra yezoensis*. There is, therefore, a strong demand to prevent the damage from Nori farmers. Since 2007, *Asteroplanus karianus* appears in the Ariake Sea almost every year. In addition, *Eucampia zodiacus* appears in Ariake Sea since 2012. There is a strong demand on estimation of relatively large size of diatoms appearance, size and appearance mechanism).

In this paper, the chlorophyll-a concentration algorithm developed for MODIS[3] is firstly validated. Then apply the algorithm to MODIS data which are acquired at the Ariake Sea areas, Japan specifically. Then a trend analysis of chlorophyll-a concentration in winter in 2011 to 2015 is made. The major influencing factor of *Eucampia zodiacus* appearance is chlorophyll-a concentration. The other environmental factors, such as sea water temperature, northern wind for convection of sea water have to be considered. Also, the relations between tidal effects and chlorophyll-a concentration as well as between ocean wind speed and chlorophyll-a concentration together with between sunshine duration a day and chlorophyll-a concentration.

In the next section, the method and procedure of the experimental study is described followed by experimental data and estimated results. Then conclusion is described with some discussions.

II. METHOD AND PROCEDURE

A. The Procedure

The procedure of the experimental study is as follows,

1) Gather the truth data of chlorophyll-a concentration measured at the observation towers in the Ariake Sea areas together with the corresponding areas of MODIS derived chlorophyll-a concentration,

2) Gather the meteorological data which includes sunshine duration a day, ocean wind speed and direction, tidal heights,

---

[1] http://en.wikipedia.org/wiki/Porphyra

[2] http://www.eos.ubc.ca/research/phytoplankton/diatoms/centric/eucampia/e_zodiacus.html

[3] http://modis.gsfc.nasa.gov/

3) Correlation analysis between the truth data and MODIS derived chlorophyll-a concentration as well as between geophysical parameters, ocean wind speed, sunshine duration a day, tidal heights and chlorophyll-a concentration is made.

B. The Intensive Study Areas

Fig.1 shows the intensive study areas in the Ariake Sea area, Kyushu, Japan.

There are three observation tower points, TW, S, and A. TW is closely situated to the Saga Ariake Airport and is situated near the river mouth. On the other hand, A is situated most closely to the coastal area while S is situated in the middle point of the Ariake Sea width and is situated most far from the coastal areas and river mouths.

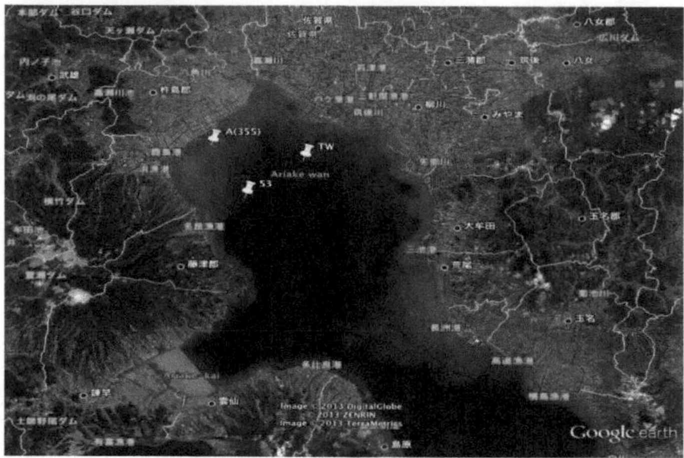

Fig. 1. Intensive study areas (Yellow pins shows the areas)

III. EXPERIMENTS

A. The Data Used

The truth data of chlorophyll-a concentration measured at the observation towers in the intensive study areas in the Ariake Sea areas together with the corresponding areas of MODIS derived chlorophyll-a concentration which area acquired for the observation period of one month during from January 1 to February 1 in 2011 to 2015 are used for the experiments. Also, the meteorological data which includes sunshine duration, ocean wind speed and direction, tidal heights which are acquired for the same time periods as MODIS acquisitions mentioned above. In particular for 2015, two months data are used for trend analysis.

Fig.2 shows an example of the chlorophyll-a concentration image which is derived from MODIS data which is acquired on 2 March 2015. The chlorophyll-a concentration is measured at the tower, TW. This is red tide (Phytoplankton) blooming period. Such this MODIS derived chlorophyll-a concentration data are available almost every day except cloudy and rainy conditions.

Blooming is used to be occurred when the seawater becomes nutrient rich water, calm ocean winds, long sunshine duration after convection of seawater (vertical seawater current from the bottom to sea surface).

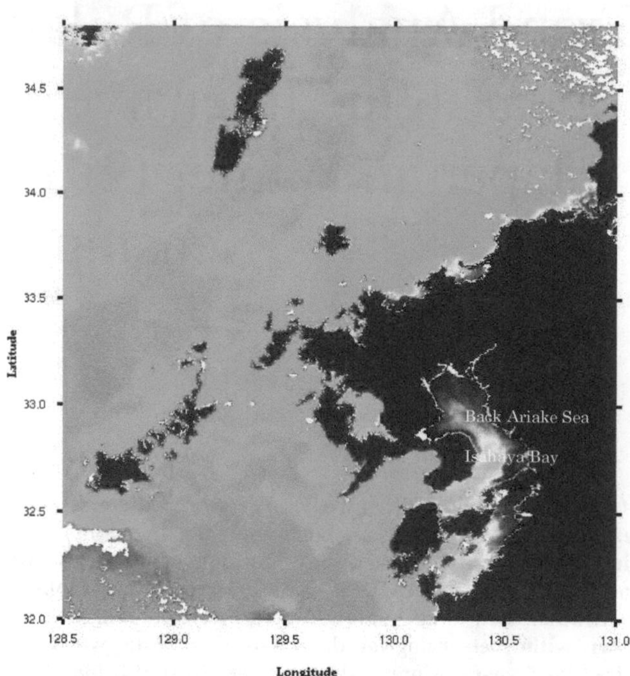

Fig. 2. Example of the chlorophyll-a concentration image which is derived from MODIS data which is acquired on 2 March 2015

Therefore, there must are relations between the geophysical parameters, ocean wind speed, sunshine duration, tidal heights and chlorophyll-a concentration. As shown in Fig.2, it is clear that the diatom appeared at the back in the Ariake Sea and is not flown from somewhere else. Also, there is relatively low chlorophyll-a concentration sea areas between Isahaya bay area and the back in the Ariake Sea area. Therefore, chlorophyll-a concentration variations are isolated each other (Isahaya bay area and the back in the Ariake Sea area).

B. The Relation Between Truth Data and MODIS Derived Colorophyl-a Concentrations (Validation of the Algorithm for Chlorophyl-a Concentration Estimation)

In order to validate the chlorophyll-a concentration estimation algorithm, the relation between truth data of Shipment data as well as Tower data and MODIS derived chlorophyll-a concentration is investigated. Before that, Tower data of chlorophyll-a concentration is compared to Shipment data. Fig.3 shows the relation between these for intensive study area of TW. Also, Fig.4 shows relation of chlorophyll-a concentration between tower data and the other two of shipment data as well as MODIS data derived chlorophyll-a concentration. The time for data collection by ship is different from MODIS data acquisition time and tower data acquisition. Spatial resolutions of MODIS data derived Chlorophyll-a Concentration is 500 m^2 while that of the shipment data and the tower data is just one point of data. Tower data is acquired every one hour. In the validation, averaged chlorophyll-a concentration a day is used because the shipment data acquisition time is varied and also MODIS data acquisition time is different by day by day. Therefore, relation between truth data and MODIS derived Chlorophyll-a Concentration is so scattered.

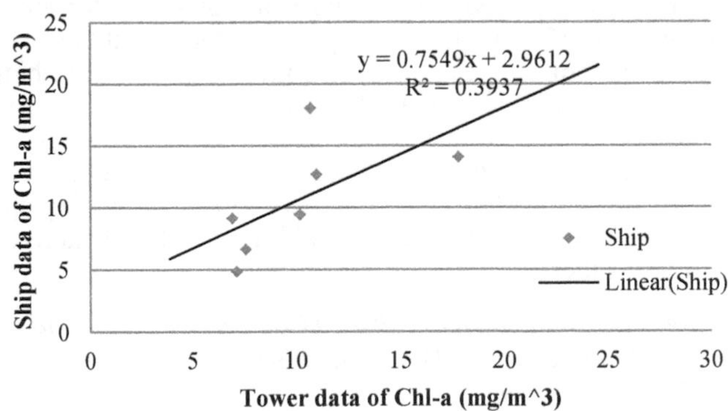

Fig. 3. Relation of the measured chlorophyll-a concentration between Tower data and Shipment data

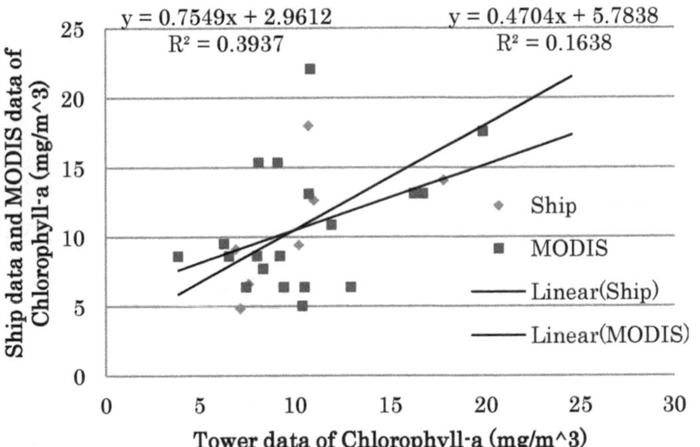

Fig. 4. Validation of Chlorophyll-a Concentration estimation algorithm

C. *Trend Analysis*

Fig.5 shows the trends of shipment data, tower data and MODIS derived chlorophyll-a concentrations measured in January to February in 2015. It seems that trends of MODIS, Tower, and Ship data derived chlorophyll-a concentration are similar. Also, it is found that high chlorophyll-a concentration is occurred on Spring tide while low chlorophyll-a concentration appears on neap time frames. Namely, the dates of neap of this time period are January 15, January 29, February 14 and February 28.

Meanwhile those of spring tide are January 7, January 23, February 7, February 21, and March 7. Chlorophyll-a concentration get up and down repeatedly on spring tides and neaps.

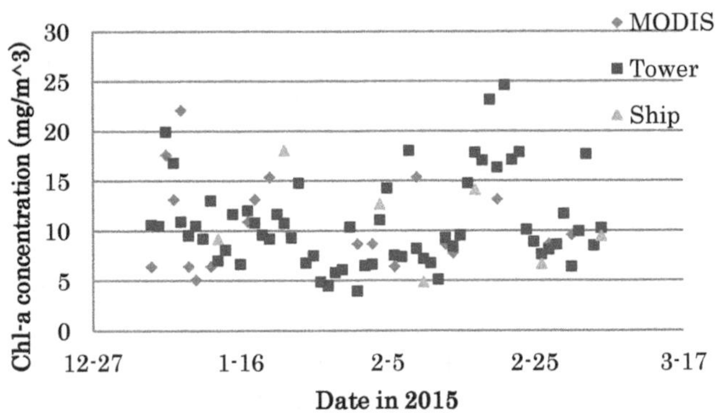

Fig. 5. Chlorophyll-a trend in January and February in 2015

On the other hand, Fig.6 (a) to (d) shows the results from trend analysis of chlorophyll-a concentrations in 2014, 2013, 2012, and 2011, respectively.

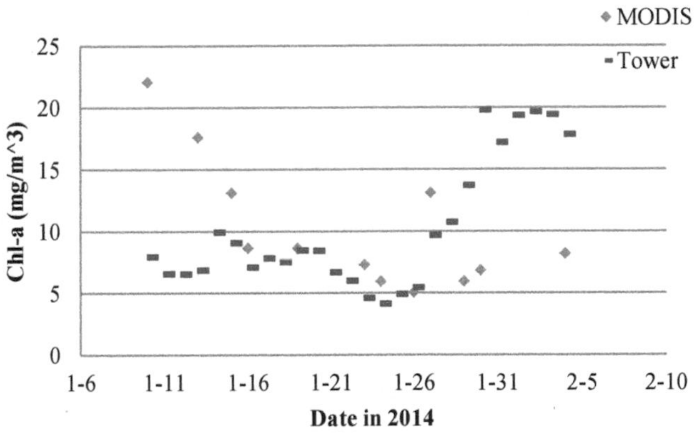

(a)January 2014

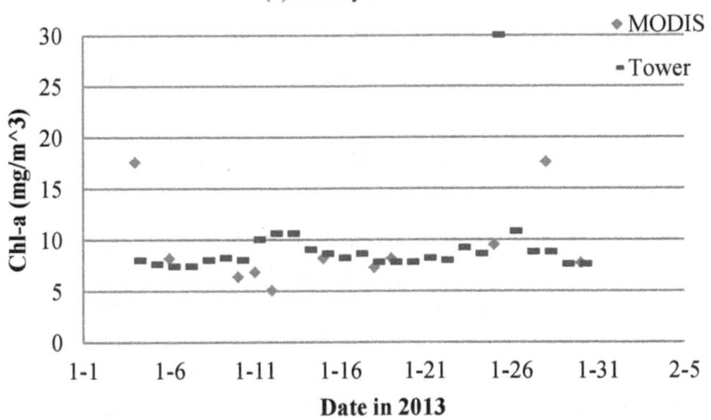

(b)January 2013

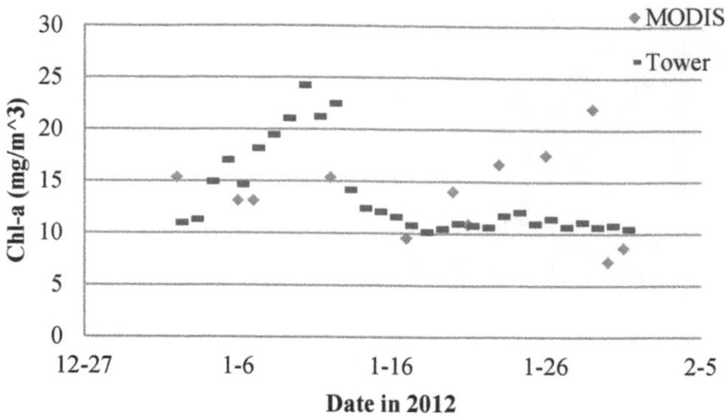

(c)January 2012

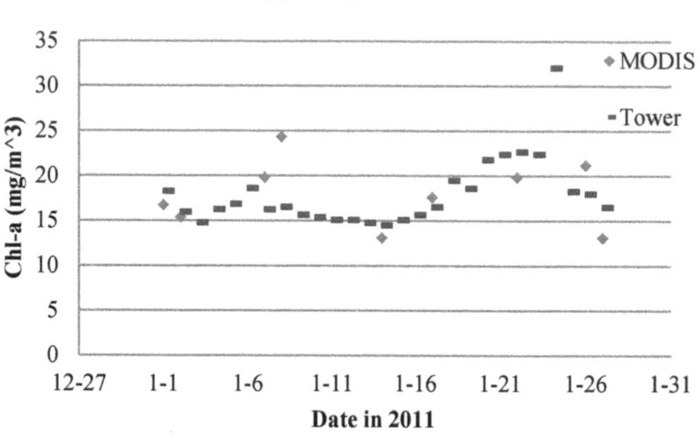

(d)January 2011

Fig. 6. Chlorophyll-a concentration trends in 2011 to 2014

Spring tides occurred on January 3, January 19 and February 2 while neaps appeared on January 10, and January 26 in 2014. Namely, chlorophyll-a concentration raised from January 26 and reached at the maximum on February 2.

It, however, is not always true. There is no peak at the spring tide on January 19 in the tower data derived chlorophyll-a concentration.

In 2013, the spring tides occurred on January 14 and January 30 while neaps appeared on January 7 and January 22, respectively. Tower data derived chlorophyll-a concentration shows the trend of the up and down of the concentration on sprig tides and neaps slightly. Meanwhile, the spring tides occurred on January 11, and January 25 while the neaps appeared on January 3, January 19 and February 2 in 2012, respectively. Although the tower data derived chlorophyll-a concentration shows the peak on January 11, there is no such peak on January 25 while the MODIS data derived chlorophyll-a concentration shows two peaks on January 11 and January 25. On the other hand, the spring tides occurred on January 6 and January 22 while the neaps appeared on January 14 and January 22, respectively. For both of the spring tide periods, both of MODIS data and tower data derived chlorophyll-a concentrations show the peaks and also show the valleys on the neaps. Therefore, it may conclude that there is strong relation between tide and chlorophyll-a concentration, it is not always true though.

In the neap period, vertical direction of sea water mixing due to tidal effect is not so large. Therefore, relatively large scale of diatoms moves to the sea bottom. Meantime, turbidity is getting down in neap period. Then the moved diatoms may be survived when transparency of sea water is getting up. After that, vertical direction of seawater mixing is occurred in spring tide period. Then the survived diatoms are getting up to sea surface. Thus blooming would occur if nutrition rich seawater is there.

D. Relation Among Chlorophyll-a Concentration and Tidal Height Difference a Day, Sun Shine Duration a Day, and Wind Speed from North

Relation among chlorophyll-a concentration (Chl-a), tidal height difference a day, sun shine duration a day and wind speed from the north is clarified. Fig.7 shows the result.

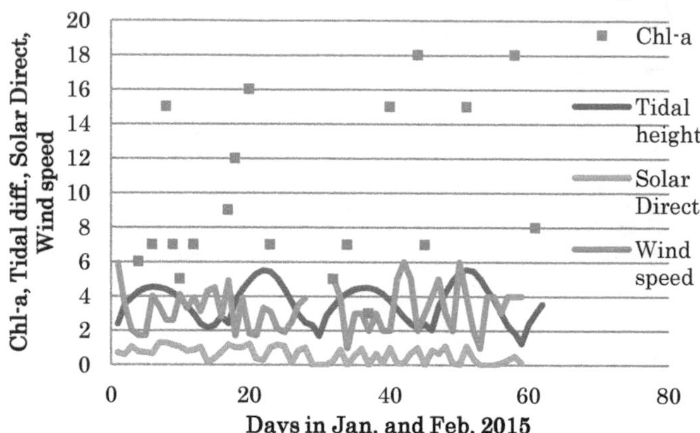

Fig. 7. Relation among chlorophyll-a, tidal height difference a day, sun shine duration a day and wind speed from the north

As shown in Fig.6, there is strong relation between tides and chlorophyll-a concentration. It, however, is not always true. Other factors, sun shine time duration a day is one of those as well as wind speed, in particular, north wind. In accordance with the other measured data of three dimensional chlorophyll-a as well as seawater temperature, salinity, Dissolved Oxygen: DO, turbidity, ph ratio, etc., it shows that poor oxygen water mass appears at the bottom of the sea and then it raised up to the sea surface (accordingly, chlorophyll-a rich seawater raised from the bottom of the sea and raised up to the sea surface). Wind speed helps to convection of the sea surface water while sun shine time duration helps to increase chlorophyll-a concentration in conjunction of warm up of the sea surface temperature.

These relations are almost same for the other year of chlorophyll-a concentration. Correlation coefficients are calculated between chlorophyll-a concentration and the other data of tidal difference a day, sun shine time duration a day and wind speed from the north. The result shows that there is a strong relation between chlorophyll-a concentration and tidal difference a day, obviously followed by wind speed from the north as shown in Fig.8.

It is not always true. The situation may change by year by year. In particular, there is clear difference between year of 2011 and the other years, 2012 to 2015. One of the specific reasons for this is due to the fact that chlorophyll-a concentration in 2011 is clearly greater than those of the other years. Therefore, clear relation between chlorophyll-a concentration and the other data of tidal difference a day, sun shine time duration a day and wind speed from the north cannot be seen. That is because of the fact that there is time delay of chlorophyll-a increasing after the nutrient rich bottom seawater is flown to the sea surface.

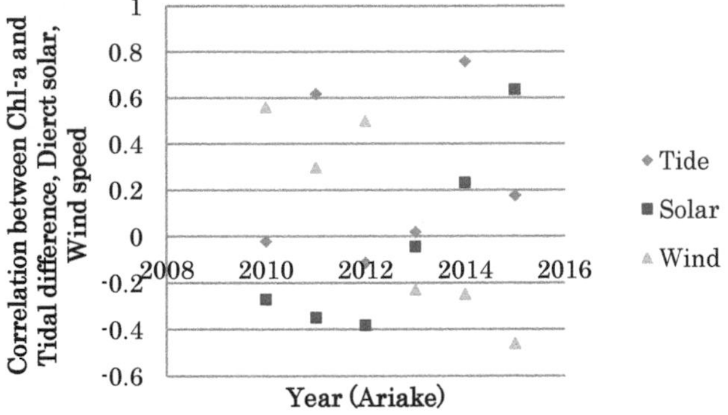

Fig. 8. Correlation coefficients between chlorophyll-a concentration and the other data of tidal difference a day, sun shine time duration a day and wind speed from the north

IV. Conclusion

Behavior of relatively large size of diatoms which appear in the Ariake Sea areas, Japan in winter based on remote sensing satellite data is clarified. Through experiments with Terra and AQUA MODIS data derived chlorophyll-a concentration and truth data of chlorophyll-a concentration together with meteorological data and tidal data which are acquired for 5 years (winter 2011 to winter 2015), it is found that strong correlation between the chlorophyll-a concentration and tidal height changes. Also it is found that the relations between ocean wind speed and chlorophyll-a concentration. Meanwhile, there is a relatively high correlation between sunshine duration a day and chlorophyll-a concentration.

An analysis on phytoplankton distribution changes monitoring for the intensive study area of the Ariake Sea, Japan based on remote sensing satellite data is conducted. Phytoplankton distribution changes in the Ariake Sea areas, Japan based on remote sensing satellite data is studied. Through experiments with Terra and AQUA MODIS data derived chlorophyll-a concentration and truth data of chlorophyll-a concentration together with meteorological data and tidal data which are acquired in January in 2011 to 2015, it is found that strong correlation between the truth data of chlorophyll-a and MODIS derived chlorophyll-a concentrations with R square value ranges from 0.677 to 0.791. Also it is found that the relations between ocean wind speed and chlorophyll-a concentration as well as between tidal difference a day and chlorophyll-a concentration. Meanwhile, there is a relatively high correlation between sunshine duration a day and chlorophyll-a concentration.

One of the knowledge raised from this study is the diatom appearance mechanism. The diatom appeared at the back in the Ariake Sea and is not flown from somewhere else. Also, there is relatively low chlorophyll-a concentration sea areas between Isahaya bay area and the back in the Ariake Sea area. Therefore, chlorophyll-a concentration variations are isolated each other (Isahaya bay area and the back in the Ariake Sea area).

Further study is required for clarification of

1) the difference between 2011 and the other years, 2012 to 2015,

2) the reason why it is not always true that chlorophyll-a concentration get high on the spring tide while that get low on the neap,

3) some other mechanism for diatoms appearance is clarified,

4) three dimensional measurements of seawater temperature, salinity, turbidity, DO, ph, and chlorophyll-a concentration have to be made,

5) nutrition rich water current from the river mouths has to be taken into account,

6) interaction between relatively large size of diatoms and the other red tide species has to be clarified and is taken into account on the appearance mechanism studies.

Acknowledgment

The authors would like to thank Dr. Yuichi Hayami, Dr. Kei Kimura, Kenji Yoshino, Naoki Fujii and Dr. Takaharu Hamada of Institute of Lowland and Marine Research, Saga University for their great supports through the experiments.

References

[1] Yuji Ito, Toshiya Katano, Naoki Fujii, Masumi Koriyama, Kenji Yoshino, and Yuichi Hayami, Decreases in turbidity during neap tides initiate late winter large diatom blooms in a macrotidal embayment, Journal of Oceanography,69: 467-479. 2013.

[2] Nishikawa T (2002) Effects of temperature, salinity and irradiance on the growth of the diatom *Eucampia zodiacus* caused bleaching seaweed *Porphyra* isolated from Harima-Nada, Seto Inland Sea, Japan. Nippon Suisan Gakk 68: 356–361. (in Japanese with English abstract)

[3] Nishikawa T (2007) Occurrence of diatom blooms and damage tocultured *Porphyra* thalli by bleaching. Aquabiology 172: 405–410. (in Japanese with English abstract)

[4] Nishikawa T, Hori Y (2004) Effects of nitrogen, phosphorus and silicon on the growth of the diatom *Eucampia zodiacus* caused bleaching of seaweed *Porphyra* isolated from Harima-Nada, Seto Inland Sea, Japan. Nippon Suisan Gakk 70: 31–38. (in Japanese with English abstract)

[5] Nishikawa T, Hori Y, Nagai S, Miyahara K, Nakamura Y, Harada K, Tanda M, Manabe T, Tada K (2010) Nutrient and phytoplankton dynamics in Harima-Nada, eastern Seto Inland Sea, Japan during a 35-year period from 1973 to 2007. Estuaries Coasts 33: 417–427.

[6] Nishikawa T, Hori Y, Tanida K, Imai I (2007) Population dynamics of the harmful diatom *Eucampia zodiacus* Ehrenberg causing bleachings of *Porphyra* thalli in aquaculture in Harima- Nada, the Seto Inland Sea, Japan. Harmful algae 6: 763–773.

[7] Nishikawa T, Miyahara K, Nagai S (2000) Effects of temperature and salinity on the growth of the giant diatom *Coscinodiscus wailesii* isolated from Harima-Nada, Seto Inland Sea, Japan. Nippon Suisan Gakk 66: 993–998. (in Japanese with English abstract)

[8] Nishikawa T, Tarutani K, Yamamoto T (2009) Nitrate and phosphate uptake kinetics of the harmful diatom *Eucampia zodiacus* Ehrenberg, a causative organism in the bleaching of aquacultured *Porphyra* thalii. Harmful algae 8: 513–517.

[9] Nishikawa T, Yamaguchi M (2006) Effect of temperature on lightlimited growth of the harmful diatom *Eucampia zodiacus* Ehrenberg,

a causative organism in the discoloration of *Porphyra* thalli. Harmful Algae 5: 141–147.

[10] Nishikawa T, Yamaguchi M (2008) Effect of temperature on lightlimited growth of the harmful diatom *Coscinodiscus wailesii*, a causative organism in the bleaching of aquacultured *Porphyra* thalli. Harmful Algae 7: 561–566.

[11] Syutou T, Matsubara T, Kuno K (2009) Nutrient state and nori aquaculture in Ariake Bay. Aquabiology 181: 168–170. (in Japanese with English abstract)

[12] Harada K, Hori Y, Nishikawa T, Fujiwara T (2009) Relationship between cultured *Porphyra* and nutrients in Harima-Nada, eastern part of the Seto Inland Sea. Aquabiology 181: 146–149. (in Japanese with English abstract)

20

Cognitive Consistency Analysis in Adaptive Bio-Metric Authentication System Design

Gahangir Hossain
Electrical & Computer Engineering
Indiana University-Purdue University Indianapolis
Indianapolis, IN USA

Habibah Khan
Instructional Design & Technology
The University of Memphis
Memphis, TN USA

Md.Iqbal Hossain
Electrical & Computer Engineering
The University of Memphis
Memphis, TN USA

***Abstract*—Cognitive consistency analysis aims to continuously monitor one's perception equilibrium towards successful accomplishment of cognitive task. Opposite to cognitive flexibility analysis – cognitive consistency analysis identifies monotone of perception towards successful interaction process (e.g., biometric authentication) and useful in generation of decision support to assist one in need. This study consider fingertip dynamics (e.g., keystroke, tapping, clicking etc.) to have insights on instantaneous cognitive states and its effects in monotonic advancement towards successful authentication process. Keystroke dynamics and tapping dynamics are analyzed based on response time data. Finally, cognitive consistency and confusion (inconsistency) are computed with Maximal Information Coefficient (MIC) and Maximal Asymmetry Score (MAS), respectively. Our preliminary study indicates that a balance between cognitive consistency and flexibility are needed in successful authentication process. Moreover, adaptive and cognitive interaction system requires in depth analysis of user's cognitive consistency to provide a robust and useful assistance.**

Keywords—Cognitive authentication; Cognitive consistency; Fingertip dynamics; Maximal Information Coefficient; Bivatiate plot

I. INTRODUCTION

With the increase of adaptive interfaces including natural gestures, touch-screen, tactile, speech enable, implicit and tangible interactions; fingertips (keystrokes, mouse click and tapping interfaces) are still dominating since their invention [1]. Users like to perform authentic interaction by a fingertip. Slip of tip might hinder the user in robust interaction and authentication process just because of inflexibility of accessibility and authentication schemes. This becomes more challenging when the users need assistance from the interaction system. Meanwhile, the system strategically requires a robust user authentication, adaptation and automation to tie its users continuously into its loop. Question arises: how to maximize authentic accessibility to benefit user in assistive interaction? The challenging accessibility problem requires a different usability engineering solution than we currently practice. Analysis of end-user's cognitive pattern of interaction and deficiencies may improve the future interface accessibility towards authentic and adaptive accessibility design. More specifically, cognitive approach of authentication may allow some flexibility in user authentication process based on users' past history of success and present consistent interactions (monotone), even though unsuccessful in authentication process.

Cognitive consistency analysis is the fundamental principle in social cognition and important factor in balancing interests through adaptive collaborative system design. The analysis uncovers three key factors that tend toward (cognitive) consistency: perception, emotion, and action. More specifically, the process helps the system to reveal user's intrinsic conscious conflict situation that leads to adaptive behavior.

In an adaptive authentication, cognitive consistency analysis is a critical research process, that aims to (1) identify progressive interests in biometric authentication, (2) develops understanding of inherent meaning, values and motives in cognitive activities, (3) study adaptive interaction trends with connection to current research, (4) construct models of the relations between cognitive capability, personal profile and participants' actions, (5) elucidate the fundamental contradiction which are developing as a result of action based on ideologically frozen understanding, (6) participate in a program to see new ways of the situation, (7) Theoretically ground the principles applied in the analysis.

With the holistic goal of adaptive authentication system development, two specific aims are analyzed through the research. First specific aim is to understand consistent or inconsistent interactions during authentic interaction. The second specific aim is to co-analyze user's past history of successful authentication record with the degree of present cognitive activity to provide an essential decision support in authentication process. This study covers the first aim with two different datasets. Moreover, the conditions with understandings are compared, ideologies are criticized, and immanent possibilities for action are discovered.

The rest of the paper is organized as following: In section 2, gives brief descriptions on cognitive approaches in fingertip dynamics and authentication with general and mixed frameworks. Section 3, explains the methods used in cognitive consistency analysis process. Exploratory analyses of cognitive effort on some benchmark biometric data sets are shown in section 4. Finally, the research is concluded with findings and future works in section 5.

II. Fingertips Dynamics Approaches

Fingertips dynamics includes finger related user interactions including keystroke, key tapping, and mouse clicking. Roman and Vanue [1] reviewed a classification of the state-of-the-art behavioral biometrics related to user's skills, style, preference, knowledge, motor-skills or strategy that users use in their everyday task accomplishment. Fingertips dynamics fall into the behavioral biometrics classification and play an importance role in conjunction to everyday hand related activities. Keystroke dynamics biometric pattern analysis and user modeling are proposed by Killourhy[3]. Subjects typed a strong password and their key pressing response times are recoded to have up key, holding and down key and their combination response analysis. Ahmed and Traore's [2] proposed new biometric fingertips with mouse clicking dynamics, which was improved in Nakkabi et al. [4] in terms of clicking response time. Poh and Tistarelli[9] customized biometric authentication systems by a novel method with discriminative score. Liang et al. [10] proposed a combined analysis of fingertips dynamics with head pose estimation. The real-time fingertip gesture tracking is proposed in Oka et al. [13] which determine an appropriate threshold value for image binarization during initialization by examining the histogram of an image of a user's hand placed open on a desk.

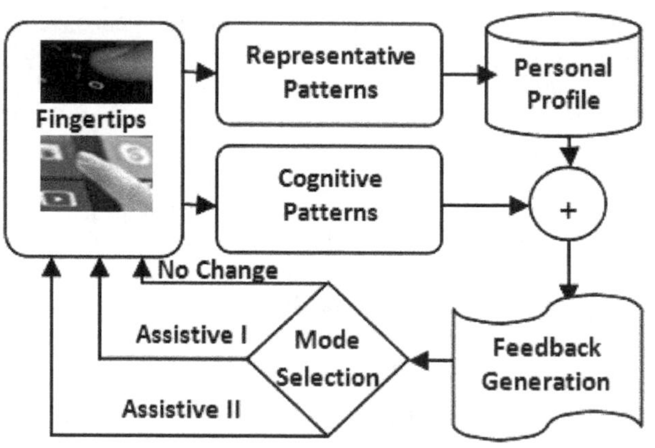

Fig. 1. Overview of proposed cognitive biometric authentication system with flexible feedbacks

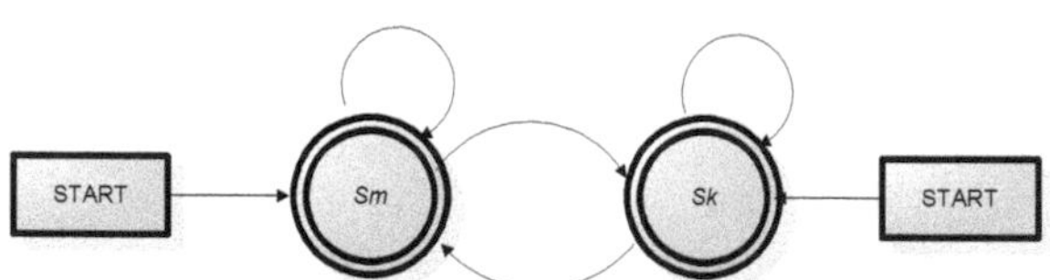

Fig. 2. Keystroke and mouse click are complementary; Sm and Sk stands for the mouse clicking state and keystroke state respectively

Cognitive issues in fingertips includes but not limited to: mental effort, consistencies (monotone) in action, dissonance, bias and more. The cognitive mechanisms of touch are proposed in some works including, Rola [13], Jone et al. [14] gives more elaboration of tactile sensory systems with brain signal analysis. Hossain et al.'s work [5, 6] proposes some relevant cognitive mechanism of cognitive workload during assistive technology interaction. These literatures prove an important grounding in cognitive analysis of fingertips dynamics. Fishel[9] proposed a direct measure of fingertip strength through a robust micro-vibration sensor for bio-metric fingertips.

A. *The General Structure*

In keystroke and tapping dynamics, user identifiable patterns are considered as representative patterns and recorded as personal profile. Cognitive patterns relates to mental activities which are related to user's task performance in terms of response time in case of fingertip dynamics. For example, the monotone in response times in every key (or dot) a user press (or tap) in authentication process. Combining both type of patterns are proposed in cognitive approach of authentication system. More specifically, the combined approach is considered to be useful in better assistance it's used in need. Figure 1 shows a functional block diagram of cognitive authentication systems with assistive modes. The assistive modes may vary based on flexibility of design principles.

B. *The Mixed or Hybrid Structure*

Keystroke and mouse clicking are still considered as primary input mechanisms. While the keystroke is considered as feed-forward interaction, the mouse click is considered as feedback interaction which is mostly replaced by tapping interaction essential in authentication process. These two interaction mechanisms works as two wings of a bird in robust interaction. It is quiet impossible for a user to accomplish a task depending only on key pressing or tapping/mouse interaction rather a combination of keyboard and mouse click/ tapping. Although keystroke and mouse clicks are sometimes complementary, their combination makes the interaction more natural. The main idea of mixed cognitive authentication is to allow the user a natural way of access rights with a combined or individual approach of fingertips and dynamics. Key objectives are: (a) to identify similarity and differences of cognitive efforts a user experiences during continuous fingertips and (b) to analyze the leverage of the differences towards adaptive authentication design. Figure 2 shows a state diagram of cognitive authentication with possible beginning and ending states of keyboard and mouse clicks use in action. Measures of cognitive effort dynamics become important and challenging research issue.

This work analyzed the keystroke response time and fingertips dynamics to have insights of cognitive efforts and their differences.

C. *Theoretical Background*

Naturally, we expect and have a preference for monotonic interaction in our lives as well as other things including in authentication process. We need consistency in the whole interaction (e.g., key pressing) process of authentication. This section explains cognitive consistency analysis with a theory and mathematical modeling techniques.

1) Cognitive consistency theory

Consistency becomes like a form of human gravity. It holds everything down and together. It helps us to understand the world and our place in it. The fundamental thrust of consistency theories is to enforce equilibrium among one's cognitions. Cognitive consistency can be defined with - cognitive consistency theory[7 -9], that focuses on the balance individuals create cognitively when inconsistencies create tensions and thus motivate our brains and body to respond. The fundamental thrust of cognitive consistency theory is to enforce equilibrium among one's cognitions. Cognitive consistency theory shows how people motivate themselves to work and adjust inconsistent measures. Three steps of consistency theory are: (1) estimation of expected consistency, (2) resolving inconsistencies that create a state of dissonance, and (3) the dissonance drives us to restore consistency. This theory is the basis for equilibrium for individuals in authentication process. However, the importance of positive and negative outcomes to reduce stressful choices relates to cognitive dissonance theory of Leon Festinger[18, 19]. Although, the cognitive consistency theory touches on both issues, it focuses on the affects of inconsistencies in interaction process motivating to react and consequent actions.

Fingertips dynamics represent finger related user interactions including keystroke, key tapping, and mouse clicking. Cognitive issues in fingertips includes but not limited to: cognitive effort, consistencies (monotone) in action, dissonance, bias and more. For example, consider in keystroke and tapping dynamics. User identifiable patterns are considered as representative patterns and recorded as personal profile. Cognitive patterns are some derived patterns from online task performance in terms of response time in fingertip dynamics. Combining both type of patterns are proposed in cognitive approach of authentication system.

Main research goal is to have an adaptive interface that can understand consistency of mental state in fingertip dynamics. More specifically, the goal is to have a robust and effective user accessibility framework which is useful in flexible user authentication process. Cognitive analysis of fingertips dynamics can be combined with users' prior interaction behavior profiles to have more effective authentication system. Cognitive effort, load, and cost are some analysis considered in consistent authentication interface design.

Figure 3, illustrates an example of lock pattern tapping interaction and cognitive analysis to understand user's consistent and dissonance interactions. Figure3, a & b shows the schematic of interaction process. Relative response time between dots or dot-to-dot transition may uncover the cognitive consistence. Main assumption behind the consistency identification is that, user with a balance consistence-dissonance value need no assistance from the system in authentication process.

III. THE CRITICAL ANALYSIS

This work analyzed two datasets: the keystroke dynamics dataset [3], which is publicly available as a benchmark dataset and the lock pattern tapping [18] dataset - used with permission from the author. Both dataset are passed through the institutional review board (IRB) for secondary evaluation. Maximal Information-based Nonparametric Exploration (MINE) tool [17] is used in consistency and dissonance analysis.

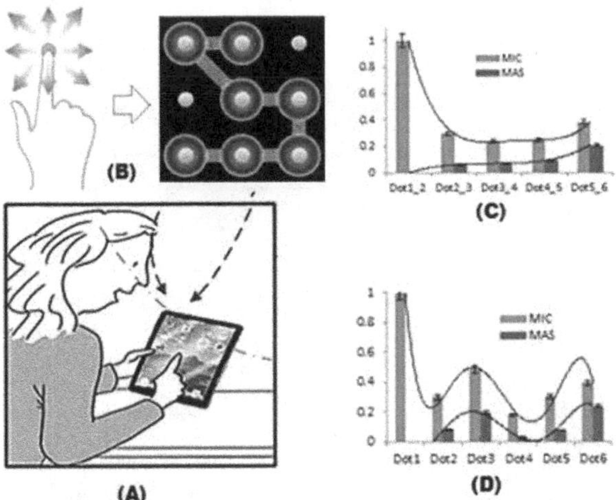

Fig. 3. Cognitive authentication in lock pattern tapping (fingertips) dynamics analysis

A. Technical Analysis

The consistency is analyzed with maximal information coefficient (MIC), which is a non-parametric measure of two-variable dependence. The MIC is widely used to identify important relationships in data sets and to characterize them. Different relationship types give rise to characteristic matrices with different properties. For instance, strong relationships yield characteristic matrices with high peaks, monotonic relationships yield symmetric characteristic matrices, and complex relationships yield characteristic matrices whose peaks are far from the origin. The MIC is used to measure relationship strength between two responses (say, dot1 and dot2) in terms of response time. Let the two response variables be defined as D and A, respectively. The MIC can be written as –

$$\mathrm{IC(M)} = \max_{\mathrm{DA<B(n)}} \mathrm{utual(M)_{D,A}}$$

$$= \max_{\mathrm{DA<B(n)}} \frac{\mathrm{I * (M, D, A)}}{\mathrm{log(min\, D, A)}}$$

where B(n) = n is the search-grid size, I(M, D, A) is the maximum mutual information over all grids D-by-A, of the distribution induced by M on a grid having D and A bins.

The Maximum Asymmetry Score (MAS)captures the deviation from monotonicity, and useful for detecting periodic relationships to have idea about cognitive dissonance. It can be defined as:

$$\mathrm{MAS(M)} = \max_{\mathrm{DA<B(n)}} |\mathrm{Mutual(M)_{D,A}} - |\mathrm{Mutual(M)_{A,D}}|$$

The Maximum Edge Value (MEV) measures the closeness to being a function is defined as:

$$\mathrm{MEV}(D) = \max_{XY<B(n)} \{M(D)_{X,Y} : X = 2\ or\ Y = 2\}.$$

Cognitive consistency visualization and analysis:

Box plot and robust bi-variant bag plots are used to visualize fingertips dynamics (consistencies) in terms of data location, spread, skewness and outliers.

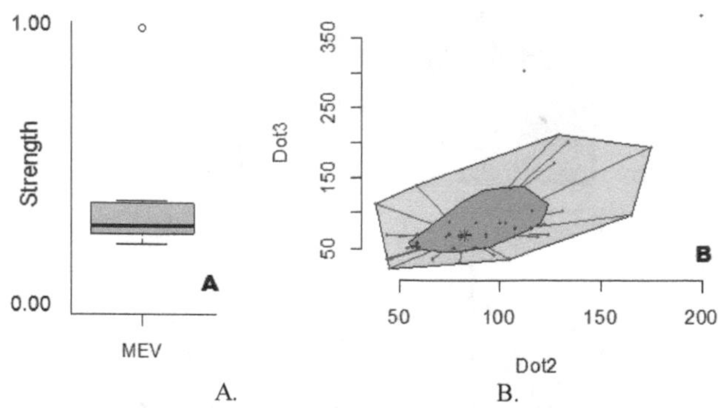

Fig. 4. Bi-variate plots illustration of two consicutive dots responsetime. *Left* - A. the box plot and *Right* - B. the bag plot

Boxplot shows the difference in cognitive levels at fingertips inconsistencies and control. Boxes show the median, 25th and 75th percentiles, error bars show 10th and 90th percentiles, and filled symbols show outliers (figure 4A).

In bag plot, the representation consists of three nested polygons: 'bag', 'fence' and 'loop' (figure 4B). The bag is the inner polygon, which is a construction of the smallest depth region containing at least 50% of the total number of observations, also known as Tukey depth around the center point of median (Tukey median). The most outer polygon known as fence (which is not drawn in the figure) but computed for the outlier points identification. Observations outside the fence are flagged as outliers. The 'fence' polygon can be obtained by inflating the bag (relative to the Tukey median) by a factor ρ. According to Rousseeuw et al. [21] ρ = 3, a recent work [22] prefers ρ = 2.58, as that allows the fence to contain 99% of the observations when the projected bivariate scores follow standard normal distributions. The convex hull of the observations that are not marked as outlier creates another polygon known as loop. The highest possible Tukey depth (median) is also marked in orange color near the center of the graph. The importance of use of bagplot is not only it's robustness against outliers, but also invariant under afine transformation.

B. Datasets

The Keystroke Dynamics - Benchmark Data Set is the accompaniment to Kevin and Roy [3]. The data consist of keystroke-timing information from 51 subjects (typists), each typing a password (.tie5Roanl) 400 times. The second dataset is – the lock pattern tapping dataset [20] – according to [20] a total of 32 different participants successfully completed lock pattern task assigned to them using the mobile application. Participants were 20 male and 12 female with different age groups (from 19 to 56 years old), cultural and educational backgrounds, and having different levels of experience interacting with touch-screen smart phones. They performed the test on different Android phones: Samsung Galaxy SII (18), Nexus S (8), HTC Legend (4) and HTC Vision (2). With email permission from [20], this dataset is used in this study for secondary evaluation.

MIC, MAS and MEV are performed in both dataset to identify underlying monotone, inconsistencies and functional closeness, respectively. Bi-variate box and bag plots are performed visual illustration of cognitive states that shows the trends in monotone and consistencies.

IV. Results and Discussions

Understanding cognitive states are important in accessible and adaptive system development. Ideas presented through this research are transformative and critical (not qualitative or quantitative) and aim to bring additive flexibilities in adaptive interaction design. Finger tapping is gaining increasing popularity; hence, the MIC-MAS trends are analyzed that is observed in finger tapping dataset. Figure 3 shows partial result of monotonic interaction and dissonance in lock pattern tapping interaction. Figure3, c & d illustrates the dot-to-dot and in dot cognitive response time relations, respectively. Figure 3C, shows that the increasing number of transition to remember increases more gap (delay). Result shows partial dot-to-dot and in-dot monotonic interaction and inconsistencies in terms of MIC and MAS. In dot-to-dot transition (Figure 3c), user shows monotonically decreasing relative responses which reaches in a steady state in between dot transitions (dot2-3, dot3-4, dot4-5) which was analyzed on all correct lock pattern response data. Moreover, user has an increase of cognitive inconsistencies (dissonance) with increase of dot-to-dot transition options (dot2-3, dot3-4, dot4-5, and dot5-6). Relative in-dot response time (figure3d) shows a stable pattern of MAS value (dissonance) always lower than consistent dot visit (MIC value), which is similar to dot-to-dot analysis, but no steady trend is observed. So, the dot-to-dot transition MIC-MAS relationship uncovers more cognitive mental aspects in authentication process.

Similar to figure 3c &3d, in keystroke data analysis, key up to next key down response time and key holding time MIC MAS are separately computed and plotted in figure 5. The cognitive consistency and inconsistency trends in key up-down and key holding response times are shown in figure5 (up) and (down), respectively. While the tapping interaction shows a frequent sinusoidal trend, keystroke interaction shows a delayed period in sinusoids.

Figure 6, shows the box plot of the complete data set with additional MINE tool, the maximum edge value (MEV) in lock pattern tapping data. According to keystroke dynamics dataset [3], users are instructed to try different log patterns. Their tapping time and performance are logged in terms of response time. This study performed an analysis of correlative activities with MIC and MAS. Box plots in figure6, represent overall analysis on MIC, MAS and MAV values. The DD analysis (the right part of the figure) indicates a consistent responses by test subjects in successfully password typing. In figure 7, holding time (H) and key-up-to-key-down (UD) time analysis shows, inconsistencies in MAS value. Whereas, more inconsistencies (dissonance) are observed in holding time (H), cognitively meaning that subject is more confused in selecting and pressing keys related to password.

Deeper lock-pattern tapping analysis is performed with a comparative tapping response time analysis in within and between dots and their transition.

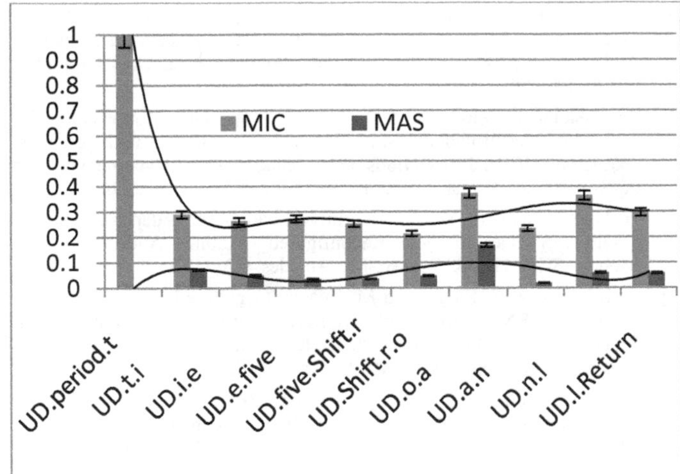

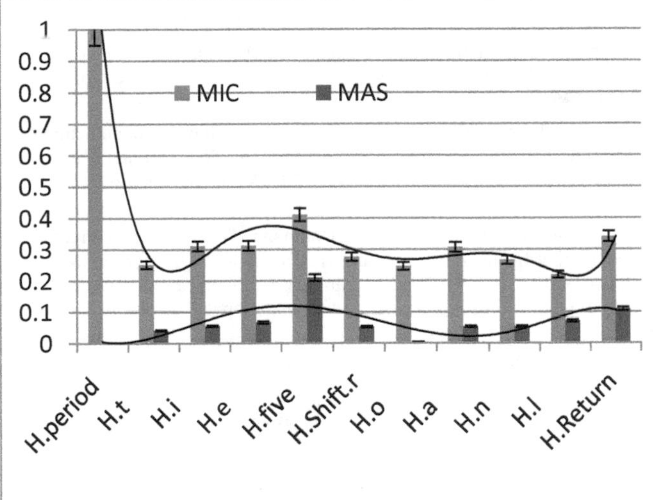

Fig. 5. Trends of user cognitive involvement in keystroke change (top) and key holding (bottom)

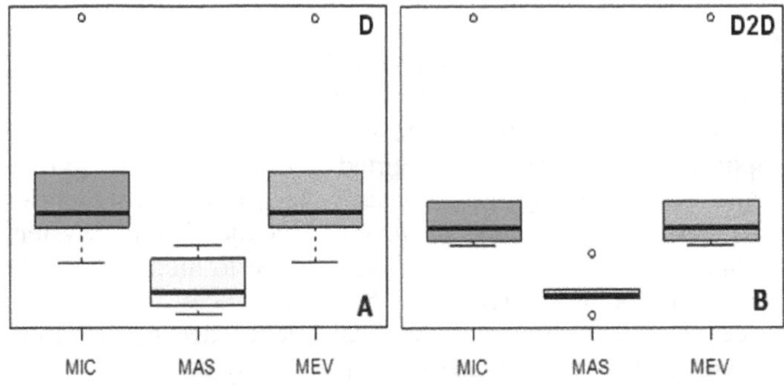

Fig. 6. Box plot of lock pattern tapping data. D in the top right corner of figure A represents the results from the response time analysis on dots. Where D2D represents the response time analysis on dot-to-dot. The Y-axis represents strength in 0.00-1.00 scale

These are illustrated with bag plots in figure 8. Similarly, in password typing (keystroke dynamics) dataset, inter key holding time and key transition time (key up and key down) and their comparisons are illustrated with bag-plots in figure 9. Both box plots and bi-variate bag plots combined illustrates the underlying cognitive factors in authentication.

Both in, in-dot and key-holding, subjects have similar cognitive states (gray dotted line on box plots figure A, and figure8A), then in dot-to-dot and key-up to key-down time. This observation, illustrates that they have more consistencies in finger tapping memory (recall) task then decision (switching) task. Existence of more sparse outliers, in later bags indicates decision inconsistencies in selecting transitions (C and D parts in figure8 and 9). For instance, the last bag plot of both cases is smaller because of sparse outlier.

Having MIC, MAS and MAV values; and outliers' situation, some derived feature values with keystroke dynamics and finger tapping datasets, a representative feature vector can be created to apply machine learning approaches towards cognitive authentication system design.

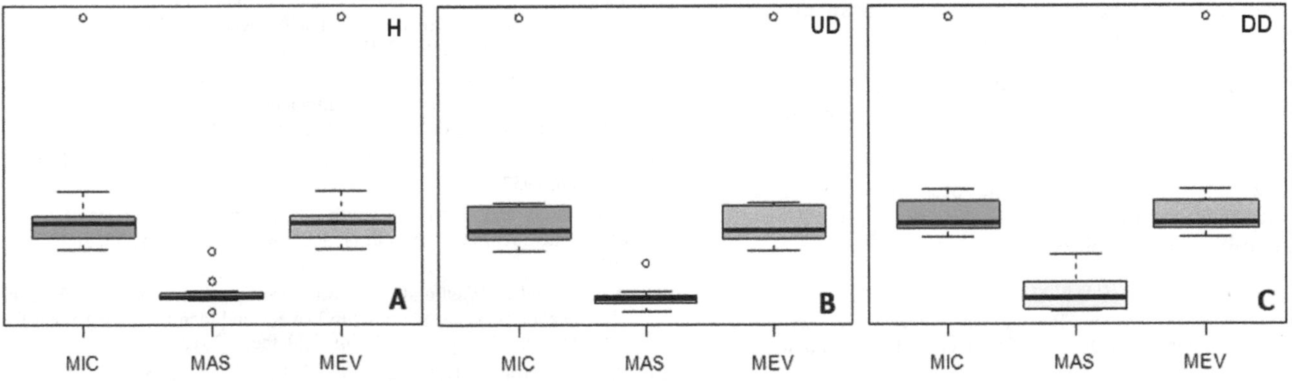

Fig. 7. Box plot of keystroke dynamics dataset[3]. DD in the top right corner in right plot C, represents the results from the key-down to key-down response time analysis. Whereas, H represents the key holding response time, and UD represents the key-up to key-down response times plot. The Y-axis represents strength in 0.00-1.00 scale

V. Conclusion

Rapid increase of accessible technology needs continuous access of user's ability based authentic interaction - an adaptive-cognitive authentication protocol becomes an important issue. This study started with the old principle of cognitive consistency and inconsistencies (dissonance) theories with novel applications in adaptive authentication system development. Exploratory analyses are performed to identify trends of cognitive activities. Some instances of cognitive processes (perception, attention, and action/decision-making) are considered in this analysis. The associated properties - emotion, intuition, collaborative action etc. can be derived with more analysis.

By fingertips dynamics, we can consider tapping, double tapping, long press, scroll, pan, flick, two finger tapping, two finger scroll, pinch, zoom, and rotate during interaction. This paper presents a compilation of user interface requirements that arise in fingertip dynamics based cognitive, assistive and adaptive interaction design. It draws implications imposed by the user interaction requirements on the architectures of cognitive consistent system. It defines a special class of cognitively defined authentication systems. Various cognitive mechanisms of providing efficient mental states can be included with similar framework. This paper also draws attention to the need for more non-parametric analysis in cognitive system design.

This paper provided a way of cognitive consistency analysis in keystroke based user interface requirements for constructing good user experience. A special user interface design needs to have assistive (helps it's user in need) and adaptive (learns and updates personal profile) properties, that can be accomplished through consistency analysis. This analysis presents opportunities to improve the system performance relying on the special structure of such interactions. Some ways to exploit this structure are pointed out through this study. Additional research could be warranted in the following areas to further exploit the nature of such problems. One may experiment with various task outcomes and cognitive load schemas for memorizing passwords and lock patterns of computations across user sessions.

Additional cognitive analysis could be gained by learning across cognitive dissonance theory [18, 19]. Such learning may involve cognitively confused patterns in test items.

More trial and time could be devoted to devising and experimenting with algorithms that are essential and connected to robust and online accessibility.

References

[1] V. Y. Roman, and G. Venu, "Behavioural biometrics: a survey and classification." International Journal of Biometrics 1, no. 1 pp. 81-113, 2008

[2] A. E. Ahmed and I. Traore, "A New Biometrics Technology based on Mouse Dynamics", IEEE Transactions on Dependable and Secure Computing, Vol. 4 No. 3, pp.165-179, 2007.

[3] K. S. Killourhy, and A. M. Roy, "Comparing anomaly-detection algorithms for keystroke dynamics," Dependable Systems & Networks, pp.125-134, 2009. IEEE Computer Society Press, Los Alamitos, California, 2009.

[4] Y. Nakkabi, I. Traore, A. A. E Ahmed, "Improving Mouse Dynamics Biometric Performance using Variance Reduction with Separate Extractors", IEEE Transactions on Systems, Man and Cybernetics—Part A: Systems and Humans, Vol. 40, No. 6, pp.1345 – 1353,2010.

[5] G. Hossain and M. Yeasin, Cognitive Load Based Adaptive Assistive Technology Design for Reconfigured Mobile Android Phone, MobiCASE 2011, pp.370-380, Los Angeles, CA, USA, 2011.

[6] G. Hossain, A. S. Shaik, and M. Yeasin, Cognitive Load and Usability Analysis of RMAP for people who are blind or visually impaired. In Proceedings of the 29th ACM International Conf. on Design of Communication (SIGDOC) ACM, Pisa, Italy, 2011.

[7] P. Tannenbaum, The Congruity Principles: Retrospective Reflection and Recent Research. In Theories of Cognitive Consistency: A Sourcebook. Abelson, R. & others (eds.) Rand McNally & Co.; Skokie, 111, 1968.

[8] M. Venkatesan, Cognitive Consistency and Novelty Seeking. In Ward, S. & Robertson, T. (eds.) Consumer Behaviour: Theoretical Sources. Prentice-Hall, Inc; Englewood Cliffs, New Jersey, 1973.

[9] T. Newcomb, Interpersonal Balance. In Abelson, R. et al Theory of Cognitive Consistency: A Sourcebook. Rand McNally & Co. Skokie, 111, 1968

[10] N. Poh and M. Tistarelli, "Customizing biometric authentication systems via discriminative score calibration." In Computer Vision and Pattern Recognition (CVPR), 2012 IEEE Conference on, pp. 2681-2686. IEEE, 2012.

[11] H. Liang, Y. Junsong, and D. Thalmann. "Hand pose estimation by combining fingertip tracking and articulated ICP." In Proceedings of the 11th ACM SIGGRAPH International Conference on Virtual-Reality Continuum and its Applications in Industry, pp. 87-90. ACM, 2012.

[12] V. Štruc, Z. Jerneja, V. Boštjan, and N. Pavešić. "Beyond parametric score normalisation in biometric verification systems." IET Biometrics 3, no. 2 (2014): 62-74.

[13] K. Oka, S. Yoichi, and K. Hideki, "Real-time fingertip tracking and gesture recognition." Computer Graphics and Applications, IEEE 22, no. 6 (2002): 64-71.

[14] E. T. Rolls, "The affective and cognitive processing of touch, oral texture, and temperature in the brain." Neuroscience & Biobehavioral Reviews 34, no. 2 pp.237-245, 2010.

[15] L. A. Jones, and A. M. Smith. "Tactile sensory system: encoding from the periphery to the cortex." Wiley Interdisciplinary Reviews: Systems Biology and Medicine 6, no. 3 pp. 279-287, 2014.

[16] J. Fishel, V. J. Santos, and G. E. Loeb. "A robust micro-vibration sensor for biomimetic fingertips." in 2nd IEEE RAS & EMBS International Conference on, pp. 659-663. IEEE, 2008.

[17] D. Reshef, Y. Reshef, H. Finucane, S. Grossman, G. McVean, P. Turnbaugh, E. Lander, M. Mitzenmacher, P. Sabeti. Detecting novel associations in large datasets. Science 334, 6062 (2011).

[18] L. Festinger, A theory of cognitive dissonance, Evanston, IL: Row & Peterson, 1957.

[19] L. Festinger and J. M. Carlsmith, "Cognitive consequences of forced compliance. Journal of Abnormal and Social Psychology, 58, 203 – 210, 1959.

[20] A. Julio and E. Wästlund. "Exploring touch-screen biometrics for user identification on smart phones." In Privacy and Identity Management for Life, pp. 130-143. Springer Berlin Heidelberg, 2012.

[21] P. Rousseeuw, I. Ruts and J.W. Tukey, `The bagplot: A bivariate boxplot', The American Statistician Vol. 53(4), pp.382-387, 1999.

[22] R.J. Hyndman, and S. H. Lin, "Rainbow plots, bagplots, and boxplots for functional data." Journal of Computational and Graphical Statistics 19, no. 1 (2010).

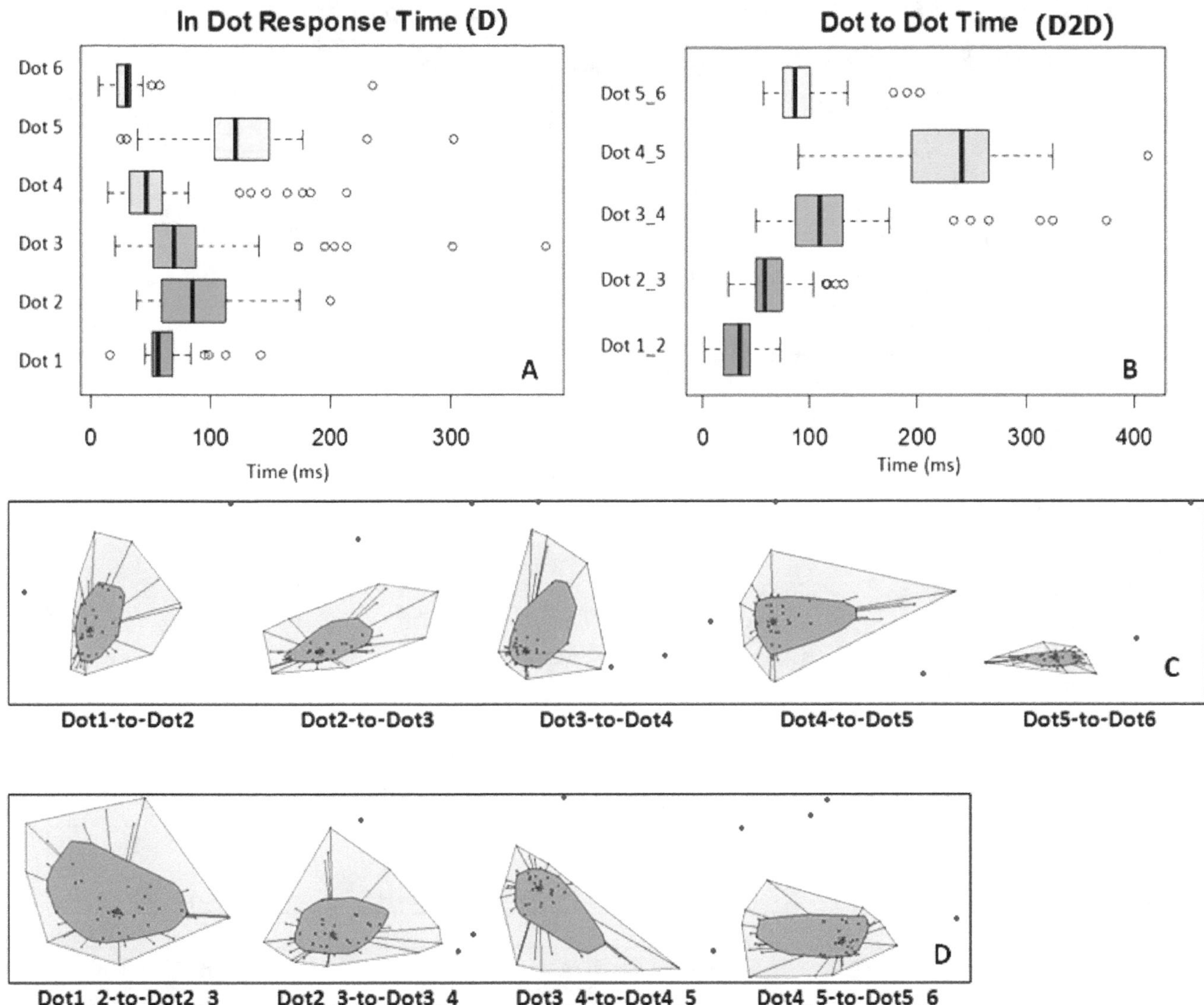

Fig. 8. Trends in Cognitive Consistency Analysis (Lock Pattern data): (A) Box plot of in-dot (D) response time; (B) Box plot of Dot-to-Dot time (D2D) and (c) Median and Outlier distributions in tapping on consecutive dots and (d) Median and Outlier distributions in consecutive dot-to-dot transitions

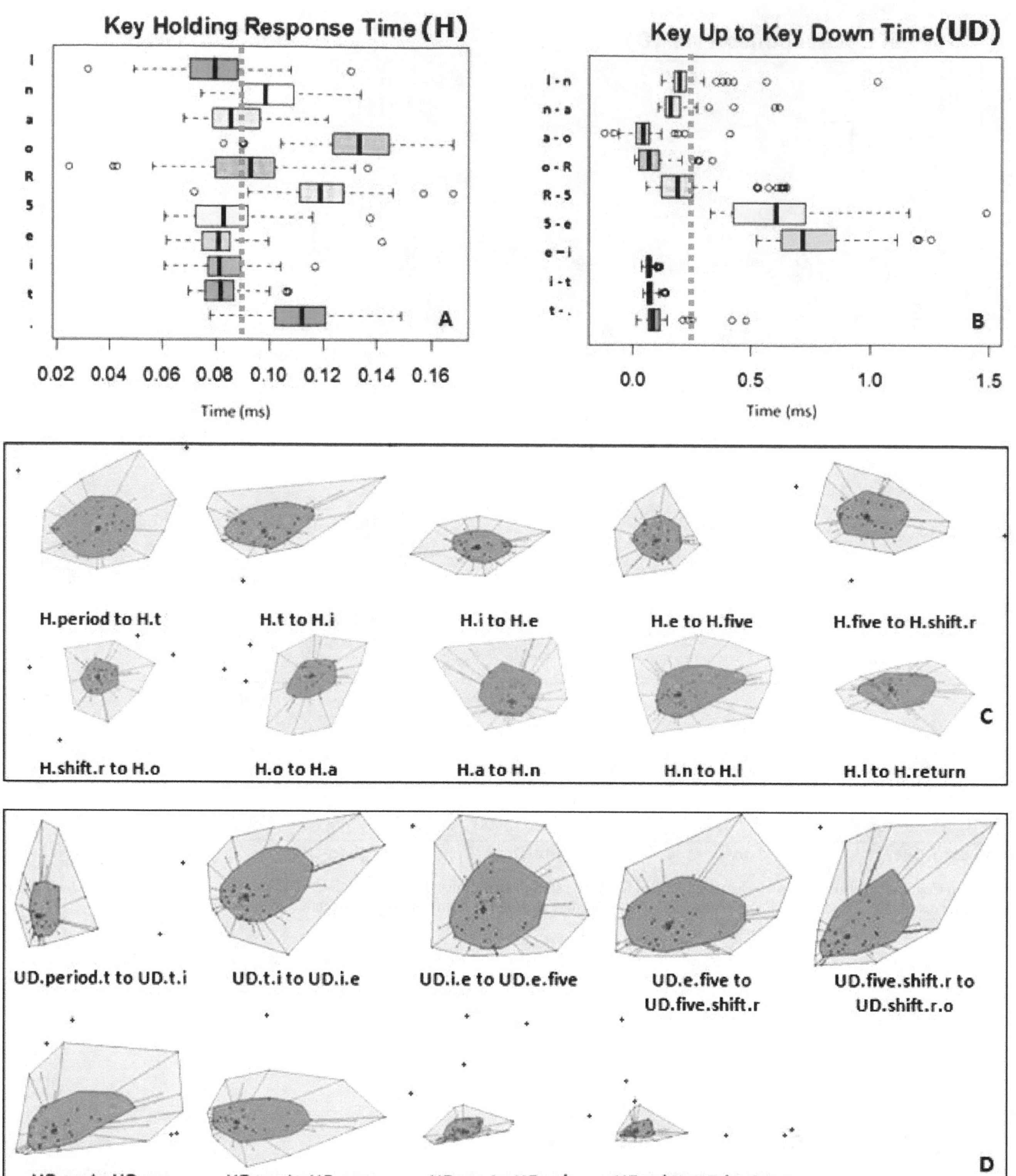

Fig. 9. Trends in Cognitive Consistency Analysis (Keystroke data): (A) Box plot of key holding (H) response time; (B) Box plot of Key Up-to-Key Down time (UD) and (c) Median and Outlier distributions in consecutive key holding and (d) Median and Outlier distributions in consecutive key up to key down transitions

21

Method for Surface Reflectance Estimation with MODIS by Means of Bi-Section between MODIS and Estimated Radiance as well as Atmospheric Correction with Skyradiometer

Kohei Arai 1
1Graduate School of Science and Engineering
Saga University
Saga City, Japan

Kenta Azuma 2
2 Cannon Electronics Inc.
Tokyo, Japan

***Abstract*—Method for surface reflectance estimation with MODIS by means of bi-section algorithm between MODIS and estimated radiance is proposed together with atmospheric correction with sky-radiometer data. Surface reflectance is one of MODIS products and is need to be improved its estimation accuracy. In particular the location near the skyradiometer or aeronet sites of which solar direct, aureole and diffuse radiance are measured, it is possible to improve the estimation accuracy of surface reflectance. The experiment is conducted at the skyradiometer site which is situated at Saga University. There is Ariake Sea near the Saga University. It is rather difficult to estimate surface reflectance of the sea surface because the reflectance is too low in comparison to that of land surface. In order to improve surface reflectance estimation accuracy, atmospheric correction is mandated. Atmospheric correction method is also proposed by using skyradiometer data. Through the experiment, it is found that these surface reflectance estimation and atmospheric correction methods are validated.**

Keywords—Sea surface reflectance; Atmospheric correction; Sky-radiometer; MODIS; satellite remote sensing

I. INTRODUCTION

Sea surface reflectance, water leaving radiance are fundamental characteristics and are importance parameters for the estimations of chlorophyll-a concentration, suspended solid, etc. Therefore, there is a strong demand to improve sea surface reflectance estimation accuracy. In order to improve surface reflectance, it is required to improve atmospheric correction accuracy. In the visible to near infrared wavelength region, the absorption components due to water vapor, ozone, aerosols, and the scattering due to atmospheric molecules, aerosols are major components. In particular, aerosol absorption and scattering (Mie scattering) is not so easy to estimate rather than scattering component due to atmospheric molecules (Rayleigh scattering). After the estimation of these components, radiative transfer equation has to be solved for the atmospheric correction. This is the process flow of the atmospheric correction [1]-[6]. Also, atmospheric component measurement, estimation, retrievals are attempted together with sensitivity analysis [7]-[17]. It is still difficult to estimate the aerosol characteristic estimation which results in difficulty on surface reflectance estimations.

The method proposed here is based on ground based Skyradiometer[1] which allows aerosol refractive index and size distribution through measurements of spectral optical depth through direct and aureole as well as diffuse solar irradiance. These are measured aerosol refractive index and size distribution, not estimated refractive index and size distribution. Therefore, it is expected that atmospheric correction can be done much precisely rather than estimation without sky radiometer data.

One of the examples are shown here for sea surface reflectance estimation with MODIS[2] data of Ariake Sea in Japan. Method for surface reflectance estimation with MODIS by means of bi-section algorithm between MODIS and estimated radiance is proposed together with atmospheric correction with sky-radiometer data. Surface reflectance is one of MODIS products and is need to be improved its estimation accuracy. In particular the location near the skyradiometer or aeronet sites of which solar direct, aureole and diffuse radiance are measured, it is possible to improve the estimation accuracy of surface reflectance. The experiment is conducted at the skyradiometer site which is situated at Saga University. There is Ariake Sea near the Saga University. It is rather difficult to estimate surface reflectance of the sea surface because the reflectance is too low in comparison to that of land surface. In order to improve surface reflectance estimation accuracy, atmospheric correction is mandated. Atmospheric correction method is also proposed by using skyradiometer data.

In the next section, the method and procedure of the experimental study is described followed by experimental data and estimated results. Then conclusion is described with some discussions.

---

[1] http://skyrad.sci.u-toyama.ac.jp/
[2] http://modis.gsfc.nasa.gov/

II. Proposed Method

A. The Proposed Method

Atmospheric correction is important for estimation of surface reflectance (Remote Sensing Reflectance [3]) in particular for estimation of sea surface reflectance estimation. The proposed atmospheric correction method is based on Skyradiometer data derived aerosol size distribution and refractive index. The aerosol refractive index and size distribution can be estimated by using SkyradPack[4] with direct and diffuse solar irradiance those are measured with skyradiometer. Scattering phase function, extinction as well as scattering and absorption coefficients and asymmetry index are then estimated by using mie2new software code with the estimated refractive index and size distribution. Meantime, geometric relation among the satellite sensor of MODIS onboard AQUA satellite is estimated with MODIS Level 1B product. These estimated values are set to the input parameters (Tape 5) of MODTRAN[5] of atmospheric radiative transfer code. Other input parameters are set at the default values. In the process of estimation of the Top of the Atmosphere: TOA Radiance, MODTRAN is used.

The well-known bi-section method is used for estimation surface reflectance because TOA radiance is getting large in accordance with sea surface reflectance. First, initial value of the sea surface reflectance is assumed. By using the initial sea surface reflectance together with the aforementioned input parameters, all the required input parameters are set for MODTRAN. Then TOA radiance can be estimated based on MODTRAN. The estimated TOA radiance is compared to MODIS Level 1B product derived at sensor radiance. The sea surface reflectance can be estimated by minimizing the difference between TOA radiance and the at sensor radiance by changing the sea surface reflectance. The proposed process flow is shown in Figure 1.

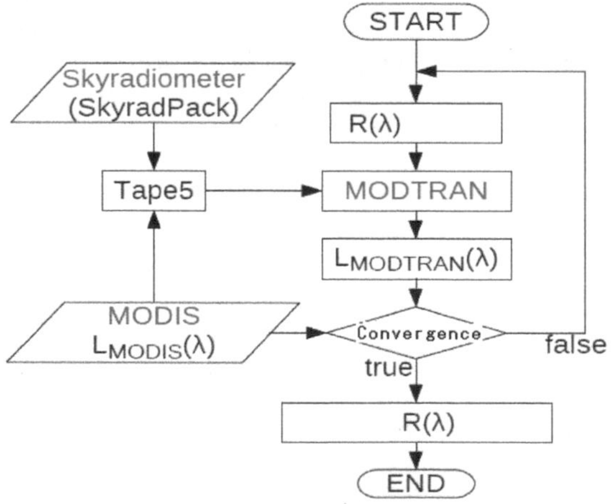

Fig. 1. Process flow of the proposed surface reflectance estimation method

R denotes the remote sensing reflectance while $L_{MODTRAN}$ denotes MODTRAN derived radiance. L_{MODIS} of SeaDAS[6] defined standard product of sea surface reflectance derived from MODIS data is used

The bi-section process is converged within 10 times of iterations because there is only one unknown parameter. The accuracy of this iterative process is around 0.0009765.

B. The Intensive Study Areas

Figure 2 shows the intensive study areas in the Ariake Sea area, Kyushu, Japan.

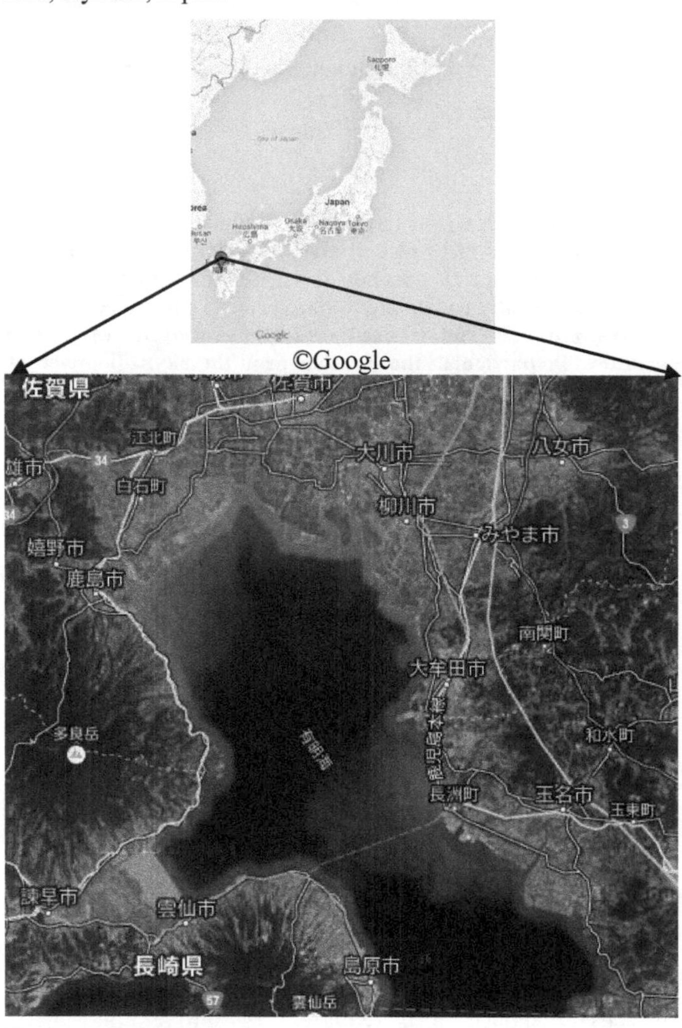

Fig. 2. Intensive study areas

III. Experiment

A. The Data Used

Terra/MODIS Level 1B of Band 8 to 16 product of Ariake Sea (Latitude: 32.82-33.25 N, Longitude: 130.05-130.65 E), Japan which is acquired at 02:20 (GMT) on May 1 2003 is used. The number of pixel data of Ariake Sea is 638 pixels (Ground resolution of MODIS is 1 km). MODIS Level 1B imagery data is shown in Figure 3.

MODIS on Terra L1B, 2003/5/1 02:20
Latitude 32.82 - 33.25 degree, Lon 130.05 - 130.65 degree

---

[3] https://books.google.co.jp/books?id=Sy_4jIcRmvUC&pg=PA36&dq=remote+sensing+reflectance+SeaDAS&hl=ja&sa=X&ved=0CBwQ6AEwAGoVChMIip6IntLyxwIVoiimCh1LTAC0#v=onepage&q=remote%20sensing%20reflectance%20SeaDAS&f=false

[4] SkyradPack is available from the University of Tokyo, Nakajima, et al., 1996

[5] http://modtran5.com/

[6] http://seadas.gsfc.nasa.gov/

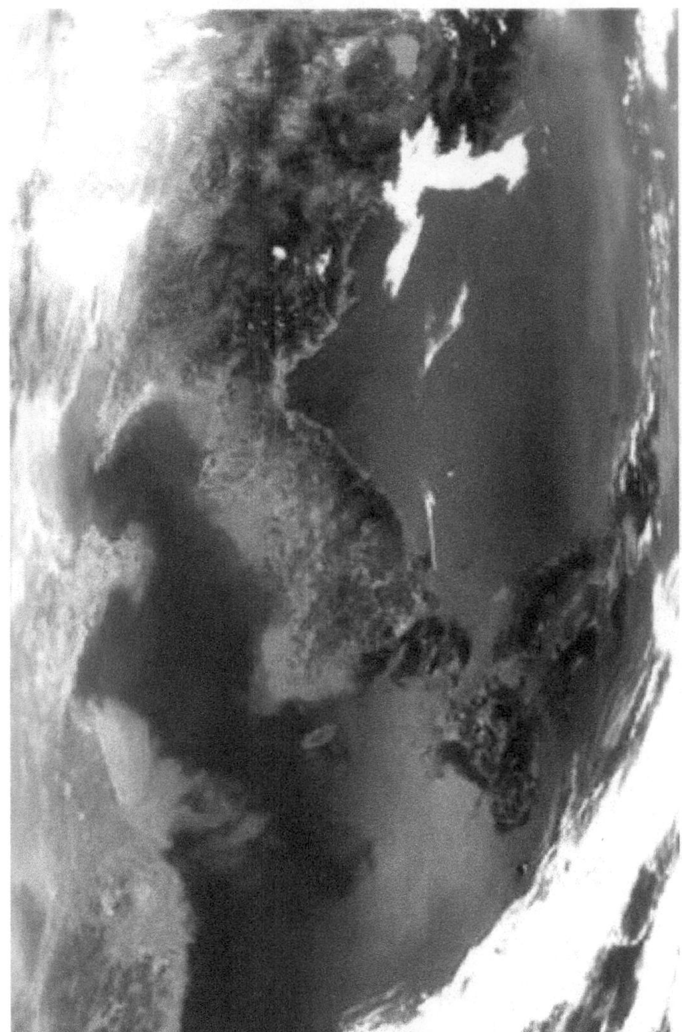

(a)Portion of MODIS image

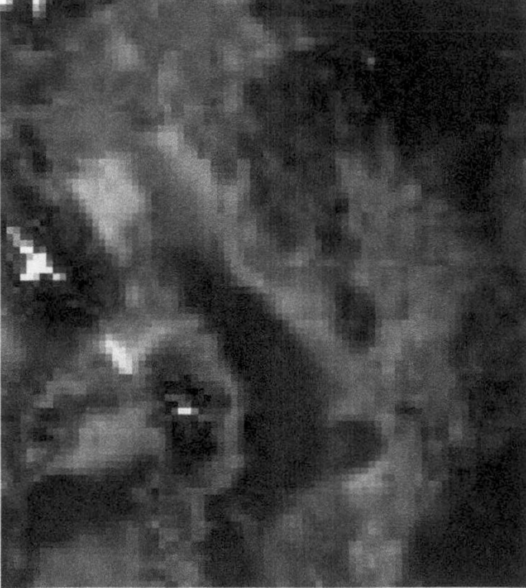

(b)MODIS image of the intensive study area of Ariake Sea

Fig. 3. MODIS image of the intensive study area of Ariake Sea acquired on May 1 2003

The locations of MODIS pixels of the intensive study area of Ariake Sea are shown in Figure 4.

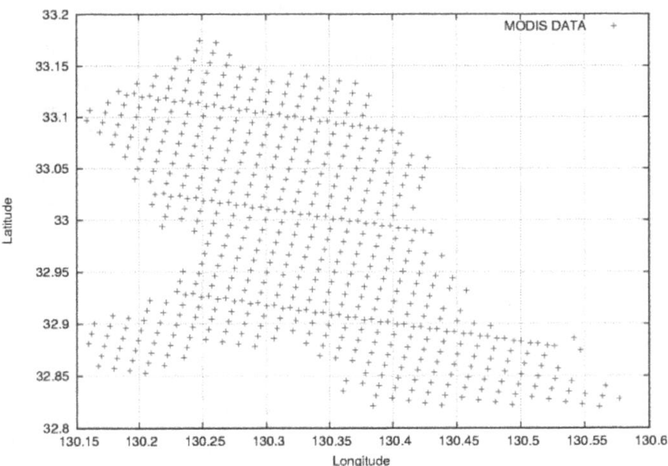

Fig. 4. Location of the MODIS pixels of the intensive study area

B. The Experimental Results

MODIS band number, center wavelength, Root Mean Square Difference: RMSD between MODIS standard product of surface reflectance and SeaDAS defined standard remote sensing reflectance, the estimated remote sensing reflectance by the proposed method with the default input parameters of the used MODTRAN (Mid-Latitude Summer), and the estimated remote sensing reflectance by the proposed method with the input parameters including phase function of aerosols are shown in Table 1.

TABLE I. ROOT MEAN SQUARE DIFFERENCE: RMSD COMPARISONS

Band	Center Wavelength(nm)	RMSD(1/str)		
		SeaDAS	Default	Proposed
8	412	0.00283	0.00257	0.00156
9	443	0.00361	0.0019	0.00144
10	488	0.00552	0.00341	0.0022
11	531	0.00692	0.00444	0.00336
12	551	0.00677	0.00396	0.003
13	667	0.00221	0.0016	0.00149
14	678	0.00216	0.00154	0.00146
15	748	0.000495	0.000651	0.000739
16	869	0.000248	0.00082	0.000831

In the Table 1, "Default" denotes the proposed method with the default input parameters of the atmosphere without using skyradiometer data while "Proposed" denotes the proposed method with using skyradiometer data. From the table, it may say that the remote sensing reflectance by the proposed method is much closer than the others to the standard product of surface reflectance, L1B product derived remote sensing reflectance in particular for shorter wavelength rages from 412 to 678 nm. Meanwhile, SeaDAS defined remote sensing reflectance is much closer than the others for the longer wavelength ranges from 748 to 869 nm (Near infrared wavelength region). Therefore, it may say that it would be better to use the measured skyradiometer data for improvement of estimation accuracy of surface reflectance. Moreover, the TOA radiance (at sensor radiance) can be estimated simultaneously for vicarious calibration, in particular.

C. Sensitivity Analysis

The relations between aerosol refractive index (Real and Imaginary parts) and extinction coefficient, scattering coefficient, absorption coefficient, and asymmetry parameter are investigated at the wavelengths, 340, 380, 400, 500, 550, 675, 870, and 1020nm (relatively transparent wavelength). Figure 5 shows the relations for the real part of the refractive index of aerosol and extinction, scattering, absorption coefficients and asymmetry parameter while Figure 6 shows those for the imaginary part of the refractive index and extinction, scattering, absorption coefficients and asymmetry parameter.

Extinction coefficient consists scattering and absorption coefficients of aerosol particles. On the other hand, asymmetry parameter is an asymmetric characteristic of aerosol scattering phase function. Rayleigh scattering phase function is symmetry while Mie scattering phase function is asymmetry (Forward scattering is dominant). Optical property of aerosol particles can be expressed with these coefficients and asymmetry parameter. Influencing components of aerosol particles to the optical property are refractive index and size distribution. Refractive index consists of real and imaginary parts, complex function. Real part represents refractive component of aerosol particles while imaginary part expresses absorptive component. There are some approximated size distribution functions of aerosol particles. Log-Normal distribution, Power Law distribution as well as Junge distribution functions are representatives. Therefore, the relations among these parameters are examined in these figures,

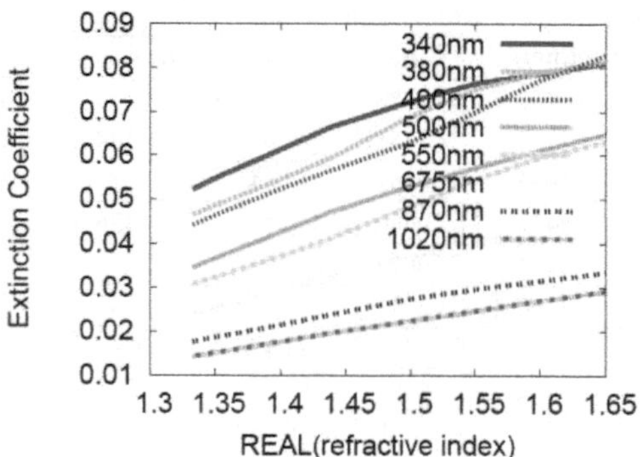

(a)Extinction Coefficient

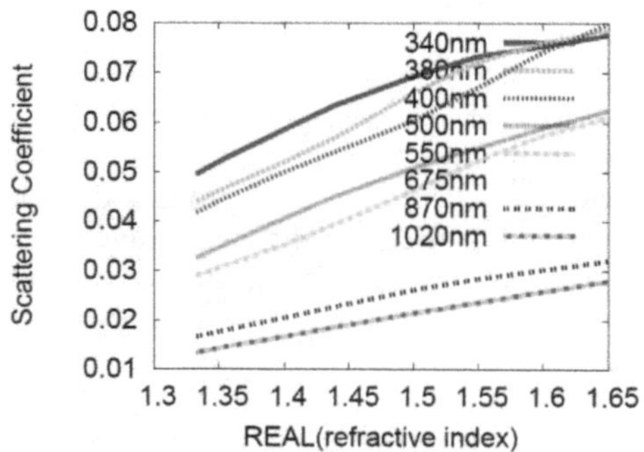

b)Scattering Coefficient

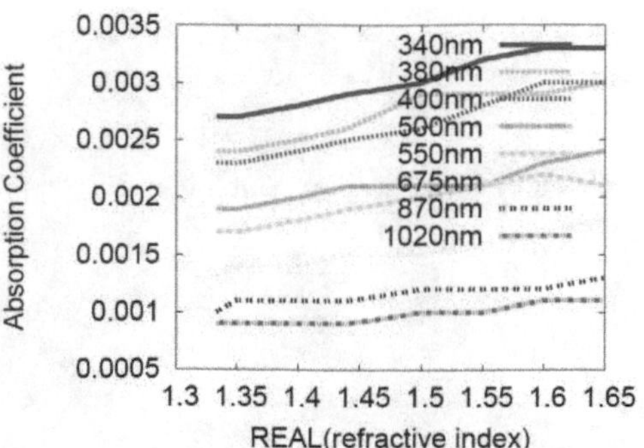

(c)Absorption Coefficient

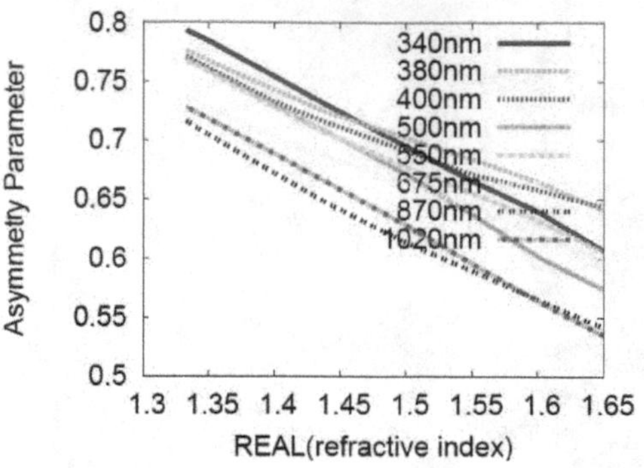

(d)Asymmetry Parameter

Fig. 5. Relations between real part of refractive index and extinction, scattering, absorption coefficients and asymmetry parameter

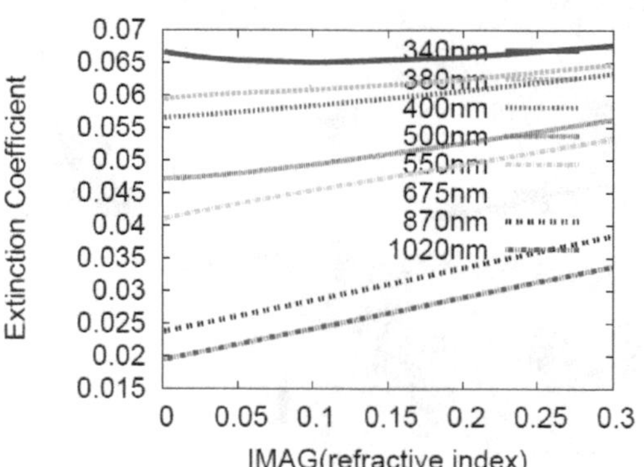

(a)Extinction Coefficient

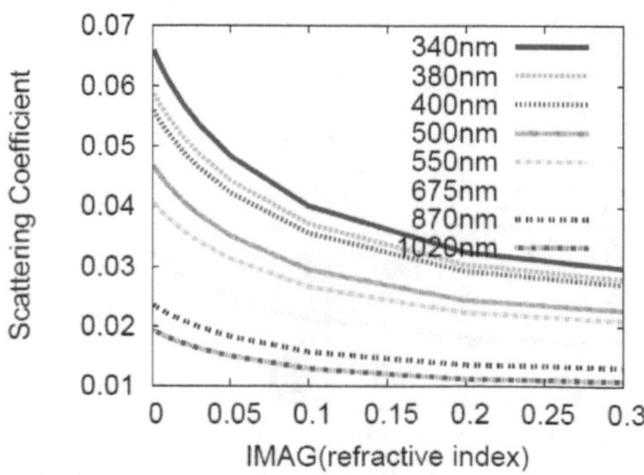

(b)Scattering Coefficient

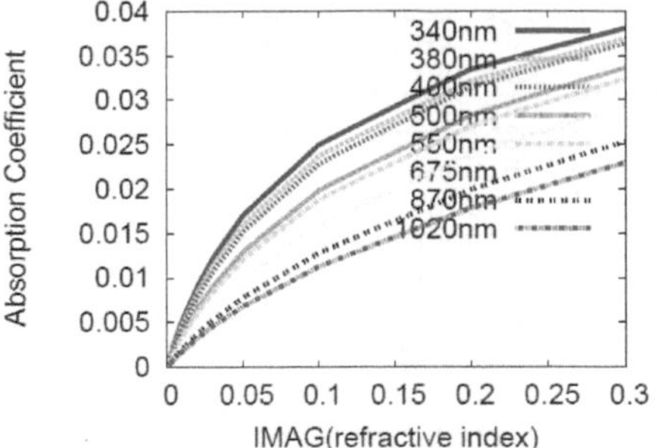

(c)Absorption Coefficient

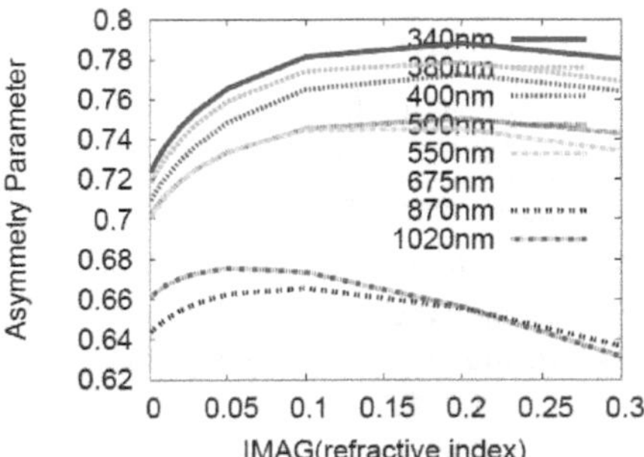

(d)Asymmetry Parameter

Fig. 6. Relations between imaginary part of the refractive index and Extinction, Scattering, Absorption coefficients, and asymmetry parameter

Figure 7 (a) shows a typical size distribution function of volume spectrum while Figure 7 (b) shows a typical size distribution of number spectrum (logarithmic function of aerosol particle number). In the figures, dark blue size distributions are measured at Saga University on October 15 2008. Red colored linear function shows Junge distribution with Junge parameter v in the equations (1) and (2). As shown in these figures, in usual, size distribution can be expressed with bi-modal function of Log-Normal function and is based on Power Law expression.

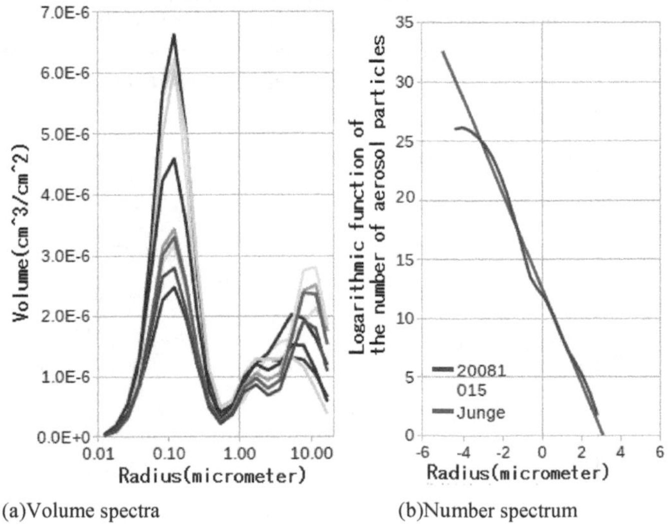

(a)Volume spectra (b)Number spectrum

Fig. 7. Typical aerosol size distributions

On the other hand, Figure 8 and 9 shows the relations between Junge parameter and extinction, scattering, absorption coefficients and asymmetry parameter as well as the coefficient "C" of the truncated Power Law Distribution function of aerosol size distribution (equations (1) and (2)) and extinction, scattering, absorption coefficients and asymmetry parameter, respectively. There are two appropriate aerosol distribution functions,

Power Law and Log-Normal Distributions. Meanwhile, there are four major atmospheric components, extinction, scattering, absorption coefficients and asymmetry parameter. Power Law Distribution function is as follows,

$$n(r) = C10^{v+1} \quad (r \leq 0.1\mu m) \quad (1)$$

$$n(r) = Cr^{-(v+1)} \quad (r > 0.1\mu m) \quad (2)$$

where n, r, C denotes the number of aerosol particles, radius of aerosol particles, and coefficient.

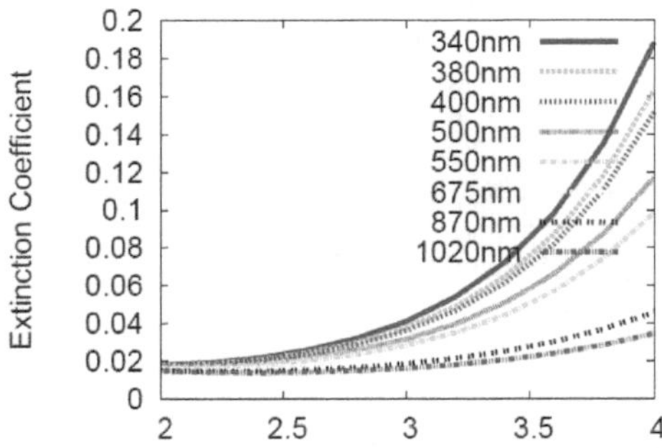

(a)Extinction Coefficient

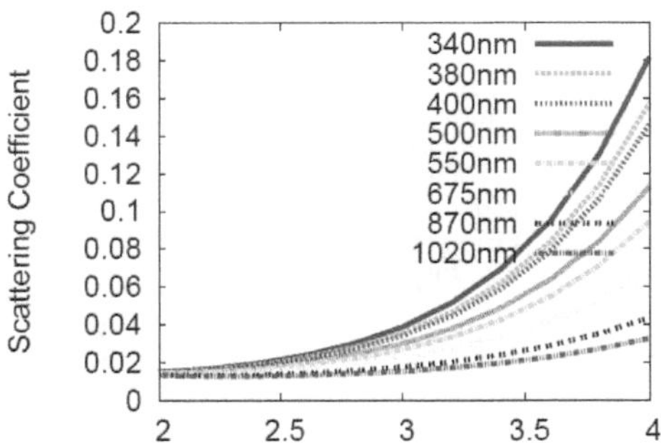

(b)Scattering Coefficient

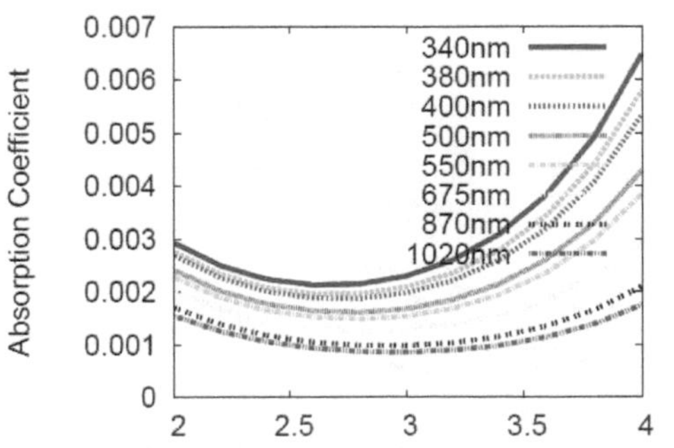

(c)Absorption Coefficient

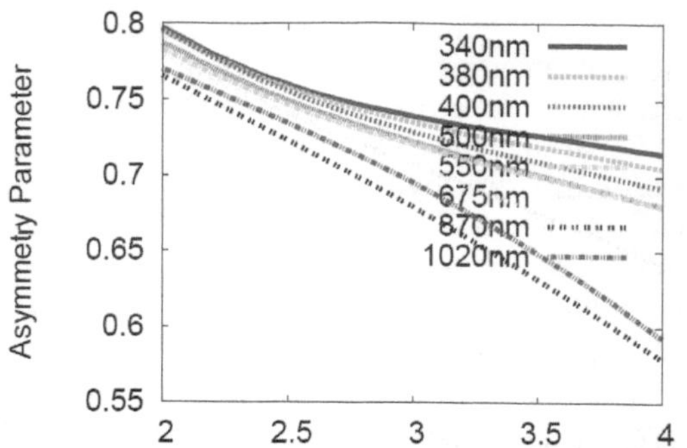

(d)Asymmetry Parameter

Fig. 8. Relation between Junge parameter for the truncated Power Law Distribution and extinction, scattering, absorption coefficients and asymmetry parameter

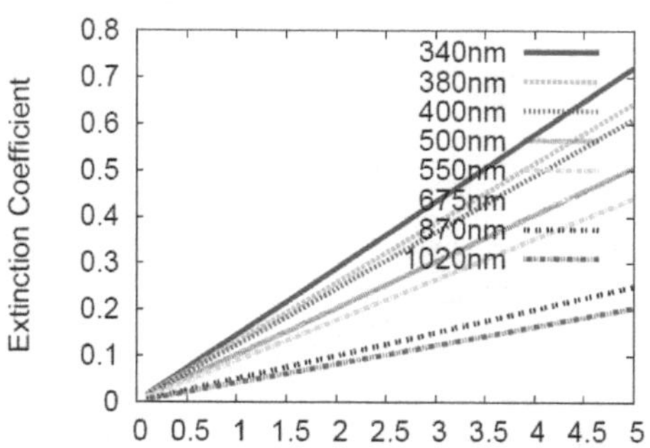

(a)Extinction Coefficient

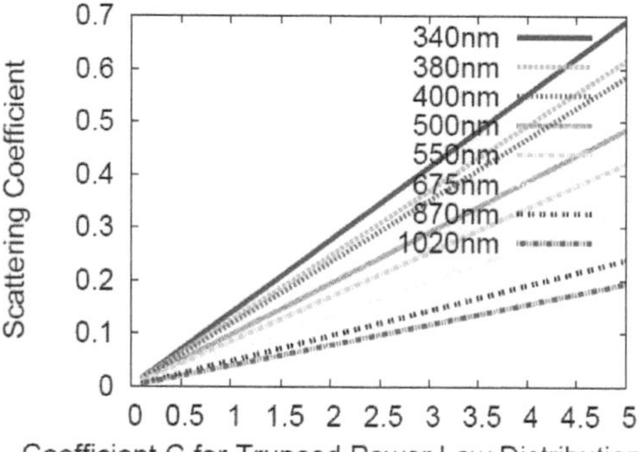

(b)Scattering Coefficient

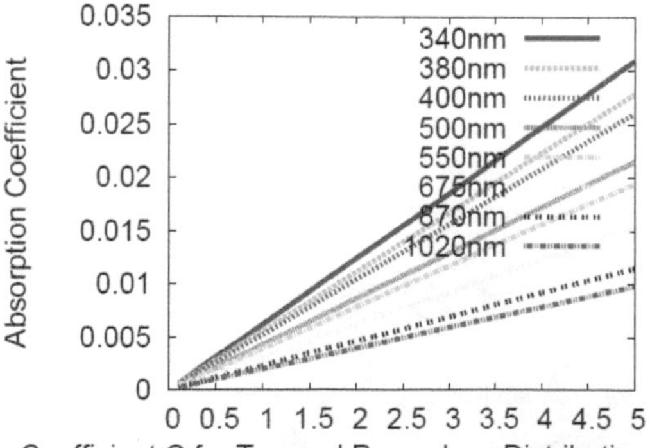

(c)Absorption Coefficient

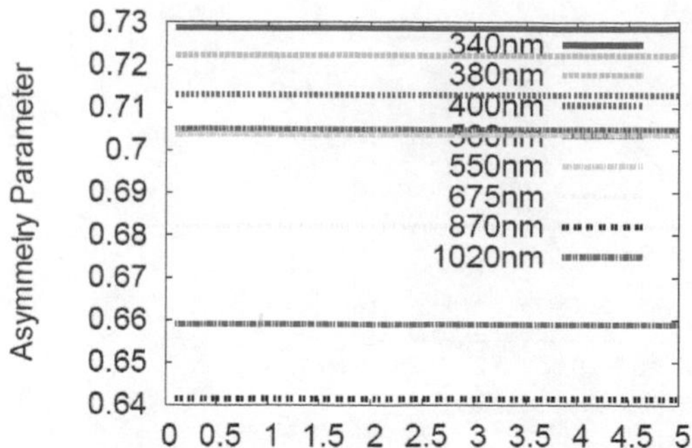

(d)Asymmetry Parameter

Fig. 9. Relation between the coefficient C for truncated Power Law Distribution and extinction, scattering, absorption coefficients and asymmetry parameter

If the Log-Normal Distribution is assumed for aerosol size distribution, then the results from the sensitivity analysis are shown in Figure 10. Log-Normal Distribution function is as follows,

$$\log \sigma_g = (\Sigma n_i(\log D_i - \log D_g)^2/(N-1))^{-1/2} \quad (3)$$

Where,

σ_g = geometric standard deviation (GSD)
D_i = midpoint particle diameter of the ith bin
n_i = number of particles in group i having a midpoint size Di
N = $\sigma_{ni,}$ the total

The parameter for the Log-Normal Distribution is as follows,

n=1.0 [cm^{-3}]
σ_g=0.4[micrometer]

There is a parameter for the Log-Normal Distribution, averaged distribution of n. The sensitivity of extinction, scattering, and absorption coefficients as well as asymmetry parameter are varied by the averaged distribution as shown in Figure 10. It is necessary to care about these sensitivity as well as selection of aerosol size distribution function for the convergence process in the proposed process flow which is shown in Figure 1.

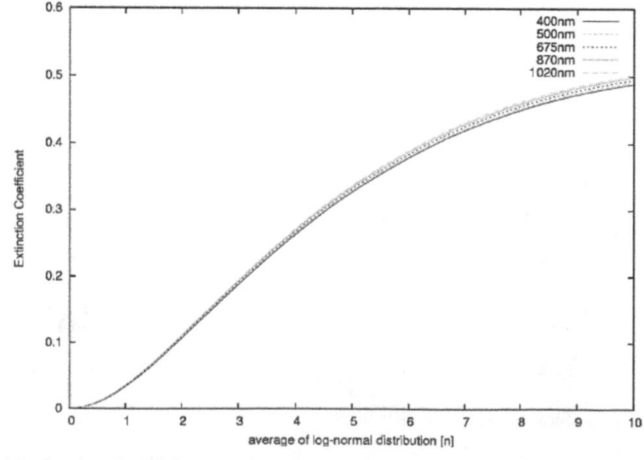

(a)Extinction Coefficient

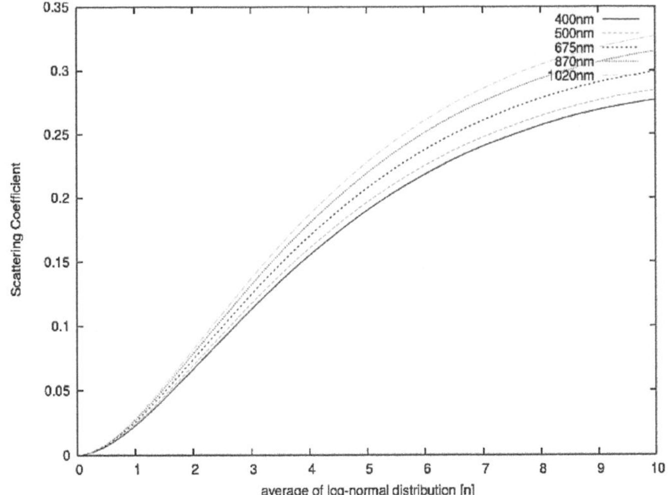

(b)Scattering Coefficient

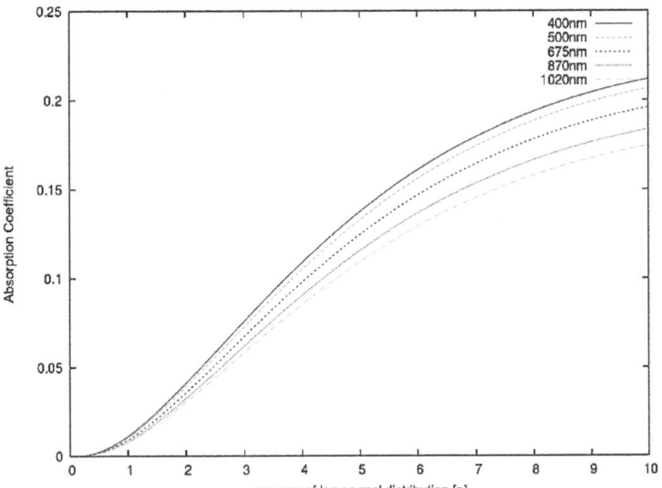

(c)Absorption Coefficient

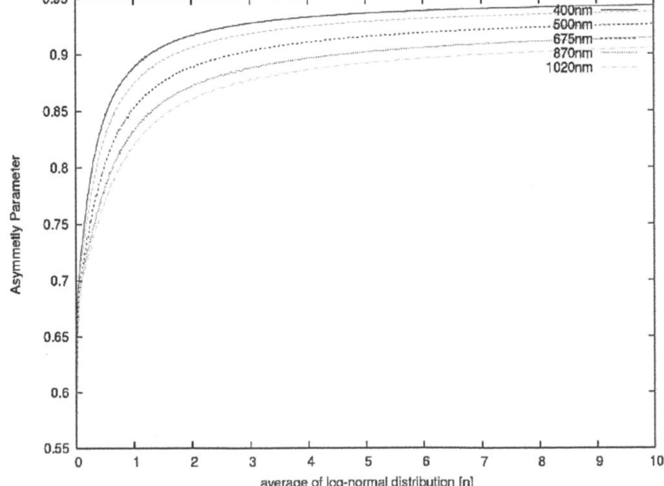

(d)Asymmetry Parameter

Fig. 10. Results from the sensitivity analysis assuming Log-Normal Distribution for aerosol size distribution

IV. Conclusion

Through experiments with the standard surface reflectance product of MODIS and the estimated remote sensing reflectance based on SeaDAS processing software, and the proposed bi-section based convergence process of estimation method with skyradiometer data derived aerosol refractive index and size distribution, it is found that the proposed method with skyradiometer data is superior to the SeaDAS derived remote sensing reflectance.

Further investigations are required for selection of appropriate aerosol size distribution function. The experiment is conducted with the assumed Junge ditribution with the parametrization of Junge parameter. It, however would better to take the other aerosol size distribution functions, Log-Normal, and Power Law distributions from the results of the sensitivity analysis.

References

[1] Ramachandran, Justice, Abrams(Edt.),Kohei Arai et al., Land Remote Sensing and Global Environmental Changes, Part-II, Sec.5: ASTER VNIR and SWIR Radiometric Calibration and Atmospheric Correction, 83-116, Springer 2010.

[2] Kohei Arai, Atmospheric correction and vicarious calibration of ADEOS/AVNIR and OCTS, Advances in Space Research, Vol.25, No.5, pp.1051-1054, 2000.

[3] Kohei Arai, Atmospheric correction and residual error in vicarious calibration of AVNIR and OCTS both onboard ADEOS, Advances in Space Research, Vol.25, No.5, pp.1055-1058, 2000.

[4] K.Arai, Atmospheric correction and vicarious calibration of ADEOS/AVNIR and OCTS, Advances in Space Research, Vol.25, No.5, pp.1051-1054, (2000).

[5] K.Arai, Atmospheric correction and residual errors in cross calibrationof AVNIR and OCTS both onboard ADEOS, Advances in Space Research, Vol.25, No.5, pp.1055-1058, (2000).

[6] Chrysoulakis,Abrams, Feidas and Kohei Arai, Comparison of Atmospheric correction methods using ASTER data for the area of Crete, Greece, International Journal of Remote Sensing, 31,24,6347-6385,2010.

[7] K.Arai, Monte Carlo simulation of polarized atmospheric irradiance for determination of refractive index of aerosols, International Journal of Research and Review on Computer Science, 3, 4, 1744-1748, 2012.

[8] O.Uchino, T.Sakai, T.Nagai, I.Morino, K.Arai, H.Okumura, S.Takubo, T.Kawasaki, Y.mano, T.Matsunaga, T.Yokota, On recent stratspheric aerosols observed by Lidar over Japan, Journal of Atmospheric Chemistry and Physics, 12, 11975-11984, 2012(doi:10.5194/acp-12, 11975-2012).

[9] K.Arai, Monte Carlo ray tracing based sensitivity analysis of the atmospheric and oceanic parameters on the top of the atmosphere radiance, International Journal of Advanced Computer Science and Applications, 3, 12, 7-13, 2012.

[10] K.Arai Error analysis on estimation method for air temperature, atmospheric pressure, and relative humidity using absorption due to CO2, O2, and H2O which situated at around near infrared wavelength regions, International Journal of Advanced Computer Science and Applications, 3, 12, 192-196, 2012.

[11] Kohei Arai, Method for estimation of aerosol parameters based on ground based atmospheric polarization irradiance measurements, International Journal of Advanced Computer Science and Applications, 4, 2, 226-233, 2013

[12] Kohei Arai, Sensitivity analysis and validation of refractive index estimation method with ground based atmospheric polarized radiance measurement data, International Journal of Advanced Computer Science and Applications, 4, 3, 1-6, 2013.

[13] O.Uchino, T.Sakai, T.Nagai, I.Morino, T.Maki, M.Deushi, K.Shibata, M.Kajino, T.Kawasaki, T. Akaho, S.Takubo, H.Okumura, Kohei Arai, M.Nazato, T.Matsunaga, T.Yokota, Y.Sasano, DIAL measurement of

lower tropospheric ozone over Saga (33.24N, 130.29E) in Japan and comparison with a chemical climate model, Journal of Atmospheric Measurement Techniques, 7, 171-194, 2014.

[14] Kohei Arai, Comparative study among least square method, steepest descent method, and conjugate gradient method for atmospheric sounder data analysis, International Journal of Advanced Research in Artificial Intelligence, 2, 9, 30-37, 2013.

[15] Kohei Arai, Sensitivity analysis for aerosol refractive index and size distribution estimation methods based on polarized atmospheric irradiance measurements, International Journal of Advanced Research in Artificial Intelligence, 3, 1, 16-23, 2014.

[16] Kohei Arai, Aerosol refractive index retrievals with atmospheric polarization measuring data, Proceedings of the SPIE, 7461-06, 1-9, 2009.

[17] Kohei Arai, Reflectance based vicarious calibration of ASTER/VNIR with aerosol refractive index and size distribution estimation using measured atmospheric polarization irradiance, Proceedings of the SPIE, 7461-08, 1-9, 2009.

22

Psychological Status Monitoring with Cerebral Blood Flow; CBF, Electroencephalogram; EEG and Electro-Oculogram; EOG Measurements

Kohei Arai 1
Graduate School of Science and Engineering
Saga University
Saga City, Japan

Abstract—**Psychological status monitoring with cerebral blood flow (CBF), EEG and EOG measurements are attempted. Through experiments, it is confirmed that the proposed method for psychological status monitoring is valid. It is also found correlations among the amplitudes of peak alpha and beta as well as gamma frequency of EEG signals and EOG as well as cerebral blood flow. Therefore, psychological status can be monitored with either EEG measurements or cerebral blood flow and EOG measurements.**

Keywords—Cerebral Blood Flow; CBF; EEG; EOG; psychological status

I. INTRODUCTION

Psychological status monitoring is getting much important for health care. Eye based psychological status monitoring is proposed and applied to a variety of fields such as Electric Wheel Chair control, e-learning system, etc.[1]-[22].

The methods and measuring instruments are proposed and well developed now a day. Relations among the psychological status monitored with cerebral blood flow (CBF), EEG (Electroencephalogram: EEG) and EOG (Electro-oculogram: EOG) measurements are not clarified. EEG sensor used to be affected by sounding noises. Insuch case EOG or CBF is useful. Furthermore, CBF is very expensive compared to the other two. If relations among CBF, EEG and EOG are clarified, then EEG and EOG can be used instead of CBF. There is no previous paper which deals with the relation among CBF, EEG and EOG.

In order to clarify the relations, experiments are conducted with patients through rhythm gaming and adding gaming. When the patients play rhythm game, in general, psychological statuses of the patients are calm and relax while psychological statuses of the patients are severe and irritated when they play adding game. Through the experiments, this paper intends to clarify the relations. Furthermore, appropriate monitoring method and system as well as measuring instruments for psychological status is clarified.

The paper is organized as follows. First, the method and procedure for psychological status monitoring is described followed by some experimental methods and procedures together with experimental results. Then some concluding remarks are described with some discussions.

II. METHOD AND PROCEDURE FOR PSYCHOLOGICAL STATUS MONITORING

EEG and EOG sensors of ZA[1] manufactured by Pro-Assist Co. Ltd. is used in experiments together with Near Infrared: NIR Spectroscopy (NIRS) of HOT 121-B manufactured by Hitachi Co. Ltd. for cerebral blood flow measurements. Table 1 and 2 show the major specifications of the EEG and EOG sensors of ZA and HOT 121-B of NIRS.

TABLE I. MAJOR SPECIFICATION OF EEG AND EOG MEASURING INSTRUMENT NAMED "ZA"

Electrodes	EEG and EOG
AD_converter	12_bit
Sampling_Frequency	128(kHz)
Band_Width	Brain_Wave:0.5-40Hz&Eye_Vol.:0.5-10Hz

TABLE II. MAJOR SPECIFICATION OF CEREBRAL BLOOD FLOW (CBF) MEASURING INSTRUMENT, HOT 121B

Sampling	100ms
Wave_Length	810nm
Repetation_Cycle	2kHz
Temp.Sensor	±1 degree C
Acceleration_Sensor	±2G
EMI	VCC-1_Class B
Application	Cerebral blood flow (left and &right), heart rate, LF/HF, Attitude

In Table 2, LF/HF denotes the ratio of Sympathetic to Parasympathetic which is called heart rate changing index. Sympathetic is dominant when patients are in irritated, active and having stress status. Therefore, LF/HF is increased in such time period. On the other hand, LF/HF is decreased when patients are in relaxing, taking a rest, and sleeping status because parasympathetic is dominant in such time period.

The experiments are conducted with 5 patients. In the experiments, each patient takes a rest for 1 minute and then plays for 1 minute with three games; adding game, rhythm game and breakout destroy game separately. It is suspected that most of patients are in relax status when they playing with rhythm game while are in irritating status when they plays with adding game and breakout destroy game. They have to have an instruction on how to play the games before getting start a set of experiment (1 minute for a rest and then 1 minute

[1] ZA is the type name of the measuring instrument.

for gaming). They have to have 15 sets of experiment each. All sets of experiment is finished within a hour.

EEG data is filtered by low pass filter with cut off frequency of 50 Hz (6dB octave) for noise removals. After that FFT is applied to the filtered EEG. One of the examples is shown in Figure 1.

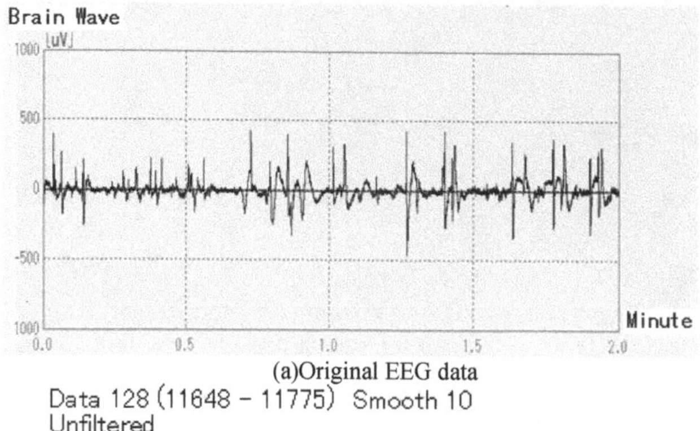

(a)Original EEG data

Data 128 (11648 - 11775) Smooth 10
Unfiltered

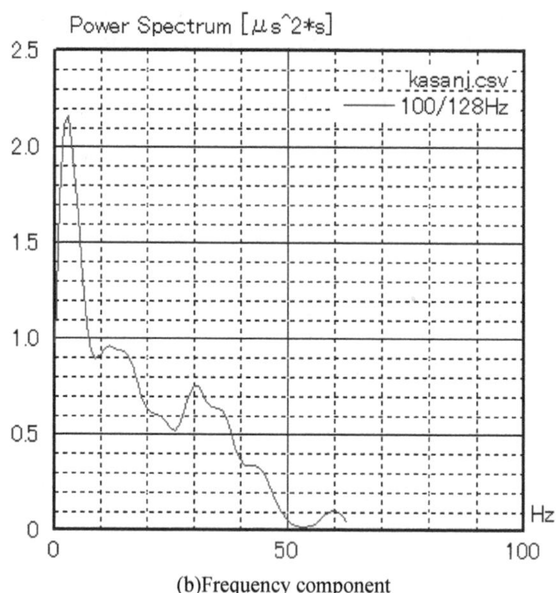

(b)Frequency component

Fig. 1. Examples of EEG and its frequency component

Usually, frequency components of 0-4 Hz, 4-8 Hz, 8-12 Hz, 12-40 Hz are named delta, theta, alpha and beta frequencies. In particular, alpha frequency component is dominant when users are in relaxing status while beta frequency component is large when they are in irritating status. When their EEG data is acquired, they have to attach electrodes on their forehead. This is the same thing for EOG measurements. They have to attach electrodes at the end of their eyes.

EOG data, on the other hand, show eye movement behavior which is reflected users' psychological status. Namely, EOG data is calm when they are in relaxing status while EOG data varied rapidly and quickly when they are in irritating status. Figure 2 shows an example of EOG data. From the data, eye movement speed can be analyzed.

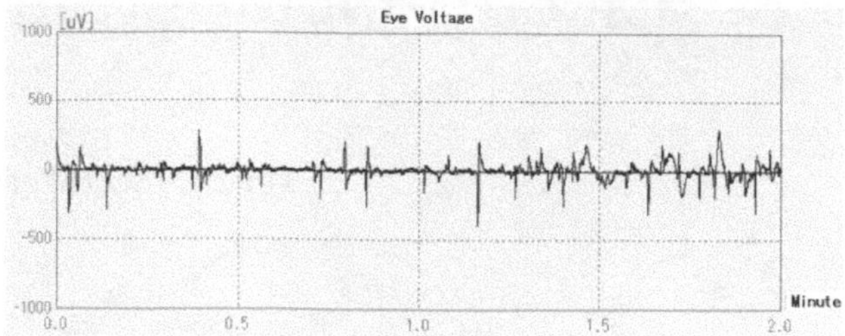

Fig. 2. Example of EOG data

Meanwhile, cerebral blood flow data shows varied rapidly and quickly when they are in irritating status while cerebral blood flow data shows calm when they are in relaxing status. It is expected that then they play with adding game, they are used to in an irritating status while they are in relaxing status when they play with rhythm game.

Figure 3 shows an example of acquired cerebral blood flow data. There are two data of cerebral blood flows, left brain (Red colored line in Figure 3) and right brain (Blue colored line in Figure 3). Also, left and right heart rate is acquired with HOT 121-B sensor. In meantime, LF/HF of right brain and left brain are also measured.

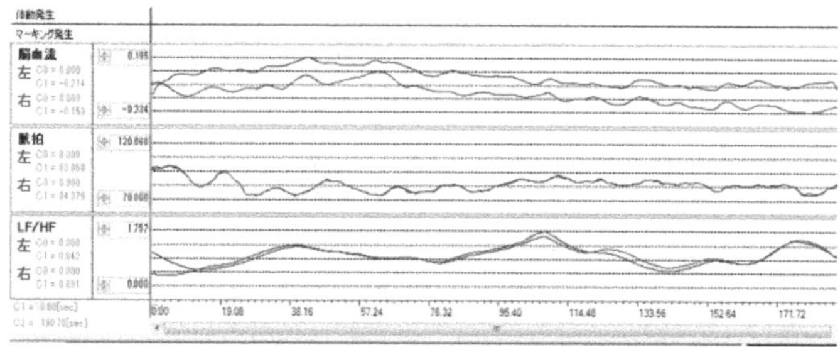

Fig. 3. shows an example of acquired cerebral blood flow data

III. Experimental Results

One of the typical measured data of cerebral blood flow, heart rate, LF/HF of one of the patients is shown in Figure 4 (a) together with EEG in Figure 4 (b) and EOG in Figure 4 (c). During the first half time, the patient takes a rest and plays rhythm game during the second half time period. As shown in Figure 4, there is not so large difference of the measured data between the first and the second half time periods. Therefore, most of the patients are in relaxing status when they play rhythm game.

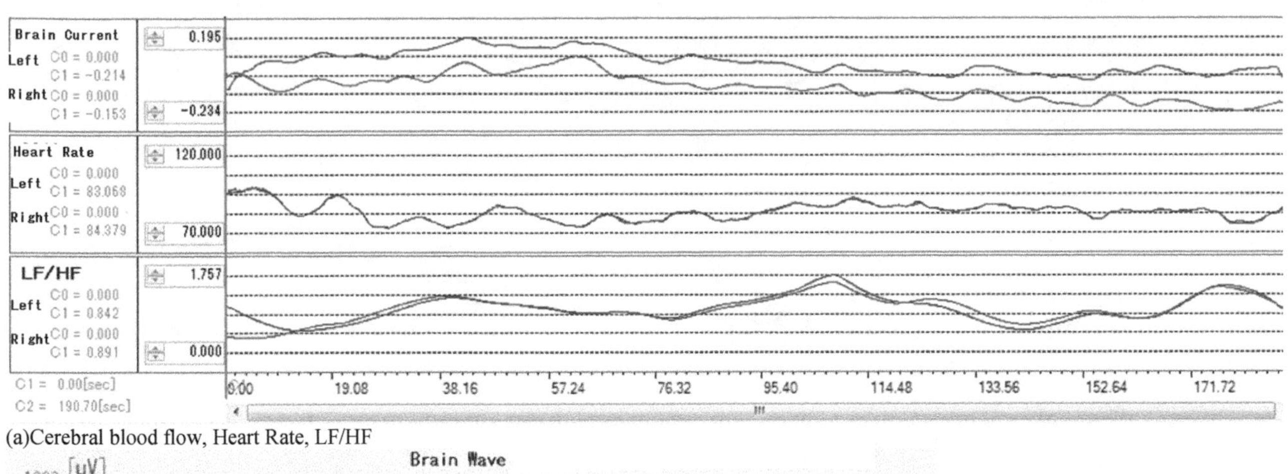

(a)Cerebral blood flow, Heart Rate, LF/HF

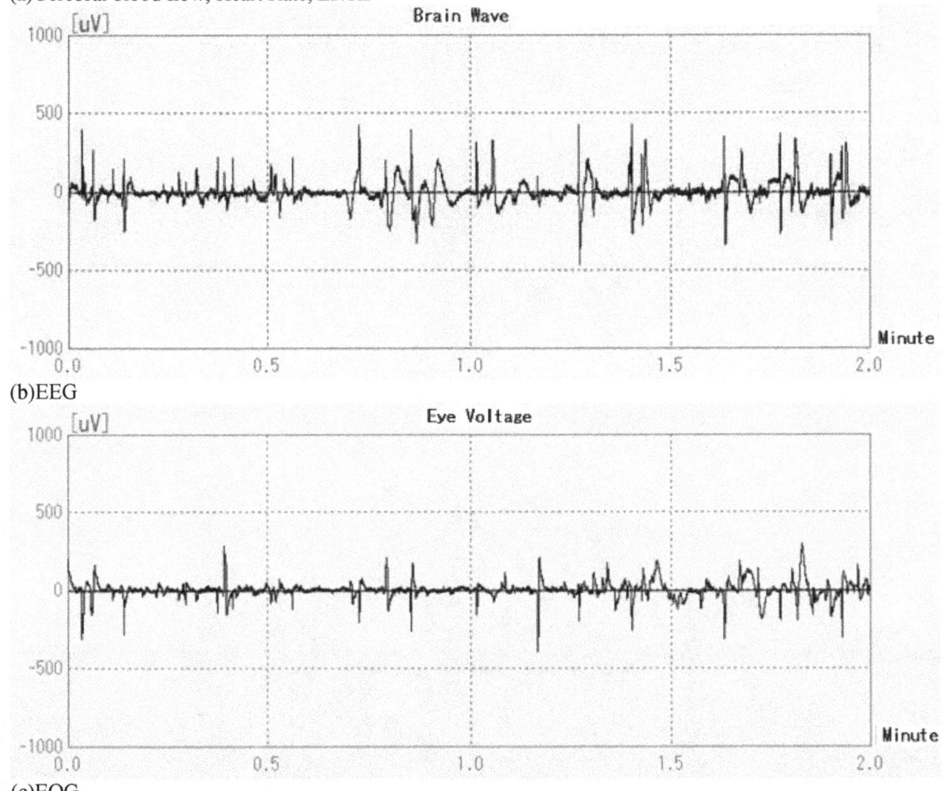

(b)EEG

(c)EOG

Fig. 4. Measured data when the typical patient plays Rhythm Game

One of the typical measured data of cerebral blood flow, heart rate, LF/HF of one of the patients is shown in Figure 5 (a) together with EEG in Figure 5 (b) and EOG in Figure 5 (c). During the first half time, the patient takes a rest and plays adding game during the second half time period. As shown in Figure 5, there is relatively large difference of the measured data between the first and the second half time periods. Therefore, most of the patients are in irritating status when they play adding game.

Meanwhile, frequency component of the measured EEGs when the patient plays rhythm game is shown in Figure 6 (a) while that for adding game is shown in Figure 6 (b).

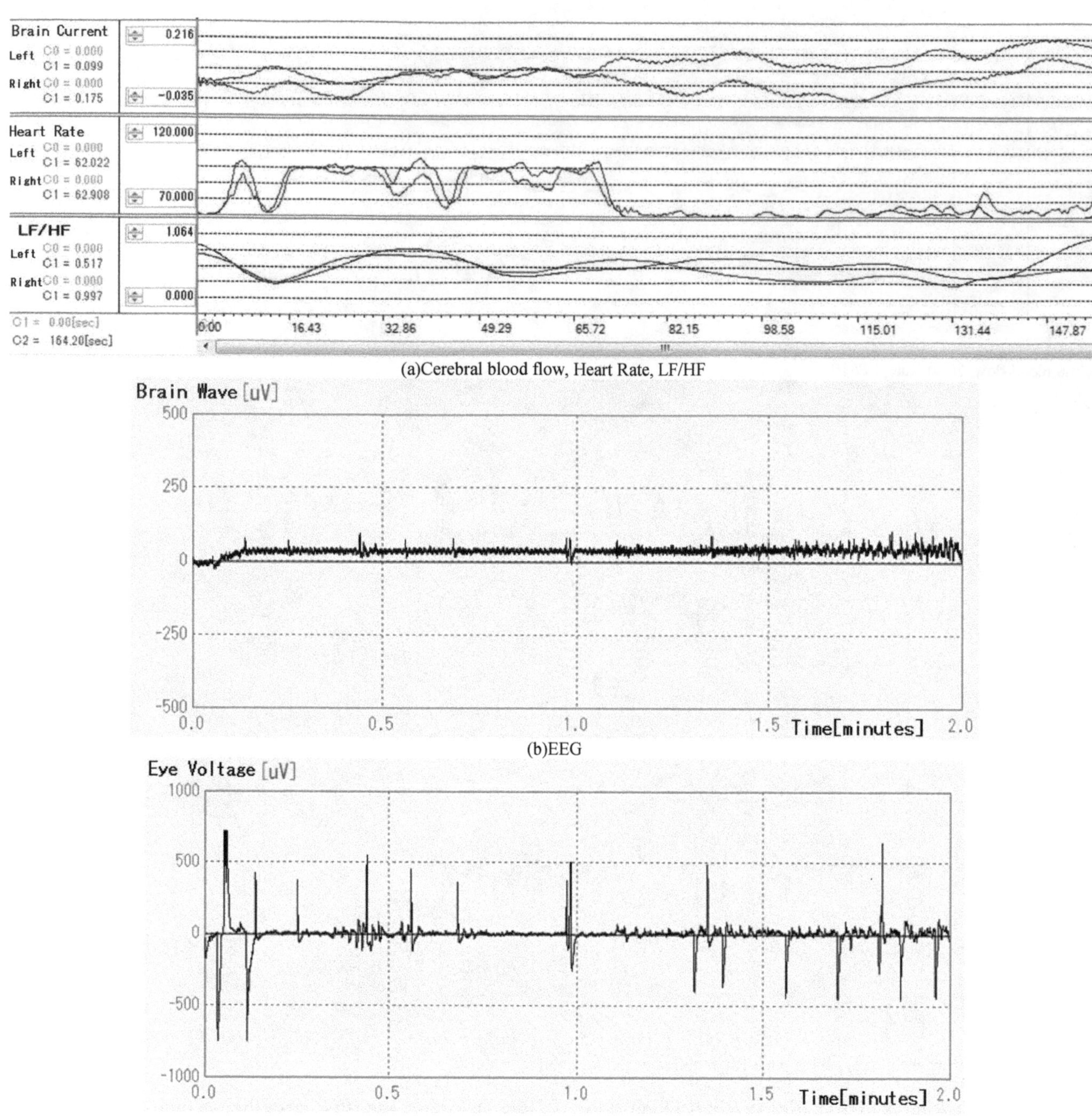

(a)Cerebral blood flow, Heart Rate, LF/HF

(b)EEG

(c)EOG

Fig. 5. Measured data when the typical patient plays Adding Game

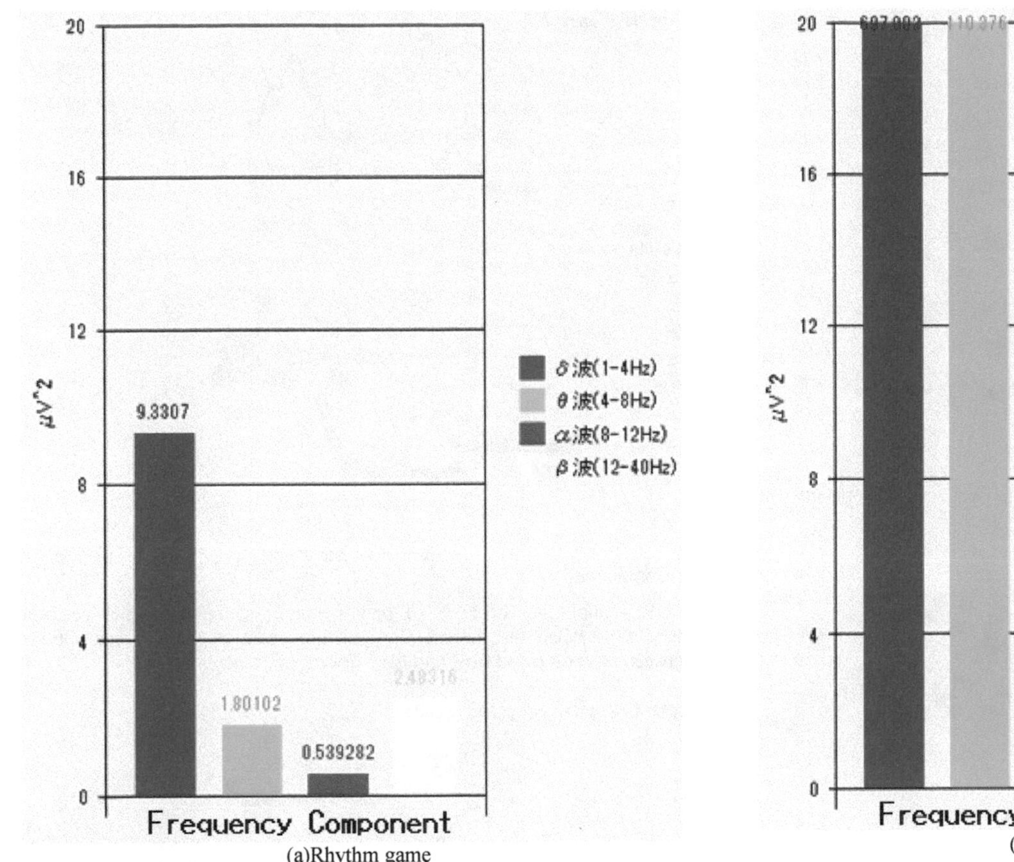

(a)Rhythm game

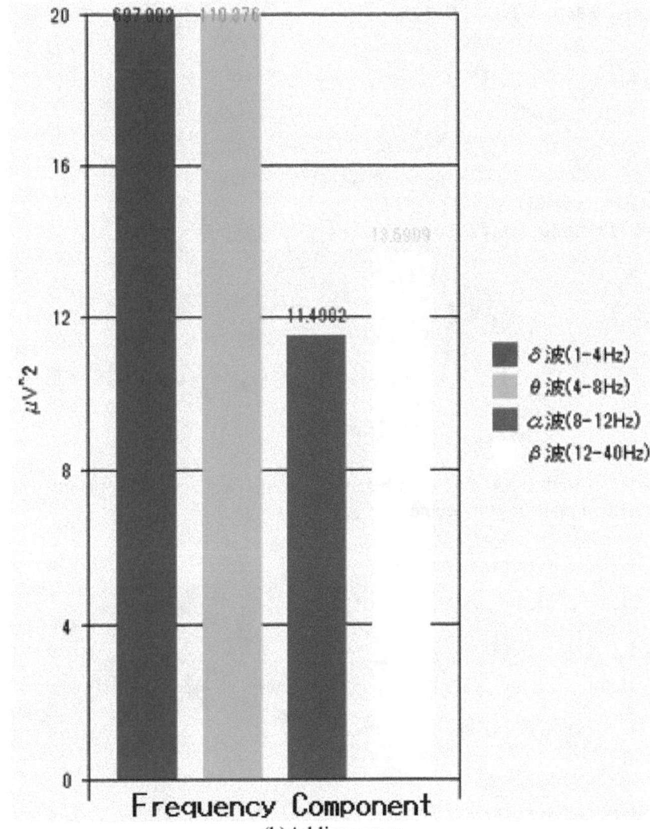

(b)Adding game

Fig. 6. Frequncy component of EEGs when the patient plays rhythm game and adding game

TABLE III. FREQUENCY COMPONENTS OF THE MEASURED EEGS WHEN THE PATIENT PLAYS THYTHM GAME AND ADDING GAME

Game	Rhythm	Adding
δ	697.003	9.3307
θ	110.376	1.80102
α	11.4002	0.53928
β	13.5909	2.48316

As shown in Table 3, alpha wave is relatively small in comparison to beta wave when the patient palys adding game rather than rhythm game.

The amplitudes of peak alpha frequency, peak beta frequncy, as well as peak gamma frequency (more than 30Hz) are used to be evaluated as psychological status indexes. These are used to be highly correlated to cerebral blood flow, heart rate, and LF/HF. The amplitudes of peak beta frequency and peak gamma frequency together with EOG of the patient when he plays rhythm game are plotted in Figure 7 (a), (b), and (c), respectively. EOG is highly correlated to the amplitudes of beta wave and gamma wave as shown in Figure 8. Correlation coefficients between EOG and beta wave amplitude is around 0.77 while that between EOG and gamma wave is 0.89, respectively. Therefore, it implies that the patient is iiritated and stressed because his eyes move so rapidly and quickly.

On the other hand, the amplitudes of peak beta frequency and right cerebral blood flow of the patient when he plays rhythm game are plotted in Figure 9 (a), and (b), respectively. Right cerebral blood flow is correlated to the amplitudes of beta wave as shown in Figure 9 (c). Correlation coefficients between right cerebral blood flow and beta wave amplitude is around 0.45.

Meanwhile, Figure 10 (a) and (b) shows amplitudes of EEG of gamma frequency components and cerebral blood flow when the patient plays breakout game as well as correlations between cerebral blood flow and EEG of gamma frequency (Figure 10 (c)). Correlation coefficients between right cerebral blood flow and beta wave amplitude of EEG is around 0.81.

For the breakout game, in general, CBF is increased together with beta and gamma waves while EOG signal amplitude is relatively large. For the adding game, beta wave is increased while EOG signal amplitude is relatively small. On the other hand, CBF is decreased together with gamma wave while EOG signal amplitude is comparatively small for the rhythm game.

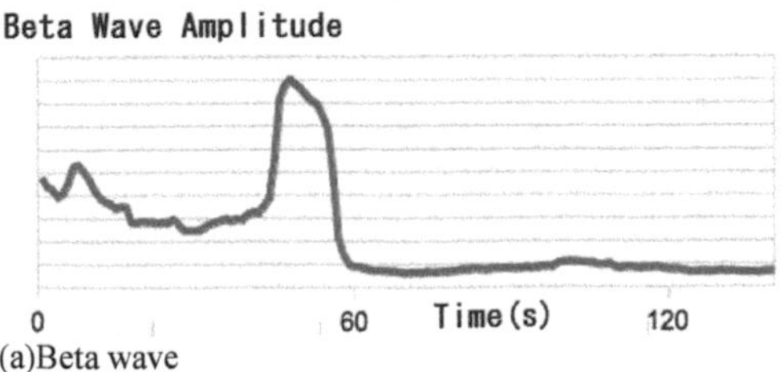

(a)Beta wave

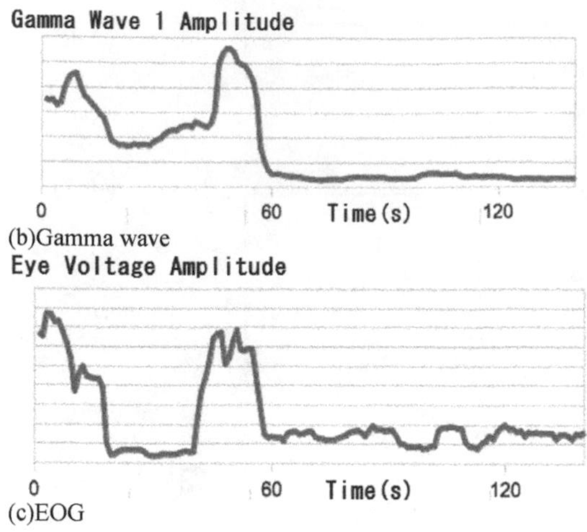

(b)Gamma wave

(c)EOG

Fig. 7. Amplitudes of EEG of beta and gamma frequency components when the patient plays rhythm game

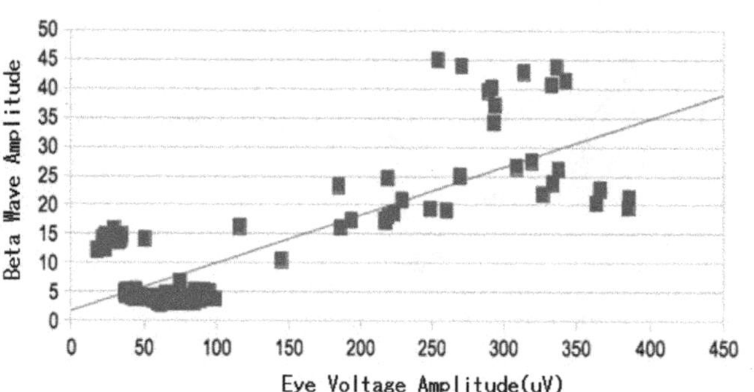

(a)Beta wave

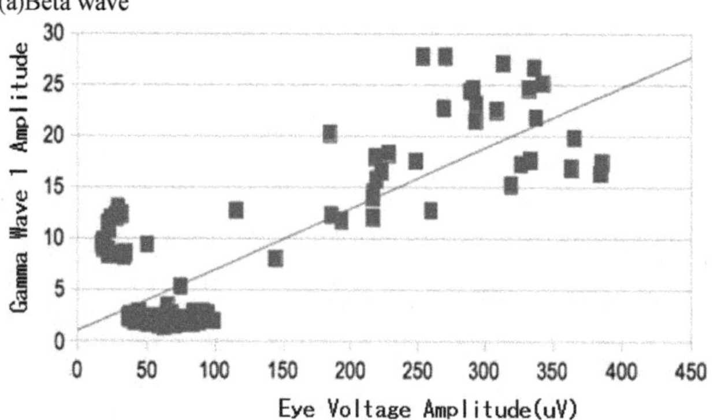

(b)Gamma wave

Fig. 8. Correlations between EOG and EEG of beta wave as well as gamma wave

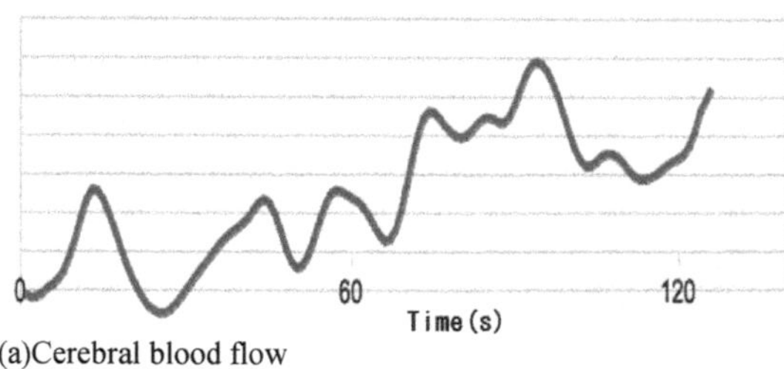

(a)Cerebral blood flow

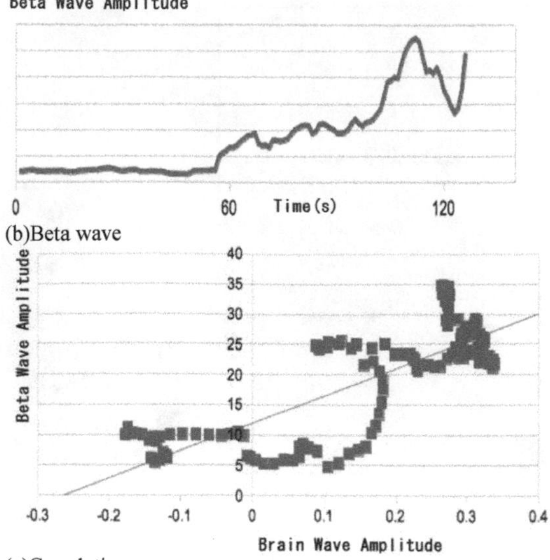

(b)Beta wave

(c)Correlation

Fig. 9. Amplitudes of EEG of beta frequency components and cerebral blood flow when the patient plays adding game as well as correlations between cerebral blood flow and EEG of beta frequency

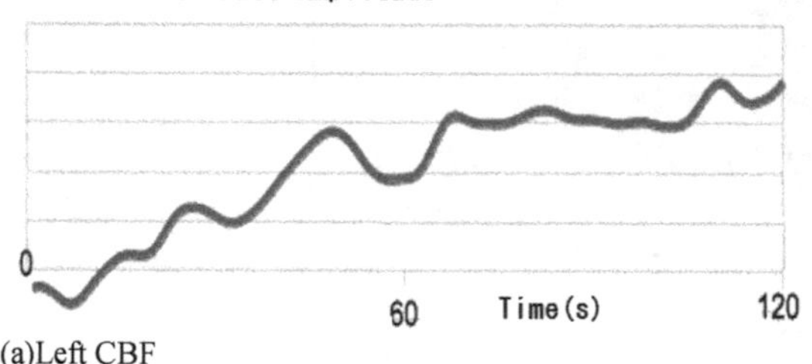

(a)Left CBF

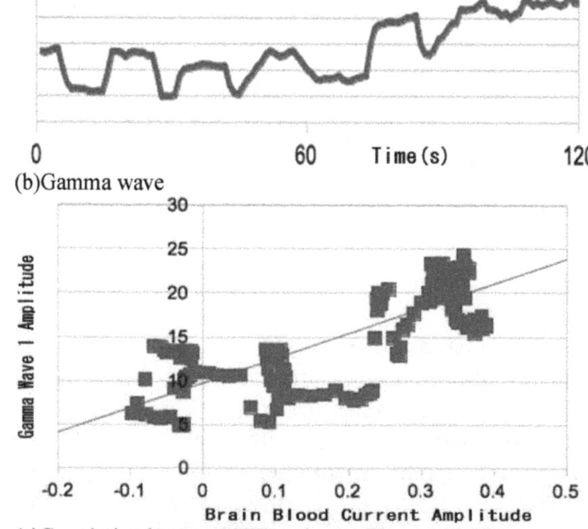

(b)Gamma wave

(c)Correlation between CBF and gamma wave of EEG

Fig. 10. Amplitudes of EEG of gamma frequency components and cerebral blood flow when the patient plays breakout game as well as correlations between cerebral blood flow and EEG of gamma frequency

As the experimental results, it is found that the followings,

A. *Breakout game:*

1) three patients out of five patients show high correlations between CBF and gamma wave of frequency component of EEG signals

2) two patients out of five patients show relatively high correlation between CBF and beta wave of frequency component of EEG signals

B. *Adding game:*

1) Two patients out of five patients show comparatively high correlation between CBF and beta wave of frequency component of EEG signals

C. *Rhythm game:*

1) Three patients out of four patients show high correlation between EOG signal amplitude and beta/gamma frequency component of EEG signals

2) Two patients out of four patients show relatively high correlation between CBF and gamma wave of frequency component of EEG

IV. CONCLUSION

Psychological status monitoring with cerebral blood flow (CBF), EEG (EEG) and EOG (EOG) measurements are attempted. Through experiments, it is confirmed that the proposed method for psychological status monitoring is valid. It is also found correlations among the amplitudes of peak alpha and beta EEGs and EOG as well as cerebral blood flow. Therefore, psychological status can be monitored with either EEG measurements or cerebral blood flow and EOG measurements.

It is found that three patients out of five patients show high correlations between CBF and gamma wave of frequency component of EEG signals for breakout game, two patients out of five patients show relatively high correlation between CBF and beta wave of frequency component of EEG signals for breakout game, two patients out of five patients show comparatively high correlation between CBF and beta wave of frequency component of EEG signals for adding game, three patients out of four patients show high correlation between EOG signal amplitude and beta/gamma frequency component of EEG signals for rhythm game, two patients out of four patients show relatively high correlation between CBF and gamma wave of frequency component of EEG for rhythm game.

From these experimental results, it may conclude that these EEG, EOG, and CBF are highly correlated. Therefore, these measurements can be used alternatively. CBF measuring instruments are relatively expensive than the others. EEG and EOG sensors are very sensitive to the surrounding noises rather than the others.

ACKNOWLEDGMENT

The author would like to thank Mr. Tsuyoshi Miyazaki for his effort to the experiment.

REFERENCES

[1] Djoko Purwanto, Ronny Mardiyanto and Kohei Arai, Electric wheel chair control with gaze detection and eye blinking, Artificial Life and Robotics, AROB Journal, 14, 694,397-400, 2009.

[2] Kohei Arai and Makoto Yamaura, Computer input with human eyes only using two Purkinje images which works in a real time basis without calibration, International Journal of Human Computer Interaction, 1,3, 71-82,2010

[3] Kohei Arai, Ronny Mardiyanto, A prototype of electric wheel chair control by eye only for paralyzed user, Journal of Robotics and Mechatronics, 23, 1, 66-75, 2010.

[4] Djoko Purwanto, Ronny Mardiyanto, Kohei Arai, Electric wheel chair control with gaze detection and eye blinking, Proceedings of the International Symposium on Artificial Life and Robotics, GS9-4, 2009.

[5] Kohei Arai and Kenro Yajima, Communication Aid and Computer Input System with Human Eyes Only, Electronics and Communications in Japan,Volume 93, Number 12, 2010, pages 1-9, John Wiley and Sons, Inc., 2010.

[6] Kohei Arai, Ronny Mardiyanto, Evaluation of users' impact for using the proposed eye based HCI with moving and fixed keyboard by using eeg signals, International Journal of Research and Reviews on Computer Science, 2, 6, 1228-1234, 2011.

[7] Kohei Arai, Kenro Yajima, Robot arm utilized having meal support system based on computer input by human eyes only, International Journal of Human-Computer Interaction, 2, 1, 120-128, 2011.

[8] Kohei Arai, Ronny Mardiyanto, Autonomous control of eye based electric wheel chair with obstacle avoidance and shortest path finding based on Dijkstra algorithm, International Journal of Advanced Computer Science and Applications, 2, 12, 19-25, 2011.

[9] Kohei Arai, Ronny Mardiyanto, Eye-based human-computer interaction allowing phoning, reading e-book/e-comic/e-learning, Internet browsing and TV information extraction, International Journal of Advanced Computer Science and Applications, 2, 12, 26-32, 2011.

[10] Kohei Arai, Ronny Mardiyanto, Eye based electric wheel chair control system-I(eye) can control EWC-, International Journal of Advanced Computer Science and Applications, 2, 12, 98-105, 2011.

[11] Kohei Arai, Ronny Mardiyanto, Electric wheel chair controlled by human eyes only with obstacle avoidance, International Journal of Research and Reviews on Computer Science, 2, 6, 1235-1242, 2011.

[12] Kohei Arai, Ronny Mardiyanto, Evaluation of users' impact for using the proposed eye based HCI with moving and fixed keyboard by using eeg signals, International Journal of Research and Reviews on Computer Science, 2, 6, 1228-1234, 2011.

[13] K.Arai, R.Mardiyanto, Evaluation of users' impact for using the proposed eye based HCI with moving and fixed keyboard by using eeg signals, International Journal of Research and review on Computer Science, 2, 6, 1228-1234, 2012.

[14] K.Arai, R.Mardiyanto, Electric wheel chair controlled by human eyes only with obstacle avoidance, International Journal of Research and review on Computer Science, 2, 6, 1235-1242, 2012.

[15] R.Mardiyanto, K.Arai, Eye-based Human Computer Interaction (HCI) A new keyboard for improving accuracy and minimizing fatigue effect, Scientific Journal Kursor, (ISSN 0216-0544), 6, 3, 1-4, 2012.

[16] K.Arai, R.Mardiyanto, Moving keyboard for eye-based Human Computer Interaction: HCI, Journal of Image and Electronics Society of Japan, 41, 4, 398-405, 2012.

[17] Kohei Arai, Ronny Mardiyanto, Eye-based domestic robot allowing patient to be self-services and communications remotely, International Journal of Advanced Research in Artificial Intelligence, 2, 2, 29-33, 2013.

[18] Kohei Arai, Ronny Mardiaynto, Method for psychological status estimation by gaze location monitoring using eye-based Human-Computer Interaction, International Journal of Advanced Computer Science and Applications, 4, 3, 199-206, 2013.

[19] Kohei Arai, Kiyoshi Hasegawa, Method for psychological status monitoring with line of sight vector changes (Human eyes movements) detected with wearing glass, International Journal of Advanced Research in Artificial Intelligence, 2, 6, 65-70, 2013.

[20] Kohei Arai, Wearable computing system with input output devices based on eye-based Human Computer Interaction: HCI allowing location based web services, International Journal of Advanced Research in Artificial Intelligence, 2, 8, 34-39, 2013.

[21] Kohei Arai Ronny Mardiyanto, Speed and vibration performance as well as obstacle avoidance performance of electric wheel chair controlled by human eyes only, International Journal of Advanced Research in Artificial Intelligence, 3, 1, 8-15, 2014.

[22] Kohei Arai Ronny Mardiyanto, Speed and vibration performance as well as obstacle avoidance performance of electric wheel chair controlled by human eyes only, International Journal of Advanced Research in Artificial Intelligence, 3, 1, 8-15, 2014.

Wildlife Damage Estimation and Prediction Using Blog and Tweet Information

Kohei Arai
Graduate School of Science and Engineering
Saga University
Saga City, Japan

Shohei Fujise
Graduate School of Science and Engineering
Saga University
Saga City, Japan

***Abstract*—Wildlife damage estimation and prediction using blog and tweet information is conducted. Through a regressive analysis with the truth data about wildlife damage which is acquired by the federal and provincial governments and the blog and the tweet information about wildlife damage which are acquired in the same year, it is found that some possibility for estimation and prediction of wildlife damage. Through experiments, it is found that R^2 value of the relations between the federal and provincial government gathered truth data of wildlife damages and the blog and the tweet information derived wildlife damages is more than 0.75. Also, it is possible to predict wildlife damage by using past truth data and the estimated wildlife damages. Therefore, it is concluded that the proposed method is applicable to estimate and predict wildlife damages.**

Keywords—Wildlife damage; Blog; Tweet; Big data analysis; Natural language recognition

I. Introduction

Wildlife damage in Japan is around 23 Billion Japanese Yen a year in accordance with the report from the Ministry of Agriculture, Japan. In particular, wildlife damages by deer and wild pigs are dominant (10 times much greater than the others) in comparison to the damage due to monkeys, bulbuls (birds), rats. Therefore, there are strong demands to mitigate the wildlife damage as much as we could. It, however, is not so easy to find and capture the wildlife due to lack of information about behavior. For instance, their routes, lurk locations are unknown and not easy to find. Therefore, it is difficult to determine the appropriate location of launch a trap. In Kyushu, Japan, wildlife damage is getting large and is one of severe problems for farmers as well as residents in the districts near the mountainous areas. The federal and provincial agricultural management organizations in the districts are surveying the wildlife damages every year. It is time consuming task and requires large budget. Also, it takes almost two years. Therefore, it is hard to make a plan for wildlife damage controls. It would be helpful to estimate and predict wildlife damages with some other methods. Meanwhile, blog and tweet information can gather with some software tools. Furthermore, it would be possible to extract some valuable information relating to wildlife damages. The method proposed here is to estimate and predict wildlife damages by using blog and tweet information. It can be done immediately after the end of the Japanese fiscal year. Therefore, wildlife damage prevention plan can be created by the end of the Japanese fiscal year.

The following section describes the proposed method for wildlife damage estimation and predictions followed by experimental data. Then, concluding remarks and some discussions are followed.

II. Literature and Related Work

According to the West, B. C., A. L. Cooper, and J. B. Armstrong, 2009, "Managing wild pigs: A technical guide. Human-Wildlife Interactions Monograph"[1], 1–551, there are the following wild pig damages, Ecological Impacts to ecosystems can take the form of decreased water quality, increased propagation of exotic plant species, increased soil erosion, modification of nutrient cycles, and damage to native plant species [1]-[5]. Agricultural Crops Wild pigs can damage timber, pastures, and, especially, agricultural crops [6]-[9]. Forest Restoration Seedlings of both hardwoods and pines, especially longleaf pines, are very susceptible to pig damage through direct consumption, rooting, and trampling [10]-[12]. Disease Threats to Humans and Livestock Wild pigs carry numerous parasites and diseases that potentially threaten the health of humans, livestock, and wildlife [13]-[15]. Humans can be infected by several of these, including diseases such as brucellosis, leptospirosis, salmonellosis, toxoplasmosis, sarcoptic mange, and trichinosis. Diseases of significance to livestock and other animals include pseudorabies, swine brucellosis, tuberculosis, vesicular stomatis, and classical swine fever [14], [16]-[18]. There also are some lethal techniques for damage managements. One of these is trapping. It is reported that an intense trapping program can reduce populations by 80 to 90% [19]. Some individuals, however, are resistant to trapping; thus, trapping alone is unlikely to be successful in entirely eradicating populations. In general, cage traps, including both large corral traps and portable drop-gate traps, are most popular and effective, but success varies seasonally with the availability of natural food sources [20]. Cage or pen traps are based on a holding container with some type of a gate or door [21]. The method and system for monitoring the total number of wild pigs in the certain district in concern is proposed [22]. All the aforementioned system is not so cheap. It requires huge resources of human-ware, hardware and software as well. Also, it is totally time consumable task. Usually, it takes two years to finalize the total number of wild animals and wildlife damages. Therefore, it is hard to plan the countermeasures for the wildlife damages.

[1] www.berrymaninstitute.org/publications,

III. Proposed Method

A. Methods for Acquisition of Blog and Tweet Information Relating to Wildlife Damages

There are some sites which allow acquisition of tweet and blog information. Fig.1 (a) shows one of the tweet information acquisition sites while Fig.1 (b) shows one of the blog information acquisition sites. For the tweet information acquisition site (https://dev.twitter.com/rest/public/search), the Search API is not complete index of all Tweets, but instead an index of recent Tweets. At the moment that index includes between 6-9 days of Tweets. Therefore, tweet information has to be acquired within 6-9 days after the event of wildlife appearance. It required some information collection robots. These examples are http://blog.ritlweb.com/ for blog information collection while http://twitter.com/ is for tweet information collections.

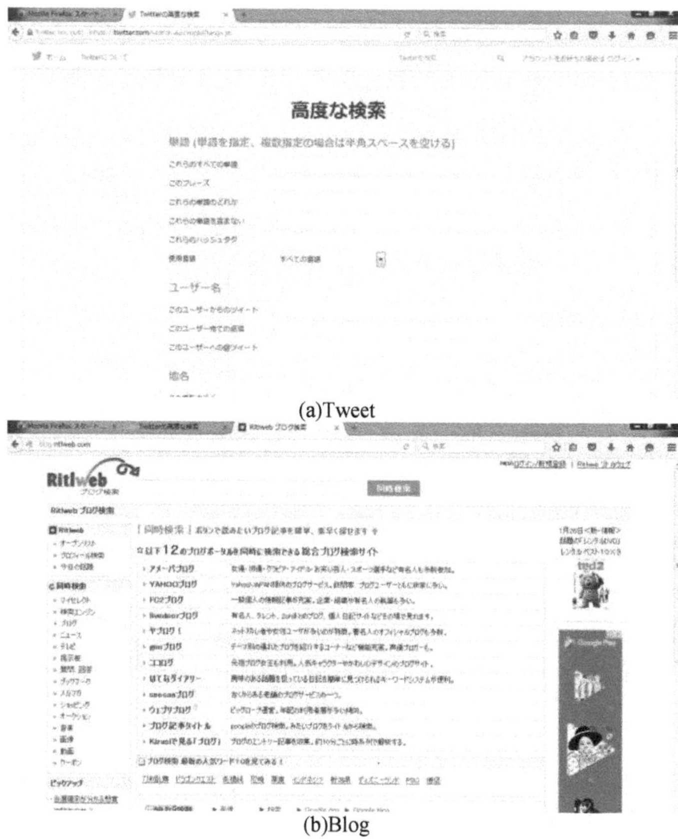

(a)Tweet

(b)Blog

Fig. 1. Examples of the tweet and the blog information acquisition sites

B. Methods for Extraction of Wildlife Damage Information from the Acquired Blog and Tweet Information

It has to be done to extract wildlife damage related information from the acquired blog and tweet information. The following set of three parameters have to be extracted, (1) the area name, (2) the types of wildlife damages, (3) the date of the wildlife damage reported. In order to extract sets of information, "Chasen" of sentence structure and words analysis software tool is used. It is morphological analysis tool. The extracted words and sentences acquired from the twitter and blog data collection sites are input to the "Chasen". Then noun and the other part of speech can be extracted as shown in Fig.2.

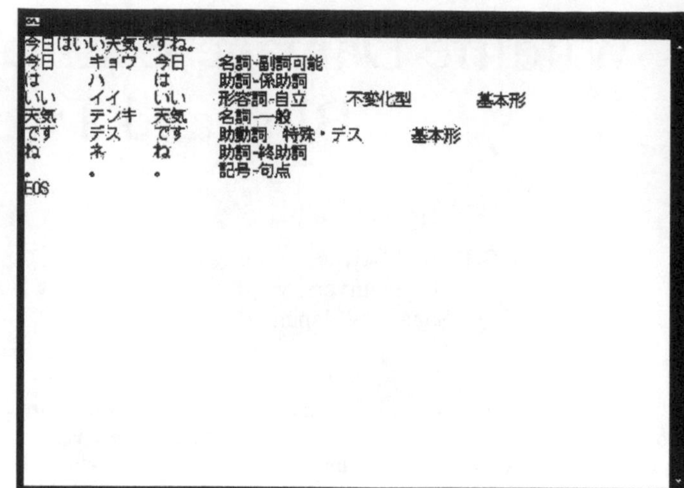

Fig. 2. Example of the screen shot of the Chasen analysis

The acquired sentence is "It is fine today" in Japanese and is appeared at the first line of the example. The first column of the second to the eighth lines "Today", "is" "Fine", "Weather", "it", and "is not it" show the words extracted from the acquired sentence. The second column shows their sounds while the forth column shows their part of speech. Thus, the words can be divided and extracted from the sentence together with their part of speech. Therefore, nouns can be extracted from the sentences. After that full text search is conducted to the extracted words.

Firstly, area names are extracted from the extracted words. In this regards, City name, Town name, and Village name in Kyushu provided by the federal and provincial governments are used in order to extract the area names. After that the names of the wildlife which is provided by the federal government of Agriculture, Forestry and Fishery ministry are extracted from the words. In this regards, combined words such as "prevention of bird damage" is recognized as the words of wildlife damage. The date of the tweet and blog information is easily extracted from the tweet and blob information because the information is dated information. Thus when, where, which wildlife can be extracted from the tweet and blog information.

C. Methods for Estimation of Wildlife Damage from the Acquired Tweet and Blob Information

The number of wildlife damage reports which are extracted from the acquired tweet and blog information in the year in concern must be proportional to the wildlife damages in that year. Therefore, linear regression would work for estimation of wildlife damage with the acquired tweet and blob information.

D. Methods for Prediction of Wildlife Damage Information from the Acquired Blog and Tweet Information

Based on the well known time series analysis method, it is possible to predict using the past wildlife damage. If the estimated wildlife damage with tweet and blob information is used for the wildlife damage in year in concern together with the past wildlife damage, then it is possible to predict future wildlife damage. In this regards, the following linear prediction is used for this,

$$y - \bar{y} = \frac{\sum_{i=1}^{n}(x_i - \bar{x})y_i}{\sum_{i=1}^{n}(x_i - \bar{x})^2}(x - \bar{x}) \quad (1)$$

where x and y denote the past wildlife damage and the current wildlife damage, respectively. xbar and ybar denote mean of the past and the current wildlife damage, respectively.

IV. EXPERIMENTS

A. Examplessof the Acquied Blog and Tweet Information Relating to Wildlife Damages

One of the examples of the tweet and blog information relating to wildlife damage is shown in Fig.3 (a). Meanwhile, the extracted words of area names and the types of wildlife are shown in Fig.3 (b) while the results from the wildlife damage estimated from the acquired tweet and blog information is shown in Fig.3 (c), respectively. The summarized results of the number of wildlife damage which are reported by twitter and blog at every province, Fukuoka, Saga, Nagasaki, Ohita, Kumamoto, Miyazaki, and Kagoshima prefectures in Kyushu in 2013.

A1 武雄「いのしし課

	A	B	C	D
1	武雄「いのしし課」猪突猛進佐賀11/24読売http://w			
2	「伊万里市では、増加するイノシシ等の有害鳥獣によ			
3	イノシシの被害は、大きい・・・佐賀では、山間に行く			
4	【イノシシ肉】イノシシによる田畑を荒らす被害に悩ま			
5	特定外来生物「アライグマ」についてアライグマは，も			
6	野生サル:小倉南出没、群れ九州最大規模120～1			
7	【#サル#福岡】北九州市小倉南区でサルによる農作			
8	サル被害対策検討会10/11NHK北九州http://www:			

(a)Tweet and blob

	A	B
1	佐賀	イノシシ
2	佐賀	イノシシ
3	佐賀	イノシシ
4	佐賀	イノシシ
5	福岡	アライグマ
6	福岡	サル
7	福岡	サル
8	福岡	サル
9		
10		

(b)Area name and types of wildlife name

1	県名	鳥獣計	鳥類計		獣類計			
2				カラス		イノシシ	サル	シカ
3	福岡	118	3	2	115	45	22	27
4	佐賀	82	8	1	74	45	5	19
5	長崎	103	2	2	101	58	12	17
6	熊本	18	0	0	18	3	0	4
7	大分	97	2	2	95	38	4	22
8	宮崎	50	2	2	48	18	6	24
9	鹿児島	0	0	0	0	0	0	0
10								

(c)Example of the results of the number of wildlife damages (for every Provinces)

Fig. 3. Examples of the acquired tweet and blog information, the area name and the types of wildlife name as well as the summarized results from the wildlife damage in Kyushu in 2013

B. True Wildlife Damage Reported by the Regional Govermental Insititude of Kyushu Agricultural Management

True wildlife damage reported by the regional governmental institute of Agricultural Management in 2013 is shown in Table 1.

TABLE I. TRUE WILDLIFE DAMAGE REPORTED BY THE REGIONAL GOVERNMENTAL INSTITUTE OF AGRICULTURAL MANAGEMENT IN 2013

	Wildlife	Birds	Crow	Animals	Wild pig	Monkey	Deer
Fukuoka	91671	36306	19551	55366	35867	2578	7986
Saga	20486	6040	4559	14446	11023	1130	0
Nagasaki	28724	3171	2194	25553	23930	1	470
Kumamoto	45531	10398	8745	35133	28031	1311	4030
Ohita	21550	1078	749	20472	14474	1355	3229
Miyazaki	72978	4242	3363	68736	33396	7287	26066
Kagoshima	43950	8848	3219	35102	17070	2183	12878

The prefecture which shows the largest wildlife damage is Fukuoka followed by Miyazaki, Kumamoto, Kagoshima. Nagasaki, Ohita and Saga. The number of reports of wildlife damage, on the other hand, is shown in Table 2. The correlation coefficient between the total numbers of the reports and the total wildlife damage is just 0.013 as shown in Table 2.

TABLE II. NUMBER OF REPORTS OF WILDLIFE DAMAGE AND TOTAL WILDLIFE DAMAGE IN KYUSHU IN 2013

	Fukuoka	Saga	Nagasaki	Kumamoto	Oita	Miyazaki	Kagoshima
Wildlife	32	7	42	12	29	8	1
Birds	1	0	0	1	0	1	0
Crow	1	0	0	0	0	0	0
Animals	31	7	42	11	29	7	1
Wild pig	8	6	35	2	3	0	0
Monkey	13	0	1	0	0	5	0
Deer	0	0	6	3	11	2	0
Wildlife	91671	20486	28724	45531	21550	72978	43950
No.of report	86	14	126	23	72	23	2

Although correlation coefficient is so poor, R=0.013, if the number of reports of wildlife damages of crow and birds, as well as monkey is deleted together with the number of report of Saga, Kumamoto and Kagoshima due to the fact that the number of reports are so small then the correlation coefficient between the total wildlife damage and the total number of the reports of wildlife damage through blog and tweet is increased R=0.538. Therefore, the relation between both is not so poor.

C. Estimation of Wildlife Damage from the Number of Reported Tweet and Blog for Every Province

Through the linear regressive analysis, it can be done to estimate wildlife damages using the reported tweet and blog information. The results from the regressive analysis are shown in Fig.5. At the top left corners of the figures in Fig.4, there are regressive equations and the R^2 values. The R^2 values range from 0.5657 to 0.9693 while slope (gain) coefficients range from 607.17 to 30686. On the other hand, the number of reports of tweet and blog (Horizontal axis of the graphs in Fig.4) range from 1 to 42. The uncertainty of the regressive analysis is totally dependent to the number of reports. Therefore, the regressive analysis results of Saga, Kagoshima, Miyazaki are not so reliable. Then the ranges of the R2 values and gain coefficients are (0.5866 – 0.9693), and (607.17 – 2893.4), respectively.

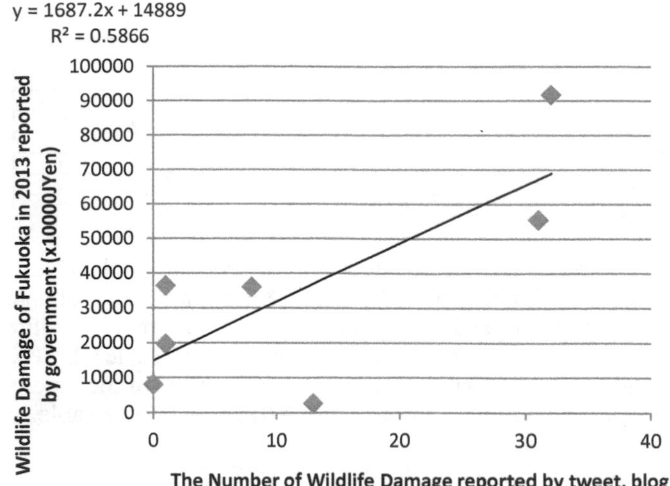

(a)Fukuoka

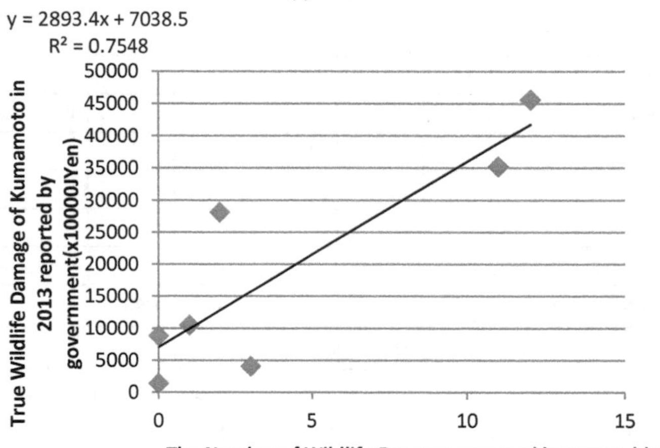

(b)Kumamoto

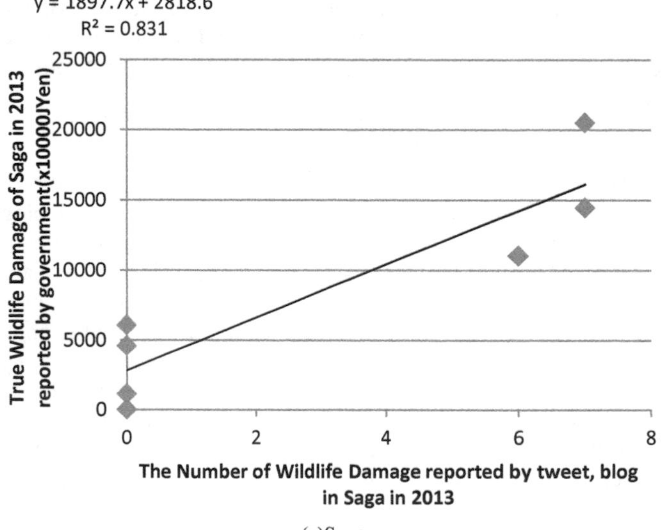

(c)Saga

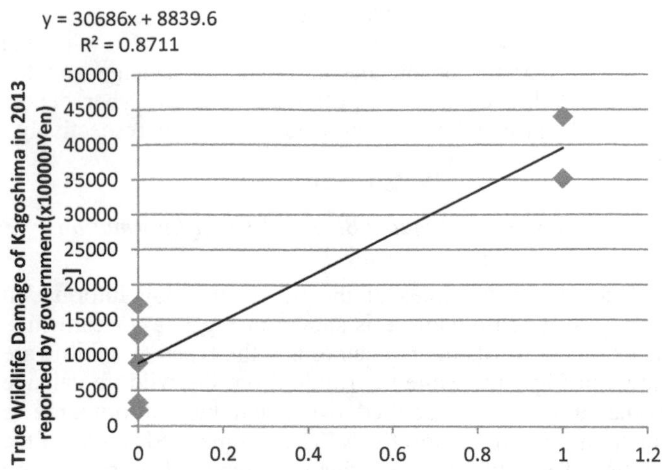

(d)Kagoshima

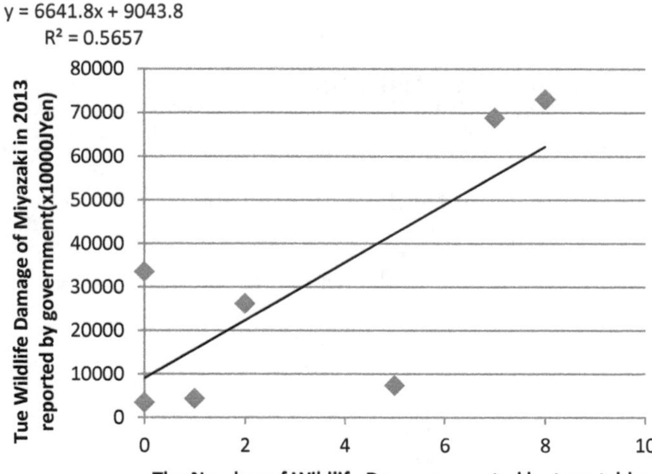

(e)Miyazaki

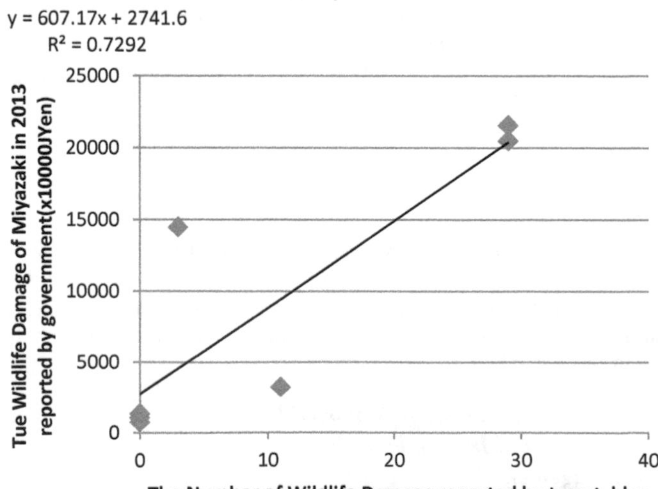

(f)Ohita

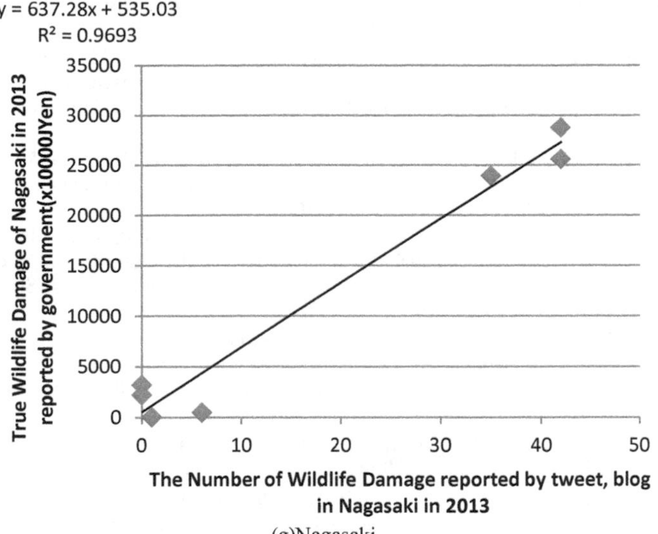

(g)Nagasaki

Fig. 4. Estimate wildlife damages for every province using the reported tweet and blog information

D. Predictions of Wildlife Damage from the Number of Reported Tweet and Blog for Every Province

The newest true wildlife damage data is 2014 which is provided by Kumamoto prefecture. There is no other prefecture of which true wildlife damage of 2014 is reported. Therefore, the wildlife damage of 2014 is predicted by using the past data of wildlife damage (2008 to 2013) based on the linear prediction which is expressed in equation (1). Table 3 shows the results from the predicted wildlife damage (in the second row of Table 3). The correlation between the wildlife damage from the true report of prefecture Kumamoto and predicted wildlife damage from the report of blog and tweet information is 0.996. By taking into account the compensation of mean and standard deviation of the predicted wildlife damage (adjusted), the difference between true wildlife damage and the predicted wildlife damage from the acquired blog and tweet information ranges from -1158 to 2944 in unit of 10,000 Japanese Yen.

TABLE III. PREDICTED WILDLIFE DAMAGE BY USING THE PAST DATA FOR 6 YEARS, 2008 TO 2013

Kumamoto	Wildlife	Birds	Crow	Animals	Wild pig	Monkey	Deer
True report	45531	10398	8745	35133	28031	1311	4030
Predicted	65270.8	14856	9520.9	52210.4	37813.5	4761.7	7181.1
Adjusted	44689.56	9399.2	5664.63	35547.28	25469.45	2333.19	4026.77
Difference	841.44	998.8	3080.37	-414.28	2561.55	-1022.19	3.23

From the relation between year and wildlife damage in Kumamoto in unit of 10,000 Japanese Yen, the wildlife damage can be calculated with the number of the tweet and the blog. Red colored number in Table 4 shows the calculated wildlife damage and the blue colored number indicates the predicted wildlife damage derived from the linear prediction with the true wildlife damage for five years (2008 – 2012) and the estimated wildlife damage in 2013. Through a comparison between true wildlife damage and the predicted one is approximately 6.0 %. Therefore, it is capable to predict wildlife damage in the next year with the past true wildlife damage reported by the local prefectural government and the relation between wildlife damage and the number of report by twitter and blog.

TABLE IV. COMPARISON OF THE WILDLIFE DAMAGES BETWEEN TRUE AND THE PREDICTION

Year	2008	2009	2010	2011	2012	2013	2014
True report	61468	70013	84516	54495	51975	45531	47235
True+Estimated	61468	70013	84516	54495	51975	58509.25	51000

Fig.5 shows the true and the predicted wildlife damages as a function of year. Therefore, it may say that wildlife damage in the next year can be predicted with the past true data of wildlife damage and the relation between the number of reports by twitter and blog.

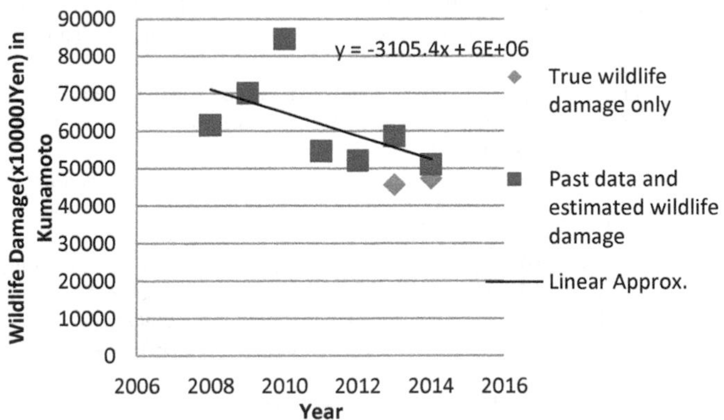

Fig. 5. True and the predicted wildlife damages as a function of year

V. CONCLUSION

Method for wildlife damage estimation and prediction using blog and tweet information relating to wildlife appearances is proposed in this paper. Through regressive analysis with the truth data about wildlife damage which is acquired by the federal and provincial governments and the blog and tweet information about wildlife damage which are acquired in the same year, it is found that some possibility for estimation and prediction of wildlife damage. Through experiments, it is found that R^2 value of the relations between the federal and provincial government gathered truth data of wildlife damages and blog tweet information derived wildlife damages is more than 0.75. Also, it is possible to predict wildlife damage by using past truth data and the estimated wildlife damages. Therefore, it is concluded that the proposed method is applicable to estimate and predict wildlife damages.

It is also found that the correlation between the wildlife damage from the true report of prefecture Kumamoto and predicted wildlife damage from the report of blog and tweet information is 0.996. By taking into account the compensation of mean and standard deviation of the predicted wildlife damage (adjusted), the difference between true wildlife damage and the predicted wildlife damage from the acquired blog and tweet information ranges from -1158 to 2944 in unit of 10,000 Japanese Yen. Therefore, future wildlife damage can be predicted by using the reports from blog and tweet information in some extent.

Further investigations are required for increasing the cases of wildlife damages for improving prediction accuracy.

ACKNOWLEDGMENT

The author would like to thank Prof. Dr. Hiroshi Okumura of Saga University for his valuable suggestions and comments for this study.

REFERENCES

[1] Patten, D. C. 1974. Feral hogs — boon or burden. Proceedings of the Sixth Vertebrate Pest Conference 6:210–234.

[2] Singer, F. J., W. T. Swank, and E. E. C. Clebsch. 1984. Effects of wild pig rooting in a deciduous forest. Journal of Wildlife Management. 48:464–473.

[3] Stone, C. P., and J. O. Keith. 1987. Control of feral ungulates and small mammals in Hawaii's national parks: research and management strategies. Pages 277–287 in C. G. J. Richards and T. Y. Ku, editors. Control of mammal pests. Taylor and Francis, London, England, and New York and Philadelphia, USA.

[4] Cushman, J. H., T. A. Tierney, and J. M. Hinds. 2004. Variable effects of feral pig disturbances on native and exotic plants in a California grassland. Ecological Applications 14:1746–1756.

[5] Kaller, M. D., and W. E. Kelso. 2006. Swine activity alters invertebrate and microbial communities in a coastal plain watershed. American Midland Naturalist 156:163–177.

[6] Bratton, S. P. 1977. The effect of European wild boar on the flora of the Great Smoky Mountains National Park. Pages 47–52 in G. W. Wood, editor. Research and management of wild hog populations. Belle W. Baruch Forest Science Institute, Clemson University, Georgetown, South Carolina, USA.

[7] Lucas, E. G. 1977. Feral hogs — problems and control on National Forest lands. Pages 17–22 in G. W. Wood, editor. Research and management of wild hog populations. Belle Baruch Forest Science Institute, Clemson University, Georgetown, South Carolina, USA.

[8] Thompson, R. L. 1977. Feral hogs on National Wildlife Refuges. Pages 11–15 in G. W. Wood, editor. Research and management of wild hog populations. Belle W. Baruch Forest Science Institute, Clemson University, Georgetown, South Carolina, USA. Kohei Arai, Preliminary Assessment of Radiometric Accuracy for MOS-1 Sensors, International Journal of Remote Sensing, Vol.9, No.1, pp.5-12, Apr.1988.

[9] Schley, L, and T. J. Roper. 2003. Diet of wild boar Sus scrofa in Western Europe, with particular reference to consumption of agricultural crops. Mammal Review 33:43–56.

[10] Whitehouse, D. B. 1999. Impacts of feral hogs on corporate timberlands in the southeastern United States. Pages 108–110 in Proceedings of the Feral Swine Symposium, June 2–3, 1999, Ft. Worth, Texas, USA.

[11] Mayer, J. J., E. A. Nelson, and L. D. Wike. 2000. Selective depredation of planted hardwood seedlings by wild pigs in a wetland restoration area. Ecological Engineering, 15(Supplement 1): S79–S85.

[12] Campbell, T. A., and D. B. Long. 2009. Feral swine damage and damage management in forested ecosystems. Forest Ecology and Management 257:2319–2326

[13] Forrester, D. J. 1991. Parasites and diseases of wild mammals in Florida. University of Florida Press, Gainesville, Florida, USA.

[14] Williams, E. S., and I. K. Barker. 2001. Infectious diseases of wild mammals. Iowa State University Press, Ames, Iowa, USA.

[15] Sweeney, J. R., J. M. Sweeney, and S. W. Sweeney. 2003. Feral hog. Pages 1164–1179 in G. A. Feldhamer, B. C. Thompson, and J. A. Chapman, editors. Wild mammals of North America. Johns Hopkins University Press, Baltimore, Maryland, USA.

[16] Nettles, V.F., J. L. Corn, G. A. Erickson, and D. A. Jessup. 1989. A survey of wild swine in the United States for evidence of hog cholera. Journal of Wildlife Diseases 25:61–65.

[17] Davidson, W. R., and V. F. Nettles, editors. 1997. Wild swine. Pages 104–133 in Field manual of wildlife diseases in the southeastern United States. Second edition. Southeastern Cooperative Wildlife Disease Study, Athens, Georgia, USA.

[18] Davidson, W. R., editor. 2006. Wild swine. Pages 105–134 in Field manual of wildlife diseases in the southeastern United States. Third . Southeastern Cooperative Wildlife Disease Study, Athens, Georgia, USA.

[19] Choquenot, D. J., R. J. Kilgour, and B. S. Lukins. 1993. An evaluation of feral pig trapping. Wildlife Research, 20:15– 22.

[20] Barrett, R. H., and G. H. Birmingham. 1994. Wild pigs. Pages D65–D70 in S. Hyngstrom, R. Timm, and G. Larsen, editors. Prevention and control of wildlife damage. Cooperative Extension Service, University of Nebraska, Lincoln, Nebraska, USA.

[21] Mapston, M. E. 1999. Feral hog control methods. Pages 117–120 in Proceedings of the Feral Swine Symposium, June 2–3, 1999, Fort Worth, Texas, USA

[22] Kohei Arai, Indra Nugraha Abdullah, Kensuke Kubo, Katsumi Sugaw, Methods for Wild Pig Identifications from Moving Pictures and Discrimination of Female Wild Pigs based on Feature Matching Method, (IJARAI) International Journal of Advanced Research in Artificial Intelligence, Vol. 4, No.7, 41-46, 2015

24

Vicarious Calibration Data Screening Method Based on Variance of Surface Reflectance and Atmospheric Optical Depth Together with Cross Calibration

Kohei Arai 1
1Graduate School of Science and Engineering
Saga University
Saga City, Japan

Abstract—**Vicarious calibration data screening method based on the measured atmospheric optical depth and the variance of the measured surface reflectance at the test sites is proposed. Reliability of the various calibration data has to be improved. In order to improve the reliability of the vicarious calibration data, some screenings have to be made. Through experimental study, it is found that vicarious calibration data screening would be better to apply with the measured atmospheric optical depth and variance of the measured surface reflectance due to the facts that thick atmospheric optical depth means that the atmosphere contains serious pollution sometime and that large deviation of the surface reflectance from the average means that the solar irradiance has an influence due to cirrus type of clouds. As the results of the screening, the uncertainty of vicarious calibration data from the approximated radiometric calibration coefficient is remarkably improved. Also, it is found that cross calibration uncertainty is poorer than that of vicarious calibration.**

Keywords—Vicarious calibration; Surface reflectance; Atmospheric Optical Depth; Sky-radiometer; Terra/ASTER; Satellite remote sensing

I. INTRODUCTION

Visible and Near Infrared mounted on earth observation satellites and the short-wavelength infrared radiation thermometer, Alternative calibration using measurement data on the ground and onboard calibration by the calibration mounting system is performed. For example, Marine Observation Satellite-1 [1], Landsat-7 Enhanced Thematic Mapper Plus [2], SeaWiFS [3], High Resolution Visible: HRV/SPOT-1 and 2 [4], Hyperion [5], POLDER [6], etc. by ASTER [7]. The calibration results have been reported. Further, report according to reciprocity with a uniform ground surface [8] over a wide area such as desert radiometer each other overlapping of the observation wavelength range have been made [9].

Vicarious calibration are conducted with consideration of the influence due to the atmosphere obviously [11]. Furthermore, the well-known cross calibration through comparisons among visible to near infrared radiometers onboard same or the different satellites is effective for calibration of the visible to near infrared radiometers in concern [12]. [13], [14]. To conduct the error analysis in the vicarious calibration of visible and near infrared radiometer, Arai et al. made it clear dominant error factors of vicarious calibration accuracy [15]. According to it, most dominant error factors are the surface reflectance measurement followed by optical depth measurement that allows estimation of aerosol property. It is still difficult to estimate the aerosol characteristic and surface reflectance estimations. In order to estimate refractive index of aerosol particles, it is strongly suggested to use skyradiometer[1] or aureole-meters [16], [17]. Since April 2003, Arai et al. have been doing the observation of aerosol by skyradiometer, POM-1 that is manufactured by PREDE Co. Ltd. [18]. The skyradiometer allows measurement of solar direct, diffuse and aureole that results in estimation of refractive index and size distribution of aerosol particles [19]. Nakajima proposes a method of estimating the volume particle size distribution and the complex refractive index [20]. Arai proposes a method for using the Simulated Annealing: SA as inverse problem-solving [21]. Furthermore, improved Modified Langley method as the calibration method of sky-radiometer and as the method for estimation of extraterrestrial solar irradiance as well as atmospheric optical depth is proposed by Arai. The method is for estimating the top of atmosphere radiance with consideration of not only down-welling but also up-welling p and s polarized irradiance and radiance [21].

Reliability of the vicarious calibration data has to be evaluated. Vicarious calibration data are used to be suffered from atmospheric conditions, existing cirrus clouds, smoke from wild fire that happens nearby test sites, enthused gasses from automobiles that situated nearby the test sites, and so on. These are invisible mostly. Therefore, vicarious calibration data are suffered from these influences even if we conducted field campaigns with great care about these. It is possible to find such these influences through careful screening test with the measured data of surface reflectance and optical depth. The method proposed here is to make a screening the vicarious calibration data suffered from the influences for improvement uncertainty of the vicarious calibration data.

In the next section, the method and procedure of the proposed screening method is described followed by experimental data and estimated results. Then conclusion is described with some discussions.

[1] http://skyrad.sci.u-toyama.ac.jp/

II. Proposed Method

A. *Vicarious Calibration Method*

Flowchart of the reflectance based vicarious calibration method is shown in Figure 1. At the test site (relatively high surface reflectance of homogeneous area of desert which is situated at comparatively high elevation (thin atmosphere) is desired, field campaign is used to be conducted. At the field campaign, atmospheric optical depth, surface reflectance, column ozone, column water vapor is measured. From the measured atmospheric optical depth, size distribution can be estimated using Angstrome exponent together with extraterrestrial solar irradiance through Langley plot. Total atmospheric optical depth can be divided into Rayleigh scattering component due to atmospheric continuant, and Mie scattering component due to aerosol particles in the scattering components. On the other hand, absorption components due to water vapor, ozone and aerosols are also estimated. Rayleigh scattering component can be estimated with air-temperature and atmospheric pressure on the ground. Absorption due to ozone can be estimated with the absorption coefficient of ozone and the measured column ozone in unit of Dobson Unit. In the visible to near infrared wavelength region, major contribution is from atmospheric continuant (O_2, N_2), water vapor, ozone, and aerosols. Therefore, these contributions in the forms of scattering and absorption have to be taken into account. Through radiative transfer equation solving process (mostly MODTRAN code is used to use) with the estimated influencing aforementioned parameters, Top of the Atmosphere: TOA radiance (at sensor radiance) can be estimated. Then the estimated TOA radiance is compared with satellite sensor data (Digital Number; DN is converted to radiance). Thus gain can be calibrated. This gain degradation is called as Radiometric Calibration Coefficient: RCC. It is referred to Vicarious Calibration Coefficient: RCCvic. On the other hands, most of visible to near infrared radiometer onboard satellites has own onboard calibration system. It provides Onboard Calibration Coefficient (OBC).

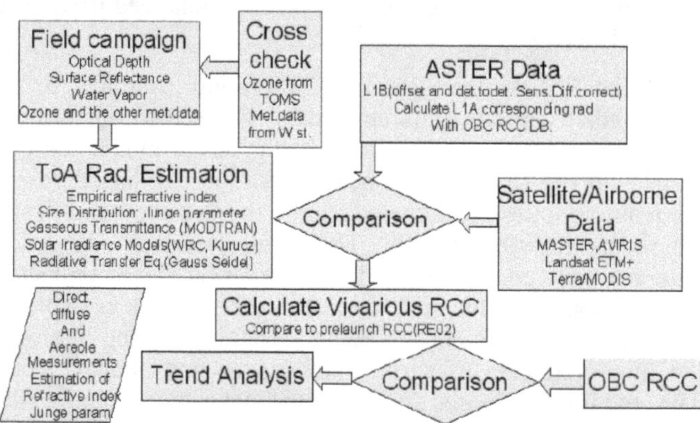

Fig. 1. Flowchart of reflectance based vicarious calibration method

Then it is possible to compare both coefficients.

B. *Error Budget Analysis*

There are 8 error sources for the vicarious calibration. The result from error budget analysis of the vicarious calibration method is shown in Table 1.

TABLE I. Error Budget for Reflectance based Vicarious Calibration Method

Error sources	Error Type	Error (%)
Optical depth	Random	1.5
Surface reflectance measurement instrument	Random	2
Standard plaque	Systematic	1
Averaging	Random	0
Refractive index	Random	1.8
Size distribution	Random	2
Radiative transfer code	Systematic	1
Registration	Random	1
RSS		4.06

Optical depth measurement has error of 1.5% while surface reflectance measurement instrument has 2% of error which includes Bi-Directional Reflectance Distribution Function: BRDF effect. Standard plaque is used as reference of reflectance and has 1% of error. Estimation accuracy of refractive index and size distribution is not high enough. 1.8 and 2 % of errors are suspected for each. Radiative transfer code has 1% of error while 1% of error is suspected due to registration of test site (location identification) in the satellite sensor image. Thus 4.06% error is suspected for vicarious calibration.

C. *Vicarious Calibration Data Screening Method*

Reliability of the vicarious calibration data has to be evaluated. Vicarious calibration data are used to be suffered from atmospheric conditions, existing cirrus clouds, smoke from wild fire which happens nearby test sites, enthused gasses from automobiles which situated nearby the test sites, and so on. These are invisible mostly. Therefore, vicarious calibration data are suffered from these influences even if we conduct field campaigns with great care about these. It is possible to find such these influences through careful screening test with the measured data of surface reflectance and optical depth. The method proposed here is to make a screening the vicarious calibration data suffered from the influences for improvement uncertainty of the vicarious calibration data. There are two major factors, optical depth and standard deviation of the measured surface reflectance. By using threshold, vicarious calibration data can be screened.

D. *Uncertainty of Vicarious Calibration*

Time series of RCCvic data can be approximated with appropriate function (Usually single exponential function of "a EXP(-b d) + c") in the sense of trend analysis. Let be RCCvic' denotes the approximated RCCvic. Then uncertainty of vicarious calibration can be expressed in equation (1).

$$U=SQRT(\Sigma(RCCvic-RCCvic')^2/n_i(n_i-p_i)) \quad (1)$$

where n and p denotes the number of vicarious calibration data and the condition number, respectively. Thus uncertainty of the vicarious calibration can be calculated.

III. Experiment

A. Trend of the Vicarious Calibration Data

One of examples of vicarious calibration data of ASTER/VNIR onboard Terra satellite is shown in Figure 2

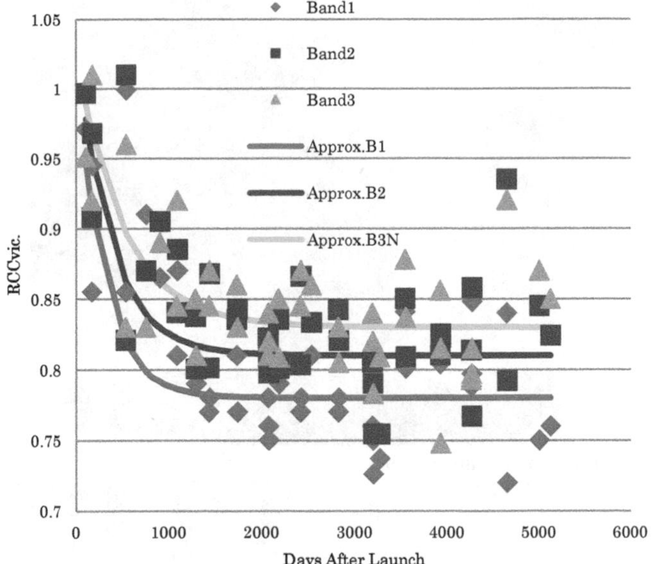

Fig. 2. Vicarious calibration data of ASTER/VNIR onboard Terra satellite

ASTER/VNIR onboard Terra satellite was launched in December 1999. This is approximately 15 years data. Solid lines are approximated function with the single exponential function. It is clear that vicarious calibration data are scattered because there are many data which are influenced by the smoke due to wild fire, exhausted gasses from automobile, cirrus, etc. VNIR has three spectral bands, 1 to 3 which are corresponding to green, red and near infrared bands.

Table 2 shows the coefficients of the approximation function of vicarious calibration data as a function of days after launch, x.

TABLE II. Coefficients of the Approximation Function of Vicarious Calibration Data as a Function of Days After Launch, X

y = b0 * exp(-b1 * x) + b2	Band1	Band2	Band3
b0	0.2374657	0.2227815	0.194918
b1	0.0033325	0.0026667	0.0019807
b2	0.7551861	0.83	0.8468567

The difference between vicarious calibration data and the approximated vicarious calibration data is shown in Table 3 while uncertainty of vicarious calibration data defined in equation (1) is shown in Table 4, respectively.

TABLE III. Difference Between Vicarious Calibration Data and the Approximated Vicarious Calibration Data

Band1	Band2	Band3
0.161	0.071	0.071

TABLE IV. Uncertainty of Vicarious Calibration Data Defined in Equation (1)

Band1	Band2	Band3
0.0101579	0.0067297	0.0067447

B. Vicarious Calibration Data After the Screening

Figure 3 shows the vicarious calibration data trend after the screening.

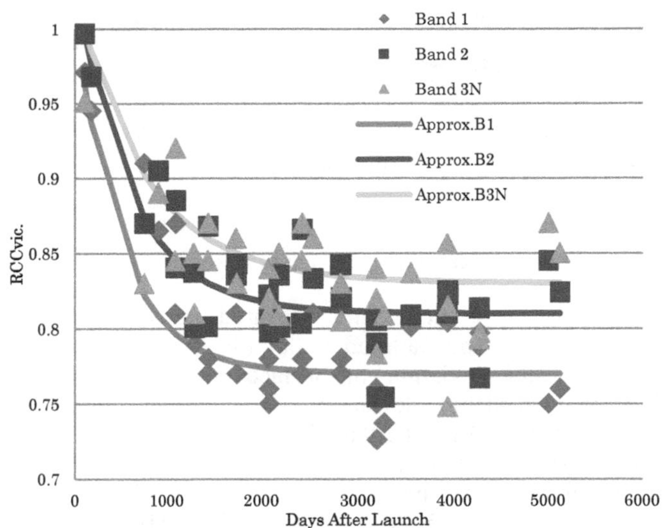

Fig. 3. Vicarious calibration data trend after the screening

One of the examples of the measured surface reflectance for test site of Railroad Valley Playa in Nevada, USA which is acquired on 25 September 2015 is shown in Figure 4 while one if examples of the measured surface reflectance without screening of the test site Ivanpah Playa in California, USA which is acquired on 18 September 2015 is shown in Figure 4 (b), respectively.

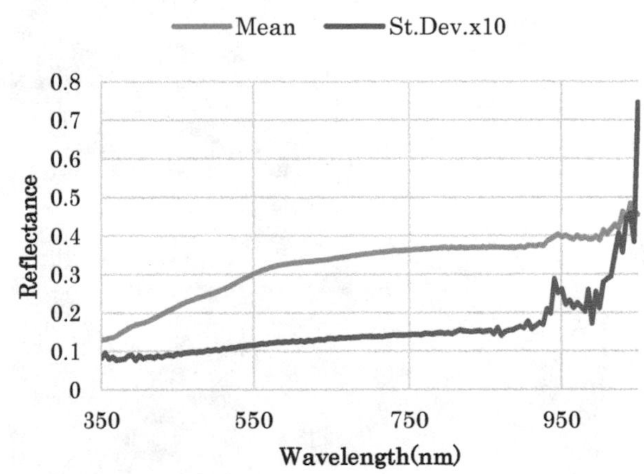

(a)Railroad Valley Playa

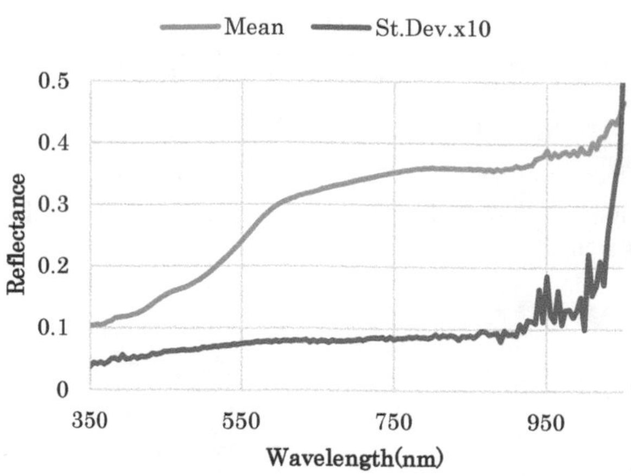

(b)Ivanpah Playa

Fig. 4. Measured surface reflectance at the test sites

As shown in Figure 4, the measured surface reflectance between Ivanpah and Railroad Valley Playas are almost same while standard deviation of surface reflectance is quite different (standard deviation of Ivanpah playa is approximately 50% greater than that of Railroad Valley playa). Figure 5 (a) shows ASTER/VNIR image of Ivanpah playa while Figure 5 (b) shows that of Railroad Valley playa. Meanwhile, Figure 5 (c) shows ASTER/TIR image of Railroad Valley playa which shows the suspected existing cirrus. Although cirrus clouds cannot be seen in the ASTER/VNIR image of Railroad Valley playa, ASTER/TIR image shows existing of cirrus clouds almost all over the test site area. During the surface reflectance measurement, solar irradiance is changed a lot due to the existing cirrus. Therefore, the standard deviation of the measured surface reflectance is 50% much greater than that of Ivanpah playa. We would better to omit such unreliable vicarious calibration data.

(a)VNIR image of Ivanpah

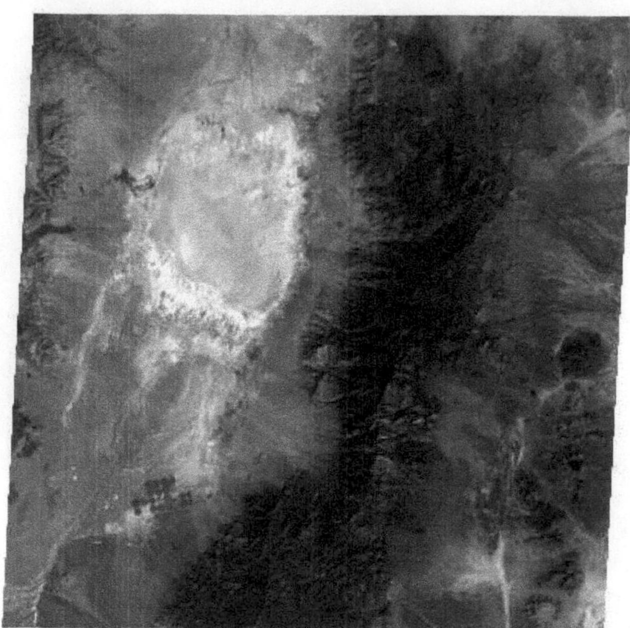

(b)VNIR image of Railroad Valley

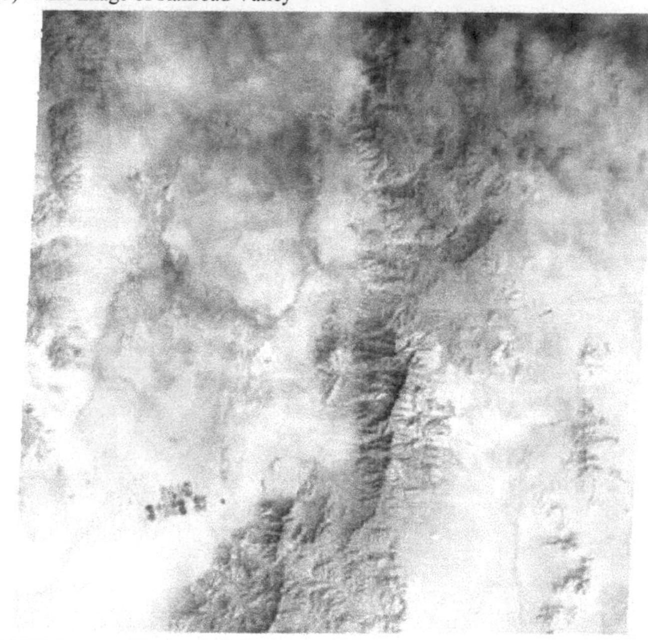

(c)TIR image of Railroad Valley

Fig. 5. ASTER/VNIR and TIR images of Ivanpah and Railroad Valley playas

Table 5 shows the coefficients of the approximation function of vicarious calibration data after the screening as a function of days after launch, x.

TABLE V. COEFFICIENTS OF THE APPROXIMATION FUNCTION OF VICARIOUS CALIBRATION DATA AFTER THE SCREENING AS A FUNCTION OF DAYS AFTER LAUNCH, X

y = b0 * exp(-b1 * x) + b2	Band1	Band2	Band3
b0	0.2374657	0.2227815	0.194918
b1	0.002	0.0016667	0.0013
b2	0.77	0.81	0.83

The difference between vicarious calibration data and the approximated vicarious calibration data is shown in Table 6 while uncertainty of vicarious calibration data defined in equation (1) is shown in Table 7, respectively.

TABLE VI. DIFFERENCE BETWEEN VICARIOUS CALIBRATION DATA AND THE APPROXIMATED VICARIOUS CALIBRATION DATA

Band1	Band2	Band3
0.040	0.024	0.036

TABLE VII. UNCERTAINTY OF VICARIOUS CALIBRATION DATA DEFINED IN EQUATION (1)

Band1	Band2	Band3
0.0061288	0.0047501	0.0058725

It is found that uncertainty of vicarious calibration can be improved remarkably in particular for Band 1.

C. Error Budget Analysis of Cross Calibration

MISR and MODIS sensors are onboard Terra satellite as well. The spectral coverage of MISR and MODIS are overlapped with ASTER/VNIR. Therefore, cross calibration can be done for VNIR and MISR (VNIR Band 1, 2, 3) and VNIR and MODIS (VNIR Band 2 and 3). Due to the fact that MODIS does not have the corresponding band for VNIR Band 1, cross calibration cannot be done. The results from error budget analysis are shown in Table 8. In the proposed cross calibration, it is conducted at the same dates for field campaigns because the vicarious calibration data can be used for cross calibration.

TABLE VIII. ERROR BUDGET FOR CROSS CALIBRATION

Error items	Error sources	Error (%)
Uncertainty of the instruments for comparison	MISR,MODIS	4.06
Registration	Uniformity of the surface reflectance	2
Spectral response	Surface reflectance	1.5
	Atmospheric effect	1
RSS		4.87

D. Cross Calibration Results with MISR

Figure 6 shows the cross calibration data trend derived RCC (RCCcross) with MISR. It is possible to approximate with the same function of the single exponential function as a function of days after launch, x. The coefficients of the approximation function are shown in Table 9 while the difference between cross calibration data and the approximated data are shown in Table 10. Uncertainty defined as equation (1) for cross calibration with MISR is shown in Table 11. In comparison to the uncertainty of the vicarious calibration, cross calibration accuracy is not better than vicarious calibration obviously. It, however, is useful to find the biases between ASTER/VNIR and MISR.

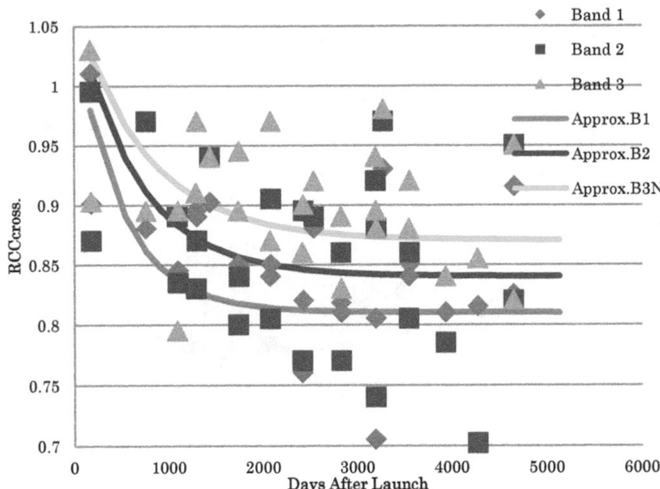

Fig. 6. Cross calibration data of ASTER/VNIR with MISR onboard Terra satellite

TABLE IX. COEFFICIENTS OF THE APPROXIMATION FUNCTION OF CROSS CALIBRATION DATA AFTER THE SCREENING AS A FUNCTION OF DAYS AFTER LAUNCH, X

y = b0 * exp(-b1 * x) + b2	Band1	Band2	Band3
b0	0.2374657	0.2227815	0.194918
b1	0.002	0.0015	0.0013
b2	0.81	0.84	0.87

TABLE X. DIFFERENCE BETWEEN CROSS CALIBRATION DATA AND THE APPROXIMATED VICARIOUS CALIBRATION DATA

Band1	Band2	Band3
8.985	9.844	10.508

TABLE XI. UNCERTAINTY OF CROSS CALIBRATION DATA DEFINED IN EQUATION (1)

Band1	Band2	Band3
0.0759393	0.0794899	0.0821237

E. Cross Calibration Results with MODIS

Figure 7 shows the cross calibration data trend derived RCC (RCCcross) with MODIS. It is possible to approximate with the same function of the single exponential function as a function of days after launch, x. The coefficients of the approximation function are shown in Table 12 while the difference between cross calibration data and the approximated data are shown in Table 13. Uncertainty defined as equation (1) for cross calibration with MODIS is shown in Table 14. In comparison to the uncertainty of the vicarious calibration, cross calibration accuracy is not better than vicarious calibration obviously. It, however, is useful to find the biases between ASTER/VNIR and MODIS.

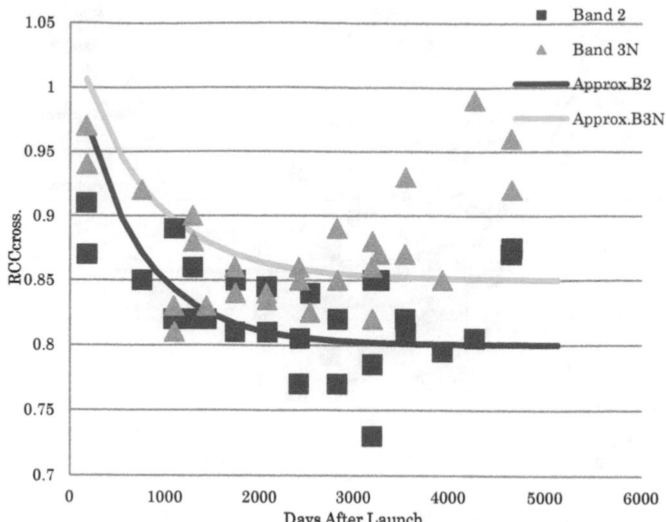

Fig. 7. Cross calibration data of ASTER/VNIR with MODIS onboard Terra satellite

TABLE XII. COEFFICIENTS OF THE APPROXIMATION FUNCTION OF CROSS CALIBRATION DATA AFTER THE SCREENING AS A FUNCTION OF DAYS AFTER LAUNCH, X

y = b0 * exp(-b1 * x) + b2	Band1	Band2	Band3
b0	-	0.222781456	0.194918032
b1	-	0.0015	0.0013
b2	-	0.8	0.85

TABLE XIII. DIFFERENCE BETWEEN CROSS CALIBRATION DATA AND THE APPROXIMATED VICARIOUS CALIBRATION DATA

Band1	Band2	Band3
-	8.884	10.035

TABLE XIV. UNCERTAINTY OF CROSS CALIBRATION DATA DEFINED IN EQUATION (1)

Band1	Band2	Band3
0	0.075513481	0.080253544

As the results from the uncertainty evaluations of the cross calibration between ASTER/VNIR and MISR as well as MODIS, it is almost same between cross calibrations of ASTER/VNIR and MISR as well as MODIS. It is also found that the uncertainty of cross calibration is poorer than that of vicarious calibration.

IV. CONCLUSION

Vicarious calibration data screening method based on the measured atmospheric optical depth and the variance of the measured surface reflectance at the test sites is proposed. Reliability of the various calibration data has to be improved. In order to improve the reliability of the vicarious calibration data, screening has to be made. Through experimental study, it is found that vicarious calibration data screening would be better to apply with the measured atmospheric optical depth and variance of the measured surface reflectance due to the facts that thick atmospheric optical depth means that the atmosphere contains serious pollution sometime and that large deviation of surface reflectance from the average means that the solar irradiance has influence due to cirrus type of clouds. As the results of the screening, the uncertainty of vicarious calibration data from the approximated radiometric calibration coefficient is remarkably improved. Also, it is found that cross calibration uncertainty is poorer than that of vicarious calibration.

ACKNOWLEDGEMENTS

Author would like to thank Dr. Akira Ono of National Institute of Advanced Industrial Science and Technology: AIST for his initiating of this research works. Also, author would like to thank Dr. Satoshi Tsuchida and his research staff, Japanese Space Systems: JSS members together with Dr. Kurtis Thome of NASA/GSFC as well as Prof. Dr. Stuart Biggar and his research staff for their contributions of the field experiments and valuable discussions.

REFERENCES

[1] Arai K., Preliminary assessment of radiometric accuracy for MOS-1 sensors, International Journal of Remote Sensing, 9, 1, 5-12, 1988.

[2] Barker, JL, SK Dolan, et al., Landsat-7 mission and early results, SPIE, 3870, 299-311, 1999.

[3] Barnes, RA, EEEplee, et al., Changes in the radiometric sensitivity of SeaWiFS determined from lunar and solar based measurements, Applied Optics, 38, 4649-4664, 1999.

[4] Gellman, DI, SF Biggar, et al., Review of SPOT-1 and 2 calibrations at White Sands from launch to the present, Proc. SPIE, Conf.No.1938, 118-125, 1993.

[5] Folkman, MA, S.Sandor, et al., Updated results from performance characterization and calibration of the TRWIS III Hyperspectral Imager, Proc. SPIE, 3118-17, 142, 1997.

[6] Hagolle, O., P.Galoub, et al., Results of POLDER in-flight calibration, IEEE Trans. On Geoscience and Remote Sensing, 37, 1550-1566, 1999.

[7] Thome, K., K. Arai, S. Tsuchida and S. Biggar, Vicarious calibration of ASTER via the reflectance based approach, IEEE transaction of GeoScience and Remote Sensing, 46, 10, 3285-3295, 2008.

[8] Cosnefroy, H., M.Leroy and X.Briottet, Selection and characterization of Saharan and Arabian Desert sites for the calibration of optical satellite sensors, Remote Sensing of Environment, 58, 110-114, 1996.

[9] Arai, K., In-flight test site cross calibration between mission instruments onboard same platform, Advances in Space Research, 19, 9, 1317-1328, 1997.

[10] Nicodemus, FE, "Directional Reflectance and Emissivity of an Opaque Surface", Applied Optic (1965), or FE Nicodemus, JC Richmond, JJ Hsia, IW Ginsber, and T. Limperis, "Geometrical Considerations and Nomenclature for Reflectance, ", NBS Monograph 160, US Dept. of Commerce (1977).

[11] Slater, PN, SFBiggar, RGHolm, RDJackson, Y.Mao, MSMoran, JMPalmer and B.Yuan, Reflectance-and radiance-based methods for the in-flight absolute calibration of multispectral sensors, Remote Sensing of Environment, 22, 11-37, 1987.

[12] Kieffer, HH and RL Wildey, Establishing the moon as a spectral radiance standard, J., Atmosphere and Oceanic Technologies, 13, 360-375, 1996.

[13] Arai, K., Atmospheric Correction and Residual Errors in Vicarious Cross-Calibration of AVNIR and OCTS Both Onboard ADEOS, Advances in Space Research, 25, 5, 1055-1058, 1999.

[14] Liu;. JI, Z. Li, YL Qiao, Y.-J. Liu, and Y.-X. Zhang, A new method for cross-calibration of two satellite sensors, Int J. of Remote Sensing, 25 , 23 5267-5281 , 2004 .

[15] Kohei Arai , error analysis of vicarious calibration of satellite visible and near infrared radiometer based KJThome, reflectance , Japan Photogrammetry Journal , Vol.39, No.2, pp.99-105, (2000) .

[16] Arai, K., Vicarious calibration for solar reflection channels of radiometers onboard satellites with deserted area of data, Advances in Space Research, 39, 1, 13-19, 2006.

[17] Arai, K. and X.Liang, Characterization of aerosols in Saga city areas, Japan withy direct and diffuse solar irradiance and aureole observations, Advances in Space Research, 39, 1, 23-27, 2006.

[18] Kohei Arai , vicarious calibration of ASTER / VNIR based on long-term observations of the optical properties of aerosols in Saga , Journal of the Remote Sensing Society of Japan , 28,3,246 over 255,2008

[19] Kohei Arai , applied linear algebra , modern science , Inc. , 2006

[20] Nakajima, T., M.Tanaka and T. Yamauchi, Retrieval of the optical properties of aerosols from aureole and extinction data, Applied Optics, 22, 19, 2951-2959, 1983.

[21] Kohei Arai , Xing Ming Liang, Estimation of complex refractive index of aerosol using direct solar direct, diffuse and aureole by simulated annealing, and polarized radiance -simultaneous estimation of particle size distribution and refractive index, Journal of the Remote Sensing Society of Japan , Vol.23, No.1, pp .11-20,2003 .

25

Location Monitoring System with GPS, Zigbee and Wifi Beacon for Rescuing Disable Persons

Kohei Arai
[1]Graduate School of Science and Engineering
Saga University
Saga City, Japan

Taka Eguchi
[1]Graduate School of Science and Engineering
Saga University
Saga City, Japan

Abstract—**Location monitoring system for rescue disable persons by switching the location estimation methods with GPS, ZigBee and WiFi beacon is proposed. Rescue system with triage using health condition monitoring together with location and attitude monitoring as well as the other data acquired with mobile devices is evaluated with the proposed location monitoring system. Through simulation study, influence due to location estimation error on rescue time is evaluated together with effect of the proposed location monitoring system. Also, it is found that the effect of triage on rescue time is clarified.**

Keywords—Rescue system; Location estimation; Attitude estimation; Health monitoring; Mobile applications; Triage; Rescue planning

I. INTRODUCTION

Most computer based simulation evacuation models are based on flow model, cellular automata model, and multi-agent-based model. Flow based model lacks interaction between evacuees and human behavior in crisis. Cellular automata model is arranged on a rigid grid, and interact with one another by certain rules [1]. A multi agent-based model is composed of individual units, situated in an explicit space, and provided with their own attributes and rules [2]. This model is particularly suitable for modeling human behaviors, as human characteristics can be presented as agent behaviors. Therefore, the multi agent-based model is widely used for evacuation simulation [1]-[4].

Recently, Geographic Information Systems: GIS is also integrated with multi-agent-based model for emergency simulation. GIS can be used to solve complex planning and decision making problems [5]-[7]. In this study, GIS is used to present road network with attributes to indicate the road conditions.

We develop a task allocation model for search and rescue persons with disabilities and simulate the rescue process to capture the phenomena and complexities during evacuations. The task allocation problem is presented by decision of volunteers to choose which victims should be helped in order to give first-aid and transportation with the least delay to the shelter. The decision making is based on several criteria such as health condition of the victims, location of the victims and location of volunteers [8]-[18].

A rescue model for people with disabilities in large scale environment is proposed. The proposed rescue model provides some specific functions to help disabled people effectively when emergency situation occurs. Important components of an evacuation plan are the ability to receive critical information about an emergency, how to respond to an emergency, and where to go to receive assistance. Triage is a key for rescue procedure. Triage can be done with the gathered physical and psychological data which are measured with a sensor network for vital sign monitoring. Through a comparison between with and without consideration of triage, it may be possible to find that the time required for evacuation from disaster areas with consideration triage is less than that without triage [19]-[20].

These studies do not taken into account location estimation accuracy. GPS utilized location accuracy is assumed to be 100% accurate (no error). There, actually, are location estimation errors. Other methods for location estimation are available, for instance, ZigBee, WiFi beacon utilizing methods. All the methods have errors which depend on the conditions, location of radio wave absorbance, weather condition, multi-path condition, etc. Some of these error sources are controllable except weather condition. A prior to location estimation, location of radio wave absorbance, and multi-path condition can be assessed. It would be possible to use the best accuracy of location estimation method can be used alternatively from among GPS, WiFi beacon, and ZigBee. In particular, ZigBee transmitter can be layout arbitrary in accordance with the required location estimation accuracy. Thus the best location estimation accuracy could be achieved.

The next section describes influence due to location estimation error on the rescue time in concern followed by the proposed location estimation method by using GPS, WiFi beacon and ZigBee alternatively. Then experimental results from rescue simulation studies which is based on triage with health condition of victims. Finally, conclusion is describes together with some discussions.

II. INFLUENCE DUE TO LOCATION ESTIMATION ERROR ON RESCUE TIME

A. Comparison of Location Estimation Methods Among GIS, WiFi Beacon and ZigBee Based Methods

The location estimation with GPS is accurate for outdoor situation with the condition without radio wave absorbance as shown in Fig.1. It, on the other hand, is poor accuracy for indoor situation with the condition with radio wave absorbance. Meanwhile, WiFi beacon based location estimation is available for both indoor and outdoor situations. Accuracy is dependent on the number of available routers and the existing radio wave absorbance.

Fig. 1. GPS based location estimation system

On the other hand, ZigBee based location estimation needs very poor power consumption as shown in Fig.2. The size of ZigBee transmitter and receiver as well as repeater is very compact so that it can be set up anywhere. Furthermore, ZigBee transmitter and receiver is relatively cheap in comparison to the WiFi router. The transmitters of WiFi beacon router and ZigBee transmitter can be set-up arbitrary depending on the required accuracy.

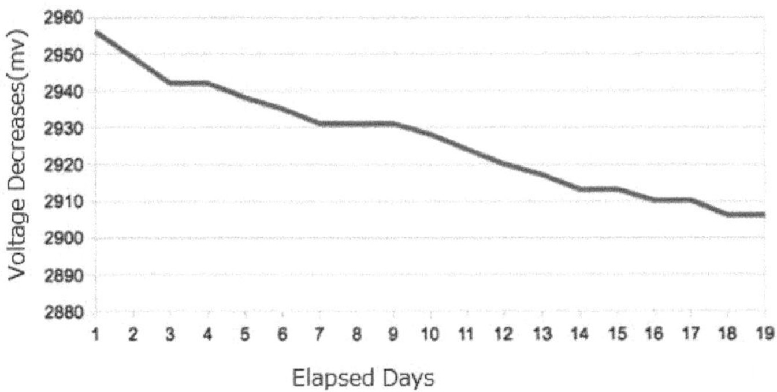

Fig. 2. Power consumption of ZigBee

Measured and theoretical signal strengths of WiFi beacon based location estimation method are shown in Fig.3. The theoretical receiving signal strength is expressed in equation (1).

$$RSSI = -(10\log_{10} d + A) \quad (1)$$

where RSSI denotes receiving signal strength while d and A denotes distance between transmitter (WiFi router) and receiver as well as signal strength for the case of the distance is 1m, respectively. Measured signal strength shows a good coincidence to the theoretical strength.

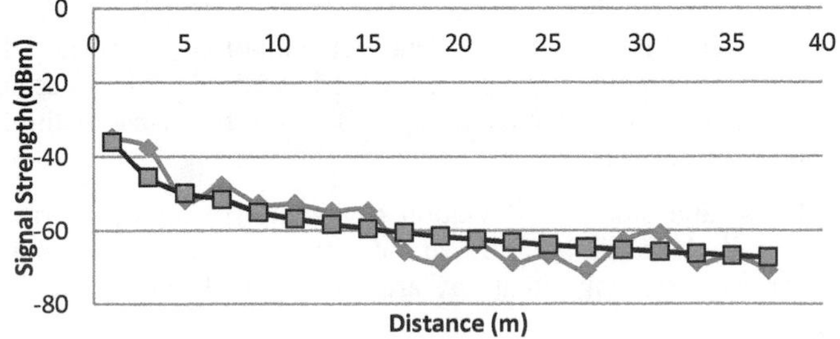

Fig. 3. Measured and theoretical signal strength as a function of distance

On the other hand, measured signal strength of ZigBee receiver as a function of distance between transmitter and receiver is shown in Fig.4 (a) for the indoor situation while that for the outdoor situation is shown in Fig.4 (b), respectively.

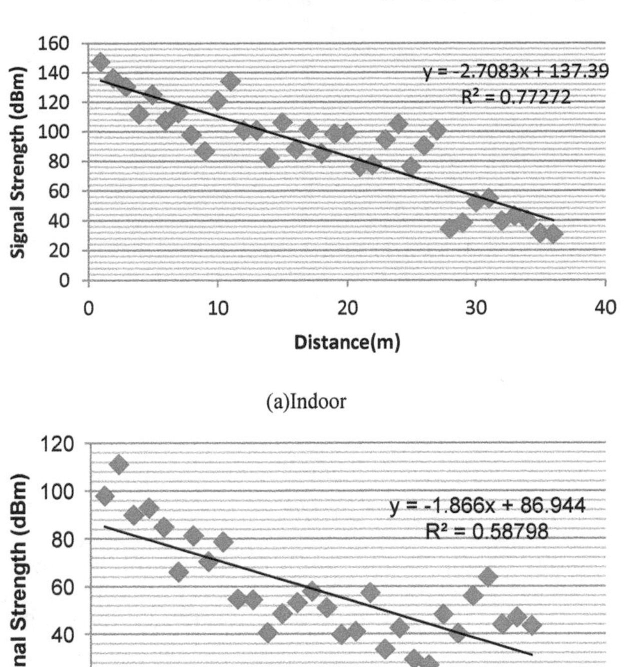

(a)Indoor

(b)Outdoor

Fig. 4. Measured signal strength of ZigBee receiver as a function of distance between transmitter and receiver

Signal strength in the case of outdoor situation is rather weak in comparison to that for indoor situation. Therefore, correlation between signal strength and distance is not so high for outdoor situation which results in relatively poor location estimation accuracy for ZigBee based location estimation method.

There is no error when the position is situated at the cross point among three circles of which the locations of WiFi routers and ZigBee transmitters are situated as shown in Fig.5 (a). On the other hand, some errors would occur in the situations which is shown in Fig.5 (b) for both WiFi router based and ZigBee based location estimation methods, Location estimate has to be made at the location as gravity center of the triangle which is formed with three circles

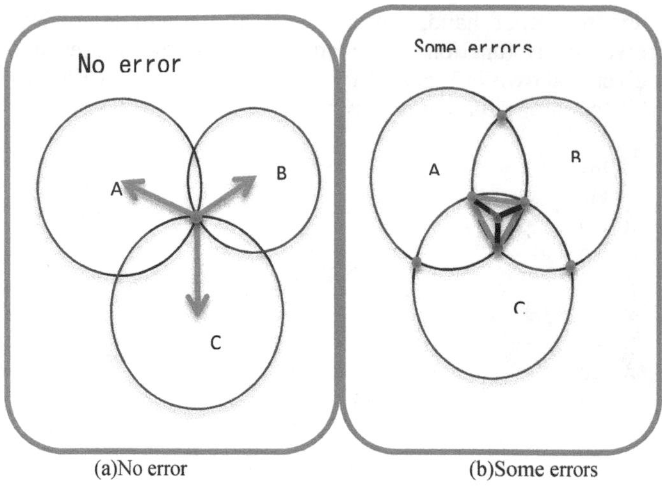

(a)No error (b)Some errors

Fig. 5. Situation when some errors occurrence

B. Location Estimation Accuracy Assessment

Location estimation accuracy of three location estimation methods with GPS, WiFi beacon and ZigBee is assessed for both indoor and outdoor situations. The actual location is the 7th building of the Science and Engineering Faculty of Saga University for indoor situation (Collider on the fourth floor). Meanwhile, outdoor is situated at the southern portion of the 7th building (Parking lot). Fig.6 (a) shows exact measured locations with GPS, WiFi beacon and ZigBee based methods while Fig.6 (b) shows probability density function of the measured location errors for WiFi beacon and ZigBee based methods.

The measured locations with GPS based method have obvious and significant bias error due to the fact that radio wave from GPS satellites comes from the left windows nearby the receiver. Even for the measured locations with WiFi beacon and ZigBee based methods have 4 to 5 meters of bias errors. In terms of mean and standard deviation, ZigBee based method is superior to WiFi beacon based method.

Meanwhile, Fig.7 (a) shows the measured locations in outdoor situation for GPS, WiFi beacon and ZigBee based methods. On the other hand, Fig.7 (b) shows probability density function of the measured location errors.

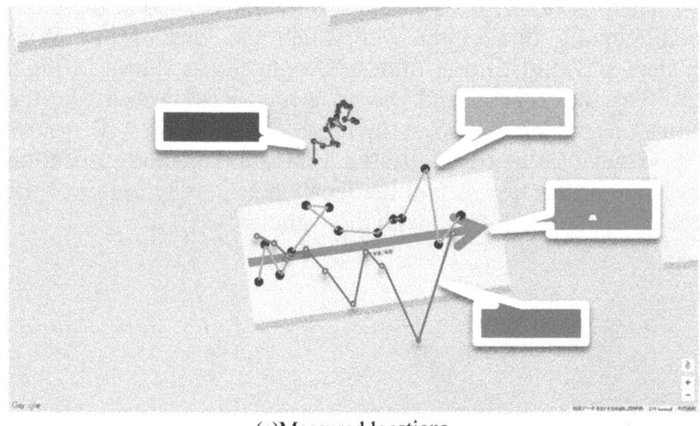

(a)Measured locations

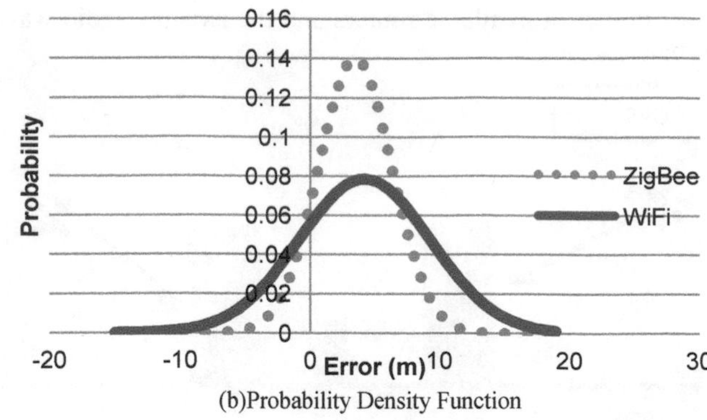

(b)Probability Density Function

Fig. 6. Location estimation accuracy in indoor and outdoor situations

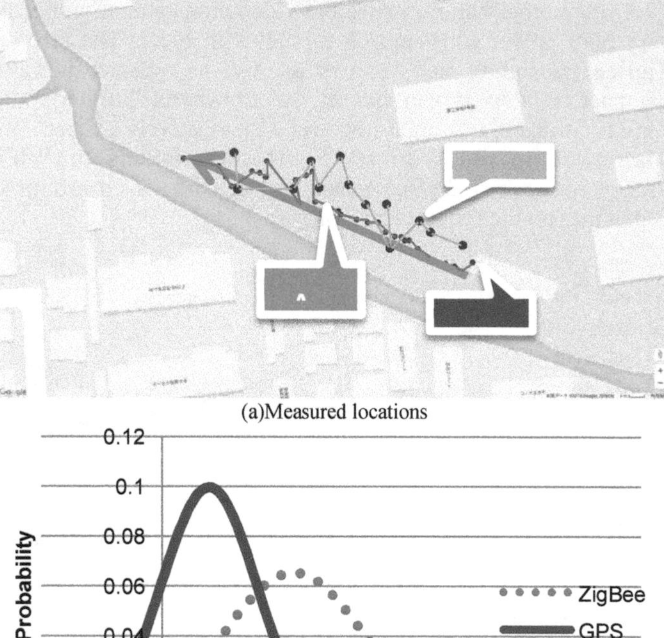

(a)Measured locations

(b)Probability Density Function

Fig. 7. Location estimation accuracy in indoor and outdoor situations

In the case of ZigBee based location estimation, the locations and layouts of transmitters are key for location estimation accuracy. For both cases of location estimations with ZigBee based method between indoor and outdoor situations, just three transmitters are used for location estimations. In the case of outdoor situation, GPS based method is superior to ZigBee based method. GPS based method shows 3m of bias error while ZigBee based method shows 11m of bias error.

Mean and standard deviation of location estimation error for both ZigBee and WiFi based location estimation methods in indoor and outdoor situations are shown in Table 1 and 2, respectively.

TABLE I. MEAN AND STANDARD DEVIATION OF LOCATION ESTIMATION ERROR IN THE CASE OF INDOOR SITUATION

	ZigBee	WiFi
Mean: μ	3.51	4.24
Standard Deviation: σ	2.85	5.08

TABLE II. MEAN AND STANDARD DEVIATION OF LOCATION ESTIMATION ERROR IN THE CASE OF OUTDOOR SITUATION

	ZigBee	GPS
Mean: μ	11	3.90
Standard Deviation: σ	6.09	4.00

C. *Relation Between Location Estimation Error and Rescue Time*

By using GAMA simulation platform, relation between location estimation error and rescue time is clarified. Fig.8 shows road network and the initial positions of victim who needs a help for evacuation (with wheel chair), shelter (green colored house), rescue people (blue colored person), and the measured location (Orange colored circle).

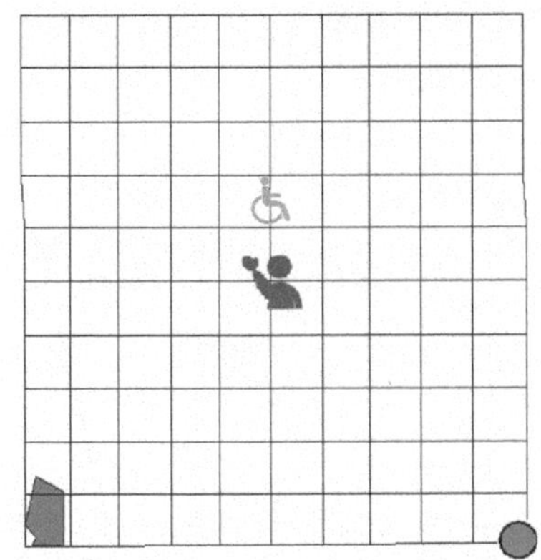

Fig. 8. Initial locations of victim, shelter, recue people and the measured location

The simulation result is shown in Fig.9.

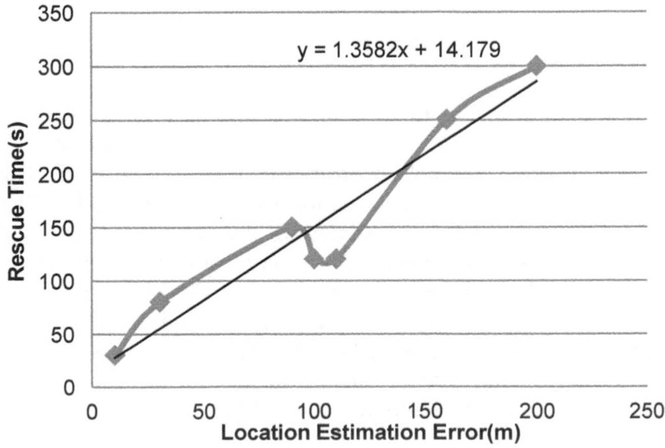

Fig. 9. Relation between location estimation error and rescue time

Even if there is no location estimation error, 14.179 of rescue time is required. In accordance with the location estimation error, rescue time is increased. Linear regressive analysis shows the following relation,

$$y = 1.3582x + 14.179 \qquad (2)$$

where x and y denotes location estimation error and rescue time, respectively.

III. RESCUE SIMULATION

A. *Rescue Simulation Model*

The centralized rescue model is presented which has three types of agent: volunteers, disabled people and route network. The route network is also considered as an agent because the condition of traffic in certain route can be changed when disaster occurs. The general rescue model is shown in Fig.10.

Fig. 10. Centralized Rescue Model

As shown in Fig.10, the concepts of the proposed rescue system. There are three major components, persons who need a help for evacuation, Information Collection Center: ICC for health, traffic, and the other conditions together with the location and attitude information of the persons who need a help and the rescue peoples. Body attached sensors allow measurements of health conditions and the location and attitude of the persons who need a help. The measured data can be transmitted to the ICC through smart-phone, or i-phone, or tablet terminals of which the persons who need a help are carrying. By using the collected health condition and the location/attitude as well as traffic condition information, most appropriate rescue peoples are determined by the person by the person.

B. *Road Network and Initial Positions of Victims, Rescue Peoples, Shelters*

The sample GIS map consists of 5 layers: road, building, rescue peoples (volunteer agents), victims (disabled persons) and shelter. Fig.11 shows the map.

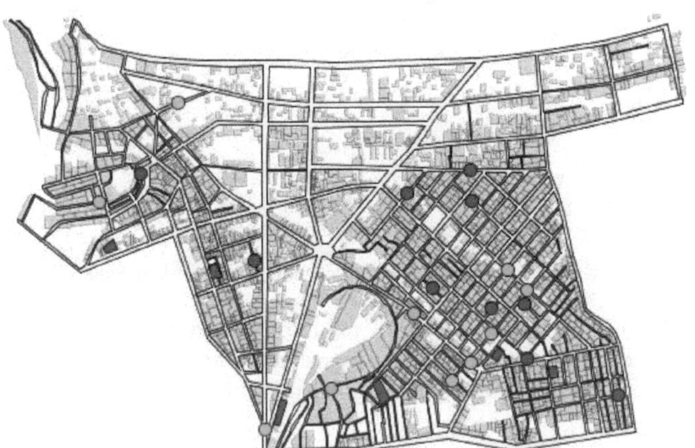

Fig. 11. Road network and the initial locations of victims, rescue peoples, shelters

The red points and green points indicate the locations of disabled persons and locations of volunteers respectively. These locations are generated randomly along the roads. Blue buildings are shelters. The initial health level of disabled persons is generated randomly between 100 and 500. Every time step of simulation, these health levels decrease by 0.5. If the health level is equal to zero, the corresponding agent is considered as dead. The movements of volunteer agents are controlled by Morimoto Traffic Simulator.

C. *Task Allocation Model*

The decision making of volunteers to help disabled persons can be treated as a task allocation problem [10]-[14]. The task allocation for rescue scenario is carried out by the central agents. The task of volunteers is to help disabled persons; this task has to be allocated as to which volunteers should help which disabled persons in order to maximize the number of survivals.

We utilize the combinatorial auction mechanism to solve this task allocation problem. At this model, the volunteers are the bidders; the disabled persons are the items; and the emergency center is the auctioneer. The distance and health level of disabled person are used as the cost for the bid. When the rescue process starts, emergency center creates a list of victims, sets the initial distance for victims, and broadcasts the information to all the volunteer agents. Only the volunteer agents whose distance to victims is less than the initial distance will help these victims. It means that each volunteer agent just help the victims within the initial distance instead of helping all the victims. The initial distance will help volunteers to reduce the number of task so that the decision making will be faster.

D. *Rescue Simulation Results for the Case Without Any Error*

With a fixed number of disabled persons and the number of volunteers increase, the correlation between number of volunteers and rescue time is shown in Fig.12.

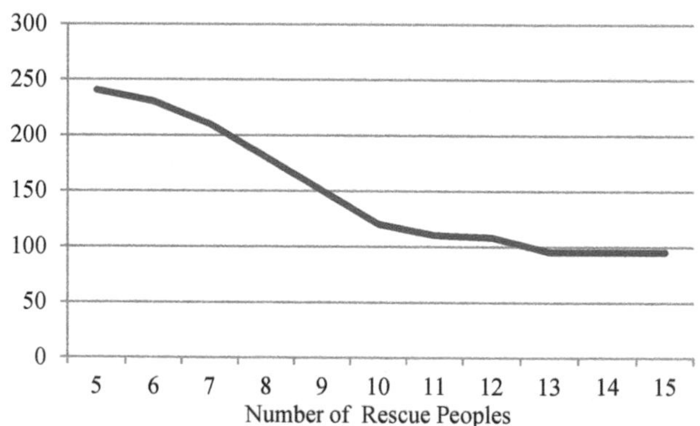

Fig. 12. Correlation between Number of Volunteers and Rescue Time

With a fixed number of volunteers and the number of disabled persons increase, the correlation between number of disabled persons and rescue time is shown in Fig.13.

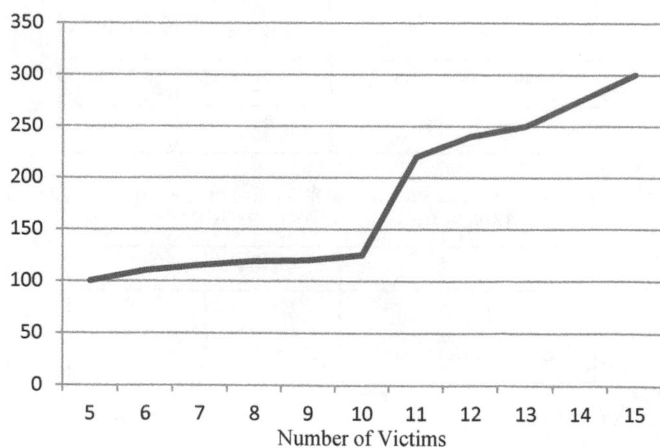

Fig. 13. Correlation between Number of Disabled Persons and Rescue Time

The number of volunteers and the number of disabled persons are fixed, whereas the number of vehicle increases. We test with the total length of road of 500 meters. The increasing number of vehicles will make traffic density higher. The correlation between number of vehicle and rescue time is shown in Fig.14.

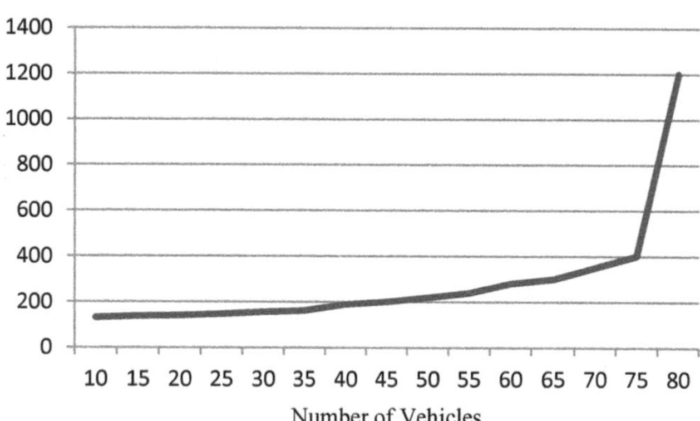

Fig. 14. Correlation between Number of Vehicles and Rescue Time

E. *Rescue Simulation Results of the Proposed Location Estimation Method*

The results of Fig.12 to 14 are for the case of no location estimation error. Meanwhile, the simulation results taking into account location estimation errors are shown in Fig.15. The basic idea of the proposed location estimation method is that three different location estimation methods are switched depending on the circumstances of the locations of radio wave absorbance, the number of access points (WiFi routers), the number of acquired GPS satellites, and the number of ZigBee transmitters. As shown in Fig.7, rescue time is increased when location estimation error is taken into account even if the best accuracy of location estimation method is selected. The rescue time for the case of which location estimation error is taken into account is evaluated. Around 10 to 20 seconds are required additionally in comparison to the rescue time without consideration of location estimation error.

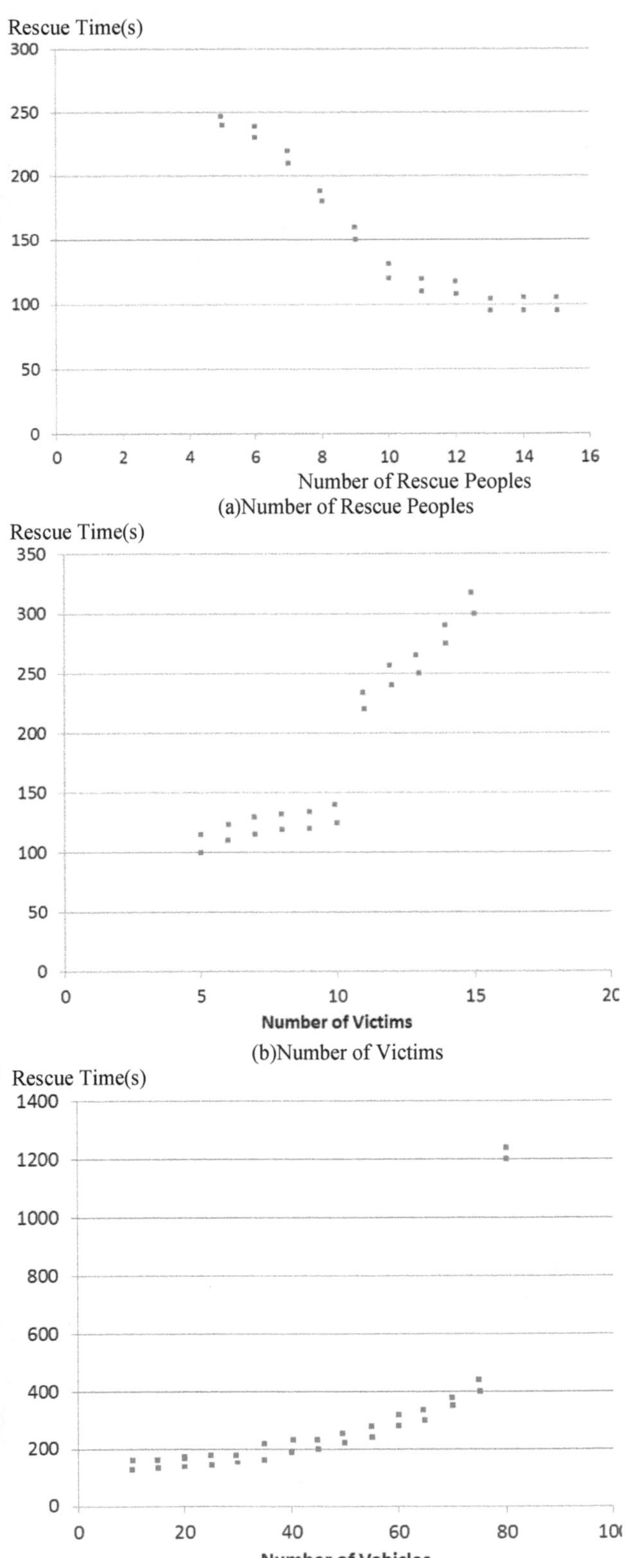

(a)Number of Rescue Peoples
(b)Number of Victims
(c)Number of Vehicles

Fig. 15. Rescue time for the case of which location estimation error is taken into account

IV. Conclusion

Location monitoring system for rescue disable persons by switching the location estimation methods with GPS, ZigBee and WiFi beacon is proposed. Rescue system with triage using health condition monitoring together with location and attitude monitoring as well as the other data acquired with mobile devices is evaluated with the proposed location monitoring system. Through simulation study, influence due to location estimation error on rescue time is evaluated together with effect of the proposed location monitoring system. Also, it is found that the effect of triage on rescue time is clarified.

Further study is required for clarifying effects of triage on reducing rescue time in actual situations.

Acknowledgment

The author would like to thank Dr. Trang Xuang Sang of Vinh University in Vietnam for his effort to conduct simulation studies.

References

[1] C. Ren, C. Yang, and S. Jin, "Agent-Based Modeling and Simulation on emergency", Complex 2009, Part II, LNICST 5, 1451 – 1461, 2009.

[2] M. H. Zaharia , F. Leon, C. Pal, and G. Pagu, "Agent-Based Simulation of Crowd Evacuation Behavior", International Conference on Automatic Control, Modeling and Simulation, 529-533, 2011.

[3] C. T. Quang, and A. Drogoul, "Agent-based simulation: definition, applications and perspectives", Invited Talk for the biannual Conference of the Faculty of Computer Science, Mathematics and Mechanics, 2008.

[4] Z. Bo, and V. Satish, "Agent-based modeling for household level hurricane evacuation", Winter Simulation Conference, 2009.

[5] J. W. Cole, C. E. Sabel, E. Blumenthal,K. Finnis, A. Dantas,S. Barnard, and D. M. Johnston, "GIS-based emergency and evacuation planning for volcanic hazards in New Zealand", Bulletin of the New Zealand society for earthquake engineering, vol. 38, no. 3, 2005.

[6] M. Batty, "Agent-Based Technologies and GIS: simulating crowding, panic, and disaster management", Frontiers of geographic information technology, chapter 4, 81-101, 2005

[7] T. Patrick, and A. Drogoul, "From GIS Data to GIS Agents Modeling with the GAMA simulation platform", TF SIM 2010.

[8] J. Kaprzy Edt., Kohei Arai, Rescue System for Elderly and Disabled Persons Using Wearable Physical and Psychological Monitoring System, Studies in Computer Intelligence, 542, 45-64, Springer Publishing Co. Ltd., 2014.

[9] K.Arai, T.X.Sang, Emergency rescue simulation for disabled persons with help from volunteers, International Journal of Research and Review on Computer Science, 3, 2, 1543-1547, 2012.

[10] K.Arai, Wearable healthy monitoring sensor network and its application to evacuation and rescue information server system for disabled and elderly person, International Journal of Research and Review on Computer Science, 3, 3, 1633-1639, 2012.

[11] K.Arai, T.X.Sang, N.T.Uyen, Task allocation model for rescue disable persons in disaster area with help of volunteers, International Journal of Advanced Computer Science and Applications, 3, 7, 96-101, 2012.

[12] Kohei Arai, Tran Xuan Sang, Decision making and emergency communication system in rescue simulation for people with disabilities, International Journal of Advanced Research in Artificial Intelligence, 2, 3, 77-85, 2013.

[13] Kohei Arai, Frequent physical health monitoring as vital sign with psychological status monitoring for search and rescue of handicapped, disabled and elderly persons, International Journal of Advanced Research in Artificial Intelligence, 2, 11, 25-31, 2013.

[14] Kohei Arai, Vital sign and location/attitude monitoring with sensor networks for the proposed rescue system for disabled and elderly persons who need a help in evacuation from disaster areas, International

Journal of Advanced Research in Artificial Intelligence, 3, 1, 24-33, 2014.

[15] Kohei Arai, Method and system for human action detection with acceleration sensors for the proposed rescue system for disabled and elderly persons who need a help in evacuation from disaster areas, International Journal of Advanced Research in Artificial Intelligence, 3, 1, 34-40, 2014.

[16] Kohei Arai, Method and system for human action detection with acceleration sensors for the proposed rescue system for disabled and elderly persons who need a help in evacuation from disaster areas, International Journal of Advanced Research in Artificial Intelligence, 3, 1, 34-40, 2014.

[17] Kohei Arai, Taka Eguchi, Realistic rescue simulation method with consideration of roadnetwork ristrictions, International Journal of Advanced Research on Artificial Intelligence, 4, 7, 21-28, 2015.

[18] K.Arai, T.X. Sang, Multi agent based rescue simulation for disabled persons with the help of volunteers in emergency situation, Proceedings of the 260th conference in Saga of Image and Electronics Engineering Society of Japan, 81-85, 2012.

[19] Kohei Arai, Rescue system with sensor network for physical and psychological health monitoring, Proceedings of the International Seminar on Intelligent Technology and Its Applications: ISITIA 2015, (Keynote Speech), 2015.

[20] Kohei Arai, Rescue system with sensor network for physical and psychological health monitoring, Proceedings of the SAI Confeernce 2015

Realistic Rescue Simulation Method with Consideration of Road Network Restrictions

Kohei Arai 1
Graduate School of Science and Engineering
Saga University
Saga City, Japan

Takashi Eguchi [1]
Graduate School of Science and Engineering
Saga University
Saga City, Japan

***Abstract*—A realistic rescue simulation method with consideration of road network restrictions is proposed. Decision making and emergency communication system play important roles in rescue process when emergency situations happen. The rescue process will be more effective if we have appropriate decision making method and accessible emergency communication system. In this paper, we propose centralized rescue model for people with disabilities. The decision making method to decide which volunteers should help which disabled persons is proposed by utilizing the auction mechanism. The GIS data are used to present the objects in a large-scale disaster simulation environment such as roads, buildings, and humans. The Gama simulation platform is used to test our proposed rescue simulation model. There are road network restrictions, road disconnections, one way traffic, roads which do not allow U-Turn, etc. These road network restrictions are taken into account in the proposed rescue simulation model. The experimental results show around 10% of additional time is required for evacuation of victims.**

Keywords—Rescue Simulation for people with disabilities; GIS MultiAgent-based Rescue Simulation; Auction based Decision Making

I. INTRODUCTION

In an emergency situation, a human tends to perform two main activities: the rescue and the evacuation. It is very difficult and costly if we want to do experiments on human rescue and or evacuation behaviors physically in real scale level. It is found that multi agent-based simulation makes it possible to simulate the human activities in rescue and evacuation process [1, 2]. A multi agent-based model is composed of individual units, situated in an explicit space, and provided with their own attributes and rules [3]. This model is particularly suitable for modeling human behaviors, as human characteristics can be presented as agent behaviors. Therefore, the multi agent-based model is widely used for rescue and evacuation simulation [1-5].

In this study, GIS map is used to model objects such as road, building, human, fire with various properties to describe the objects condition. With the help of GIS data, it enables the disaster space to be closer to a real situation [5-10]. Kisko et al. (1998) employs a flow based model to simulate the physical environment as a network of nodes. The physical structures, such as rooms, stairs, lobbies, and hallways are represented as nodes which are connected to comprise a evacuation space. This approach allows viewing the movement of evacuees as a continuous flow, not as an aggregate of persons varying in physical abilities, individual dispositions and direction of movement [11]. Gregor et al. (2008) presents a large scale microscopic evacuation simulation. Each evacuee is modeled as an individual agent that optimizes its personal evacuation route. The objective is a Nash equilibrium, where every agent attempts to find a route that is optimal for the agent [12]. Fahy (1996; 1999) proposes an agent based model for evacuation simulation. This model allows taking in account the social interaction and emergent group response. The travel time is a function of density and speed within a constructed network of nodes and arcs [13, 14]. Gobelbecker et al. (2009) presents a method to acquire GIS data to design a large scale disaster simulation environment. The GIS data is retrieved from a public source through the website OpenStreetMap.org. The data is then converted to the Robocup Rescue Simulation system format, enabling a simulation on a real world scenario [15]. Sato et al. (2011) also proposed a method to create realistic maps using the open GIS data. The experiment shows the differences between two types of maps: the map generated from the program and the map created from the real data [2]. Ren et al. (2009) presents an agent-based modeling and simulation using Repast software to construct crowd evacuation for emergency response for an area under a fire. Characteristics of the people are modeled and tested by iterative simulation. The simulation results demonstrate the effect of various parameters of agents [3]. Cole (2005) studied on GIS agent-based technology for emergency simulation. This research discusses about the simulation of crowding, panic and disaster management [6]. Quang et al. (2009) proposes the approach of multi-agent-based simulation based on participatory design and interactive learning with experts' preferences for rescue simulation [9]. Hunsberger et al. (2000), Beatriz et al. (2003) and Chan et al. (2005) apply the auction mechanism to solve the task allocation problem in rescue decision making. Christensen et al. (2008) presents the BUMMPEE model, an agent-based simulation capable of simulating a heterogeneous population according to variation in individual criteria. This method allows simulating the behaviors of people with disabilities in emergency situation [23].

Our study will focus mainly on proposing a rescue model for people with disabilities in large scale environment. This rescue model provides some specific functions to help disabled people effectively when emergency situation occurs. Important components of an evacuation plan are the ability to receive critical information about an emergency, how to respond to an emergency, and where to go to receive assistance. We propose a wearable device which is attached to body of disabled people. This device measures the condition of the disabled persons such as their heart rate, body temperature and attitude; the device can also be used to trace the location of the disabled persons by GPS. That information will be sent to emergency center automatically. The emergency center will then collect that information together with information from volunteers to assign which volunteer should help which disabled persons.

The rest of the paper is organized as follows. Section 2 describes the centralized rescue model and the rescue decision making method. Section 3 provides the experimental results of different evacuation scenarios. Finally, section 4 summarizes the work of this paper.

II. PROPOSED RESCUE SYSTEM

A. *Proposed Rescue Model*

Important components of an evacuation plan are the ability to receive critical information about an emergency, how to respond to an emergency, and where to go to receive assistance. We proposes a wearable device which is attached to body of disabled people. This device measures the condition of the disabled persons such as their heart rate, body temperature and attitude; the device can also be used to trace the location of the disabled persons by GPS. Those information will be sent to emergency center automatically. The emergency center will then collect those information together with information from volunteers to assign which volunteer should help which disabled persons. The centralized rescue model presented has three types of agents: volunteers, disabled people and route network. The route network is also considered as an agent because the condition of traffic in a certain route can be changed when a disaster occurs. The general rescue model is shown in Figure 1.

Fig. 1. Centralized Rescue Model

Before starting the simulation, every agent has to be connected to the emergency center in order to send and receive information. The types of data exchanged between agents and emergency center are listed as below.

Message from agent
A1: To request for connection to the emergency center
A2: To acknowledge the connection
A3: Inform the movement to another position
A4: Inform the rescue action for victim
A5: Inform the load action for victim
A6: Inform the unload action for victim
A7: Inform the inactive status
Message from emergency center
K1: To confirm the success of the connection
K2: To confirm the failure of the connection
K3: To send decisive information

Before starting the simulation, every agent will send the command A1 to request for connection to the emergency center. The emergency center will return the response with a command K1 or K2 corresponding to the success or failure of their connection respectively. If the connection is established, the agent will send the command A2 to acknowledge the connection. The initial process of simulation is shown in Figure 2.

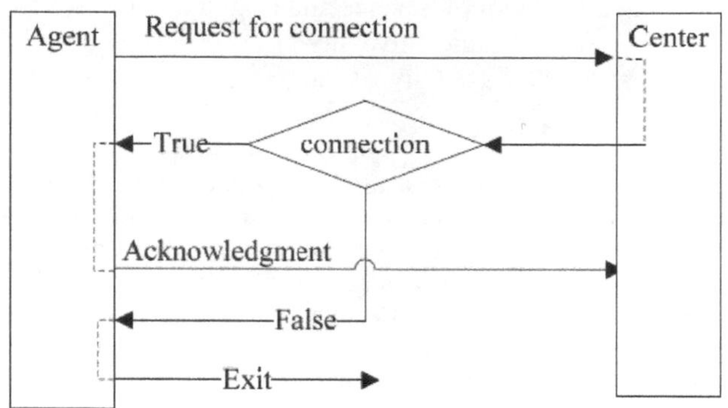

Fig. 2. Initial Process

After the initial process, all the connected agents will receive the decisive information such as the location of agents and health level via command K3; after that the rescue agents will make a decision of action and submit to the center using one of the commands from A3 to A7. At every cycle in the simulation, each rescue agent receives a command K3 as its own decisive information from the center, and then submits back an action command. The status of disaster space is sent to the viewer for visualization of simulation. The repeating steps of simulation are shown in Figure 3.

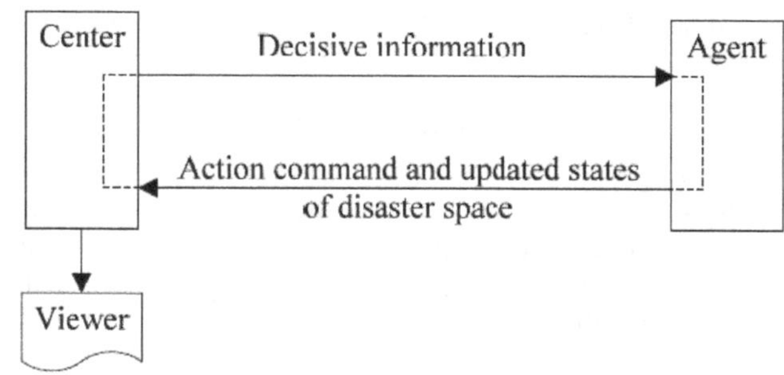

Fig. 3. Simulation Cycles

B. *Disaster Area Model*

The disaster area is modeled as a collection of objects: Nodes, Buildings, Roads, and Humans. Each object has properties such as its positions, shape and is identified by a unique ID. Table 1 to Table 7 presents the properties of Nodes,

Buildings, Roads and Humans object respectively. These properties are derived from RoboCup rescue platform with some modifications.

The topographical relations of objects are illustrated from Figure 4 to Figure 7. The representative point is assigned to every object, and the distance between two objects is calculated from their representative points.

TABLE I. PROPERTIES OF NODE OBJECT

Property	Unit	Description
x,y	–	x-yCordinate
Edges	ID	Connected_road_and_Building

TABLE II. PROPERTIES OF BUILDING OBJECT

Property	Description
x,y	x-y coordinate of the representative point
Entrance	Node_connecting_building_and_road

TABLE III. PROPERTIES OF ROAD OBJECT

Property	Unit	Description
Start_Point_and_End_Point	ID	Point_to_enter_the_Road_
Length_and_Width	m	Length_and_width_of_Road
Lane	Lane	Number_of_lanes
Blocked_road	Lane	Number_of_blocked_lanes
Clear_Cost	Cycle	Cost_required_for_clearing_blocks

TABLE IV. PROPERTIES OF VICTIM AGENT

Property	Unit	Description
Position	ID	Object_where_victim_is_on
Position_in _road	m	Lecgth_from_Start_Point_when_victim_is_on_ro ad,_Otherwise_it_is_zero
Health_Le vel	Health_P oint	Health_level_of_victim
Damage_P oint	Health_P oint	Health_level_dwindles_by_Damage
Disability_ Type	Type(1-7)	Disability_types
Disability_ Level	High_or _Low	High_disability_level_means_Highest_Damage_P oint

TABLE V. PROPERTIES OF VOLUNTEER AGENT

Property	Unit	Description
Position_on_road	ID	Object=that_victim_is_on
Position_on_road	m	Length_from_Start_Point
Current_Action	Type(1-3)	See_in_Table_VII
Energy	Level(1-5)	Empty_level_of_vehicle_gasoline
Panic_Level	Level(0-9)	Hesitance_level_of_decision

TABLE VI. ACTION OF VOLUNTEER AGENT

Action_ID	Action	Description
1	Stationary	Rescue_person_Stays
2	Move_to_Victim	Rescue_person_moves_to_victim
3	Move_to_Shelter	Rescue_person_carry_victim_to_shelter

TABLE VII. TYPE OF DISABILITY

Type	Description
1	Cognitive_Impaiment
2	Dexterity_Impairment
3	Mobility_Impairment
4	Elderly
5	Hearing_Impairment
6	Speech_and_Language_Impairment
7	Visdual_Impairment

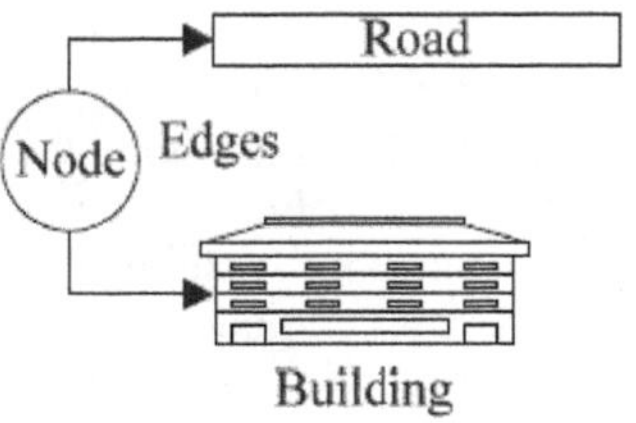

Fig. 4. Node object

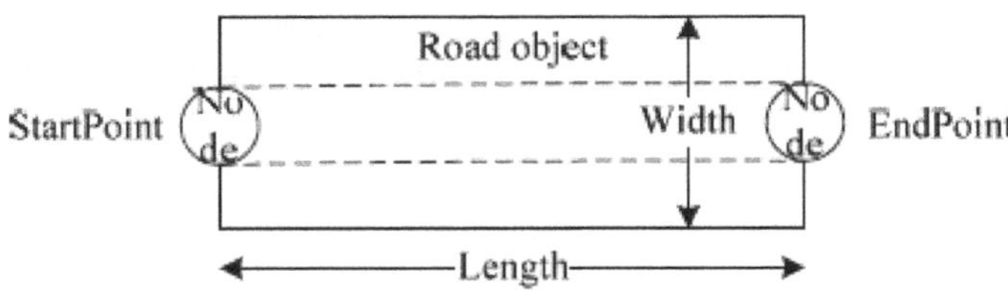

Fig. 5. Road object

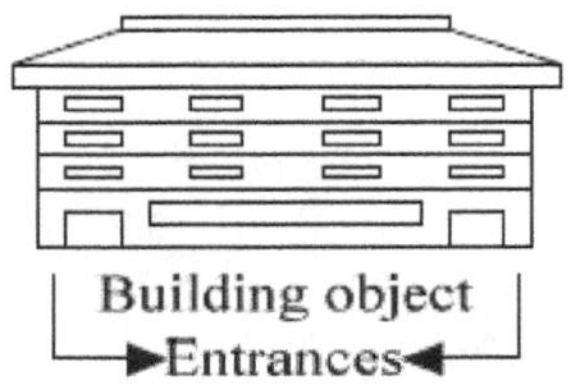

Fig. 6. Building object

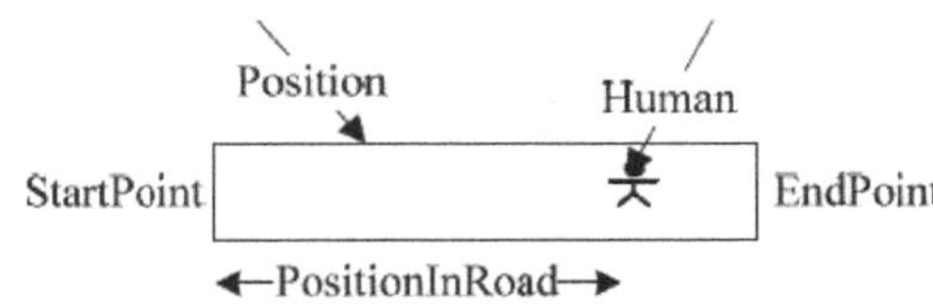

Fig. 7. Human object

C. *Path finding in Gama Simulation Platform*

After a volunteer makes the decision to help a certain victim, the path finding algorithm is used to find the route from volunteer agent to victim agent. The GIS data presents roads as a line network in graph type. Figure 8 shows an example of graph computation. The Dijkstra algorithm is implemented for the shortest path computation [8].

In this section, we present experimental studies on different scenarios. We show the experimental results with traditional rescue model which not considering the updated information of victims and volunteers such as health conditions, locations, traffic conditions.

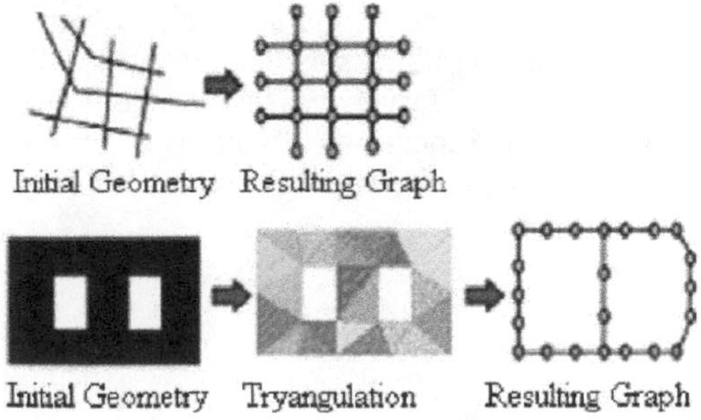

Fig. 8. Example of Graph Computation [8]

The traditional rescue model provides fixed mission for which volunteers should help which victims. Whereas, our rescue model provides flexible mission for which volunteers should help which victims. The targets of volunteers can be changed dynamically according to current situation. The experimental results of our proposed rescue model are also presented to show the advantages comparing to traditional model.

The evacuation time is evaluated from the time at which the first volunteer started moving till the time at which all saved victims arrive at the shelters. The simulation model is tested using the Gama simulation platform [8, 10].

We consider the number of volunteers, number of disabled persons, panic level of volunteer, disability level of victim and the complexity of traffic as parameters to examine the correlation between these parameters with rescue time. The traffic complexity is function of the number of nodes and links in a road network.

Figure 9 presents the sample GIS map consisting of 4 layers: road, volunteer, disabled person and shelter. The initial health levels of disabled persons are generated randomly between 100 point and 500 point. If the health level is equal or less than zero, the corresponding agent is considered as dead.

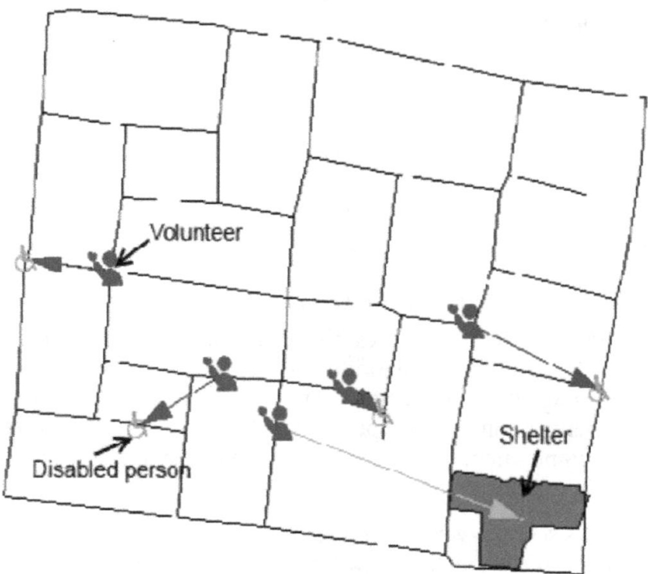

Fig. 9. Sample GIS Map of Disaster Space

D. Realistic Simulation Designating to Saga City, Japan

Figure 10 (a) shows the rescue simulation site which is situated at Saga University and the surrounding areas. The shelter around this site is Saga University. Meanwhile, Figure 10 (b) shows road network and locations of victims and rescue peoples as initial conditions. At the top right corner, entrance gate is situated at the shelter.

(a)Topographic map of the rescue simulation site

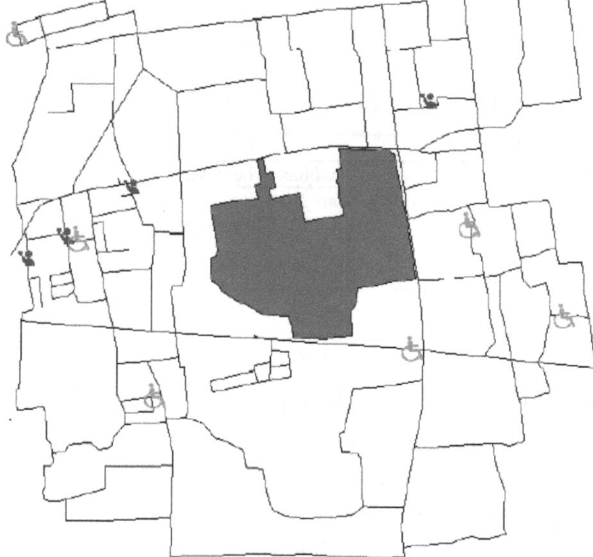

(b)Road network of the rescue simulation site

Fig. 10. Rescue simulation site of Saga University and its surrounding areas

III. Experiments

Simulations are conducted with the aforementioned initial conditions on the road network. There are two victims. ①The rescue person goes the victim #1, firstly, and ②,③he goes to the victim #2 after that. Then he goes to the shelter together with the victims #1 and #2. It is expected the rescue route length would be 4.34 km and the expected rescue time would be 10 min. and 40 sec. as shown in Figure 11 (a). Figure 11 (b) shows the route for rescue the victim #3. In this case, the number of rescue victims is just one. Therefore, the route length is 2.66 km and it takes 5 min. and 16 sec. Another possible route for rescue the victim #4 is shown in Figure 11 (c). There is only one victim. Therefore, the route length is 2.36 km and the time required for rescue is 4min. and 18 sec.

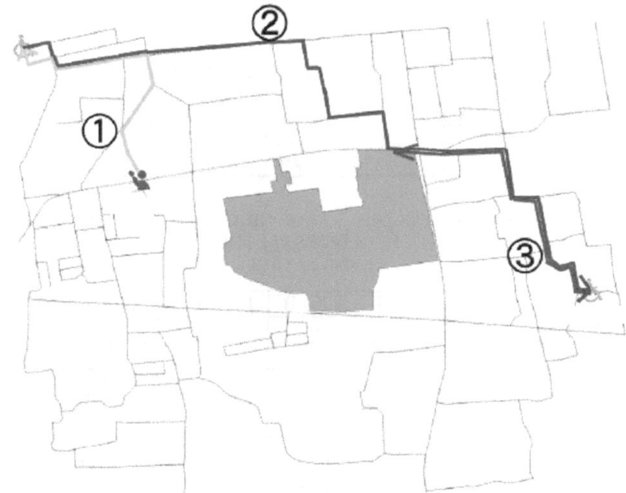

(a)Possible route for rescue the victims #1 and #2

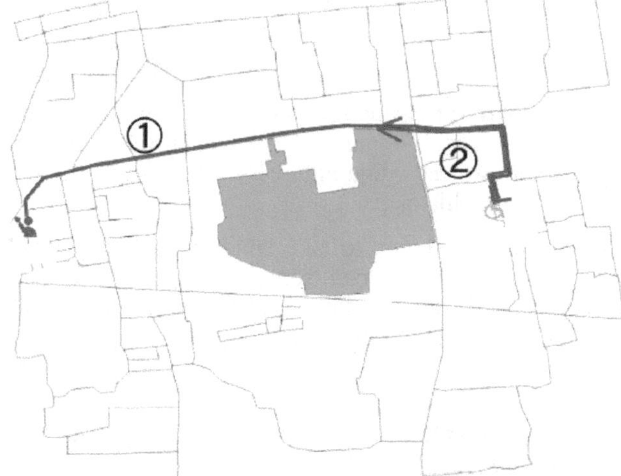

(b)Possible route for rescue the victim #3

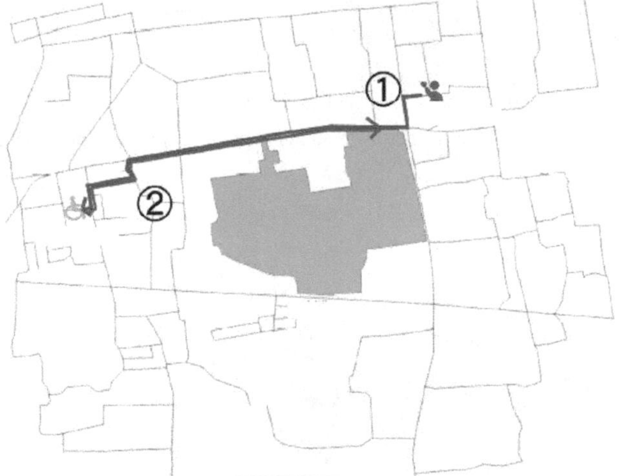

(c)Possible route for rescue the victim #4

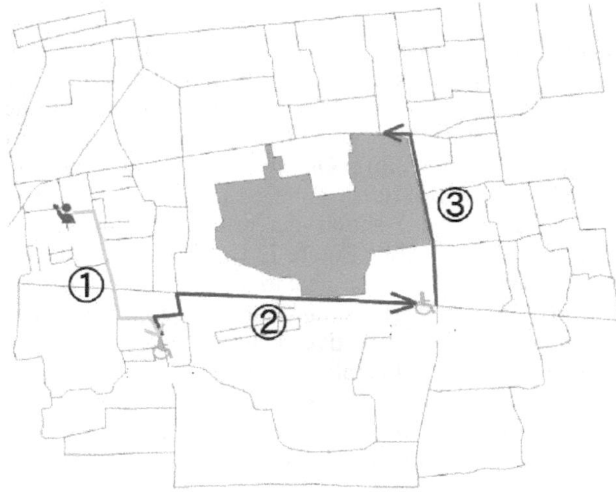

(d)Possible route for rescue the victims #5 and #6

Fig. 11. Possible routes for rescue the victims #1-#6

On the other hand, the possible route for rescue the victim #5 and #6 is shown in Figure 11 (d). In this case, the number of victims is two for the rescue person. The route length is 2.385 km and it takes 6 min. and 20 sec. The time required for rescue for each case is summarized in Table 8. These are simulation results. The experiments are conducted in real world twice. The time required for rescue is shown in Table 9 (the first trial) and 10 (the second trial).

TABLE VIII. THE TIME REQUIRED FOR RESCUE (SIMULATION)

	Rescue A	Rescue B	Rescue C	Rescue D
First victim	2:26	3:58	2:50	2:10
Second victim	8:00	---	---	4:18
Rescue time	10:40	5:16	4:18	6:20

TABLE IX. THE TIME REQUIRED FOR RESCUE (REAL WORLD TRIAL #1)

	Rescue A	Rescue B	Rescue C	Rescue D
First victim	2:56	5:36	3:36	2:36
Second victim	11:41	---	---	6:10
Rescue time	16:54	7:47	6:37	8:37

TABLE X. THE TIME REQUIRED FOR RESCUE (REAL WORLD TRIAL #2)

	Rescue A	Rescue B	Rescue C	Rescue D
First victim	1:45	5:54	4:12	1:49
Second victim	10:22	---	---	4:51
Rescue time	14:57	8:13	6:59	7:13

Most of the rescue simulation software does not care about the one way roads and the roads of which u-turn is impossible. There, however, are one way traffic roads and the u-turn impossible roads in the real world situation. Therefore, it takes much time for rescue when these realistic road conditions are taken into account in comparison to the simulations and the experiments which do not taken into account the conditions.

Figure 12 (a) shows an example of the one way traffic road in Saga city. Black cross mark shows one way traffic road from right to left. Therefore, the rescue person cannot take this one way traffic road (Figure 12 (b)). On behalf of this road, the rescue person takes another alternative road (Second shortest pass shown in Figure 12(c)). On the other hand, there are narrow roads of which u-turn cannot be done as is shown in Figure 13. Figure 13 (a) shows the route for conventional rescue simulation result while Figure 13 (b) shows the route for the proposed rescue simulation. Due to the fact that the road is narrow which does not allow make any u-turn, the rescue person have to take the alternative route of Figure 13 (b). As the results of these considerations on the proposed rescue simulation, the rescue person A, B, C, D takes the rescue time and the route length which are shown in Table 11. Meanwhile, Table 12 shows rescue time in the case of the considerations of one way traffic road and u-turn impossible road in the proposed rescue simulation. On the other hand, Table 13 shows effects of rescue time increasing in the case of the considerations of one way traffic road and u-turn impossible road in the proposed rescue simulation. In case of the consideration of one way traffic road and u-turn impossible road, the alternative route of the second shortest pass has to be taken. Therefore, rescue time is increased.

(a)One way traffic road

(b)Shortest pass (c)Second shortest pass

Fig. 12. Consideration of one way traffic road in the proposed rescue simulation

(a)U-turn traffic road (b)Alternative route

Fig. 13. Consideration of U-turn impossible roads in the proposed rescue simulation-

TABLE XI. THE TIME REQUIRED FOR RESCUE AND THE ROUTE LENGTH

	Rescue A	Rescue B	Rescue C	Rescue D
Route length	4.34	2.66	2.36	2.38
Rescue time	2’26”	3’58”	2’50”	2’10”

TABLE XII. RESCUE TIME IN THE CASE OF THE CONSIDERATIONS OF ONE WAY TRAFFIC ROAD AND U-TURN IMPOSSIBLE ROAD IN THE PROPOSED RESCUE SIMULATION

	Rescue A	Rescue B	Rescue C	Rescue D
First Victim	2’35”	4’10”	3’40”	2.10
Second victim	9’14”	---	---	4’30”
Overall Rescue time	12’15”	5’44”	6’1”	6’50”

TABLE XIII. EFFECT OF THE CONSIDERATIONS OF ONE WAY TRAFFIC ROAD AND U-TURN IMPOSSIBLE ROAD IN THE PROPOSED RESCUE SIMULATION

	Rescue A	Rescue B	Rescue C	Rescue D
First Victim	+9”	+12”	+50”	0”
Second victim	+1’14”	---	---	+12”
Overall Rescue time	+1’35”	+28”	+1’43”	+30”

It is much hard to pass the route if the road is much narrow road which is shown in Figure 14. Figure 14 (a) shows the photo of the narrow road while Figure 14 (b) shows the narrow road on the topographic map.

In such case, it is better to take another alternative much wider road than the narrow road from the rescue time point of view. Therefore, it is better to take the route #2 (Figure 15 (b)) rather than the route #1 (Figure 15 (a)).

(a)Photo of the narrow road (b)The narrow road on map

Fig. 14. Consideration of the narrow road in the proposed rescue simulation

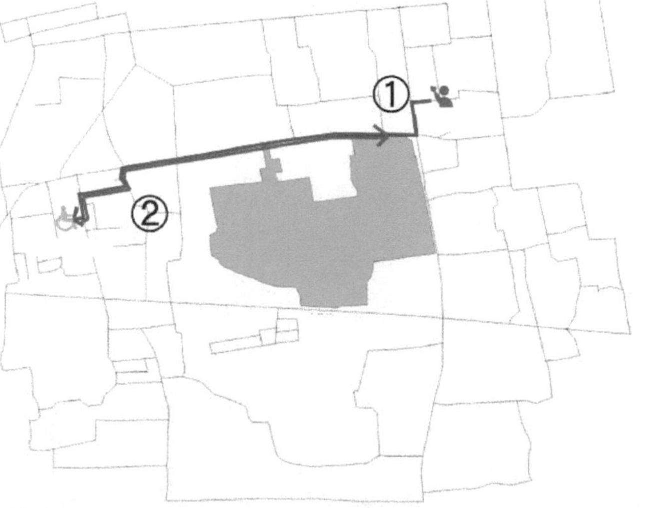

(a)Route #1

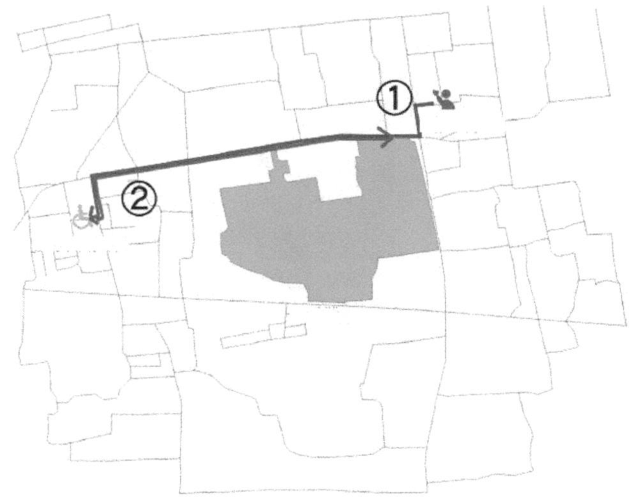

(b)Route #2

Fig. 15. Consideration of priority of wide roads rather than narrow roads in the proposed rescue simulation

Meanwhile, there are the routes which take much longer time for crossing very wide boulevards than narrow roads as shown in Figure 16. Figure 16 (a) shows an example of photo of the route which takes much longer time for crossing very wide boulevard than narrow roads while Figure 16 (b) shows the route which takes much longer time for crossing very wide boulevard than narrow roads on map.

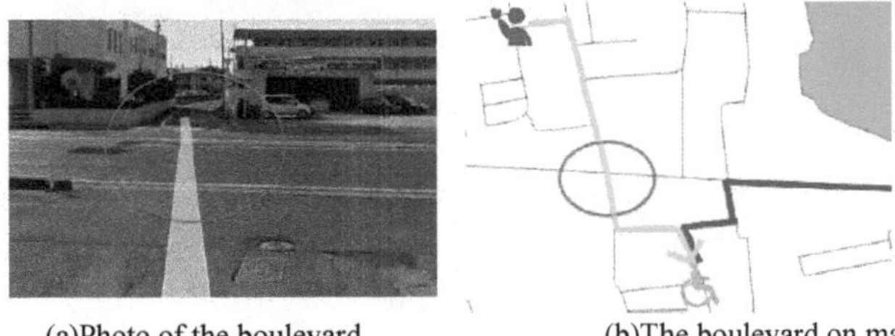

(a)Photo of the boulevard (b)The boulevard on map

Fig. 16. Routes which take much longer time for crossing very wide boulevards than narrow roads

IV. CONCLUSION

A realistic rescue simulation method with consideration of road network restrictions is proposed. Decision making and emergency communication system play important roles in rescue process when emergency situations happen. The rescue process will be more effective if we have appropriate decision making method and accessible emergency communication system.

In this paper, we propose centralized rescue model for people with disabilities. The decision making method to decide which volunteers should help which disabled persons is proposed by utilizing the auction mechanism. The GIS data are used to present the objects in a large-scale disaster simulation environment such as roads, buildings, and humans.

The Gama simulation platform is used to test our proposed rescue simulation model. There are road network restrictions, road disconnections, one way traffic, roads which do not allow U-Turn, etc. These road network restrictions are taken into account in the proposed rescue simulation model.

The experimental results show around 10% of additional time is required for evacuation of victims in maximum. In order to reduce rescue time, considerations of priority of wide roads rather than narrow roads in the proposed rescue simulation are taking into account.

REFERENCES

[1] Obelbecker G., & Dornhege M., "Realistic cities in simulated environments - an Open Street Map to Robocup Rescue converter", Online-Proceedings of the Fourth International Workshop on Synthetic Simulation and Robotics to Mitigate Earthquake Disaster, 2009.

[2] Sato, K., & Takahashi, T., "A study of map data influence on disaster and rescue simulation's results", Computational Intelligence Series, vol. 325. Springer Berlin / Heidelberg, 389–402, 2011.

[3] Ren C., Yang C., & Jin S., "Agent-Based Modeling and Simulation on emergency", Complex 2009, Part II, LNICST 5, 1451 – 1461, 2009.

[4] Zaharia M. H., Leon F., Pal C., & Pagu G., "Agent-Based Simulation of Crowd Evacuation Behavior", International Conference on Automatic Control, Modeling and Simulation, 529-533, 2011.

[5] Quang C. T., & Drogoul A., "Agent-based simulation: definition, applications and perspectives", Invited Talk for the biannual Conference of the Faculty of Computer Science, Mathematics and Mechanics, 2008.

[6] Cole J. W., Sabel C. E., Blumenthal E., Finnis K., Dantas A., Barnard S., & Johnston D. M., "GIS-based emergency and evacuation planning for volcanic hazards in New Zealand", Bulletin of the New Zealand society for earthquake engineering, vol. 38, no. 3, 2005.

[7] Batty M., "Agent-Based Technologies and GIS: simulating crowding, panic, and disaster management", Frontiers of geographic information technology, chapter 4, 81-101, 2005.

[8] Patrick T., & Drogoul A., "From GIS Data to GIS Agents Modeling with the GAMA simulation platform", TF SIM 2010.

[9] Quang C. T., Drogoul A., & Boucher A., "Interactive Learning of Independent Experts' Criteria for Rescue Simulations", Journal of Universal Computer Science, Vol. 15, No. 13, 2701-2725, 2009.

[10] Taillandier T., Vo D. A., Amouroux E., & Drogoul A., "GAMA: a simulation platform that integrates geographical information data, agentbased modeling and multi-scale control", In Proceedings of Principles and practice of multi-agent systems, India, 2012.

[11] Kisko, T.M., Francis, R.L., Nobel C.R., "EVACNET4 User's Guide", University of Florida, 1998.

[12] Gregor, Marcel R., & Nagel K., "Large scale microscopic evacuation simulation", Pedestrian and Evacuation Dynamics, 547-553, 2008.

[13] Fahy R. F., "User's Manual, EXIT89 v 1.01, An Evacuation Model for High-Rise Buildings, National Fire Protection Association", Quincy, Mass, 1999.

[14] Fahy R. F., "EXIT89 – High-Rise Evacuation Model –Recent Enhancements and Example Applications" Interflam '96, International Interflam Conference – 7th Proceedings; Cambridge, England, pg. 1001-1005, 1996.

[15] Gobelbecker M., & Dornhege C., "Realistic cities in simulated environments- an open street map to RoboCup Rescue converter". In 4th Int'l Workshop on Synthetic Simulation and Robotics to Mitigate Earthquake Disaster (SRMED 2009), Graz, Austria, July 2009.

[16] Nair R., Ito T., Tambe M., & Marsella S., "Task allocation in the rescue simulation domain: A short note", Volume 2377 of Lecture Notes in Computer Science. Springer, Berlin 751–754, 2002.

[17] Boffo F., Ferreira P. R., & Bazzan A. L., "A comparison of algorithms for task allocation in robocup rescue", Proceedings of the 5th European workshop on multiagent systems, 537–548, 2007.

[18] Hunsberger L.,& Grosz B., "A combinatorial auction for collaborative planning", Proceedings of the fourth international conference on multiagent systems, 2000.

[19] Beatriz L., Silvia S., & Josep L., "Allocation in rescue operations using combinatorial auctions", Artificial Intelligence Research and Development, Vol. 100, 233-243, 2003.

[20] Chan C. K., & Leung H. F., "Multi-auction approach for solving task allocation problem", Lecture Notes in Computer Science, Vol 4078,

240-254, 2005.

[21] Sandholm T., "Algorithm for optimal winner determination in combinatorial auctions", Artificial Intelligence, Vol 135, 1-54, 2002.

[22] Arai K., & Sang T. X., "Multi Agent-based Rescue Simulation for Disable Persons with the Help from Volunteers in Emergency Situations", International Journal of Research and Reviews in Computer Science, Vol. 3, No. 2, April 2012.

[23] Arai K., & Sang T. X., "Fuzzy Genetic Algorithm for Prioritization Determination with Technique for Order Preference by Similarity to Ideal Solution", International Journal of Computer Science and Network Security, Vol. 11, No. 5, pp. 229-235, May 2011.

[24] Christensen K. M., & Sasaki Y., "Agent-Based Emergency Evacuation Simulation with Individuals with Disabilities in the Population", Journal of Artificial Societies and Social Simulation 11(3)9, 2008.

Yahoo!Search and Web API Utilized Mashup based e-Learning Content Search Engine for Mobile Learning

Kohei Arai 1
Graduate School of Science and Engineering
Saga University
Saga City, Japan

Abstract—**Mashup based content search engine for mobile devices is proposed. Mashup technology is defined as search engine with plural different APIs. Mash-up has not only the plural APIs, but also the following specific features, (1) it enables classifications of the contents in concern by using web 2.0, (2) it may use API from the different sites, (3) it allows information retrievals from both sides of client and server, (4) it may search contents as an arbitrary structured hybrid content which is mixed content formed with the individual content from the different sites, (5) it enabling to utilize REST, RSS, Atom, etc. which are formed from XML conversions. The mash-up should be a flexible search engine for any purposes of content retrievals. The proposed search system allows 3D space display of search menus with these peculiarities on Android devices. The proposed search system featuring Yahoo!search BOSS and Web API is also applied for e-learning content retrievals. It is confirmed that the system can be used for search a variety of e-learning content in concern efficiently.**

Keywords—Mashup; API; web 2.0; mobile devices; e-learning content; content retrieval; Yahoo!saerch BOSS; Web API

I. Introduction

E-learning content retrievals with mobile terminals have not been done easily due to the fact that display size is not good enough for search as well as there are a variety of contents types, video, document, etc. On the other hand, mashup technology [1], [2] supported web 2.0 [3] allows gathering information of e-learning contents efficiently and effectively. Meanwhile, visualization of web contents has been well developed. Therefore, e-learning content search engine can be improved in terms of efficiently and effectively based on the aforementioned mashup technology and visualization tools1.

Web API [4] based e-learning content search engine is developed with Yahoo!search BOSS (**B**uild your **O**wn **S**earch Service) [5] which is called ELDOXEA: E-Learning Content Search Engine [6]-[10]. Not only major keyword, but also minor keywords derived from the descriptions and keywords in the header information of URL which is hit by the first major keyword are used for search. The search engine proposed here is ELDOXEA for mobile devices. One of the disadvantages of mobile devices is relatively small size of display. Therefore, in particular, representation method for search results has to be considered for fitting the search result contents efficiently and effectively.

There are many web visualization tools. In particular, UNIX software of Natto view [11] and mashup based search engine with www visualization of Flowser[2] are sophisticated tools. With the reference to these software tools, the proposed e-learning content search engine is designed.

E-learning content search engine with web API based approach is proposed in the following section. Bing and Google API are additionally featured. Furthermore, thesaurus based search results prioritized method is featured. Application software tool for the search engine is described followed by implementation and some experimental results. Then concluding remarks are described with some discussions.

II. Proposed Method and System

A. Mashup Technology

Mashup is defined as a method for providing web services and applications based on combination of more than two web APIs. It allows improve usability by combining search results which are obtained from the different web APIs separately. Search results from web APIs, in general, in the form of XML or JSON [12]. Therefore, it is easy to combine the search results into one. The following four web APIs are used for the proposed search engine,

1) Yahoo!search: web URL search API
http://search.yahooapis.jp/PremiumWebSearchService/V1/webSearch

2) Yahoo!search: Image retrieval API
http://search.yahooapis.jp/PremiumImageSearchService/V1/imageSearch

3) Youtube Data API developed by Google Inc.
http://gdata.youtube.com/feeds/api/videos

4) Product Advertising API developed by Amazon.com, Inc.
http://ecs.amazonaws.jp/onca/xml

[1] https://supportcenter.checkpoint.com/supportcenter/portal?eventSubmit_doGoviewsolutiondetails=&solutionid=sk30765

[2] http://www.flowser.com/

B. Search Result Display

In order to display search results in a comprehensive manner, the following four candidate models, helix, star, star helix and star slide are trade-off.

One of important roles of the displaying search results is to show the priority of the results. From the top to bottom priority, "helix model" of displaying method allows priority representation from the top to the bottom of the display positions in a spiral order. It, however, is not appropriate displaying method because there are many types of the search results, document, URL, image, etc. Therefore, it is complicated and overlapped so much among the search result types.

Another model of the search result representation is "star model". This model is adopted by Flowser. The search results are classified by content types of nodes. From the top node to the node in concern, the length of rink between both shows priority. It, however, is not so easy to see because the nodes are located in the 3D domain of the display positions.

Star-Helix model combines the aforementioned two models, star and helix models. It, however, is still difficult to see the priority. Star-Slide model of search result display method is the most appropriate model as shown in Fig.1.

Fig. 1. Star-Slide Model

In the Fig.1, star shows search result types while smile marks shows contents (URLs). From the top to bottom priority, "star-slide model" of displaying method allows priority representation from the top to the bottom of the display positions. It has to be displayed onto mobile device displays in a 3D CG representation. Away3D [9] is used for this purpose. There are a plenty of 3D CG representation software tools, OpenGL, ADOBE AIR, etc. under the programming language of Java, Objective-C, etc. In particular, Away3D is used on the ADOBE AIR (Stage3D) of Action Script as a library. The geometric relation between camera and object is shown in Figure 2 as well as rendering of object together with staging device.

C. LEDOXEA

LEDOXEA is mobile devices version of ELDOXEA: E-Learning Content Search Engine. After the activation of LEDOXEA, initial image of start-up is displayed onto mobile device display as shown in Fig.3 (a). Immediately after this, key-in is available with screen keyboard as shown in Fig.3 (b).

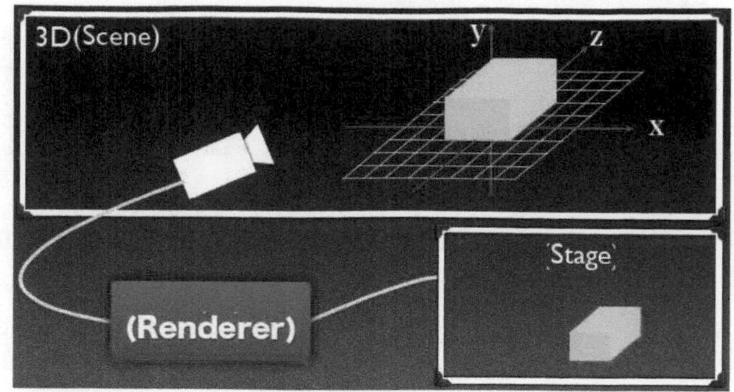

Fig. 2. Geometric relation among 3D view port, camera, object and 3D area definition for rendering

Title
Discription
(a) Initial image

(b) Key-in image

Fig. 3. Display image

When users key-in their keyword in the dialog box as shown in Fig.4 (a), then users press the search button for retrievals. After that, users get their search results as shown in Fig.4 (b).

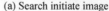

(a) Search initiate image

(b) Search results

Fig. 4. Display image

In this case, five content types and icons of search results of URLs appear in order of their priorities. Five content types are (1) search engines such as Google, Yahoo search, (2) moving picture contents such as YouTube, (3) document contents, (4) image contents and (5) general purpose content search such as Amazon.com which are shown in Fig.5.

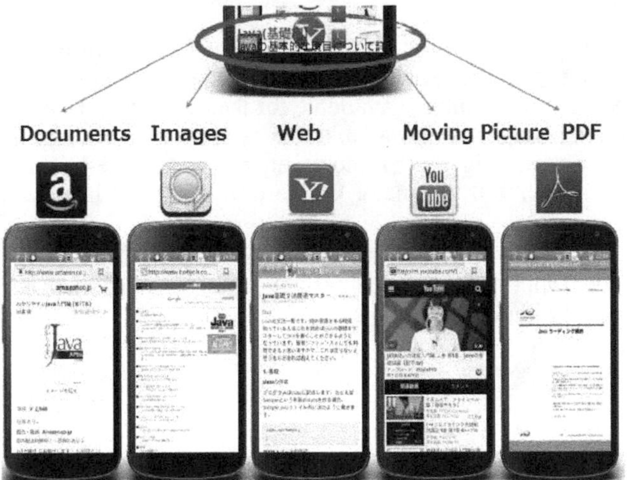

Fig. 5. Content types for search

Search results can be moved up and down when users swipe in the vertical direction as shown in Fig.6 (a). Therefore, search result content can be seen one by one. When users would like to see other types of search results of contents, they have to swipe the display in the horizontal direction as shown in Fig.6 (b).

(a)vertical swipe (b)horizontal swipe

Fig. 6. Move up and down as well as ration with vertical swipe and horizontal swipe

The icon sizes are different by content type for representation of the search results in 3D space on the display as shown in Fig.7. When users tap the title summary which is located the bottom of the display as shown in Fig.8, then users may take a look at the search result of contents through browser or external applications. There is "back button" for getting back to the previous display image after the referencing the search result contents.

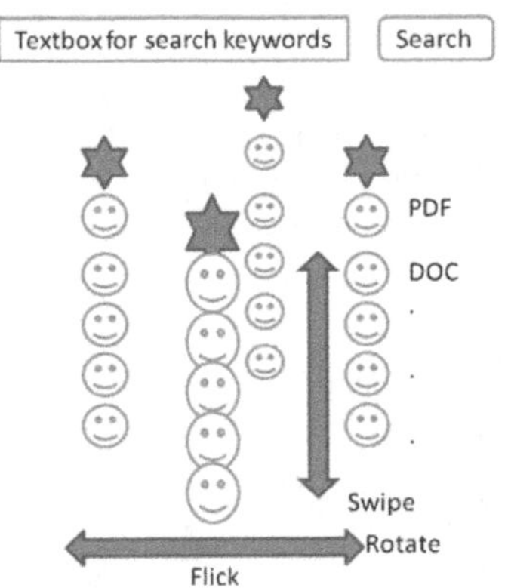

Fig. 7. 3D display of the icon during flick, swipe, and rotational operations

Search content types are rotated when users swipe the content type icons. These rotation angles can be calculated as shown in Fig.8.

Icon operations with swipe actions are illustrated in Fig.9. Swipe action in the vertical direction allows search results candidate selections. Meanwhile, swipe action in the horizontal direction allows content type selections. These can be done through camera position changes in 3D space in the CG space of Away 3D.

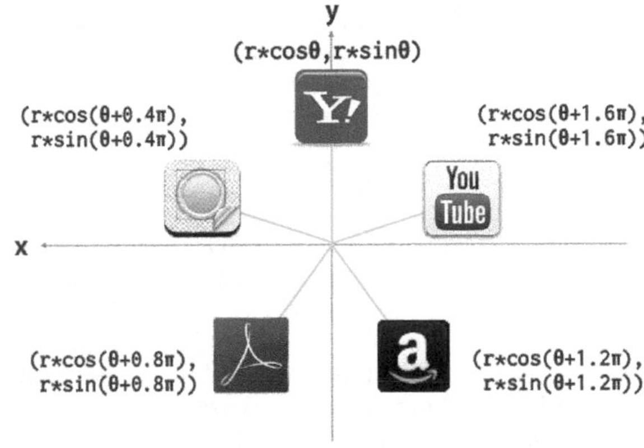

Fig. 8. Rotation angle of the five icons

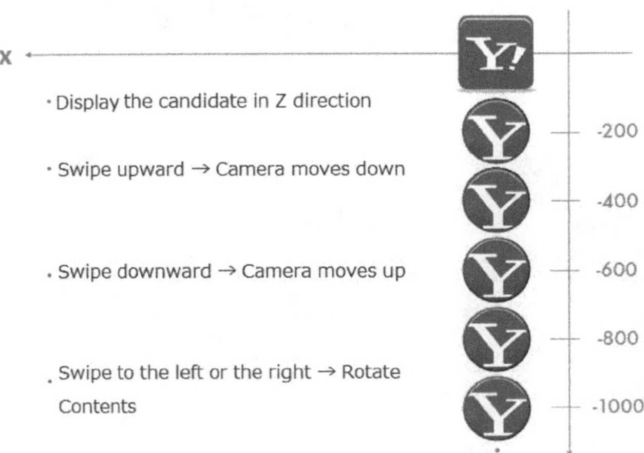

Fig. 9. icon manipulations utilizing swipe operations

A. LEDOXEA version 2

From the web site stability or availability points of view, the proposed search engine employed Bing API instead of Yahoo!Search. Display image layout is shown in Fig.10.

Fig. 10. Example of search result display with Bing API based LEDOXEA

YouTube (moving pictures), Amazon site, ADOBE content (documents), and Image contents are available for search. These are common to the previous LEDOXEA featuring Yahoo!Search. Search results are aligned in the order of their priority. When users choose one of the icon of search result, then the hyperlink of the search result appears on the bottom of the screen. Also, users get the contents when they click the hyperlinked search result. This is referred to the version 1 of LEDOXEA hereafter.

B. Thesaurus Approach

In order to improve effectiveness of search engine, not only major keyword but also minor keywords can be used for e-learning content search. In addition to the dialog box, thesaurus search is available for the LEDOXEA version 2. Example of the keyword input image display is shown in Fig.11 (a). There is the dialog box for thesaurus search which is situated just under the dialog box for the major keyword. Example of the image of thesaurus search results is shown in Fig.11 (b).

(a)Thesaurus search engine (b)Example of thesaurus search result

Fig. 11. Thesaurus search featured LEDOXEA

In this example, "Wikipedia" of thesaurus search engine is selected. Appropriate minor keywords are available as the thesaurus search results. Therefore, much appropriate e-learning content search can be done with major and minor keywords.

Fig.12 (a) shows the other thesaurus search result by keyword types abstract, history, word, footnote, reference, related site, other link are available for thesaurus search. Also, several languages are available for search by clicking the bottom button. Fig.12 (b) shows an example of e-learning content search with the major keyword of "Java". When users key-in the minor keyword of "文法" (grammar in English) in the dialog box of thesaurus search engine, then they can get the much appropriate search results rather than LEDOXEA version 1 (without thesaurus engine).

(a)Thesaurus search keyword (b)Example for "Java"

Fig. 12. Search keyword types selection for thesaurus search and example of e-learning content search with the major keyword of "Java"

Fig.13 (a) shows an example of search image when users key-in the major keyword "Java" and the minor keyword "grammar" for the thesaurus search engine while Fig.13 (b) shows an example of search image when users key-in the major keyword "Java" and the minor keyword "question and answer" for the thesaurus search engine.

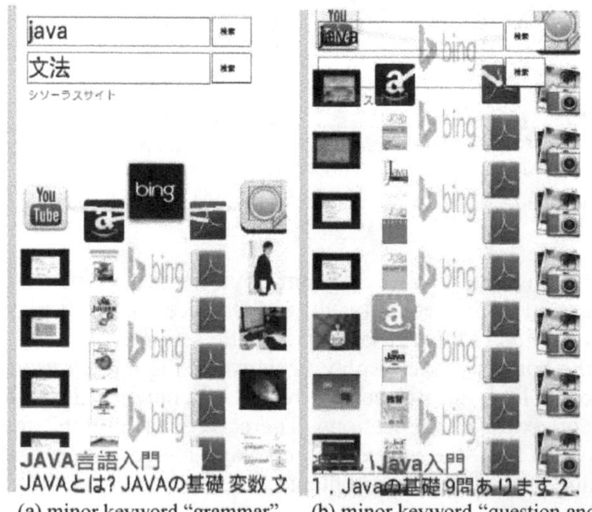

(a) minor keyword "grammar" (b) minor keyword "question and answer"

(c)minor keyword "lecture"

Fig. 13. Examples of e-learning content search results with major keyword of "Java" and minor keywords, "grammar" and "Question and answer" as well as "Lecture" for thesaurus search results

On the other hand, Fig.13 (c) shows an example of search image when users key-in the major keyword "Java" and the minor keyword "question and answer" for the thesaurus search engine.

Thus users can take a look at the search results from the top to the bottom priority of the search results by content type by type. Also, the highest priority of search result appears at the bottom of the image with hyperlink. In this example, the book on Java language specifically programming grammar is greatly featured is shown as search result. Users may input minor keywords on their own. Users also may use thesaurus search engine. It is totally up to users.

III. Experiments

A. Implementation

Implementation of the proposed software is done with FlashProfessionalCS6 on Android 4.0.3. In order to publicize of the implemented source code based on Action Script, the following library, framework, image files are executed,

1) Away3D 4.1.0 Alpha (away3d-core-fp11 4 1 0 Alpha.swc)[8]
2) Bulk Loader (bulk loader.swc)[13]
3) As3 Crypto (as3crypto.swc)[14
4) norm-top.png (128x128px of png file)
5) norm.png (128x128px of png file)
6) img-top.png (128x128px of png file)
7) img.png (128x128px of png file)
8) doc-top.png (128x128px of png file)
9) doc.png (128x128px of png file)
10) mv-top.png (128x128px of png file)
11) ama-top.png (128x128px of png file)
12) ama.png (128x128px of png file)
13) defImgXML.xml (XML file which are acquired with Y! Image Retrieval API)

These source code package are provided by the URL of Github

https://github.com/legnoh/ledoxea

A.1 Main.as(ActionScript3.0)

```
package {
import flash.display.BitmapData;
import flash.display.Sprite;
import flash.display.SimpleButton;
import flash.events.*;
import flash.filesystem.StorageVolumeInfo;
import flash.geom.Matrix;
import flash.geom.Vector3D;
import flash.text.*;
import flash.utils.ByteArray;
import flash.utils.escapeMultiByte;
import flash.net.URLLoader;
import flash.net.URLRequest;
import flash.ui.Keyboard;
import flash.desktop.NativeApplication;
import away3d.containers.*;
import away3d.entities.*;
import away3d.materials.TextureMaterial;
import away3d.primitives.*;
```

On the other hand, XML file is created which is defined Android application software. Meanwhile, used APIs are Yahoo!JAPAN Web Retrieval API,Yahoo!JAPAN Image Retrieval API,YouTube Data API, and Amazon Product Advertising API.

B. Experiments

Fig.14 shows an example of display image of smartphone of which the proposed search engine is implemented (github.com/legnoh/ledoxea).

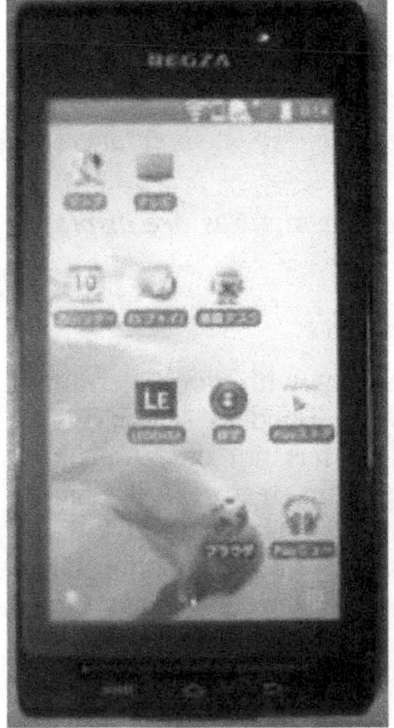

Fig. 14. Example of display image of smartphone of which the proposed search engine is implemented (github.com/legnoh/ledoxea)

"LE" of icon denotes LEDOXEA.

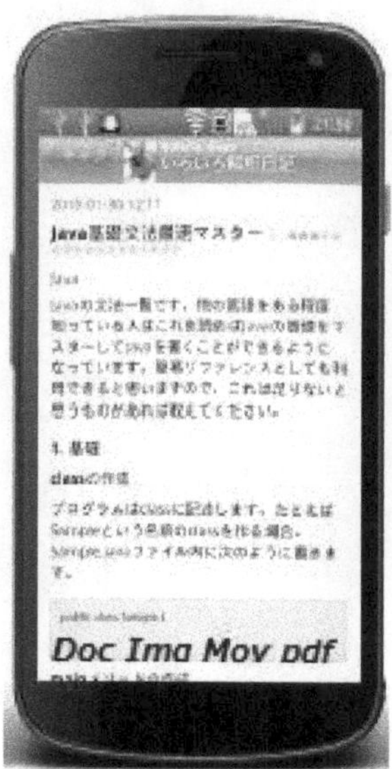

Fig. 15. Example of linked retrieved results (Merged contents of Yahoo search results with the other retrieved contents)

Twenty of university students (information science department) participate the experiments. They use the LEDOXEA and then they evaluate the LEDOXEA. Their comments and suggestions are as follows,

1) It is comfortable to use the LEDOXEA because operability using swipe operations, smooth rendering, usability of Andoroid tablet terminals, WWW visualizations are excellent,

2) Search results can be seen simultaneously,

3) PDF data and YouTube of search results can be accessible from the LEDOXEA directly,

4) Appropriate e-learning contents can be retrieved efficiently with a single keyword.

5) Prioritizing the search results is excellent because the most appropriate e-learning content can be found easily,

6) Five e-learning contents can be seen simultaneously with a single keyword,

7) Still pictures as well as moving pictures are displayed their thumbnail, it is comprehensive.

IV. Conclusion

Mashup based content search engine with mobile terminals is proposed. Mashup technology is defined as search engine with plural different APIs. Mash-up has not only the plural APIs, but also the following specific features. it enables classifications of the contents in concern by using web 2.0, it may use API from the different sites, it allows information retrievals from both sides of client and server, it may search contents as an arbitrary structured hybrid contents which is mixed contents formed with the individual contents from the different sites, it enabling to utilize REST, RSS, Atom, etc. which are formed from XML conversions.

Although mashup allows content search which is same as portal, mashup has the aforementioned different features from portal. Therefore, mash-up is possible to create more flexible search engine for any purposes of content retrievals. The search system which is proposed here is that make it possible to control the graph in the 3D space display with these peculiarity on Android devices. The proposed search system is applied for e-learning content retrievals. It is confirmed that the system enables to search a variety of content in concern efficiently.

It is found that users can take a look at the search results from the top to the bottom priority of the search results by content type by type. Also, the highest priority of search result appears at the bottom of the image with hyperlink. In this example, the book on Java language specifically programming grammar is greatly featured is shown as search result. Users may input minor keywords on their own. Users also may use thesaurus search engine. It is totally up to users.

Acknowledgment

The author would like to thank Mr. Ryoma Kai and Mr. Yohei Nishiguchi for their effort to the implementation and the experiment.

References

[1] Ahmet Soylu, Felix Mödritscher, Fridolin Wild, Patrick De Causmaecker, Piet Desmet. 2012 . "Mashups by Orchestration and Widget-based Personal Environments: Key Challenges, Solution Strategies, and an Application." Program: Electronic Library and Information Systems 46 (4): 383–428.

[2] Endres-Niggemeyer, Brigitte ed. 2013. Semantic Mashups. Intelligent Reuse of Web Resources. Springer. ISBN 978-3-642-36402-0

[3] O'Reilly, T., 2005. What is Web 2.0. Design Patterns and Business Models for the Next Generation of Software, p. 30

[4] Benslimane, D.; Dustdar, S.; Sheth, A. (2008). "Services Mashups: The New Generation of Web Applications". *IEEE Internet Computing* **10** (5): 13–15. doi:10.1109/MIC.2008.110

[5] Tuesday, August 17th, 2010 (2010-08-17). "As Bing Takes Over Yahoo Search, SearchMonkey Dies, BOSS Is No Longer Free, But Site Explorer Still Works". TechCrunch. Retrieved 2012-09-18.

[6] Kohei Arai, Mash-up based content search engine for mobile devices, International Journal of Advanced Research in Artificial Intelligence, 2, 5, 39-43, 2013.

[7] Kohei Arai and Tolle Herman, Module based content adaptation of composite e-learning content for delivering to mobile learners, International Journal of Computer Theory and Engineering, 3, 3, 382-387, 2011.

[8] Kohei Arai, Herman Tolle, Efficiency improvements of e-learning document search engine for mobile browser, International Journal of Research and Reviews on Computer Science, 2, 6, 1287-1291, 2011.

[9] K.Arai, T.Herman, Efficiency improvement of e-learning document search engine for mobile browser, International Journal of Research and review on Computer Science, 2, 6, 1287-1291, 2012.

[10] K.Arai, T.Herman, Video searching optimization with supplemental semantic keyword for e-learning video searching, International Journal of Research and Review on Computer Science, 3, 3, 1640-1644, 2012.

[11] H. Shiozawa, K. Okada, Y. Matsushita: 3D Interactive Visualization for Inter-Cell Dependencies of Spreadsheets, Proc. IEEE InfoVis '99, pp. 79-82, Oct. 1999.

[12] Douglas Crockford (July 2006). "IANA Considerations". *The application/json Media Type for JavaScript Object Notation (JSON).* IETF. sec. 6. RFC 4627. https://tools.ietf.org/html/rfc4627#section-6. Retrieved October 21, 2009.

Multiple-Language Translation System Focusing on Long-distance Medical and Outpatient Services

Rena Aierken and Li Xiao
The Xinjiang Technical Institute of Physics & Chemistry Academy of Sciences, Urumqi, China
University of Chinese Academy of Science,

Su Sha and Dawa Yidemucao
Key Laboratory of Xinjiang Multi-Language Technology, Xinjiang University, Urumqi, China
Department of information science and engineering, Xinjiang University, Urumqi, China

Abstract—**For people living in the countryside, an effective long-distance medical and health service is very important. People living in western China, especially, require convenient communication in their native language with doctors working in a modern city. To address this problem, a multiple-language translation system for long-distance medical and outpatient services is discussed. This system initially provides a table containing basic information including disease names and symptoms for different medical classifications, and then translates the sentences selected from the table automatically using a machine translation system. Finally, a PDF file is created for the doctor and the patient. In this paper, the system construction and evaluation of the machine translation are introduced.**

Keywords—questionnaire for outpatient cases; Chinese Uyghur language; medical Chinese-Minority language parallel corpus; statistical machine translation.

I. INTRODUCTION

Long-distance medical and health services via internet is extremely beneficial for people living in the countryside. A large percentage of people living in the countryside of Xinjiang cannot speak Mandarin or other languages, while most medical practitioners or doctors do not know the local language. Thus, there is a significant communication-gap between the doctors and the patients. Although a manual translation system could be a possible solution, it costs time and money, besides increasing the burden on patients [1].

A study is going for overcoming the communication problems existing in the medical and health field by building a translation system. The system was built in three steps: first, we built a high-quality Chinese and Uyghur language corpora focusing on the various medical terms in collaboration with medical and health institutions. The corpora were carefully collected by hospital clinics and professionals from medical universities. Next, a set of parallel sentences (PS) including a larger volume of words used in the medical and health field was created. Finally, a translation system that searches the PS and a statistical machine translation (SMT) method were implemented.

When using the translation system, a patient first answers questions in his/her native language. The results of the translation obtained by searching through the PS followed by an SMT are then displayed on the table. Next, the patient selects the most suitable result from the translation result by clicking on the table buttons. Finally, a PDF file that will be used as an electronic medical record (EMR) is generated by the system automatically.

The remainder of this paper is organized as follows. The related works are presented in section 2, and discuss the overall translation system in section 3. Section 4 shows the experimental setup and analyzes the experimental results. The conclusions and future work will be presented in section 5.

II. RELATED WORKS

Until now, there have been only limited reports on domestic multilingual translation systems for health care research and development. In our previous study, we have reported an electronic medical record translation system that used the strategy of statistical and example-based machine translation for the Chinese and Uyghur languages [2]. The research group of the Ministry of Japan and Ikeda, reported a multi-language support system using a practical text-set for outpatient services [3,4]. They also confirmed that the system performance deteriorates with a larger number of unknown words. Collection of all the relevant words and dialogs in a native language is considerably difficult. In this study, we report a high performance machine translation system focused on the outpatient service. Our system is built based on bilingual parallel sentences in Chinese and Uyghur, relevant to the medical field and the statistical machine translation method.

III. PREPARE YOUR PAPER BEFORE STYLING

A. Questionnaire for the Outpatient Service

An outpatient doctor may provide a patient's condition survey table, as shown in Figure 1, to the patient. The patient answers each question in his/her native language. For demonstration illustration purposes, a Uyghur Latin alphabet-based questionnaire is presented.

To use the system, a patient first answers the question as prompted, entry, depicted in Figure 1 (2) and Figure 1 (3). Next, when the patient clicks on the appropriate buttons, depicted in Figure 1 (2) or Figure 1 (4), the system will display an interface, as shown in Figure 2, automatically, according to patient's input, in two languages (the patient's language and the doctor's language). Column (3) of Figure 2 shows the PS search results and column (4) of Figure 2, shows

Sponsor: National Natural Science foundation of China

the MT result for an entry in column (2) of Figure 2. The PS search results are shown in column (3) of Figure 2, obtained by an N-gram (2 gram or 4 gram) statistical language model and a maximum likelihood criterion. In this case, the system can provide three matching results. If the instance or the input string does not exist in the PS, the system does not display the results. Column (4) of Figure 2 shows the result of the MT. the most suitable result from the translation result by clicking Common machine translation systems may generate mistranslation results. Therefore, column (5) of Figure 2 is highlighted with a red string display, reminding the user that the machine translation results may have a translation error.

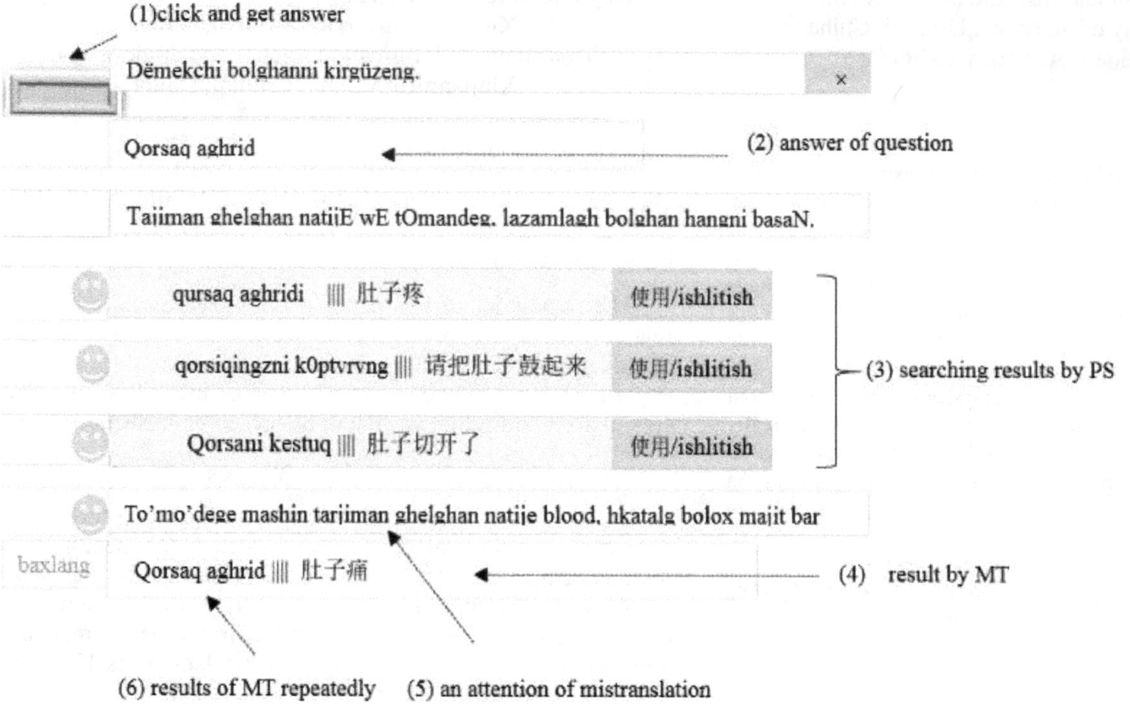

Fig. 1. The complete interface

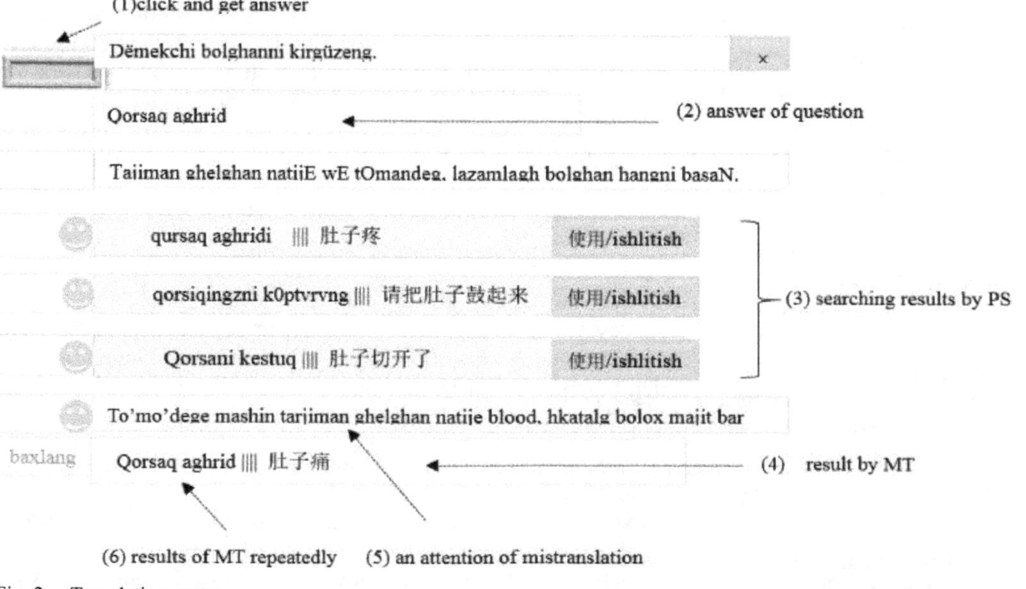

Fig. 2. Translation system

In addition, column (4) of Figure 2 shows the results of the original input strings that are repeatedly translated. Thus, the patient can repeatedly modify the original input, namely column (2) of Figure 2, and choose the most satisfactory translation result to improve the accuracy of the machine translation. The MT provides only one result and the PS gives three, hence, the translation system in Figure 2 can display four different translation results. Finally, the patient can select and click column (3) or column (4) of Figure 2 according to the actual situation, using the right button "use." Then the system prints a PDF file in two languages. Finally, both the doctor and the patient can save the PDF file for electronic medical records.

B. *Parallel Sentence and Statistical Machine Translation*

This section describes the PS system and the decoder of the machine translation. 240K Chinese sentences including medical, medicine, and drug-names were developed in our previous research. Then, sentences in the Uyghur and Kazak languages corresponding to these Chinese sentences were added manually. Finally, a multi-lingual, aligned medical corpus was created for the system test [5]. We applied a machine translation software based on the general Moses software, developed by the Key Laboratory of multiple language information technology of the Xinjiang University [6,7,8].

IV. EXPEXPERIMENTAL RESULTS

In order to confirm the effectiveness of the questionnaire, we investigated the test results of the PS and the SMT. Then, we tested the performance using an integrated system (PS and SMT).

A. *Test Conditions*

Ten students, interested in science and engineering, and proficient in Chinese and Uyghur, participated in the experiment. They were asked to enter their own experience of the pathology and relevant medical information. Six students, who specialized in medicine, checked the results of the translation.

To increase the usability of the system, the source language is set to Uyghur and target language is set to Chinese. Additionally,

1) We guaranteed confidentiality in this study. Private information was not revealed.

2) Each participant filled both the investigation form and the system-provided form. These two forms were filled with the same content and in the same order.

3) Taking into account the user's familiarity with the system operation, the two forms were repeated twice. Table 1 shows examples of the questions. Answers were filled in both a paper-form and a system-provided form. The experiment was carried out with native speakers of the local language (Uyghur). The system assumes that 258 pairs of sentences can be used in the experiment.

TABLE I. QUESTIONNAIRE CONTENTS

No.	So'al sorash mazmi (Question content)	o'al sorash shekli (Question form)
1	qandaq Alamet kõrulwatidu?	ixtiyari kirgüzüsh
2	qachandin bashlap?	
3	iqki takxurux ainiki bilen tegxarganma?	qemni yaki kunupkini chĕkip kirgüzüng
4	Qandaq dora yedingiz?	ixtiyari kirgüzüsh
5	dorini hazirmu yemsiz?	

B. *Test by Searching the PS only*

In this experiment, we investigate two kinds of test results. Figure 3 shows a PS set search result and the MT for question No. 1 in Table I . It is clear that the three results shown in (3) of Figures 2, 3, and 4, by searching the PS alone, are not better than the result of the MT shown in (4) of Figures 2, 3, and 4, for the answer string /*axqazan mijaz yaqxiemes* (meaning: my stomach is upset) /. This is because the answer string contains commonly used words and does not contain medical terms. We can observe anther translation result for question No. 3 in Table I , shown in figure 4. Here, an answer for question No. 3, /iqki takxurux ainiki tegxarixni qlghan/, contains medical terms, hence, the PS search results, (3) of Figures 4 is considerably better than that of the MT, (4) of Figures 3 and 4.

C. *Tests using MT only*

In this experiment, we assume that the machine translation is an independent implementation. Each sentence that appears in MT is compared with the results. Table II shows the results of the 10 sentences using the PS search. The column, results of the PS (Chinese), includes results of translation by the PS for the source language (Uyghur).

The column, results of the MT, includes results by the MT for the source language (Uyghur). The column, indicating correctness, shows a manual evaluation of the MT for the source language (Uyghur). In this experiment, a total of 20 sentences were selected for the test and 10 sentences were evaluated after excluding repetitive sentences. From Table II, it can be seen that there are four mistranslations among the 10 sentences or four of the translation results do not match the source language. In addition, for a sentence with ID:1 in Table II, although the result using MT and the result of the PS are close in meaning, the result by MT does not answer the question "What are the symptoms?", hence, it is judged as an incorrect result. From the above discussions, it is obvious that

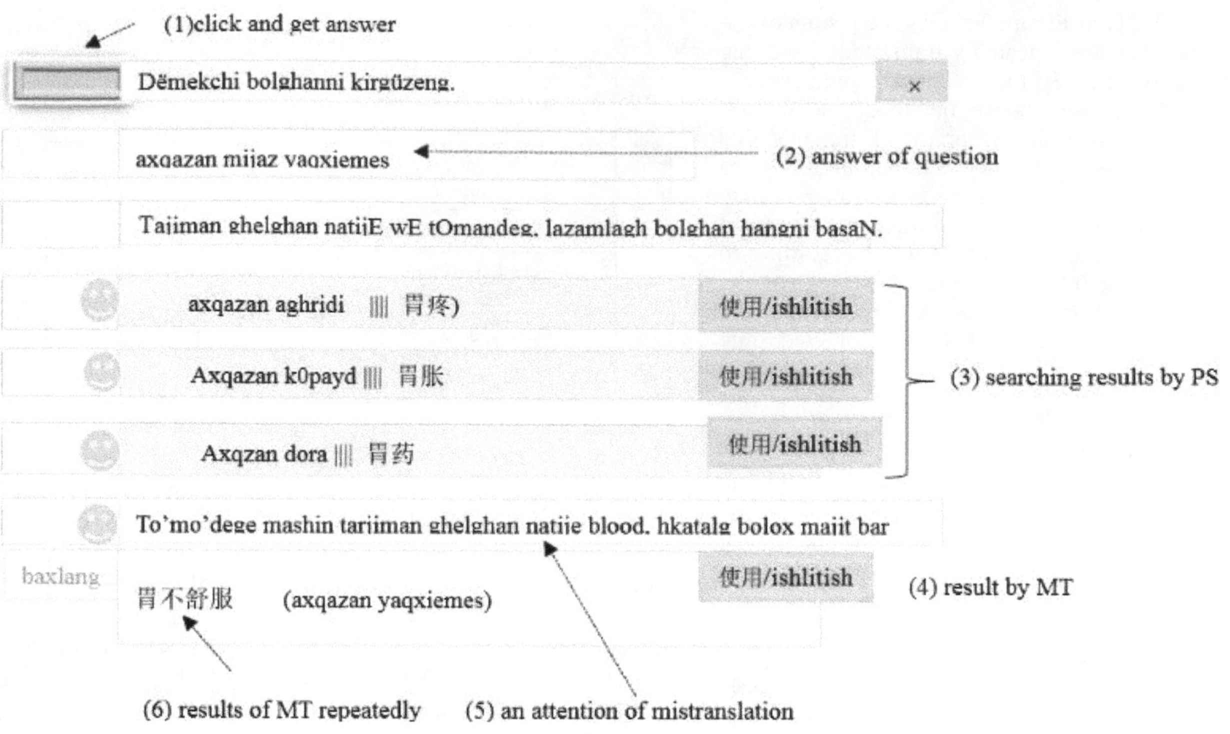

Fig. 3. A translation result for question No. 1

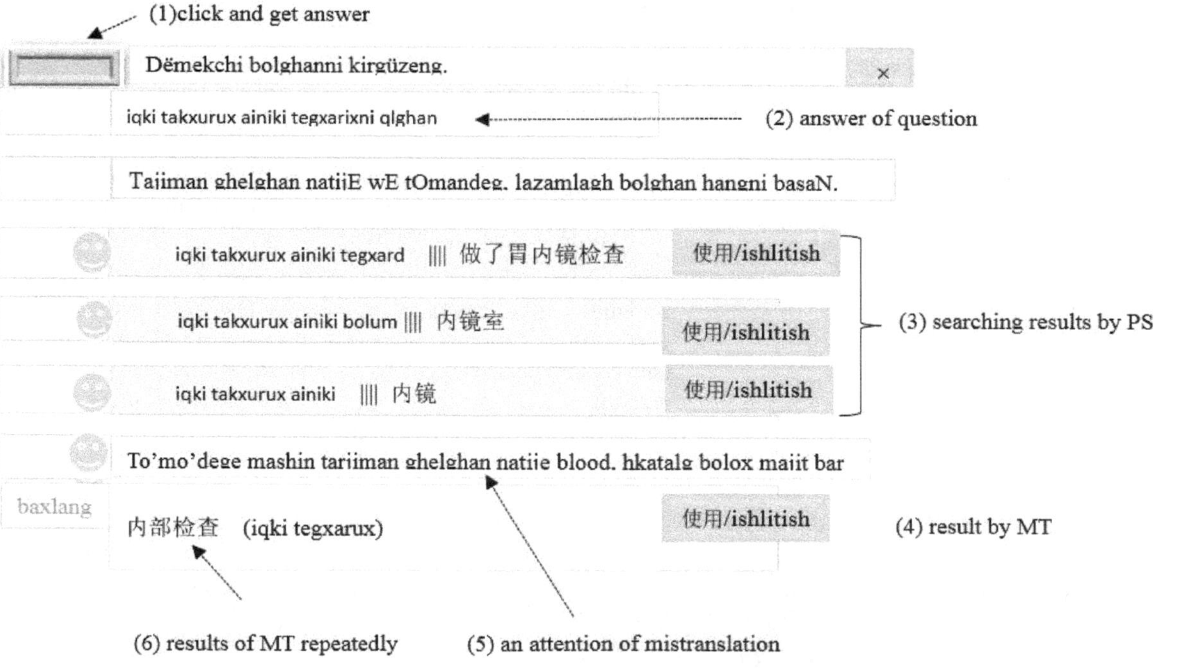

Fig. 4. Translation result for question No. 3

it is difficult to collect all the native language information used in the medical field and the machine translation technique cannot ensure the translation accuracy for the special terms used in the medical and health fields. If the mistranslation by machine translation is to be corrected using the PS search, then, the proposed method of merging both the PS and the SMT is feasible and complete. A real test EMR is shown in Appendix A.

TABLE II. EXAMPLES OF THE SENTENCES USED IN THE EXPERIMENT

ID	source language (Uyghur)	results of PS (Chinese)	results of MT (Chinese)	correctness
1	Qizitma örlesh	发烧 (fever)	有发热 (have hot)	x
2	Bash aghrish	头疼 (headache)	头疼 (headache)	○
4	Müre sirqirap aghrish	肩膀酸疼 (shoulder ache)	肩膀很疼(shoulder is very painful)	x
8	Zukam	感冒 (have a cold)	感冒 (have a cold)	○
21	Qorsaq aghrish	肚子疼(collywobbles)	肚子疼 (collywobbles)	○
47	Dora rë'aksiye qilish	食物过敏(food allergy)	食品有过敏(food allergies)	x
50	Bash qëyish	头昏(dizzy)	头昏(dizzy)	○
60	Ishtiha yoq	没有食(loss of appetite)	没有食欲 (loss of appetite)	○
72	Nepeste qëlinish	呼吸困难(dyspnea)	呼吸困难 (dyspnea)	○
83	Ashqazan qattiq aghris	胃疼的厉害(have a stomachache)	有时胃疼(sometimes a stomachache)	

V. CONCLUSIONS

In this paper, we have presented a multiple-language translation system that focuses on long-distance medical and outpatient services. The system merges the PS and the SMT approaches and translates the dialogues of the doctors and patients automatically, to create an electronic medical record (EMR). Through experiments, we confirmed that the proposed method is feasible and practical.

In the future, we will further improve the system accuracy and extend the system to service more languages, such as languages along Silk Road.

REFERENCES

[1] People government of Xinjiang Uyghur Autonomous Region, China. Study and publicize the article implement the second central work conference.http://www.xj.xinhuanet.com/

[2] Dawa Idomucao. A Study on Chinese Minority Medical Document Translation Based on Hybrid-strategy [J]. Journal of Xinjiang University (Natural Science Edition). 2015. Vol. 32(1). 123-128.

[3] G. Zhenyi. Establishment multi-language Medical Outpatient registered system for foreign patients [J], Jornal of the Japan Institute of electronic information and communication. 2009, Vol.192-D No.6 708-718.

[4] Ikeda, T, Ando, S, Satoh, K Okumura, A. and Watanabe., T, Aoutomatic interpretation system integrating Free-style Sentence translation and Parallel text based Translation[C], Prec.Workshop on speech-to-speech translation 2002,85-92.

[5] Dawa Idomucao. Research based on multi-language machine translation system with electronic medical record [M], 2012, National Natural Science and Technology Fund: Number:61163030.

[6] Hieu Hoang and Philipp Koehn. Design of the moses decoder for statistical machine translation. n Proceedings of ACL Workshop on Software engineering, testing, and quality assurance for NLP, pp. 58–65, 2008.

[7] Philipp, Koehn, Statistical Machine Translation[M], Uk, Cambridge University,20

[8] YANG Pan, LI Miao,ZHANG Jian. Chinese-Uyghur ranslation system for phrase-based statical translation [J], Journal of Computer Application 2009，29(7),2022-2025.

Appendix A:

医院/： دوختۇرخانا　　　　病类科/： بۆلۈم

<<电子病历 EMR>>　　　　2015年10月07日

==

患者姓名/: بىمار ئىسمى

出生年月日/: تۇغۇلغان يىل ئاي كۈن

民族/: مىللىتى

住址/： ئادرېس

联系电话/: تېلفون

病情调查表/： كېسەللىك ئەھۋالى تەكشۈرۈش جەدۋىلى

医生提问? دوختۇر سورىغان سوئال؟	患者选择回答 بىمار تاللىغان جاۋاب	机器翻译结果/ ماشىنا تەرجىمە نەتىجىسى	医生确认结果/ دوختۇر تەكشۈرۈش نەتىجىسى
1. 你有什么症状？/ سىزدە قانداق ئەھۋال كۆرۈلدى؟	بېشىم ئاغرىيدۇ	头疼||| باش ئاغرىش	√
2.什么时候开始/ قاچان باشلانغان؟	2-3 كۈن	2-3 天|||? 2-3 كۈن	√
3.你有高血压吗/ سىزدە يۇقىرى قان بېسىم بارمۇ؟	بار	有/بار	√
4.你现在还服降压药物吗？/ سىز ھازىر قان بېسىم چۈشۈرۈش دورىسى ئىچەمسىز؟	يېقىندىن بېرى دورا ئىچمىگەن	近期没有服药||| دورا ئىچمىگەن	×
5.其他还有什么症状？/ يەنە قانداق ئالامەتلەر بار	دىئابېت كېسىلىم بار	患有糖尿病||| دىئابېت كېسىلىگە گىرىپتار بولغان	√
6.对于糖尿病服用了什么药物？ قەنت سىيىش كېسىلىگە نىمە دورا ئىشلەتتىڭىز؟	ئىنسۇلىن	胰岛素||| ئىنسۇلىن	√
7. 你睡眠好吗？/ ئۇيقىڭىز ياخشىمۇ؟	ئانچە ياخشى ئەمەس	不怎么好||| ياخشى ئەمەس	√
8.现在有降压药物吗?/ قان بېسىم چۈشۈرۈش دورىسى بارمۇ؟	ھازىر يوق	目前没有||| ھازىرچە يوق	√
9.服用几天的降压药物好吗/ بىرنەچچە كۈنلۈك قان بېسىم چۈشۈرۈش دورىسى ئىشلەتسىڭىز بولامدۇ؟	بولىدۇ	吃吧||| دورا يەي	√
10.吃完降压药物再来检查可以吗?/ قان بېسىم چۈشۈرۈش دورىسى ئىچىپ بولۇپ يەنە تەكشۈرۈپ باقسىڭىز قانداق؟	بولىدۇ	可以||| بولىدۇ	√

医生门诊结果填写/： دوختۇر ئامبۇلاتورىيە نەتىجىسى تولدۇرۇلىدۇ

1） 患者症状/： كېسەل ئەھۋال

2） 门诊结果/： دىئاگنوز نەتىجىسى

3） 用药情况/： دورا ئىشلىتىش ئەھۋالى

医生签字（盖章）

دوختۇر ئىمزاسى(تامغا)/

29

Relation between Rice Crop Quality (Protein Content) and Fertilizer Amount as Well as Rice Stump Density Derived from Helicopter Data

Kohei Arai 1
Graduate School of Science and Engineering
Saga University
Saga City, Japan

Masanoori Sakashita [1]
Information Science, Saga University
Saga, Japan

Osamu Shigetomi [2]
Saga Prefectural Agricultural Research Institute
Saga Prefectural Government, Japan

Yuko Miura [2]
Saga Prefectural Agricultural Research Institute
Saga Prefectural Government, Japan

***Abstract*—Relation between protein content in rice crops and fertilizer amount as well as rice stump density is clarified with a multi-spectral camera data mounted on a radio-wave controlled helicopter. Estimation of protein content in rice crop and total nitrogen content in rice leaves through regression analysis with Normalized Difference Vegetation Index: NDVI derived from camera mounted radio-controlled helicopter is already proposed. Through experiments at rice paddy fields which is situated at Saga Prefectural Research Institute of Agriculture: SPRIA in Saga city, Japan, it is found that total nitrogen content in rice leaves is linearly proportional to fertilizer amount and NDVI. Also, it is found that protein content in rice crops is positively proportional to fertilizer amount for lower fertilizer amount while protein content in rice crop is negatively proportional to fertilizer amount for relatively high fertilizer amount.**

Keywords—Rice Crop; Rice Leaf; Total nitrogen content; Protein content; NDVI; Fertilizer amount; Rice stump density

I. INTRODUCTION

Vitality monitoring of vegetation is attempted with photographic cameras [1]. Grow rate monitoring is also attempted with spectral reflectance measurements [2]. Bi-Directional Reflectance Distribution Function: BRDF is related to the grow rate for tealeaves [3]. Using such relation, sensor network system with visible and near infrared cameras is proposed [4]. It is applicable to estimate total nitrogen content and fiber content in the tealeaves in concern [5]. Therefore, damage grade can be estimated with the proposed system for rice paddy fields [6]. This method is validated with Monte Carlo simulation [7]. Also Fractal model is applied to representation of shapes of tealeaves [8]. Thus the tealeaves can be asse3ssed with parameters of the fractal model. Vitality of tea trees are assessed with visible and near infrared camera data [9]. Rice paddy field monitoring with radio-control helicopter mounting visible and NIR camera is proposed [10] while the method for rice quality evaluation through total nitrogen content in rice leaves is also proposed [10]. The method which allows evaluation of rice quality with protein content in rice crop estimated with NDVI1 which is acquired with visible and NIR camera mounted on radio-control helicopter is also proposed [10]. The fact that protein content in rice crops is highly correlated with NDVI which is acquired with visible and Near Infrared: NIR camera mounted on radio-control helicopter is well reported [10]. It also is reported that total nitrogen content in rice leaves is correlated to NDVI as well.

Protein content in rice crop is negatively proportional to rice taste. Therefore, rice crop quality can be evaluated through NDVI observation of rice paddy field. Relation among total nitrogen content in rice leaves, amount of fertilizer amount, NDVI and protein content in rice crops as well as stump density has to be clarified in this paper.

The proposed method is described in the next section followed by experiments. The experimental results are validated in the following section followed by conclusion with some discussions.

II. PROPOSED METHOD FOR ESTIMATION OF PROTEIN CONTENT IN RICE CROPS

A. Radio Controlled Helicopter Based Near Infrared Cameras Utilizing Agricultural Field Monitoring System

The helicopter used for the proposed system is "GrassHOPPER" [2] manufactured by Information & Science Techno-Systems Co. Ltd. The major specification of the radio controlled helicopter used. Canon Powershot S1002 [3] (focal length=24mm) is mounted on the GrassHOPPER. It allows acquire images with the following Instantaneous Field of

[1] Normalized Difference Vegetation Index: NDVI is expressed with the following equation,
NDVI=(NIR-R)/(NIR+R)
where NIR, R denotes Near Infrared and Red wavelength region of reflectance

[2] http://www.ists.co.jp/?page_id=892

[3] http://cweb.canon.jp/camera/dcam/lineup/powershot/s110/index.html

View: IFOV at the certain altitudes, 1.1cm (Altitude=30m) 3.3cm (Altitude=100m) and 5.5cm (Altitude=150m) .

Spectral response functions of filters attached to the camera used are sensitive to Green, Red and Near Infrared bands.

In order to measure NIR reflectance, standard plaque whose reflectance is known is required. Spectralon[4] provided by Labsphere Co. Ltd. is well known as well qualified standard plaque. It is not so cheap that photo print papers are used for the proposed system. Therefore, comparative study is needed between Spectralon and the photo print papers.

The system consist Helicopter, NIR camera, photo print paper. Namely, photo print paper is put on the agricultural plantations, rice leaves in this case. Then farm areas are observed with helicopter mounted Visible and NIR camera. Total nitrogen content in rice leaves, protein content in rice crops can be estimated with the camera data based on the previously established regressive equation [12].

B. *Rice Paddy Field at Saga Prefectural Agricultural Research Institute: SPARI*

Specie of the rice crop is Hiyokumochi4 which is one of the late growing types of rice species. Hiyokumochi[5] is one of low amylase (and amylopectin rich) of rice species (Rice No.216).

Paddy fields of the test site of rice paddy field at SPARI[6] which is situated at 33°13'11.5" North, 130°18'39.6"East, and the elevation of 52 feet. The paddy field C4-2 is for the investigation of water supply condition on rice crop quality. There are 14 of the paddy field subsections of which water supply conditions are different each other.

There are two types of water supply scheduling, short term and standard term. Water supply is stopped in the early stage of rice crop growing period for the short term water supply subsection fields while water supply is continued comparatively longer time period comparing to the short term water supply subsection fields.

Meanwhile, there are three types of water supply conditions, rich, standard, and poor water supply subsection fields. On the other hand, test sites C4-3 and C4-4 are for investigation of total nitrogen of chemical fertilizer amount dependency on rice crop quality. There are two types of paddy subsections, densely and sparsely planted paddy fields. Hiyokumochi rice leaves are planted 15 to 20 fluxes per m_2 on June 22 2012. Rice crop fields are divided into 10 different small fields depending on the amount of nutrition including total nitrogen ranges from zero to 19 kg/10 a/total nitrogen.

Total nitrogen of chemical fertilizer amount is used to put into paddy fields for five times during from June to August. Although rice crops in the 10 different small fields are same species, the way for giving chemical fertilizer amount are different. Namely, the small field No.1 is defined as there is no chemical fertilizer amount at all for the field while 9, 11, and 13 kg/10a/ total nitrogen of after chemical fertilizer amount are given for No.2 to 4, respectively, no initial chemical fertilizer amount though. Meanwhile, 9, 11, 13 kg/10 a/total nitrogen are given as after chemical fertilizer amount for the small field No.5, 6, and 7, respectively in addition to the 3 kg/10 a/total nitrogen of initial chemical fertilizer amount. On the other hand, 12, 14, and 16 kg/10 a /total nitrogen are given for the small fields No.5, 6, 7, respectively as after chemical fertilizer amount in addition to the initial chemical fertilizer amount of 3 kg/ 10 a/ total nitrogen for the small field No. 15, 17, 19, respectively. Therefore, rice crop grow rate differs each other paddy fields depending on the amount of total nitrogen of chemical fertilizer amount.

III. EXPERIMENTS

A. *Acquired Near Infrared Camera Imagery Data*

Radio wave controlled helicopter mounted near infrared camera imagery data is acquired at C4-2, C4-3, C4-4 in SPARI on 18 and 22 August 2013 with the different viewing angle from the different altitudes. In the acquired camera images, there is Spectralon of standard plaque as a reference of the measured reflectance in between C4-3 and C4-4. Just before the data acquisition, some of rice crops and leaves are removed from the subsection of paddy fields for inspection of total nitrogen content in rice leaves. Using the removed rice leaves, total nitrogen content in rice leaves is measured based on the Keldar method and Dumas method[7] (a kind of chemical method) with Sumigraph NC-220F [8] of instrument. The measured total nitrogen content in rice leaves and protein content in rice crops are compared to the NDVI.

The camera images are acquired on 18 August and 22 August. Meanwhile, these images have influences due to shadow and shade of rice leaves and water situated under the rice leaves as well as narrow roads between rice paddy fields. In order to eliminate the influences, thresholding process is applied to the acquired images.

Measured total nitrogen contents in rice leaves of rice paddy fields of partitioned A1 to A8 and B1 to B8 on 14 and 22 August 2013 are shown in Table 1 and 2, respectively.

Before estimation of total nitrogen content in rice leaves, geometric correction is applied to the acquired camera image after extraction of intensive study areas. Also it is found that total nitrogen of chemical fertilizer amount; water management as well as plant density is different from each other partitioned rice paddy fields as aforementioned.

B. *Total nitrogen Content in Rice Leaves*

Total nitrogen content in the rice leaves seem to reflect the fact of chemical fertilizer amount of total nitrogen, water supply management, and plantation density, obviously. Fig.1 shows the relation between fertilizer amount and total nitrogen content in the rice leaves.

---

[4] http://www.labsphere.com/products/reflectance-standards-and-targets/reflectance-targets/spectralon-targets.aspx

[5] http://ja.wikipedia.org/wiki/%E3%82%82%E3%81%A1%E7%B1%B3

[6] http://www.pref.saga.lg.jp/web/shigoto/_1075/_32933/ns-nousisetu/nouse/n_seika_h23.html

[7] http://note.chiebukuro.yahoo.co.jp/detail/n92075

[8] http://www.scas.co.jp/service/apparatus/elemental_analyzer/sumigraph_nc-220F.html

TABLE I. MEASURED TOTAL NITROGEN CONTENT IN RICE LEAVES ON 14 AUGUST 2013

Farm Area	Total nitrogen (%)
A1	2.61
A3	2.85
A5	2.84
A8	2.77
B1	2.82
B3	2.74
B5	3.16
B8	2.78

TABLE II. MEASURED TOTAL NITROGEN CONTENT IN RICE LEAVES ON 22 AUGUST 2013

Farm Area	Total nitrogen (%)
A1	2.46
A2	2.88
A4	2.97
A5	2.89
A6	2.67
A8	3.22
B1	2.33
B2	2.79
B4	2.84
B5	2.85
B6	2.96
B8	3.14

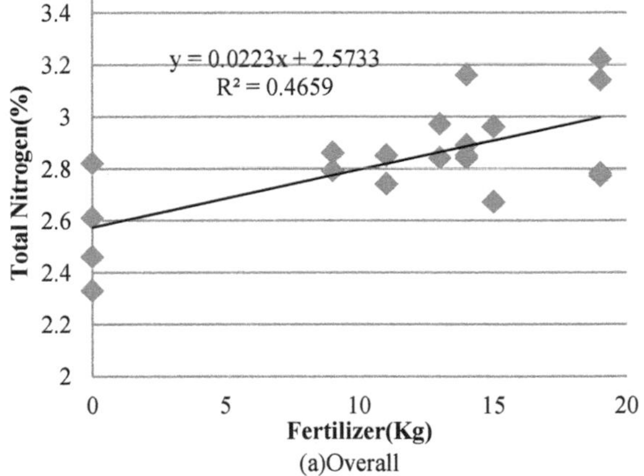

(a)Overall

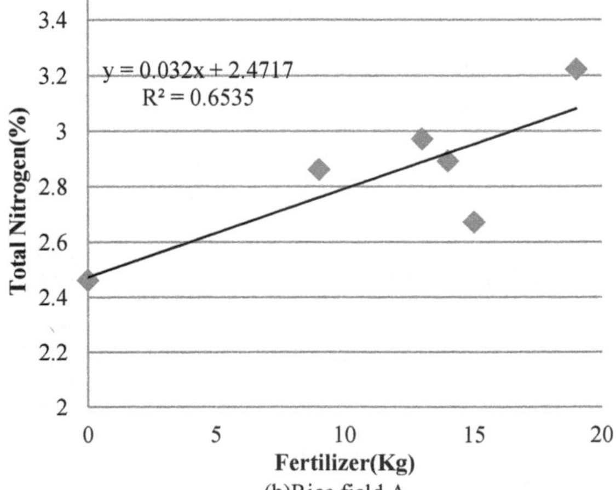

(b)Rice field A

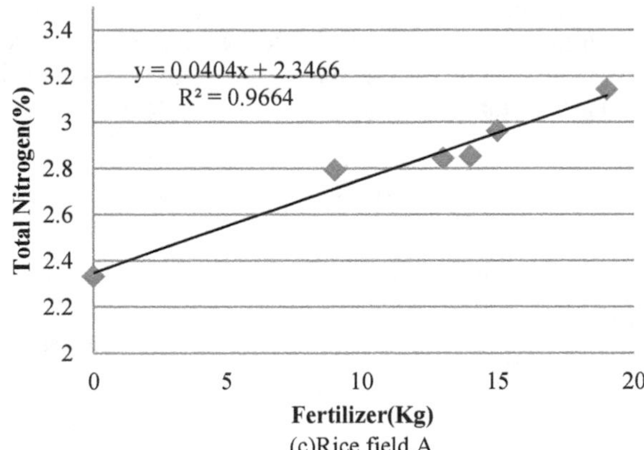

(c)Rice field A

Fig. 1. Relation between fertilizer amount and total nitrogen content in the rice leaves

The chemical fertilizer amount is put into paddy fields four times. Fertilizer amount denotes total chemical fertilizer amount. Fig.1 (a) shows overall relation between fertilizer amount and total nitrogen content in the rice leaves while Fig.1 (b) and (c) shows the relation for measured total nitrogen content in rice leaves on August 18 and August 22, respectively. Rice crops are harvested in the begging of October. Therefore, rice crops and rice leaves are grown up a little bit for four days. These figures show the proportional relation between both. The regressive equations for rice paddy field A is expressed in equation (1) while that for rice paddy field B is represented in equation (2), respectively.

$$y = 0.032x + 2.4717 \quad (1)$$
$$R^2 = 0.6535$$
$$y = 0.0404x + 2.3466 \quad (2)$$
$$R^2 = 0.9664$$

These regressive coefficients are very similar (the difference of proportional coefficients for both rice paddy fields is just 20.79 % while that of bias coefficients is 5.06 %). Therefore, the relationship between fertilizer amount and total nitrogen content in the rice leaves is clarified. In total, regressive equation between fertilizer amount and total nitrogen content in rice leaves is expressed with equation (3)

$$y = 0.0223x + 2.5733 \quad (3)$$
$$R^2 = 0.4659$$

R square value (determination value[9]) is around 0.5. This implies that the fertilizer amount is proportional to the total nitrogen in rice leaves.

C. NDVI and Protein Content in Rice Crops

Meanwhile, protein content in rice crops shows different relation against the relation between total nitrogen in rice leaves and fertilizer amount. Namely, there is proportional relation between protein content in rice crops and fertilizer amount ranged from zero to 12 Kg while there is negatively proportional relation between both for the fertilizer amount ranged from 12 to 16 Kg. Rice taste depends on protein content in the rice crops. Namely, protein rich rice crops taste

[9] $R^2 = 1 - \frac{\sum_{i=1}^{n}(y_i - f_i)^2}{\sum_{i=1}^{n}(y_i - m)^2}$ where y, f, and m denotes observed value, predicted value and mean value, respectively.

bad. Therefore, in accordance with increasing of fertilizer amount, total nitrogen content in rice leaves and NDVI of rice leaves are increased while protein content in rice crops is decreased as shown in Fig.2.

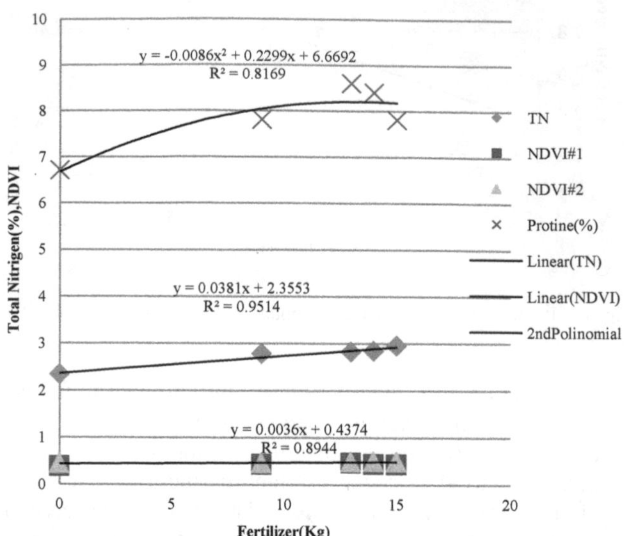

Fig. 2. Relation among fertilizer amount, total nitrogen content in rice leaves and NDVI of rice leaves as well as protein content in rice crops

Through the regressive analysis, it is found that there is proportional relation between fertilizer amount and NDVI with the following regressive equation,

$$y = 0.0036x + 0.4374 \quad (4)$$
$$R^2 = 0.8944$$

Meanwhile, there must is the proportional relation between fertilizer content and total nitrogen content in rice leaves as shown in equation (5).

$$y = 0.0381x + 2.3553 \quad (5)$$
$$R^2 = 0.9514$$

On the other hand, it seems that the relation between fertilizer amount and protein content in rice crop is different from the above relations. It is most likely the second order polynomial relation of equation (6) rather than proportional relation.

$$y = -0.0086x^2 + 0.2299x + 6.6692 \quad (6)$$
$$R^2 = 0.8169$$

The determination index (R square value) is greater than 0.8. Therefore, it may say that there is saturation level of fertilizer amount. One of the reasons for this is due to the fact that the relation between NDVI and total nitrogen content in rice leaves differ from the relation between NDVI and protein content in rice crops as shown in Fig.6 (An Example).

On the other hand, remote controlled helicopter mounted visible and near infrared camera data are acquired on 18 and 22 August, total nitrogen content in rice leaves is measured on August 22 only though. Therefore, August 22 data show much reliable than August 18 data. For four days, rice leaves and rice crops are grown a little bit as shown in Fig.4.

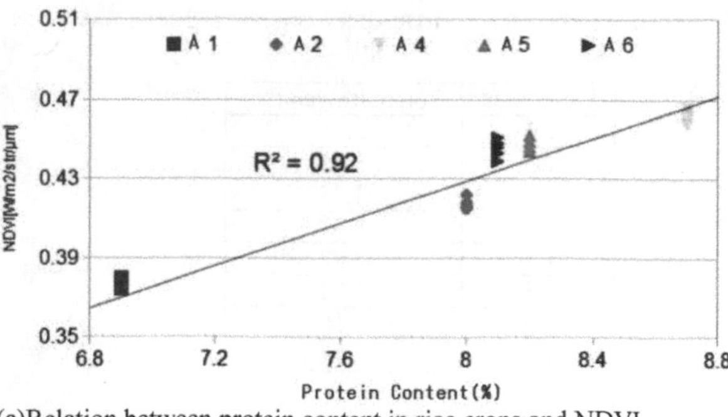

(a)Relation between protein content in rice crops and NDVI

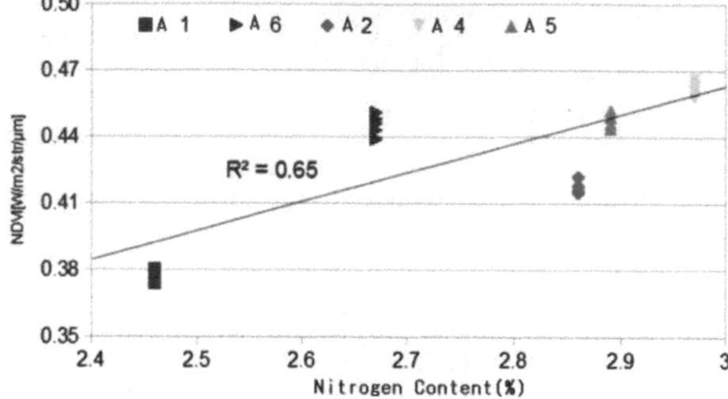

(b)Relation between total nitrogen content in rice leaves and NDVI

Fig. 3. Different relations between NDVI and total nitrogen content in rice leaves as well as protein content in rice crops

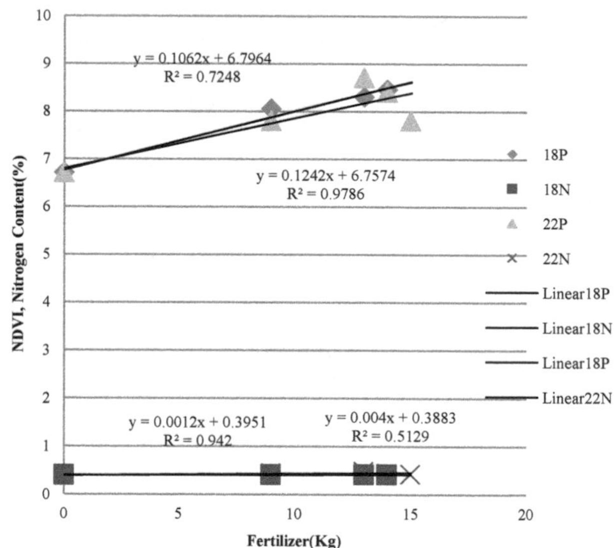

Fig. 4. Difference between August 18 and 22 data derived total nitrogen content and NDVI of rice leaves

D. Four Day Changes of NDVI and Protein Content in Rice Crops

NDVI and Total nitrogen content in rice leaves which are measured on 18 and 22 August are shown in Fig.5 (a) and (b), respectively. Both are increased a little bit. Also, the variances of these NDVI and total nitrogen content in rice leaves are increased, in particular, for NDVI.

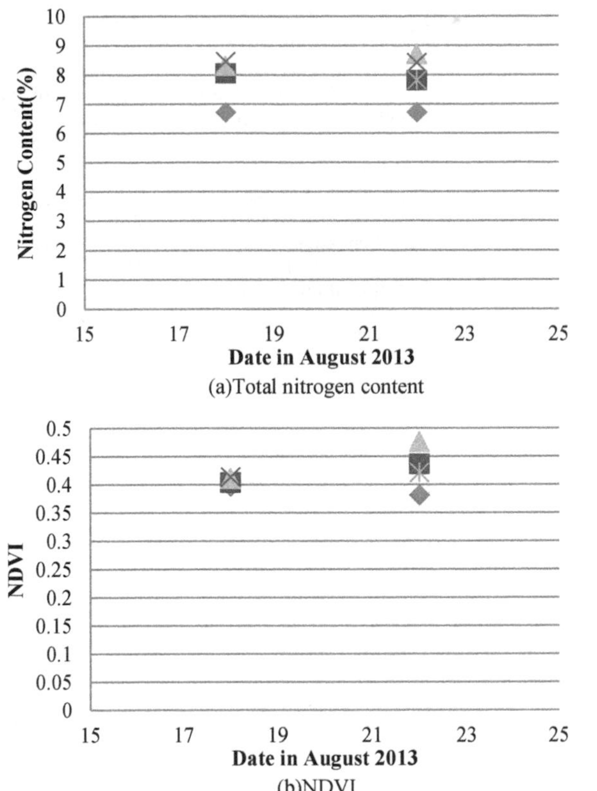

(a)Total nitrogen content

(b)NDVI

Fig. 5. NDVI and Total nitrogen content in rice leaves which are measured on 18 and 22 August

E. Rice Stump Density

On the other hand, there are two rice stump densities, 15.15 and 21.21 stumps/ m^2 in the rice paddy fields A and B. NDVI, total nitrogen content in rice leaves and protein content in rice crops for each rice stump density are shown in Fig.6 (a), (b) and (c), respectively. NDVI and total nitrogen content in rice leaves are decreased in accordance with increasing of the density while protein content in rice crops is decreased with increasing of the density. This implies that rice crop quality of relatively high density of rice stump is better than that of poorly dense rice paddy field.

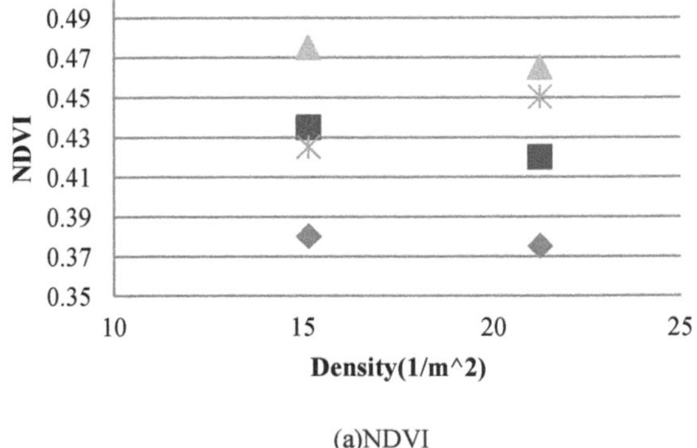

(a)NDVI

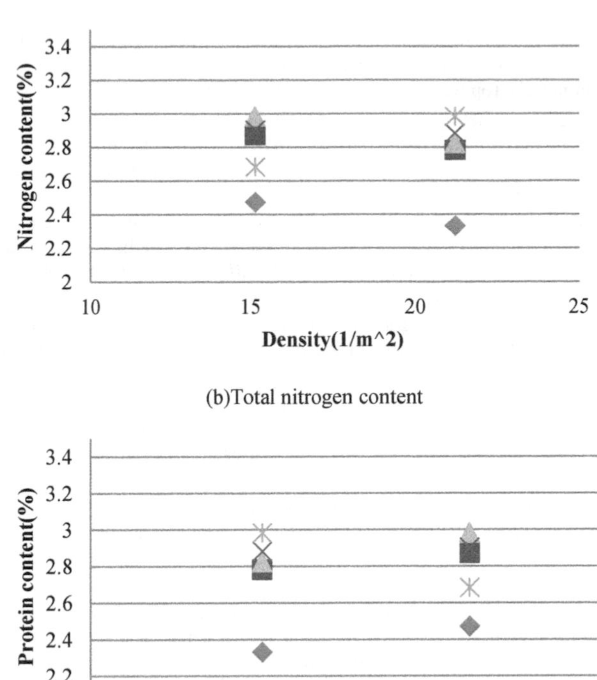

(b)Total nitrogen content

(c)Protein content

Fig. 6. NDVI, total nitrogen content in rice leaves and protein content in rice crops for each rice stump density

IV. Conclusion

Estimation of protein content in rice crop and total nitrogen content in rice leaves through regression analysis with Normalized Difference Vegetation Index: NDVI derived from camera mounted radio-control helicopter is proposed. Through experiments at rice paddy fields which is situated at Saga Prefectural Research Institute of Agriculture: SPRIA in Saga city, Japan, it is found that protein content in rice crops is highly correlated with NDVI which is acquired with visible and Near Infrared: NIR camera mounted on radio-control helicopter. It also is found that total nitrogen content in rice leaves is highly correlated to NDVI as well.

Protein content in rice crop is negatively proportional to rice taste. Therefore rice crop quality can be evaluated through NDVI observation of rice paddy field. It is found that total nitrogen content in rice leaves is linearly proportional to fertilizer amount and NDVI. Also, it is found that protein content in rice crops is positively proportional to fertilizer amount for lower fertilizer amount while protein content in rice crop is negatively proportional to fertilizer amount for relatively high fertilizer amount. It is also found that rice crop quality of low density of rice stump is better than that of highly dense rice paddy field.

REFERENCES

[1] Wiegand, C., Shibayama, M, Yamagata, Y, Akiyama, T., 1989. Spectral Observations for Estimating the Growth and Yield of Rice, Journal of Crop Science, 58, 4, 673-683, 1989.

[2] Kohei Arai, Method for estimation of grow index of tealeaves based on Bi-Directional reflectance function:BRDF measurements with ground based netwrok cameras, International Journal of Applied Science, 2, 2, 52-62, 2011.

[3] Kohei Arai, Wireless sensor network for tea estate monitoring in complementally usage with Earth observation satellite imagery data based on Geographic Information System(GIS), International Journal of Ubiquitous Computing, 1, 2, 12-21, 2011.

[4] Kohei Arai, Method for estimation of total total nitrogen and fiber contents in tealeaves with grond based network cameras, International Journal of Applied Science, 2, 2, 21-30, 2011.

[5] Kohei Arai, Method for estimation of damage grade and damaged paddy field areas sue to salt containing sea breeze with typhoon using remote sensing imagery data, International Journal of Applied Science,2,3,84-92, 2011.

[6] Kohei Arai, Monte Carlo ray tracing simulation for bi-directional reflectance distribution function and grow index of tealeaves estimation, International Journal of Research and Reviews on Computer Science, 2, 6, 1313-1318, 2011.

[7] Kohei.Arai, Fractal model based tea tree and tealeaves model for estimation of well opened tealeaf ratio which is useful to determine tealeaf harvesting timing, International Journal of Research and Review on Computer Science, 3, 3, 1628-1632, 2012.

[8] Kohei.Arai, H.Miyazaki, M.Akaishi, Determination of harvesting timing of tealeaves with visible and near infrared cameradata and its application to tea tree vitality assessment, Journal of Japanese Society of Photogrammetry and Remote Sensing, 51, 1, 38-45, 2012

[9] Kohei Arai, Osamu Shigetomi, Yuko Miura, Hideaki Munemoto, Rice crop field monitoring system with radio controlled helicopter based near infrared cameras through total nitrogen content estimation and its distribution monitoring, International Journal of Advanced Research in Artificial Intelligence, 2, 3, 26-37, 2013.

[10] Kohei Arai, Masanori Sakashita, Osamu Shigetomi, Yuko Miura, Estimation of Protein Content in Rice Crop and Total nitrogen Content in Rice Leaves Through Regression Analysis with NDVI Derived from Camera Mounted Radio-Control Helicopter, International Journal of Advanced Research in Artificial Intelligence, 3, 3, 13-19, 2013

30

Relations between Psychological Status and Eye Movements

Kohei Arai
Graduate School of Science and Engineering
Saga University
Saga City, Japan

Abstract—**Relations between psychological status and eye movements are found through experiments with readings of different types of documents as well as playing games. Psychological status can be monitored with Electroencephalogram: EEG sensor while eye movements can be monitored with Near Infrared: NIR cameras with NIR Light Emission Diode: LED. EEG signals are suffred from noises while eye movement can be acquired without any influence from nise. Therefore, psychlogical status can be monitored with eye movement detection instead of EEG signal acquisition if there is relation between both. Through the experiments, it is found strong relation between both. In particular, relation between the number of rapid changes of line of sight directions and relatively high frequency components of EEG signals is found. It is also found that the number of rapid eye movement is counted when the users are reading the documents. The rapid eye movement is defined as 10 degrees of look angle difference for one second. Not only when the users change the lines in the document, but also when the users feel a difficulty for reading words in the document, the users' line of sight direction moves rapidly.**

Keywords—EEG; eye movement; psychological status; alpha wave; beta wave

I. INTRODUCTION

Nowadays, the eye-based HCI has been widely used to assist not only handicap person but also for a normal person. In handicap person, especially paraplegic, they use eye-based HCI for helping them to self-sufficient in the daily life such as input text to compute [1], communication aids [2], controlling wheelchair [3], [4] , fetch a meal on table using robot arm [5], etc. The eye key-in system has been developed by many researchers [1]. The commercial available system provided by Tobii Tracker Company has been used by many researchers for developing text input, customer interest estimator on the business market, etc. [6]. Technology has been successful in rehabilitating paraplegics` personal lives. Prof Stephen Hawking, who was diagnosed with Athe proposedotrophic Lateral Sclerosis (ALS), uses an electronic voice synthesizer to help him communicate with others [7]. By typing the text through aid of a predictive text entry system, approximating his voice, he is able to make coherent speech and present at conferences. To give another example, a paraplegic patient wearing a head-mounted camera is able to draw figures, lines, and play computer games [1]. Clearly, through the use of assistive technology, handicapped people are able to do feats on par with non-handicapped people.

The published papers discussing eye-based HCI system are categorized into: (1) vision-based and (2) bio-potential-based. The vision-based method utilized camera to capture an image and estimate the user sight. The key issue here is how the method/system could be deal with environment changing. Lighting changing, user movement, various types of user, etc have to cooperate with the system. The vision-based system could be explained as follow.

Eye mouse based on sight is developed [8]. The method searches and tracks the eye by using projection of difference left-right eye after it is success to detect the face. It obtains left and right direction that could be used for controlling mouse pointer without upward and downward directions. They implemented the method for controlling application such as "Block Escape" game and spelling program. Eye mouse by utilizing sight obtained from pupil location and detected by using Haar Classifier (OpenCv function) is developed [9]. It used blinking for replacing left click of mouse event. The system enabling it to control mouse by utilizing face movement and blinking is developed [10]. The Adaboost face detection method detects center position of face and tracks it by using Lucas-Kanade Optical flow. The user used the resulted location for controlling mouse cursor and will execute left mouse click by detecting the blinking. Human-computer interface by integrating eye and head position monitoring devices is developed [11]. It controlled the interface based on sight and blinking. The user command could be translated by system via sight and blinking. Also, they modified calibration method for reducing visual angle between the center of target and the intersection point (derived by sight). They reported that the modification allowed 108 or higher number of command blocks to be displayed on 14 inch monitor. Also, they reported that it has hit rate of 98% when viewed at the distance of 500mm. For triggering, the blinking invokes commands.

The bio-potential-based method estimated user behavior (eye behavior) by measuring user's bio-potential. The bio-potential measurement instrument such as Electrooculograph (EOG), Electroencephalograph (EEG), etc could be used for measuring eye behaviors. The example of bio-potential-based has been applied in application of electric wheelchair controlled using EOG analyzed user eye movement via electrodes directly on the eye to obtain horizontal and vertical eye-muscle activity.

Signal recognition analyzed Omni directional eye movement patterns [12]. I search pupil location on an eye image by using the proposed method that has been published [13]. I estimate the sight by converting the obtained position of pupil to sight angle. After the sight angle is estimated, I control the mouse cursor by using this sight.

In the following section, eye movement detection method is described followed by psychological status monitoring with EEG sensor. Then relations between eye movement and psychological status are described followed by conclusion with discussions.

II. Eye Movement Detection System

A. Hardware Configuration

The system utilizes Infrared Camera NetCowBoy DC NCR-131 to acquire user image. I modified the position of 8 IR LED for adjusting illumination and obtaining stable image even illumination of environment changes as shown in Fig. 1.

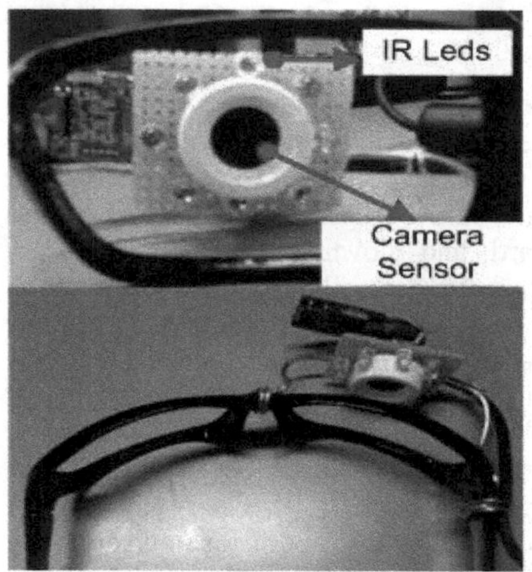

Fig. 1. Modified Camera Sensor

The use of IR camera will solve problem of illumination changes. I put the camera on user's glasses. Distance between camera and eye is 2 cm. I use Netbook Asus Eee PC 1002 HA with Intel Atom N270 CPU (1.6 GHz), 1GB Memory, 160 GB SATA HDD, and have small screen display 10 inch as main processing device. The software is developed under C++ language of Visual Studio 2005 and OpenCv, image processing library, which can be downloaded as free at their website. The hardware configuration is shown in Fig.2.

B. Eye Detection

The eye detection is handled by adaptive thresholds and pupil knowledge. I do not need the tracking method because the camera is on the user's glasses, so the next eye location has the same position as the previous one. Pupil location is detected by using pupil knowledge such as color, size, shape, sequential location, and movement.

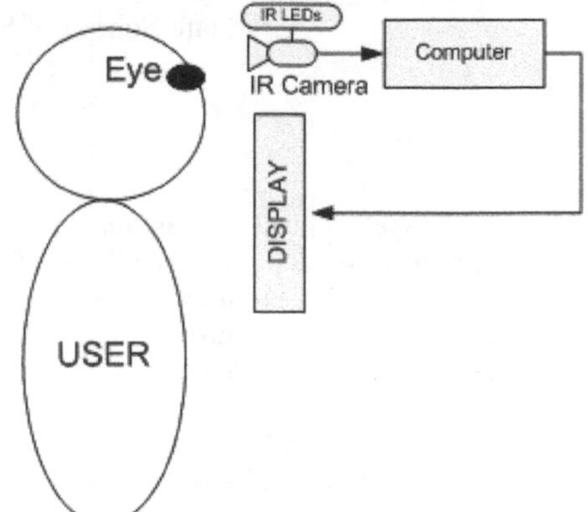

Fig. 2. Hardware Configuration

In pupil detection, pupil location is estimated as stated earlier using pupil knowledge extracted using an adaptive threshold to separate the pupil from other eye components. The threshold value T is 0.27% below the mean μ of eye image I and is derived from adjusting illumination intensity of 150 lux. The proposed threshold is suitable only for this condition but enables the camera to automatically adjust illumination as it changes.

$$\mu = \frac{1}{N}\sum_{i=0}^{N-1} I_i \qquad (1)$$

$$T = 0.27\mu \qquad (2)$$

Output from the adaptive threshold is black pixels representing the pupil in the image. To eliminate noise, I use a median filter. Widely adaptive threshold output is divided into three categories: (1) case 1, in which noise free black pixels clearly represent the pupil, (2) case 2, in which noise appears and is the same size and shape as pupil, and (3) case 3, when no pupil properties can be used to locate pupil. Cases are shown in Fig. 3.

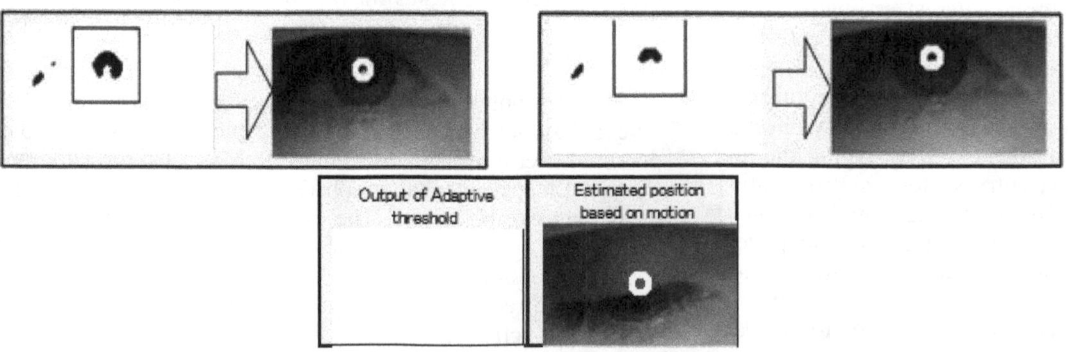

Fig. 3. Three cases in pupil detection that to be solved by using pupil knowledge

Once adaptive threshold output is classified, I estimate pupil location in three steps based on pupil knowledge. In case 1, the pupil is easily located by shape and size, even with noise in the image. In case 2, noise appears as almost the same size and shape as the pupil, e.g., when the adaptive threshold cannot be separated from other eye components such as the eyelid or the corner of the eye. To minimize these problems I recorded pupil movement histories assuming that the true pupil is closest to its previous location.

$$P(t-1)-C<P(t)<P(t-1)+C \quad (3)$$

Reasonable pupil location *P(t)* is within the surrounding of previous location *P(t−1)* with area *C*. In case 3, when features cannot be found to locate the pupil as when the user is looking from the corner of the eye, I estimate pupil location based on the movement.

C. EEG Sensor

The EEG sensor used for getting EEG signal from users' forehead in the experiment is Brain Catcher BCG-1S manufactured by TECHNOS Co. Ltd. Outlook of the sensor is shown in Fig.4. The sensor consists of two subsystems, EEG sensor and EEG signal receiver. Major specification is shown in Table 1.

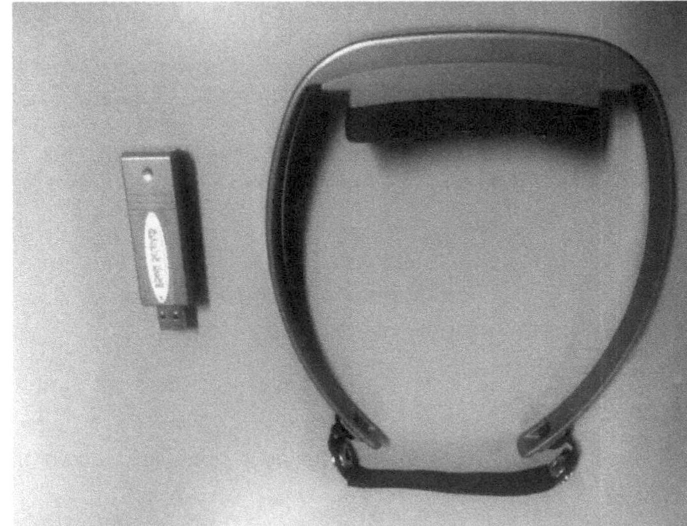

Fig. 4. Outlook of the EEG sensor used

TABLE I. SPECIFICATION OF BRAIN CATCHER BCG-1S MANUFACTURED BY TECHNOS CO. LTD.

Size	W:23 × D:79 × H:15(mm)
Weight	16g

(a)EEG signal receiver (USB terminal)

Size	W:160 × D:180 × H:37(mm)
Weight	85g

(b)EEG sensor

III. EXPERIMENTS

A. Eye detection Accuracy

In order to measure performance of pupil detection, I involved five different users who have different race and nationality (Indonesian, Japanese, Sri Lankan, and Vietnamese). The uses of many samples will prove that the proposed method work when it is used for all types of users. The eye movement data was collected from each user while was making several eye movement.

Eye images of three Indonesian show that even though images were taken from same country, each person has different eye shape. The number of one of the Indonesian images 882 samples. One of them has slated eye and the other two Indonesians have narrow eye and clear pupil. The number of samples of these two Indonesian is 552 and 668, respectively. The collected data from Sri Lankan consists of the number of images is 828 samples. His skin color is black and eyelid is thick. Collected data from Japanese consists of the number of images is 665 samples. His skin color is bright and eye is slant. The collected data from Vietnamese is almost same of the other five.

This experiment evaluates pupil detection accuracy and variance against different user by counting the success sample and the fail one. After accuracy of pupil detection has been counted, I compared the proposed method with adaptive threshold method and also template matching method. The compared adaptive threshold method uses combination between adaptive threshold itself and connected labeling method. Also, another compared method uses pupil template as reference and matched with the pupil images. The accuracy of the proposed pupil detection method against different users is shown in TABLE II. . The result data show that the proposed pupil detection method is accurate and robust against different users with variance value is 16.27.

TABLE II. ACCURACY OF THE PROPOSED PUPIL DETECTION METHOD AGAINST DIFFERENT USERS, THIS TABLE SHOWS THAT THE PROPOSED METHOD IS ROBUST AGAINST DIFFERENT USER AND ACCURATE

User Types	Nationality	Adaptive Threshold (%)	Template Matching (%)	Proposed Method (%)
1	Indonesian	99.85	63.04	99.99
2	Indonesian	80.24	76.95	96.41
3	Sri Lankan	87.8	52.17	96.01
4	Indonesian	96.26	74.49	99.77
5	Japanese	83.49	89.1	89.25
6	Vietnamese	98.77	64.74	98.95
Average		91.07	70.08	96.73
Variance		69.75	165.38	16.27

Line of sight direction is detected with NIR camera and NIR LED. Therefore, eye color and skin color are not matter for the eye movement detections.

B. Relation between Eye Movements and EEG Signals

Four Japanese users are selected for the experiment of eye movement detections and EEG signal acquisitions. Through the experiment, two documents (simple document [1] and complicated document[2] which are shown in Fig.5) are read by the users. It is expected that brain is in very active for the complicated document while users are in relax status for the simple document.

むかし、あるところに、おじいさんが ひとりで すんでいました。
あるひ、ひるごはんのあと いっぷくしていると、つばめが いちわ どまに おちてきました。
つばめは、はねを バタバタさせて、まいあがろうとするのですが、すぐにまた、どまに おち て、もがいています。

おじいさんが、どうしたのかとおもって みてみると、あしが おれていました。
「おぅおぅ、かわいそうに。これでは、とぶことはできんな。わしが なおしてやろうなぁ」と、くす りをつけて、こえだを そえて ほうたいを まいてやりました。
おじいさんは、まいにち、つばめを かいほうしてやりました。

しばらくして つばめのあしは、なおって、とべるようになりました。
おじいさんは、「よかった、よかった、もう だいじょうぶ!」と、いって はなしてやりました。
つばめは、おじいさんのいえのうえを、ぐるりと とんでから、とおくのそらへ とんでいきまし た。
「これから、きをつけて、げんきでくらせよー。そうして、らいねんも また、こいよー」つばめが みえなくなっても、おじいさんは、こえを かけつづけました。

それから、いちねん たちました。
おじいさんが、まえにわで いっぷくしていると、つばめが やってきました。
「おぅ、きたかや。おまえは きょねん、たすけてやった やつか。よう きたなぁ」と、はなしかけ ました。

つばめは、しばらく おじいさんの あたまのうえを とびまわっ て いましたが、まっくろけな おおきな つぶを、ひとつぶ ポ テーン と おとして、とんでいってしまいました。

「なんだ これは! フンを おとしていったのかい」と、いいながら よく みると、おおきな すいか の たねでした。
「あれ あれ、すいかのたねを、おとしていったのか。そうか、おれに これをうえて、すいかが なったら たべろということかな。きょねん、たすけてやったおかえしに、すいかのたねを、もって きたというわけか」

おじいさんは、つばめが もってきた すいかのたねを、はたけに まきました。
みずをやったり、こやしをやったり、だいじに せわをしていると、やがて おおきな つるが の びてきて、みが ひとつ なりました。
おじいさんは、まいにち はたけにいって、すいかを なでて やりました。

すいかの みは、どんどん どんどん おおきくなっていきました。
あるひ、おじいさんが、すいかをたたくと、ちょうど たべごろの おとがしました。
おじいさんは、おおきな すいかを うんとこしょ どっこいしょ と、やっとのことで、いえにもちかえりました。

おじいさんが、すいかを まっぷたつに わったとたん、たねの ひとつぶ ひとつぶが、だいくさ んやこびきさんになって、どっと でてきました。

(a)Simple document

脳波（のうは、Electroencephalogram：EEG）は、ヒト・動物の脳から生じる電気活動を、頭皮上、蝶形骨底、鼓膜、脳表、脳深部などに置いた電極で記録したものである。
英語のElectroencephalogramの忠実な訳語として、脳電図、EEGという呼び方もあり、中国語ではこちらの表現法を取っている。
本来は、脳波図と呼ぶべきであるが、一般的には「脳波」と簡略化して呼ばれることが多い。
脳波を測定、記録する装置を脳波計(Electroencephalograph：EEG)と呼び、それを用いた脳波検査(Electroencephalography：EEG)は、医療での臨床検査として、また医学、生理学、心理学、工学領域での研究方法として用いられる。
検査方法、検査機械、検査結果のどれも略語はEEGとなるので、使い分けに注意が必要である。
個々の神経細胞の発火を観察する単一細胞電極とは異なり、電極近傍あるいは遠隔部の神経細胞集団の電気活動の総和を観察する（少数の例外を除く）。
近縁のものに、神経細胞の電気活動に伴って生じる磁場を観察する脳磁図（のうじず、Magnetoencephalogram：MEG）がある。
直接記録する方法はしばしば臨床検査として用いられる。 背景脳波（基礎律動）や突発活動（てんかん波形など）を観察する。
各種のてんかん、ナルコレプシー、変性疾患、代謝性疾患、神経系の感染症、脳器質的疾患、意識障害、睡眠障害、精神疾患などの診断の補助・状態把握などに用いられる。
波形の加工の方法として、主なものに加算平均法、双極子推定法、周波数解析、コヒーレンス法、主成分分析、独立成分分析などがあり、一部は臨床でも用いられている。
脳波を直接記録する方法はしばしば臨床検査として用いられる。 背景脳波（基礎律動）や突発活動（てんかん波形など）を観察する。
各種のてんかん、ナルコレプシー、変性疾患、代謝性疾患、神経系の感染症、脳器質的疾患、意識障害、睡眠障害、精神疾患などの診断の補助・状態把握などに用いられる。
波形の加工の方法として、主なものに加算平均法、双極子推定法、周波数解析、コヒーレンス法、主成分分析、独立成分分析などがあり、一部は臨床でも用いられている。

(b)Complicated document

Fig. 5. Documents used for the experiment

Fig.6 shows EEG signals when the users read the simple document.

Superimposed EEG signals for four users are shown in Fig.7 (The simple document).

Meanwhile, Fig.8 shows those when the users read the complicated document.

---

[1] Simple document in Japanese is written in "Hiragana" character only. The content of the document is story for children. Therefore, it is very easy to read.
[2] Complicated document in Japanese is written in "Kanji" and "Hiragana" characters. The content of the document is text book for psychology for university students. Therefore, it is rather difficult to read.

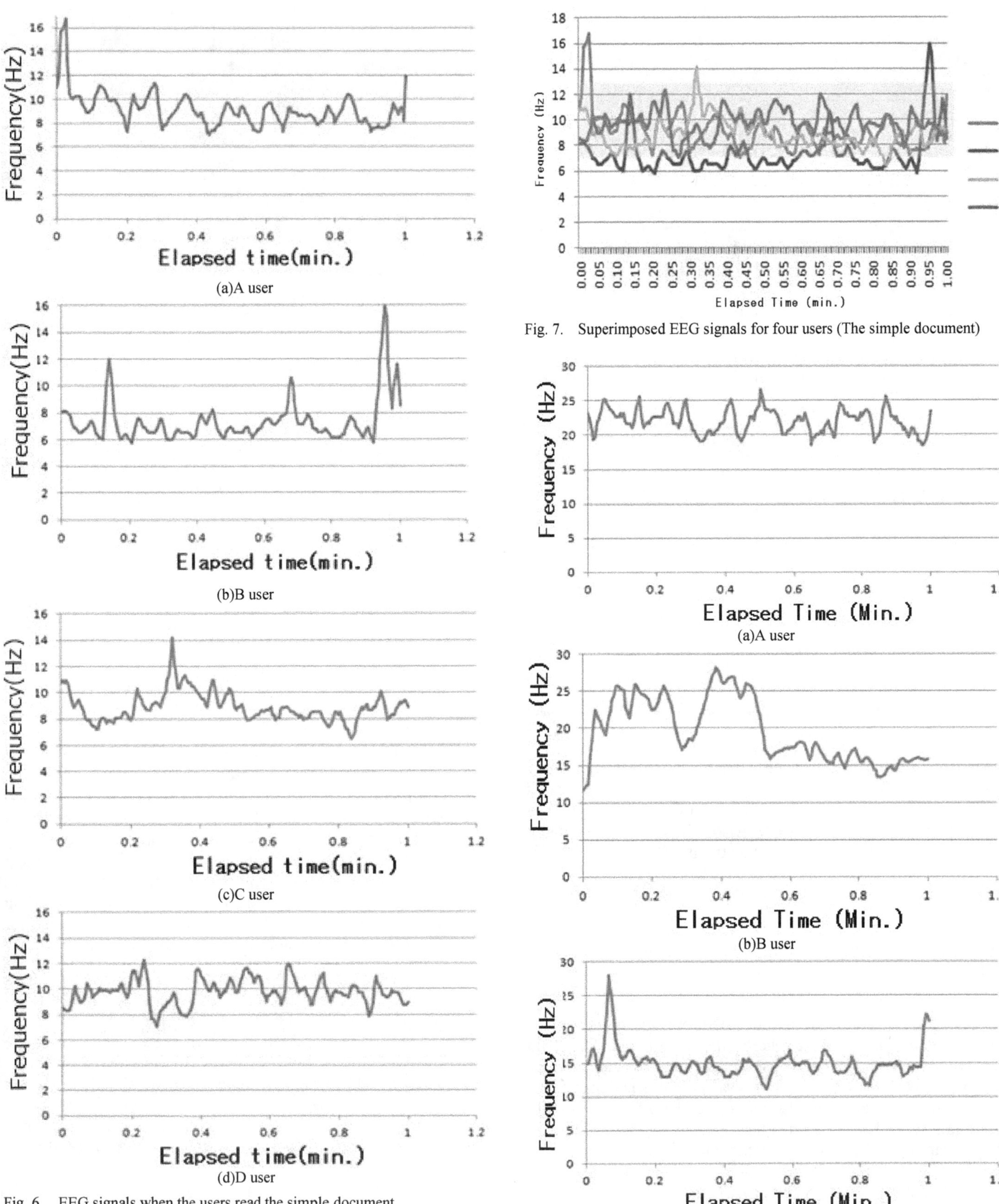

(a)A user

(b)B user

(c)C user

(d)D user

Fig. 6. EEG signals when the users read the simple document

Fig. 7. Superimposed EEG signals for four users (The simple document)

(a)A user

(b)B user

(c)C user

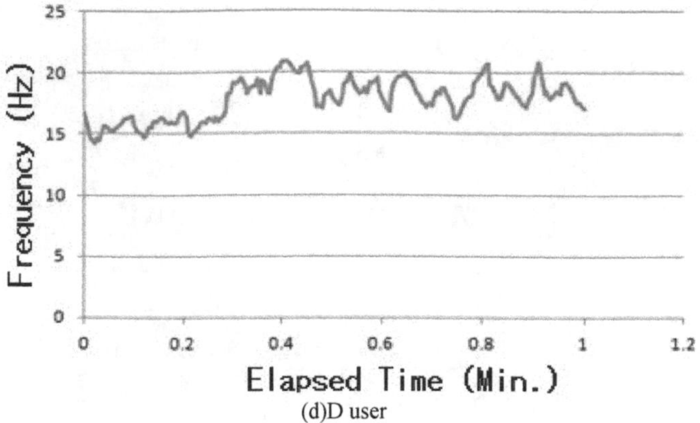

(d)D user

Fig. 8. EEG signals when the users read the complicated document

Superimposed EEG signals for four users are shown in Fig.9 (The complicated document).

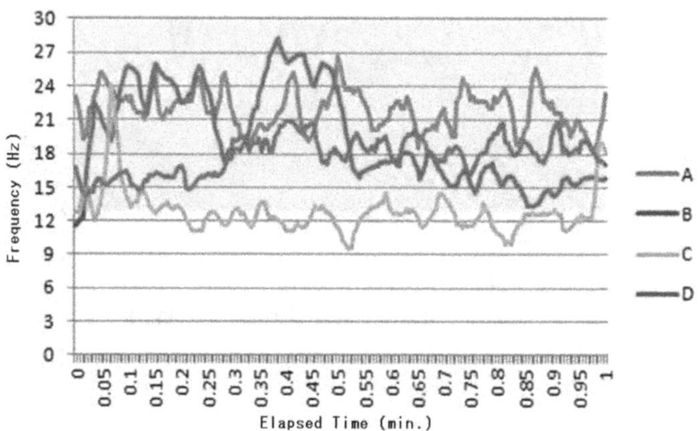

Fig. 9. Superimposed EEG signals for four users (The complicated document)

It is found that the frequency components for the simple document are much lower than those for the complicated document through a comparison between Fig.7 and Fig.9. The time variation is not so large. Also, there is not so large discrepancy among four users except user C. The user C is good at reading documents and has a plenty of knowledge and experiences on the contents of the documents.

From time to time, frequency component is getting high when users are reading complicated portions of the documents while is getting down when users are reading simple portions of the documents as shown in Fig.7 and Fig.9.

On the other hand, the number of rapid eye movement is counted when the users are reading the documents. The rapid eye movement is defined as 10 degrees of look angle difference for one second. Not only when the users change the lines in the document, but also when the users feel a difficulty for reading words in the document, the users' look direction moves rapidly. The number of the rapid eye movements is shown in Fig.10. The timing of the rapid eye movements is synchronized to raising the EEG frequency component. Namely, when the users feel a difficulty of reading the documents and line changes of the documents, EEG frequency components are raising and eye moves rapidly as shown in Fig.11 (Red circle indicates such moments).

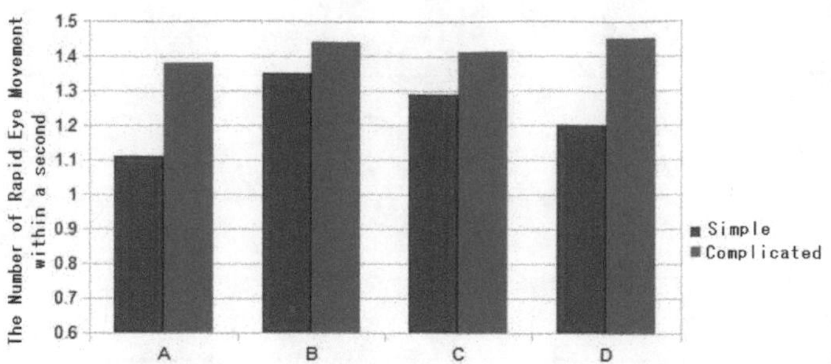

Fig. 10. The number of rapid eye movements

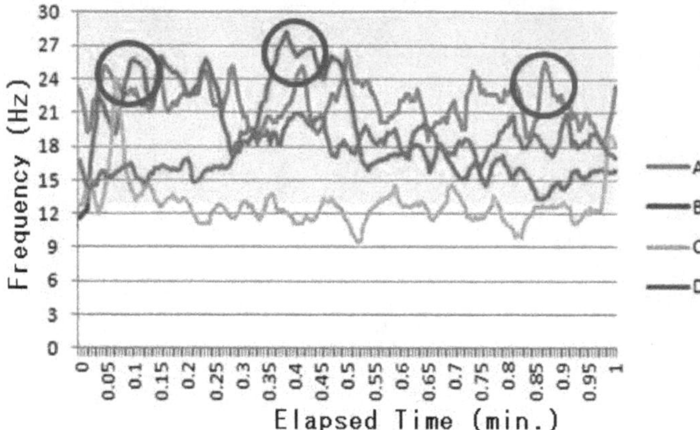

Fig. 11. EEG frequency components are raised when look direction changes rapidly

IV. CONCLUSION

Relations between psychological status and eye movements are found through experiments with readings of different types of documents as well as playing games. Psychological status can be monitored with Electroencephalogram: EEG sensor while eye movements can be monitored with Near Infrared: NIR cameras with NIR Light Emission Diode: LED. It is found that strong relation between both. In particular, the number of rapid changes of looking directions and relatively high frequency components of EEG signals. Therefore, it is possible to detect psychological status with monitoring eye movements. Due to the fact that EEG signals used to be affected by surrounding noises, eye movement detection is much useful for psychological status monitoring instead of EEG signals.

It is found that the number of rapid eye movement is counted when the users are reading the documents. The rapid eye movement is defined as 10 degrees of look angle difference for one second. Not only when the users change the lines in the document, but also when the users feel a difficulty for reading words in the document, the users' look direction moves rapidly. It is also found that the timing of the rapid eye movements is synchronized to raising the EEG frequency component. Namely, when the users feel a difficulty of reading the documents and line changes of the documents, EEG frequency components are raising and eye moves rapidly.

Future research works are required for typical psychological status monitoring, hatred, beloved, joy, anger, happy, surprise etc. with eye movement detection. Then the patients without mimic muscle function can represent their emotions. Also, psychological status monitored with the proposed eye movement detection is useful for triage of victims when they evacuate from disaster occurred areas to safe areas. Because vital sign which consist heart rate, blood pressure, the number of breath, body temperature and copiousness is required for triage. If patients in hospitals, group homes, etc. wear the glass with eye movement detection function and physical status monitoring sensor and network system for transmission of these data, then the patients can be saved their life because their status are known together with their location and attitude.

ACKNOWLEDGMENT

The author would like to thank Dr. Ronny Mardiyanto for his effort to the experiment on eye detection accuracy evaluation.

REFERENCES

[1] EYEWRITER: low-cost, open-source eye-based drawing system. Disponivel em: <http://www. crunchgear.com/2009/08/25/%20eyewriter-low-cost-%20open-source-eye-%20based-drawing-system/>. Acesso em: 7 October 2011.

[2] LEFF, R. B.; LEFF, A. N. 4.954.083, 1990.

[3] ARAI, K.; MARDIYANTO, R. A Prototype of ElectricWheelchair Controlled by Eye-Only for Paralyzed User. Journal of Robotic and Mechatronic, p. 66-74, 2011.

[4] DJOKO PURWANTO, R. M. A. K. A. Electric wheel chair control with gaze detection and eye blinking. Artificial Life and Robotics, AROB Journal. 2009. p. 397-400.

[5] ARAI, K.; YAJIMA, K. Robot Arm Utilized Having Meal Support System Based on Computer Input by Human Eyes Only. International Journal of Human Computer Interaction (IJHCI), p. 1-9, 2011

[6] JOHN J, M. et al. EyeKeys: A Real-Time Vision Interface Based on Gaze Detection from a Low-Grade Video Camera. 2004 Conference on Computer Vision and Pattern Recognition Workshop. 2004. p. 159.

[7] ARAI, K.; YAJIMA, K. Robot Arm Utilized Having Meal Support System Based on Computer Input by Human Eyes Only. International Journal of Human Computer Interaction (IJHCI), p. 1-9, 2011

[8] JOHN J, M. et al. EyeKeys: A Real-Time Vision Interface Based on Gaze Detection from a Low-Grade Video Camera. 2004 Conference on Computer Vision and Pattern Recognition Workshop. 2004. p. 159.

[9] CHANGZHENG, L.; CHUNG-KYUE, K.; JONG-SEUNG, P. The Indirect Keyboard Control System by Using the Gaze Tracing Based on Haar Classifier in OpenCV. 2009 International Forum on Information Technology and Applications.. 2009. p. 362-366.

[10] ZHU, H.; QIANWEI, L. Vision-Based Interface: Using Face and Eye Blinking Tracking with Camera. 2008 Second International Symposium on Intelligent Information Technology Application. 2008. p. 306-310.

[11] PARK K.S., L. K. T. Eye-controlled human/computer interface using the line-of-sight and the intentional blink. Computers and Industrial Engineering. 1993. p. 463-473.

[12] BAREA, R. et al. System for Assisted Mobility using Eye Movements based on Electrooculography. IEEE Transaction on Neural System and Rehabilitation Engineering, v. 10, p. 209-218, 2002.

[13] ARAI, K.; MARDIYANTO, R. Improvement of gaze estimation robustness using pupil knowledge. Proceedings of the International Conference on Computational Science and Its Applications (ICCSA2010). 2010. p. 336-350.

31

Defending Grey Attacks by Exploiting Wavelet Analysis in Collaborative Filtering Recommender Systems

Zhihai Yang, Zhongmin Cai* and Aghil Esmaeilikelishomi
Ministry of Education Key Lab for Intelligent Networks and Network Security,
Xi'an Jiaotong University, Xi'an, 710049, China

***Abstract*—"Shilling" attacks or "profile injection" attacks have always major challenges in collaborative filtering recommender systems (CFRSs). Many efforts have been devoted to improve collaborative filtering techniques which can eliminate the "shilling" attacks. However, most of them focused on detecting push attack or nuke attack which is rated with the highest score or lowest score on the target items. Few pay attention to grey attack when a target item is rated with a lower or higher scores than the average score, which shows a more hidden rating behavior than push or nuke attack. In this paper, we present a novel detection method to make recommender systems resistant to such attacks. To characterize grey ratings, we exploit rating deviation of item to discriminate between grey attack profiles and genuine profiles. In addition, we also employ novelty and popularity of item to construct rating series. Since it is difficult to discriminate between the rating series of attacker and genuine users, we incorporate into discrete wavelet transform (DWT) to amplify these differences based on the rating series of rating deviation, novelty and popularity, respectively. Finally, we respectively extract features from rating series of rating deviation-based, novelty-based and popularity-based by using amplitude domain analysis method and combine all clustered results as our detection results. We conduct a list of experiments on the Book-Crossing dataset in diverse attack models. Experimental results were included to validate the effectiveness of our approach in comparison with benchmarked methods.**

Keywords*—*recommender system; grey attack; discrete wavelet transform; shilling attack

I. INTRODUCTION

Collaborative filtering recommender systems (CFRSs) have become a popular and effective tool for information retrieval especially when users facing information overload. CFRSs also have played an important role in many popular web services such as Netflix, Amazon etc., which are designed to recommend items based on relevant information for the specific user [3], [5], [11], [14], [30], [33]. However, CFRSs are particularly vulnerable to "shilling" attacks or "profile injection" attacks in which an attacker signs up as a number of "puppet" users and rates fake scores in an attempt to promote or demote the recommendations of specific items by using knowledge of the recommender algorithms [2], [20], [21], [25], [26]. In such attacks, the attackers deliberately insert attack profiles into genuine profiles to change the prediction results which would reduce the trustworthiness of recommendation. The attack profiles indicate the attacker's intention that he wishes a particular item can be rated highest score (called push attack) or lowest score (called nuke attack) [4], [6], [7], [9], [10], [12], [16], [18], [19]. In addition, to avoid being detected easily by traditional detection techniques, the attackers may rate a higher score or lower score on the target items, which generates relatively hidden attack intents in comparison with push attacks or nuke attacks [24], we also call them grey attacks. Of course, they belong to the "shilling" attacks. Therefore, constructing an effective system to defend the attackers and remove them from the CFRSs is crucial.

Although existing work in this area have focused on detecting and preventing the "shilling" attacks or "profile injection" attacks, it has not reached a fully acceptable level of detection performance. We can briefly summarize that it is difficult to improve detection performance for detecting such attacks when filler size or attack size is small. Moreover, few pay attention to the grey attack detection. As an attacker demotes (nuke attack) the target items by rating lowest score or promotes (push attack) the target items by rating highest score, he also can demote or promote the target items by rating lower or higher scores. In fact, the rating behavior of an attacker is very similar to the behavior of a genuine user if the rating of target item is close to the actual rating. For the nuke attack, an attacker is simply shifting the rating given to the target item from the minimum rating to a rating one step higher, for the push attack, and vice versa [24]. Any profile that includes these ratings is likely to be less suspect. Although a minor change, this has a key effect. Thus, a challenging detection method should not only perform well when attack size or filler size is small, but also effectively defend the grey attack profiles.

In this paper, we propose an unsupervised attack detection method to make recommender systems resistant to such attacks, which combines discrete wavelet transform (DWT) and EM-based (Expectation-maximization based) clustering method. Since the attackers mimic some rating details of genuine users in shilling attacks, the rating behavior between attackers and genuine users will become more similar, especially for the grey attacks. Although existing features extracted from user profiles can characterize the shilling attacks to some extent, it's difficult to fully discriminate between attack profiles and genuine profiles. Moreover, the above challenges are also significant in grey attacks. Our basic assumption is that we can use DWT to amplify the differences between attack profiles and genuine profiles. In addition, to characterize the features of grey ratings, we use rating deviation of item to address this crucial problem. To construct input series for DWT, we create a list of transformed rating series to address this problem, which exploits the novelty, popularity and rating deviation of item for each user profiles, respectively. Moreover, we employ the empirical model decomposition (EMD) method to extract intrinsic mode functions (IMFs) from the rating series [17]. These can be seen that there are some but not obvious difference between the attack profiles and genuine profiles (as shown in Figures 4-6). To amplify the difference, we further use DWT to transform these series. In essence, a rating series is a non-stationary random series. Therefore, it is very suitable to be processed by DWT which performs well for non-stationary data [17]. After DWT, the differences between attack profiles and genuine profiles become more obvious (as shown in Figures 7-9). Based on the output series of DWT, we extract a list of effective features by using amplitude domain analysis method. And then exploiting EM clustering method to discriminate jointly attackers and genuine users based on the extracted features. In addition, the effectiveness of our proposed approach is validated and benchmark methods are briefly discussed. Experimental results show that our approach performs well for detecting the grey attacks in comparison with the benchmarked methods.

The remaining parts of this paper are organized as follows: Section 2 reviews some related work. Section 3 describes the attack model and introduces the theory of discrete wavelet transform. Our proposed detection method is introduced in Section 4. Experimental results and analysis are presented and discussed in Section 5. Finally, we conclude the paper with a brief summary and directions for future work.

II. RELATED WORK

Although existing detection techniques have focused on detecting and preventing the "shilling" attacks or "profile injection" attacks, it has not reached a fully acceptable level of detection performance. To name only a few, Burke et al. [3] proposed and studied several attributes derived from user profiles for their utility in attack detection. They employed kNN as their classification approach. However, it was unsuccessful when detecting attacks with small filler size and also suffered from low classifier precision. Then, Williams et al. [15], [24], [28] used several trained classifiers to detect shilling attacks based on extracted features of user profiles. Although, [24] used the higher or lower ratings instead of the maximum or minimum ratings to the target item, discussion of detecting such attacks was limited. Moreover, the detection performance was limited when filler size is small. Mobasher et al. [29] employed signatures of attack profiles and were moderately accurate. But, the method suffered from low accuracy in detecting shilling attack. They just focused on individual users and mostly ignored the combined effect of such attackers. In addition, the detection performance was limited when the attack profiles are obfuscated. Zhang et al. [31] proposed an ensemble approach to improve the precision of detection by using meta-learning technique. Their proposed method performs better detection performance than the bench marked methods. He et al. [32] employed rough set theory to detect shilling attacks though taking features of user profiles as the condition attributes of the decision table. However, their method also suffered from low precision. F. Zhang et al. [17] proposed an online method to detect profile injection attacks based on HHT and SMV. Zhou et al. [1] proposed a detection technique for identifying group attack profiles, called DeR-TIA, which combines an improved metric based on Degree of Similarity with Top Neighbors (DegSim) and Rating Deviation from Mean Agreement (RDMA). Zhang et al. [19] proposed a spectral clustering method to make recommender systems resistant to the shilling attacks in the case that the attack profiles are highly correlated with each other. Their experimental results reported good performance in random, average and bandwagon attacks. However, it also performed poor precision and recall in AOP attack when attack size is small.

III. PRELIMINARIES

In this section, we firstly describe the attack profiles and attack models. Then, we introduce the theory of discrete wavelet transform to facilitate discussions later.

A. Attack profiles and attack models

In the literature, "shilling" attacks are classified into two ways: nuke attack and push attack [3]. In order to nuke or push a target item, the attacker should be clearly known the form of an attack profile. The general form of an attack profile is shown in Table 1. The details of the four sets of items are described as follows:

I_S: The set of selected items with specified rating by the function $\sigma(i_k^S)$ [13];

I_F: A set of filler items, received items with randomly chosen by the function $\rho(i_l^F)$;

I_N: A set of items with no ratings;

I_T: A set of target items with singleton or multiple items, called single-target attack or multiple-targets attack. The rating is $\gamma(i_j^T)$, generally rated the maximum or minimum value in the entire profiles.

In this paper, we utilize 8 attack models to generate attack profiles. The involved attack profiles and corresponding explanations are listed in Table 2. The details of these attack models in our experiments are described as follows:

1) AOP attack: A simple and effective strategy to obfuscate the Average attack is to choose filler items with equal probability from the top x% of most popular items rather than from the entire collection of items [22].

2) Random attack: $I_S = \emptyset$ *and* $\rho(i) \sim N(\bar{r}, \bar{\sigma}^2)$ *[13];*

3) Average attack: $I_S = \emptyset$ *and* $\rho(i) \sim N(\bar{r}_i, \bar{\sigma}_i^2)$ *[13];*

4) Bandwagon (average): I_S *contains a set of popular items. Then, we use these items as* I_S *,* $\sigma(i) = r_{max}$ *or* r_{min} *or* r_{grey} *(push or nuke or grey) and* $\rho(i) \sim N(\bar{r}_i, \bar{\sigma}_i^2)$ *[13];*

5) Bandwagon (random): I_S *contains a set of popular items,*

TABLE I. GENERAL FORM OF ATTACK PROFILES

I_T			I_S			I_F			I_N		
i_1^T	...	i_j^T	i_1^S	...	i_k^S	i_1^F	...	i_l^F	i_1^N	...	i_v^N
$\gamma(i_1^T)$	...	$\gamma(i_j^T)$	$\sigma(i_1^S)$	...	$\sigma(i_k^S)$	$\rho(i_1^F)$	...	$\rho(i_l^F)$	null	...	null

TABLE II. ATTACK MODELS

Attack Model	I_S		I_F		I_N	I_T (push or nuke or grey)
	Items	*Rating*	*Items*	*Rating*		
AOP	null		x-% popular items, ratings set with normal dist around item mean.		null	r_{max} or r_{min} or r_{grey}
Random	null		randomly chosen	system mean	null	r_{max} or r_{min} or r_{grey}
Average	null		randomly chosen	item mean	null	r_{max} or r_{min} or r_{grey}
Bandwago (average)	popular items	r_{max} or r_{min}	randomly chosen	item mean	null	r_{max} or r_{min} or r_{grey}
Bandwagon (random)	popular items	r_{max} or r_{min}	randomly chosen	system mean	null	r_{max} or r_{min} or r_{grey}
Segment	segmented items	r_{max} or r_{min}	randomly chosen	r_{min} or r_{max}	null	r_{max} or r_{min} or r_{grey}
Reverse Bandwagon	unpopular items	r_{min} or r_{max}	randomly chosen	system mean	null	r_{max} or r_{min} or r_{grey}
Love/Hate	null	null	randomly chosen	r_{min} or r_{max}	null	r_{max} or r_{min} or r_{grey}

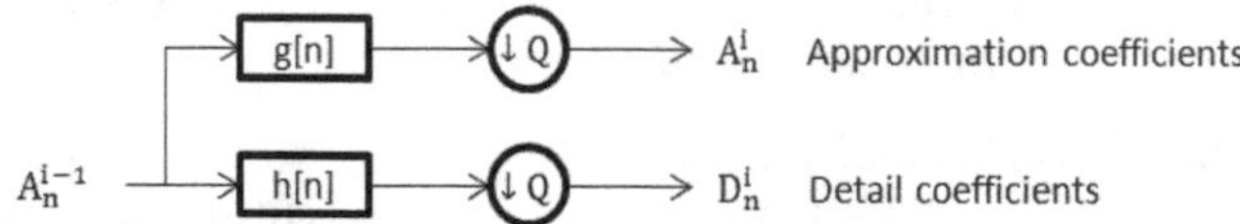

Fig. 1. Block diagram of filter analysis

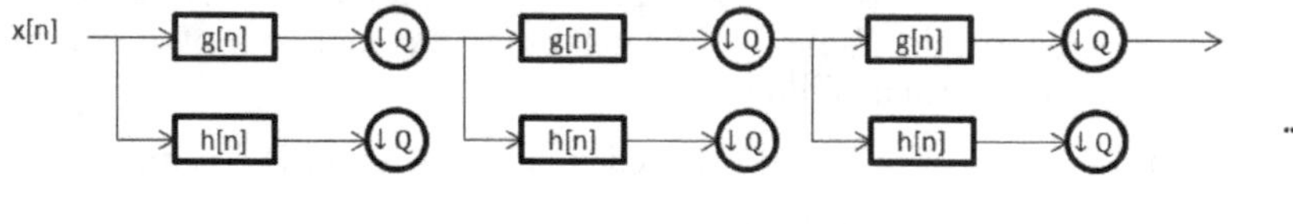

Fig. 2. K (k greater than or equal to 1) levels of filter bank

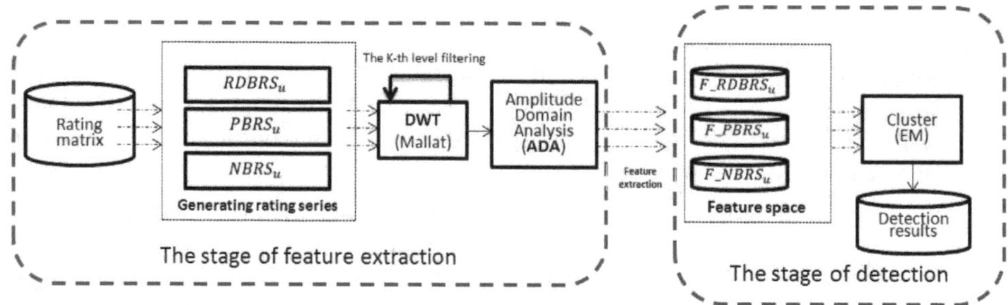

Fig. 3. The framework of our proposed method which consists of two stages: the stage of feature extraction and the stage of detection

$\sigma(i) = r_{max}$ or r_{min} or r_{grey} *and* $\rho(i) \sim N(\bar{r}, \bar{\sigma}^2)$ *(nuke or grey) [13];*

6) Segment attack: I_S *contains a set of segmented items,* $\sigma(i) = r_{max}$ or r_{min} or r_{grey} *and* $\rho(i) = r_{min}$ or r_{max} or r_{grey} *(push or nuke or grey) [8];*

7) Reverse Bandwagon attack: I_S *contains a set of unpopular items,* $\sigma(i) = r_{min}$ or r_{max} or r_{grey} *(push or nuke or grey) and* $\rho(i) \sim N(\bar{r}, \bar{\sigma}^2)$ *[9];*

8) Love/Hate attack: $I_S = \emptyset$ *and* $\rho(i) = r_{max}$ or r_{grey} *(nuke or grey) [9].*

B. Discrete wavelet transform

Discrete wavelet transform (DWT) has been recognized as a natural wavelet transform for discrete time signals. Both time and scale parameters are discrete. For a discrete-time sequence $x[n], n \epsilon Z$, DWT is defined by discrete-time multi-resolution decomposition which could be computed by Mallat pyramidal decomposition algorithm (as shown in Equations (1)-(3)) [23]. However, since half the frequencies of the signal have now been removed, half the samples can be discarded according to Nyquist's rule. The filter outputs are then sub-sampled by 2 (Mallat's and the common notation is the opposite, g- high pass and h- low pass):

$$A_n^0 = x[n],\ n \epsilon N \quad (1)$$

$$A_n^i = \sum_{k \epsilon Z} g(k-2n) A_k^{i-1},\ i = 1,2,\dots,L \quad (2)$$

$$D_n^i = \sum_{k \epsilon Z} h(k-2n) A_k^{i-1},\ i = 1,2,\dots, \quad (3)$$

where h and g are impulse responses of high-pass filter H and low-pass filter G, respectively. $\{A_n^i\}$ and $\{D_n^i\}$ are scale sequence and wavelet sequence of 2^{-i} scale. L is the maximum possible scale of the discrete signal x[n]. The signal is also decomposed simultaneously using a high-pass filter. The outputs give the detail coefficients (from the high-pass filter) and approximation coefficients (from the low-pass) as shown in Figure 1. It is important that the two filters are related to each other and they are known as a quartered mirror filter.

DWT of a signal is calculated by passing it through a series of filters. The decomposition is repeated to further increase the frequency resolution and the approximation coefficients decomposed with high and low pass filters and then down-sampled (see Figure 2). This is represented as a binary tree with nodes representing a sub-space with different time-frequency localization. And the tree is known as a filter bank.

IV. Our Proposed Approach

In this section, we firstly introduce the framework of our proposed approach. And then we give several definitions of rating series used in this paper. Finally, we briefly describe our detection method.

A. The framework

As shown in Figure 3, our proposed algorithm consists of two stages: the stage of feature extraction and the stage of detection. At the stage of feature extraction, the feature is extracted one by one from user profiles by using the proposed feature extraction method (see subsection 4.2). Inspired from previous studies (Zhang et al. [17]), we incorporate into two concepts: Empirical Mode Decomposition (EMD) and Intrinsic Mode function (IMF). EMD is an adaptive and highly efficient decomposition method and is also a necessary step to reduce any given data into a collection of intrinsic mode functions (IMF) where the DWT analysis can be applied. As we all know, DWT is a method for analyzing non-stationary data, since the rating series are non-stationary data. The IMF is defined as a function that satisfies the following requirements: (a) In the whole data set, the number of extreme and zero-crossings must either be equal or differ at most by one; (b) At any point, the mean value of the envelope defined by the local maxima and the envelope defined by the local minima is zero.

With this method, rating series can be decomposed into a finite signal and regard the signal as the input of discrete wavelet transform [17], [27]. In our proposed approach, we decompose respectively each user profiles into novelty-based, popularity-based and rating deviation-based rating series as the input signals. And then, the input signals are passed through the series of filters (including low-pass and high-pass filter, as shown in Figure 3.) to generate corresponding output signals. In the process of DWT, we perform one level transformation to get the output signals. Then, by using amplitude domain analysis method to extract features from the output signal. At the stage of detection, based on the extracted features, we respectively use EM method to cluster two groups. Finally, combing the three parts of clustering results to return our detection result.

B. Feature extraction

Previous studies [17] have disclosed that using the novelty and popularity of items to construct rating series for user profiles implies useful information. Inspired from this research, we investigate using rating deviation of items to construct rating series in order to extract features from grey attack profiles. Novelty [1] in recommendation is focusing on recommending the log-tail items (i.e., less popular items) which is generally considered to be particularly valuable to users. Popularity of items usually reflects the genuine users' tastes or preferences in collaborative recommender system. By sorting the items according to their novelty, popularity and rating deviation, we can create respectively the rating deviation-based, novelty-based and popularity-based rating series for the user profiles. Firstly, two definitions of the rating deviation are described in the following:

Definition 1 (Rating Deviation of Items, RDoI).

The $RDoI_i$ (rating deviation of item i) is defined as follows:

$$RDoI_i = \begin{cases} |r_{ui} - \overline{r_i}|, & r_{ui} \neq \perp, u \in R_g \\ 0, & r_{ui} = \perp \end{cases}, \quad (4)$$

where r_{ui} denotes the rating of user u on item i. $\overline{r_i}$ is the mean rating of item i in the system. $r_{ui} \neq \perp$ denotes item i is rated by user u, $r_{ui} = \perp$ denotes item i is not rated by user u. R_g denotes the set of genuine users in dataset.

Definition 2 (Rating Deviation-based Rating Series, RDBRS).

Let $RDoI_i$ denotes the rating deviation of item i. Sort all items in set I (a set of the entire items in the recommender system.) according to $RDoI_i$ in descending order and let $i = 1,2,\dots,|I|$ denotes the order of items after sorting, where $|I|$ denotes total number of items in the recommender system. The $RDBRS_u(i)$ [2] is defined as follows:

$$RDBRS_u(i) = \begin{cases} 1, & r_{u,i} \neq \perp \text{ and } (i = 1 \text{ or } RDNRS_u(i-1) \neq 1), \\ -1, & r_{u,i} = \perp \text{ and } (i = 1 \text{ or } RDNRS_u(i-1) \neq -1), \\ 0, & \text{otherwise.} \end{cases} \quad (5)$$

where zero value is used to meet the requirements of extreme for DWT. $r_{u,i} \neq \perp$ denotes item i is rated by user u. $r_{u,i} = \perp$ denotes item i is not rated by user u.

Novelty of Items, NoI

The NoI_i(novelty of item i) is defined as follows:

$$NoI_i = \frac{1}{|R_g|} \sum_{u \in R_g, r_{u,i} \neq \perp} NoI_{u,i}, \quad (6)$$

---

[1] The novelty of an item refers to the degree to which it is unusual with respect to the user's normal tastes.

[2] The rating deviation-based rating series of user u.

where $NoI_{u,i}$ denotes the novelty of item i for user u [17].

$$NoI_{u,i} = \frac{1}{|N_j|}\sum_{u\in R_g, r_{u,j}\neq\perp}(1 - simi(i,j)) \quad (7)$$

where N_j denotes the number of users who rate on item j. R_g denotes the set of genuine users in dataset. $simi(i,j)$ (Jaccard coefficient) denotes the similarity between item i and item j, which can be calculated as follows:

$$simi(i,j) = \frac{|V_i \cap V_j|}{|V_i \cup V_j|} \quad (8)$$

Where V_i is set of users rated by item i, V_j is the set of users rated by item j. If both V_i and V_j are empty, we define $simi(i,j) = 0$. Clearly, $0 \leq simi(i,j) \leq 1$.

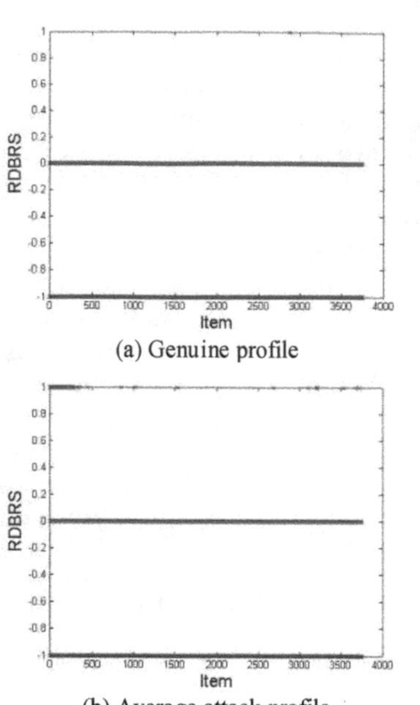

(a) Genuine profile

(b) Average attack profile

Fig. 4. Rating Deviation-based rating series. (a) The signal of a genuine profile before DWT; (b) The signal of an average attack profile before DWT

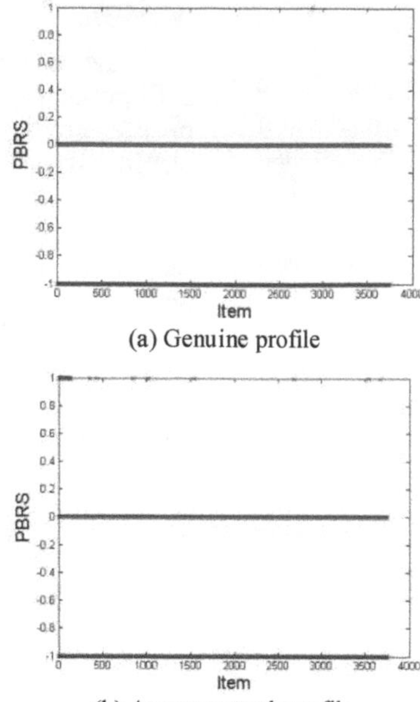

(a) Genuine profile

(b) Average attack profile

Fig. 5. Popularity-based rating series. (a) The signal of a genuine profile before DWT; (b) The signal of an average attack profile before DWT

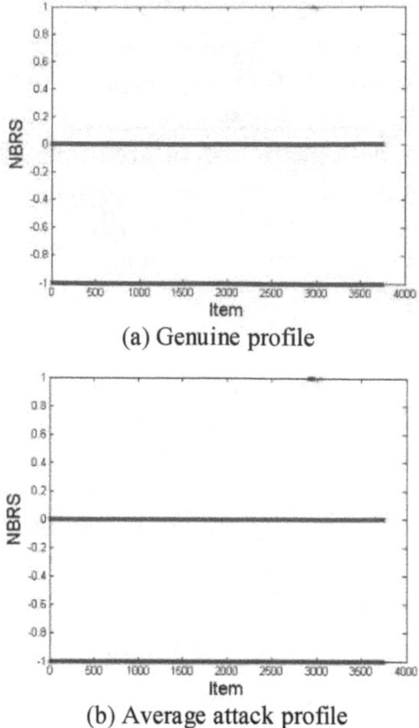

(a) Genuine profile

(b) Average attack profile

Fig. 6. Novelty-based rating series. (a) The signal of a genuine profile before DWT; (b) The signal of an average attack profile before DWT

Novelty-based Rating Series, NBRS

Let NoI_i denotes the novelty of item i. Sort all items in set I according to NoI_i in descending order and let $i = 1,2,\dots,|I|$ denotes the order of items after sorting. The novelty-based rating series of user u, $NBRS_u(i)$ is defined as follows:

$$NBRS_u(i) = \begin{cases} 1, & r_{u,i} \neq \perp \text{ and } (i = 1 \text{ or } NBRS_u(i-1) \neq 1), \\ -1, & r_{u,i} = \perp \text{ and } (i = 1 \text{ or } NBRS_u(i-1) \neq -1), \\ 0, & \text{otherwise.} \end{cases} \quad (9)$$

where zero value is used to meet the requirements of extreme for DWT [17].

Popularity of Items, PoI

The popularity of item i, PoI_i, is defined as the number of ratings given to item i by genuine users in data set [17].

Popularity-based Rating Series, PBRS

Let PoI_i denotes the popularity of item i. Sort all items in set I according to PoI_i in descending order and let $i = 1,2,\dots,|I|$ denotes the order of items after sorting. The popularity-based rating series of user u, $PBRS_u(i)$, is defined as follows:

$$PBRS_u(i) = \begin{cases} 1, & r_{u,i} \neq \perp \text{ and } (i = 1 \text{ or } PBRS_u(i-1) \neq 1), \\ -1, & r_{u,i} = \perp \text{ and } (i = 1 \text{ or } PBRS_u(i-1) \neq -1), \\ 0, & \text{otherwise.} \end{cases} \quad (10)$$

where zero value is used to meet the requirements of extreme for DWT [17].

To show the difference between genuine and attack profiles in rating series, we give examples of the novelty-based, popularity-based and rating deviation-based rating series in Figures 4-6. These rating series are constructed by the genuine profiles and the average attack profiles (take average attack for example). The genuine profiles are selected from the Book-Crossing dataset. As shown in Figures 4-6, there are very little difference between the genuine and average attack profiles in rating series. We can observe that the RDBRS for the genuine profile barely changed from starting position to ending position in compared to the RDBRS of the average attack profile decreased gradually for the rating deviation-based rating series. For the popularity-based rating series, the PBRS for the genuine profile barely changed with the item increased while the PBRS of the average attack profile decreased gradually. And for the novelty-based rating series, the NBRS for genuine profile also almost remain unchanged with the item increased, while the NBRS of the average attack profile show characteristics of more concentrated. As mentioned above, it is difficult to discriminate between genuine profiles and attack profiles regardless of using Rating Deviation-based, Popularity-based and Novelty-based rating series. To amplify the difference between genuine profiles and attack profiles, we use DWT to transform the rating series in order to extract features from output signal by using amplitude domain analysis method.

After K (k greater than or equal to 1) level discrete wavelet transform (as shown in Figure 2), we can get the local properties, which passes a series low-pass filters to obtain an approximation coefficients. As shown in Figures 7-9, we can observe that there is a more significant difference between genuine profiles and average attack profiles on rating series than before using DWT. In Figure 7, the strength of oscillations of genuine profiles show characteristics of more concentrated with the item increased while the strength of oscillations of average attack profile decreased gradually from starting position to ending position. For the popularity-based rating series, the same observations are also clear in Figure 8. And for the novelty-based rating series, we can observe that there is a little difference between the genuine profiles and average attack profiles, although they show characteristics of more concentrated similarly as illustrated in Figure 9.

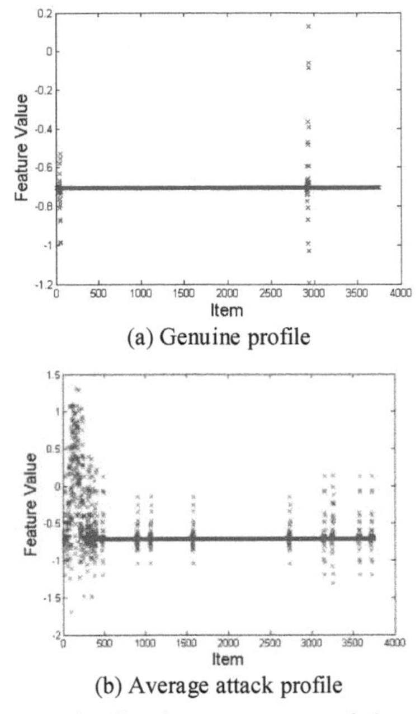

Fig. 7. The first low-pass output of the rating deviation-based rating series. (a) The signal of a genuine profile after DWT; (b) The signal of an average attack profile after DWT.

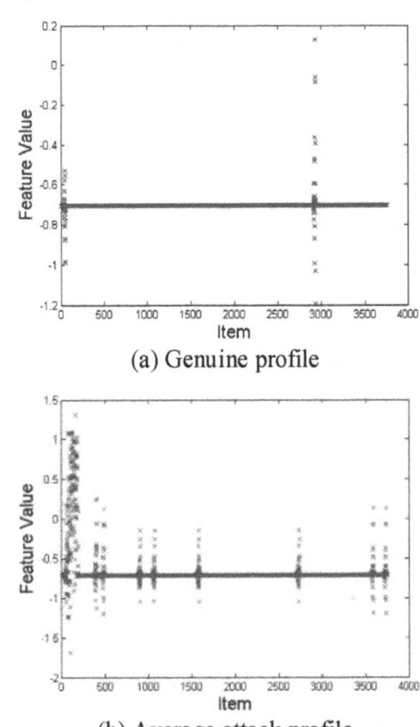

Fig. 8. The first low-pass output of the popularity-based rating series. (a) The signal of a genuine profile after DWT; (b) The signal of an average attack profile after DWT

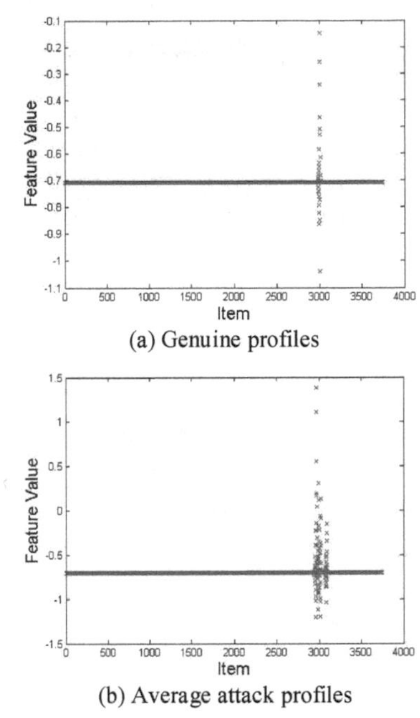

Fig. 9. The first low-pass output of the novelty-based rating series. (a) The signal of a genuine profile after DWT; (b) The signal of an average attack profile after DWT

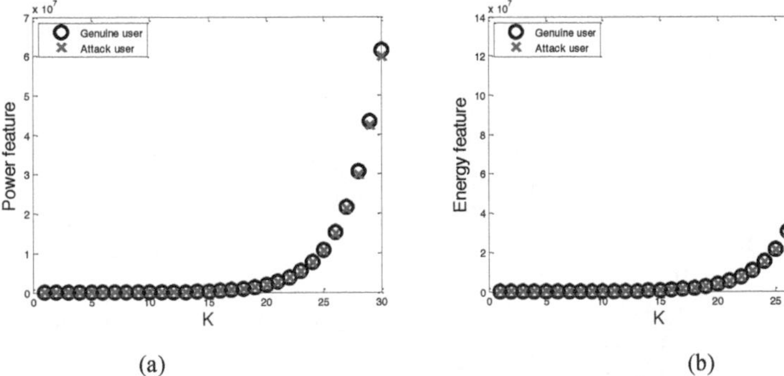

Fig. 10. The power feature and the energy feature in different K levels output of discrete wavelet transforms for a genuine user and an attacker. (a) Power features; (b) Energy features

Let F_RDBRS_u, F_PBRS_u and F_NBRS_u denotes the feature vector of user u on the rating deviation-based, novelty-based and popularity-based after DWT, respectively. The proposed feature extraction algorithm is described in algorithm 1. In algorithm 1, from step 1 to step 3 create the rating deviation-based, novelty-based and popularity-based rating series for user u respectively. Step 4 is the process of DWT. Step 5 extract features from approximation parts of rating deviation, popularity and novelty rating series, termed A_RD_k, A_P_k and A_N_k by using amplitude domain analysis method. The last step generates a feature space for the stage of detection.

Algorithm 1: Feature extraction algorithm for user profiles

Input: Rating Matrix;

Output: F_RDBRS_u, F_PBRS_u and F_NBRS_u;

Step 1: Create rating series $RDBRS_u(i)$ of u by using rating matrix and Equations (4)-(5);

Step 2: Create rating series $NBRS_u(i)$ of u by using rating matrix and Equations (6)-(9);

Step 3: Create rating series $PBRS_u(i)$ of u by using rating matrix and Equation (10);

Step 4: Generate approximation parts A and detail parts D by exploiting Mallat (discrete wavelet transform) algorithm on the rating series of $RDBRS_u(i)$, $PBRS_u(i)$ and $NBRS_u(i)$ by using Equations (1)-(3), respectively;

Step 5: Take the K level approximation parts A_RD_k, A_N_k and A_P_k from Step 4's output, respectively. And extract features from the approximation parts by using amplitude domain analysis method on A_RD_k, A_N_k and A_P_k respectively;

Step 6: Generate and return the feature space F_RDBRS_u, F_PBRS_u and F_NBRS_u respectively.

TABLE III. THE FEATURES OF THE SIGNAL AMPLITUDE DOMAIN AND THEIR DESCRIPTION

Features	Equations	Descriptions
Minimum value	$x_{min} = \min(X)$	The minimum value of the amplitude of the signal.
Maximum value	$x_{max} = \max(X)$	The maximum value of the amplitude of the signal.
Mean value	$\overline{X} = \text{mean}(X)$	The average value of the amplitude of the signal.
Peak value	$x_p = \max(\text{abs}(X))$	The maximum of the absolute value of the amplitude of the signal.
Root mean square value	$X_{rms} = \sqrt{\frac{1}{N}\sum_{i=1}^{N} x_i^2}$	The root mean square value of the amplitude of the signal.
Root mean square amplitude value	$X_r = \left(\frac{1}{N}\sum_{i=1}^{N}\sqrt{\lvert x_i\rvert}\right)^2$	Represent the energy size of the signal.
Absolute mean	$\lvert\overline{X}\rvert = \frac{1}{N}\sum_{i=1}^{N}\lvert x_i\rvert$	Absolute mean value of the amplitude of the signal.
Variance	$\sigma_x^2 = X_{rms}^2 - \overline{X}^2$	Represent the degree of dispersion of the signal.
Skewness	$\alpha = \frac{1}{N}\sum_{i=1}^{N} x_i^3$	Represent the asymmetry of amplitude probability density function on the vertical axis.
Kurtosis	$\beta = \frac{1}{N}\sum_{i=1}^{N} x_i^4$	Represent the steep degree of the signal curve.
Shape factor	$S_f = X_{rms}/\lvert\overline{X}\rvert$	A shape factor refers to a value that is affected by an object's shape but is independent of its dimensions
Crest factor	$C_f = X_{max}/X_{rms}$	Crest factor is a measure of a waveform, showing the ratio of peak values to the average value.
Impulse factor	$I_f = X_{max}/\lvert\overline{X}\rvert$	Non-dimensional parameter in amplitude domain.
Clearance factor	$CL_f = X_{max}/X_r$	Non-dimensional parameter in amplitude domain.
Kurtosis value	$K_v = \beta/X_{rms}^4$	Non-dimensional parameter in amplitude domain.

For different types of signal, there are different analysis methods such as time domain analysis, frequency domain analysis and amplitude domain analysis. As shown in Figure 10, we can observe that these are no significant difference between genuine user and attacker with the K (the K level output of DWT) increased, regardless of using the power features or energy features. In this paper, we use amplitude domain analysis to extract features from signals. The details of signal features in amplitude domain are showed in Table 3. We have 15 features to characterize the signal which extracts from the K level (we set K equal to 1 in our work) output of DWT.

C. Detection algorithm

In order to get better detection performance as far as possible, we combine the rating deviation-based, novelty-based and popularity-based methods to distinguish between genuine profiles and attack profiles. And then, we utilize EM (Expectation-maximization) clustering method (Clustering results and EM clustering method were created using Weka [3]) to separate attackers from genuine users as far as possible. Let D denotes the set of detection result. The proposed method for detecting grey attacks is described in algorithm 2. In algorithm 2, from step 1 to 3 perform EM algorithm on feature vector F_RDBRS_u, F_PBRS_u and F_NBRS_u, respectively. Step 4 obtains the set of attackers decided by using the smaller cluster, since the number of attackers less than the number of genuine users in the recommender system. In step 5, we exploit the intersection of the set D_RD, D_P and D_N, and then the detection result D was generated.

[3] http://www.cs.waikato.ac.nz/ml/weka/

Algorithm 2: Detection algorithm

Input: The set of users' feature space F_RDBRS_u, F_PBRS_u and F_NBRS_u; The number of clusters k;

Output: The detected result D;

Step 1: $\{C\_RD_1, C\_RD_2\} \leftarrow \text{EM}(F\_RDBRS_u)$;

Step 2: $\{C\_P_1, C\_P_2\} \leftarrow \text{EM}(F\_PBRS_u)$;

Step 3: $\{C\_N_1, C\_N_2\} \leftarrow \text{EM}(F\_NBRS_u)$;

Step 4: $D\_ARD = \min(C\_RD_1, C\_RD_2)$, $D\_P = \min(C\_P_1, C\_P_2)$, $D\_N = \min(C\_N_1, C\_N_2)$;

Step 5: $D \leftarrow \{D | D\_RD \cap D\_P \cap D\_N\}$;

Return D.

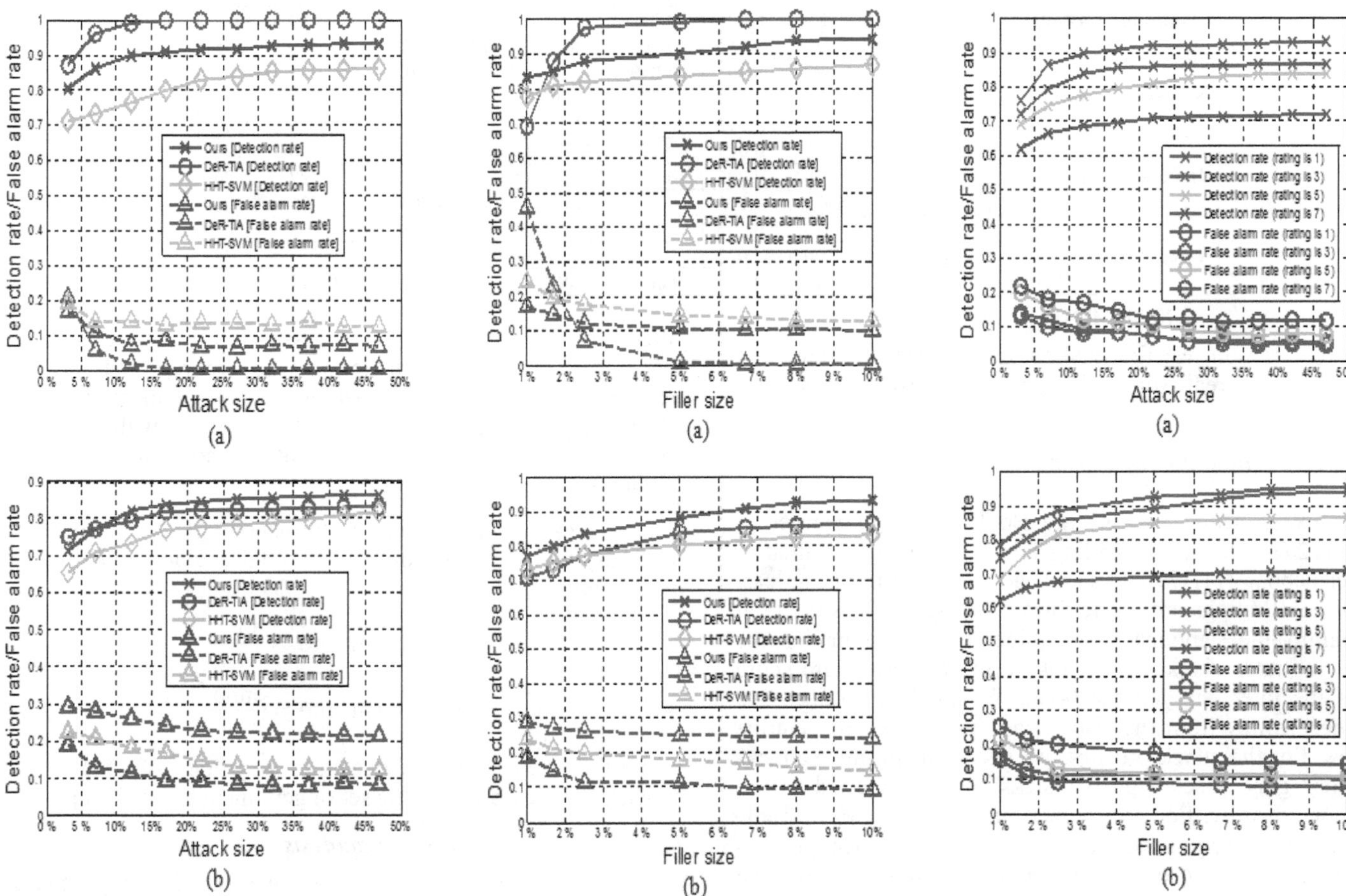

Fig. 11. The comparison of detection rate and false alarm rate in different attack sizes. (a) Grey rating is 1, filler size is 5%, single-target bandwagon (random) attack; (b) Grey rating is 3, filler size is 5%, single-target bandwagon (random) attack

Fig. 12. The comparison of detection rate and false alarm rate in different filler sizes. (a) Grey rating is 1, attack size is 17%, single-target bandwagon (random) attack; (b) Grey rating is 3, attack size is 17%, single-target bandwagon (random) attack

Fig. 13. The comparison of detection rate and false alarm rate with different grey ratings in single-target attack. (a) Filler size is 5%, attack size varies in bandwagon (average) attack. (b) Attack size is 17%, filler size varies in bandwagon (average) attack

TABLE IV. COMPARISON OF THE DETECTION PERFORMANCE OF OUR METHOD WITH TWO BENCHMARKED METHODS

Attack models	Methods	Rating							
		1		*3*		*5*		*7*	
		DR	***FAR***	***DR***	***FAR***	***DR***	***FAR***	***DR***	***FAR***
AOP	HHT-SVM	0.845	0.095	0.819	0.15	0.79	0.177	0.673	0.21
	DeR-TIA	1.0	0.005	0.715	0.185	0.734	0.225	0.707	0.275
	Ours	0.911	0.0785	**0.835**	**0.093**	**0.813**	**0.102**	**0.702**	**0.11**
Random	HHT-SVM	0.819	0.12	0.765	0.15	0.7345	0.14	0.68	0.21
	DeR-TIA	1.0	0.0025	0.735	0.175	0.727	0.195	0.731	0.265
	Ours	0.904	0.081	**0.834**	**0.086**	**0.801**	**0.093**	**0.707**	**0.11**
Average	HHT-SVM	0.873	0.1091	0.782	0.13	0.759	0.158	0.665	0.182
	DeR-TIA	1.0	0.0025	0.763	0.165	0.750	0.205	0.752	0.195
	Ours	0.907	0.085	**0.837**	**0.090**	**0.805**	**0.079**	**0.703**	**0.125**
Bandwagon (average)	HHT-SVM	0.906	0.09	0.8279	0.14	0.7869	0.16	0.675	0.19
	DeR-TIA	1.0	0.005	0.755	0.18	0.734	0.25	0.752	0.285
	Ours	0.935	0.0615	**0.852**	**0.0713**	**0.823**	**0.0682**	**0.705**	**0.115**
Bandwagon (random)	HHT-SVM	0.910	0.095	0.8179	0.13	0.8069	0.18	0.67	0.21
	DeR-TIA	1.0	0.005	0.747	0.165	0.735	0.205	0.750	0.27
	Ours	0.934	0.055	**0.868**	**0.075**	**0.83**	**0.069**	**0.718**	**0.115**
Segment	HHT-SVM	0.897	0.0891	0.819	0.13	0.7869	0.167	0.667	0.193
	DeR-TIA	1.0	0.0055	0.752	0.15	0.730	0.185	0.731	0.25
	Ours	0.915	0.075	**0.846**	**0.08**	**0.815**	**0.086**	**0.70**	**0.11**
Reveres bandwagon	HHT-SVM	0.895	0.087	0.8179	0.125	0.796	0.145	0.66	0.195
	DeR-TIA	1.0	0.005	0.739	0.175	0.754	0.185	0.727	0.26
	Ours	0.933	0.065	**0.868**	**0.075**	**0.815**	**0.0775**	**0.705**	**0.125**
Love/Hate	HHT-SVM	0.849	0.105	0.807	0.135	0.7569	0.175	0.67	0.205
	DeR-TIA	1.0	0.0025	0.752	0.16	0.727	0.195	0.750	0.24
	Ours	0.917	0.075	**0.845**	**0.065**	**0.81**	**0.0785**	**0.717**	**0.135**

V. EXPERIMENTS AND ANALYSIS

In this section, we firstly show the experimental data and settings on a real-world dataset. Then, we discuss our experimental results.

A. Experimental data and settings

In our experiments, we use the Book-Crossing [4] dataset. It contains 278,858 users providing 1,149,780 ratings (explicit or implicit) about 271,379books and each rater had to rate at least 1 books. All ratings are in the form of integral values between minimum value 1 and maximum value 10. The minimum score means the rater dislikes the book, while the maximum score means the rater enjoyed the book. We randomly select 800 genuine profiles from the dataset as the samples of genuine profiles. For the attack profiles, we just focus on nuke attacks and their grey attacks, push attacks can be detected in the analogous manner. For each attack model (as shown in Table 2), we respectively generate nuke and grey attack profiles according to the corresponding attack models with diverse attack sizes [5] {3%, 7%, 12%, 17%, 22%, 27%, 32%, 37%, 42%, 47%} and filler sizes [6] {1%, 1.7%, 2.5%, 5%, 6.7%, 8%, 10%}. In addition, to ensure the rationality of the results, the target item is randomly selected for these attack profiles. Especially in Table 2, the r_{grey} is the grey rating on target items rated by lower scores such as 1, 3, 5 and 7.

The generated attack profiles are respectively inserted into the sampled genuine profiles to construct our test datasets. Therefore, we have 560 (8 × 10 × 7) test datasets including 8 diverse attack models, 10 different attack sizes and 7 different filler sizes. Notice that, these process is repeated 10 times and the average value of detection results are reported for the experiments. All numerical studies are implemented using MATLAB R2012a on a personal computer with Intel(R) Core(TM) i7-4790 3.60GHz CPU, 16G memory and Microsoft windows 7 operating system.

To measure detection performance of the proposed methods, we use detection rate and false alarm rate in our experiments.

$$\text{detection rate} = \frac{|D \cap A|}{|A|} \quad (11)$$

$$\text{false alarm rate} = \frac{|D \cap G|}{|G|} \quad (12)$$

where D is the set of the detected user profiles, A is the set of attacker profiles, and G is the set of genuine user profiles [11].

B. Experimental results and analysis

To validate the detection performance of our proposed method, we employ two benchmarked methods including HHT-SVM [17] and DeR-TIA [1] to demonstrate the outperformance of our method. Take bandwagon (random) attack for example, Figures 11 and 12 demonstrate how each method performs under varying attack sizes and filler sizes, respectively. In the bandwagon (random) attack, a group isolated attackers always provide maximal or minimal or grey rating on a set of items when they are selected as the selected items or the filler items. As shown in Figures 11(a) and 12(a), the detection rate increased gradually and false alarm rate decreased gradually when the attack size increased and the filler size is fixed with 5% (in Figure 11 (a)) and filler size increased and attack size is 17% (in Figure 12 (a)). In addition, we can observe that our method shows significantly better

[4] http://www.informatik.uni-freiburg.de/~cziegler/BX/

[5] The ratio between the number of attackers and genuine users.

[6] The ratio between the number of items rated by user u and the number of entire items in the recommender systems.

detection performance than HHT-SVM with the attack size increased. This might be attributed to the combination of novelty-based, popularity-based and rating deviation-based rating series adopted by our proposed algorithm. The rating deviation-based strategy calculates a rating offset on a target item which can identify between the genuine profiles and attack profiles. The second observation is that DeR-TIA shows the best performance among the three algorithms. With the attack size increasing, the detection rate almost keeps maximum 100% and the false alarm rate almost keeps minimum 0, except for the early stages (attack size < 17%) as illustrated in Figure 11 (a). The same observations are also clear in Figure 12(a). However, for grey rating, as shown in Figures 11 (b) and 12 (b), we set a grey rating equal to 3 (integer rating from 1-10 in the datasets). Our method shows the best detection performance among the three methods, although the detection rate of our method shows lower than DeR-TIA in the early stage (attack size < 12%) as illustrated in Figure 11 (b). To compare with our proposed method and HHT-SVM, DeR-TIA shows higher false alarm rate than the others. Moreover, the detection rate of DeR-TIA almost remained unchanged with the attack size increased, and similar results can be observed in Figure 12 (b). The results might be attributed to grey rating. The first phase of DeR-TIA can filter out a part of genuine users by using similarity threshold, but it is difficult to capture the suspected profiles which rate grey ratings in their second phase. They defend and remove the suspected users almost depend on the similarity threshold, so they perform lower detection performance. For our proposed method, we pay more attention to the details of the all ratings rated by a user and explore the top-N items which has sorted by the rating deviation of item in order to characterize the grey ratings.

To examine the detection performance of our method in bandwagon (random) attack with different grey ratings (take bandwagon (random) attack for example), we conduct a list of experiments with diverse attack sizes and filler sizes. As shown in Figure 13, we perform 4 different grey ratings including 1, 3, 5 and 7 on the target items. One observation is that the detection rate gradually increased and false alarm rate gradually decreased with the attack size increasing (in Figure 13 (a)) or filler size increasing (in Figure 13 (b)). The other observation is that the detection performance gradually performs poor when the grey rating increased from 1 to 7, regardless of different attack sizes and filler sizes. The results may indicate that the grey ratings are close to average rating in the entire system with the grey rating on the target items increasing. The attackers rate a mean rating may show a rating behavior like genuine users, which is difficult to discriminate between attackers and genuine users and shows higher false alarm rate.

To further illustrate the detection performance of our proposed method under different attack models with different grey ratings, we conduct a list of experiments in 8 attack models for comparing the performance of our proposed method with HHT-SVM and DeR-TIA. We use 4 different ratings including 1, 3, 5 and 7 score when filler size is 5% and attack size is 17%. As shown in Table 4, we can observe that the detection rate (DR) of our method reports higher than other two benchmarked methods when the grey rating increasing, except for the grey rating is 1. Similarly, the false alarm rate (FAR) of our method reports lower than others. In addition, the second observation is that the proposed method reports better detection performance under bandwagon (both random and average) and reverse bandwagon attacks in comparison with the other attack models, especially for grey ratings (such as 3, 5 and 7 score). These results may indicate that we combine the rating deviation-based, novelty-based and popularity-based rating series in our method is useful to discriminate difference between grey attack profiles and genuine profiles. The rating deviation-based rating series may easily characterize the grey attacks in comparison with the other two methods.

VI. CONCLUSIONS AND FUTURE WORK

In this paper, we highlighted the challenges faced by the grey attacks, and then we develop an unsupervised detection approach based on discrete wavelet transform by combing the rating deviation-based, novelty-based and popularity-based rating series. Extensive experiments on the Book-Crossing dataset have demonstrated the effectiveness of the proposed approach. One of the limitations of our proposed method directly comes from the time consumption, which constructs the signals of rating series. In our future work, we intend to extend and improve grey attack detection in the following directions: 1) Considering more attack models such as Power users attack or Power items attack, etc.; 2) We will explore specific and simple method to detect grey attacks and develop better approach to construct the rating series. 3) Extracting more simpler and effective features to characterize grey attack profiles is still an open issue.

ACKNOWLEDGMENT

The research is supported by NFSC (61175039, 61221063), 863 High Tech Development Plan (2012AA011003), Research Fund for Doctoral Program of Higher Education of China (20090201120032), International Research Collaboration Project of Shaanxi Province (2013KW11) and Fundamental Research Funds for Central Universities (2012jdhz08).

REFERENCES

[1] W Zhou, Y. S. Koh, J. H. Wen, S Burki and G Dobbie. Detection of abnormal profiles on group attacks in recommender systems. Proceedings of the 37th international ACM SIGIR conference on Research & development in information retrieval, Pages 955-958, 2014.

[2] D. Jia, F. Zhang and S. Liu. A robust collaborative filtering recommendation algorithm based on multidimensional trust model. Journal of Software, vol. 8, no. 1, 2013.

[3] R. Burke, B. Mobasher and C. Williams. Classification features for attack detection in collaborative recommender systems. In Proceedings of the 12th International Conference on Knowledge Discovery and Data Mining, pages 17–20, 2006.

[4] B. Mobasher, R. Burke and J. Sandvig. Model-based collaborative filtering as a defense against profile injection attacks. AAAI. 1388, 2006.

[5] K. Bryan, M. O'Mahony and P. Cunningham. Unsupervised retrieval of attack profiles in collaborative recommender systems. In RecSys'08: Proceedings of the 2008 ACM conference on Recommender systems, pages 155–162 , 2008.

[6] H. Hurley, Z. Cheng and M. Zhang. Statistical attack detection. In: Proceedings of the Third ACM Conference on Recommender Systems (RecSys'09), pages 149–156 , 2009.

[7] B. Mehta. Unsupervised shilling detection for collaborative filtering. AAAI, 1402-1407, 2007.

[8] C Li and Z Luo. Detection of shilling attacks in collaborative filtering recommender systems. In: Proceedings of the international conference of soft computing and pattern recognition, Dalian, China, pages 190–193, 2011.

[9] I Gunes, C Kaleli, A Bilge and H Polat. Shilling attacks against recommender systems: A comprehensive survey. Artificial Intelligence Review, pages 1-33, 2012.

[10] N Giseop, Y. Kang and C. Kim. Ecsy-Recsy: Considering Sybil attack with time dynamics and economics in recommender system. International Conference on Information Networking (ICOIN), pages 566 - 571, 2013.

[11] C. Chung, P. Hsu and S. Huang. βP: A novel approach to filter out malicious rating profiles from recommender systems. Journal of Decision Support Systems, pages 314–325, April 2013.

[12] X. Zhang, T. Lee and G Pitsilis. Securing recommender systems against shilling attacks using social-based clustering. Journal of Computer Science and Technology (JCST), pages 616-624, July 2013.

[13] Z Zhang and S. Kulkarni. Graph-based detection of shilling attacks in recommender systems. IEEE International Workshop on Machine Learning for Signal Processing (MLSP), pages 1-6, 2013.

[14] B. Mehta, T. Hofmann and P. Fankhauser. Lies and propaganda: detecting spam users in collaborative filtering. In: IUI '07: Proceedings of the 12th International Conference on Intelligent User Interfaces, pages 14–21, 2007.

[15] M Morid and M Shajari. Defending recommender systems by influence analysis. Information Retrieval, pages 137-152, April 2014.

[16] Z. Wu, J Cao, B Mao and Y. Zhang. Semi-SAD: Applying semi-supervised learning to shilling attack detection. Proceedings of the 5th International Conference on Recommender Systems. New York: ACM, pages 289–292, 2011.

[17] F. Zhang and Q. Zhou. HHT–SVM: An online method for detecting profile injection attacks in collaborative recommender systems, Knowl. Based Syst. 2014.

[18] J Zou and F Fekri. A belief propagation approach for detecting shilling attacks in collaborative filtering. Proceedings of the 22nd ACM international conference on Conference on information & knowledge management (CIKM), pages 1837-1840, 2013.

[19] Z Zhang and SR Kulkarni, Detection of Shilling Attacks in Recommender Systems via Spectral Clustering. 2014 17th International Conference on Information Fusion (FUSION). Page(s):1-8, 7-10 July 2014.

[20] Fidel Cacheda, Victor Carneiro, Diego Fernandez and vreixo Formoso. Comparison of Collaborative Filtering Algorithms: Limitations of Current Techniques and Proposals for Scalable, High-Performance Recommender Systems. ACM Transactions on the Web (TWEB), Volume 5, Issue 1, February 2011.

[21] B. Mobasher, R. Burke, B. Bhaumil and C. Williams. Towards trustworthy recommender systems: an analysis of attack models and algorithm robustness. ACM Transactions on Internet Technology, 7 (4), pages 23–38, 2007.

[22] C. E. Seminario and D. C. Wilson. Attacking item-based recommender systems with power items. RecSys'14, October 6-10, 2014.

[23] M. J. Shensa, Wedding the a trous and Mallat algorithms, IEEE Trans. Signal Process. 40 (1992), 24642482.

[24] Williams, C., Mobasher, B., Burke, R., Sandvig, J., Bhaumik, R. Detection of obfuscated attacks in collaborative recommender systems. In: Workshop on Recommender Systems, ECAI, 2006.

[25] J. S. Lee, D. Zhu, Shilling attack detection: a new approach for a trustworthy recommender system, JNFORMS J. Comput. 24 (1) , pages 117–131, 2011.

[26] B. Mehta, W. Nejdl, Unsupervised strategies for shilling detection and robust collaborative filtering, User Model. User-Adap. Inter. 19 (1–2), pages 65–79, 2009.

[27] Mohamed Hamdi, Noureddine Boudriga. Detecting denial-of-service attacks using the wavelet transform. Computer Communications, 30 (16) (2007), pp. 3203–3213.

[28] C.A. Williams, B. Mobasher, R. Burke, R. Bhaumik, Detecting profile injection attacks in collaborative filtering: a classification-based approach, in: Proceedings of the 8th Knowledge Discovery on the Web International Conference on Advances in Web Mining and Web Usage Analysis (Lecture Notes in Computer Science), Springer-Verlag, 2007, pp. 167–186.

[29] B. Mobasher, R. Burke, R. Bhaumik, and C. Williams, "Toward trustworthy recommender systems: An analysis of attack models and algorithm robustness," ACM Transactions on Internet Technology (TOIT), Volume 7 , Issue 4 (October 2007), 2007.

[30] Z. A. Wu, J. J. Wu, J. Cao, D. C. Tao, HySAD: a semi-supervised hybrid shilling attack detector for trustworthy product recommendation, in: 18th ACM SIGKDD Conference on Knowledge Discovery and Data Mining, Beijing, China, August, 2012, pp. 985–993.

[31] F. Zhang, Q. Zhou, A meta-learning-based approach for detecting profile injection attacks in collaborative recommender systems, J. Comput. 7 (1) (2012) 226-234.

[32] F. He, X.Wang, B. Liu, Attack detection by rough set theory in recommendation system, in: Proceedings of 2010 IEEE International Conference on Granular Computing, 2010, pp. 692-695.

[33] W. Zhou, J. Wen, Y. S. Koh, Q. Xiong, M. Gao, G. Dobbie, and S. Alam. Shilling attacks detection in recommender systems based on target item analysis. PloS one, 2015.

32

An Arabic Natural Language Interface System for a Database of the Holy Quran

Khaled Nasser ElSayed
Computer Science Department, Umm AlQura University

Abstract—**In the time being, the need for searching in the words, objects, subjects, and statistics of words and parts of the Holy Quran has grown rapidly concurrently with the grow of number of Moslems and the huge usage of smart mobiles, tablets and lab tops. Because, databases are used almost in all activities of our life, some DBs have been built to store information about words and surah of Quran. The need for accessing Quran DBs became very important and wide uses, which could be done through database applications or using SQL commands, directly from database site or indirectly by a special format through LAN or even through the WEB. Most of peoples are not experienced in SQL language, but they need to build SQL commands for their retrievals. The proposed system will translate their natural Arabic requests such as questions or imperative sentences into SQL commands to retrieve answers from a Quran DB. It will perform parsing and little morphological processes according to a sub set of Arabic context-free grammar rules to work as an interface layer between users and Database.**

Keywords—Natural Language Processing (NLP); Arabic Question Answering System; Morphology; Arabic Grammar; Database; SQL

I. INTRODUCTION

Language obeys regularities and exhibits useful properties at a number of somewhat separable "levels". Suppose that a database user has some requests that he wishes to convey to database. His requests impose linearity on the signal. All you can play with is the properties of a sequence of tokens. A meaning gets encoded as a sequence of tokens, each of which has some set of distinguishable properties, and is then interpreted by figuring out what meaning corresponds to those tokens in that order.

The properties of the tokens and their sequence somehow "elicit" an understanding of the meaning. Language is a set of resources to enable us to share meanings, but isn't best thought of as a means for *encoding* meanings. This is a sort of philosophical issue perhaps, but if this point of view is true, it makes much of the AI approach to NLP somewhat suspect, as it is really based on the "encoded meanings" view of language.

The expression "natural" language refers to the spoken languages, such as English, Arabic, and French as opposed to artificial languages like languages of programming. NLP systems are programs perform some processes on natural language in some way or another .NLP is considered as one of the most important subfields of AI. It draws on techniques of logical and probabilistic knowledge representation and reasoning, as well as on ideas from philosophy and linguistics. It requires an empirical investigation of actual human behavior, so it is complex and interesting [1].

The main function of NLP is to extract information from the natural input sentences with no care of method of inputting sentences to the computer. It could be used in many applications like: User interfaces (just tell the computer what to do in a textual interface), Knowledge-Acquisition (programs that could read books and manuals or the newspaper, with no need to explicitly encode all of the knowledge), Information Retrieval (find articles about a given topic and to determine whether the articles match a given query), and Translation (machines could automatically translate from one language to another) [2].

Because most of persons have no knowledge of database language, they find it difficult to access database. Recently, there is a rising demand for non-expert users to query relational database in a more natural language. Therefore the idea of using natural language instead of SQL triggered the development of new method of processing named: Natural Language Interface to Database (NLIDB) [3]. The advantages of NLIDB over formal query language and form based interfaces are ; No need to know the physical data structure, No need to learn AI, and Easy to use. In the other side, the disadvantages of NLIDB are; Difficult to decide success or failure of a query, Limited dealing with natural language, and Wrong assumption by users [4].

Mobile applications of the Arabic language are going to grow in the time being and near future. There is an increasing of the need of enriching the Arabic digital content. Almost, there is no study have had its focus on identifying the challenging aspects of developing mobile applications in mobile applications in Arabic. Many studies emphasized that there is a need of considering the identified challenges by interaction designers, developers, and other stakeholders in the early stages of the software life cycle [5]. Arabic Language understanding is an important field of AI. This field can be used to build an intelligent system for translating the natural Arabic request to SQL commands.

The proposed system performs parsing and interpreting of the natural Arabic input such as a question or an imperative sentence. It applied morphology and context-free parsing techniques on context-free grammar of Arabic Language. Then, the system produces an SQL command, which could retrieve the suitable answer from the database of Quran statistics. It uses an approach that lets the computer accepts natural language sentences, but extract only the essential information from that command. Also, it enables users to learn how to build their SQL commands.

II. Related Work

Daoud introduced in [6] a SMS system named CATS, for posting and searching through free Arabic text using a technology of information extraction. This system can handle structured data stored in relational database and unstructured free Arabic SMS text. He used Arabic interaction language between sellers and buyers through SMS in a classified domain.

Al-Johar and McGregor proposed in [7] developing an Arabic natural language interface to database systems in prolog. They used the approach of intermediate meaning representation in building LMRA notation as a representative for this approach for the Arabic language. This notation divides common nouns into two classes: A mammal common noun (more than one possible gender), and a non-mammal common noun (one possible gender). It has logical formulas to represent a number of Arabic words and phrases.

Mohammad, Nasser, and Harb produced in [8] a Knowledge Based Arabic Question Answering System (AQAS) in prolog. Their system has a knowledge base of a radiation diseases domain. It divided the Arabic query into two parts: the required part (the information requested) and the known part (the thing asked about). Its parser converts the input query into internal meaning representation (IMR), and then it is processed to locate and retrieve the answer for the user. Its IMR is looking for certain words in the query to specify the required information about certain thing.

El-Mouadib, Zubi, Almagrous, and El-Feghi introduced in [9] the design and implementation of an English natural language interface to a database system. Its name is Generic Interactive Natural Language Interface to Databases (GINLIDB). It has two types of semantic grammars parser to supports a wide range of natural language statements: The first is a single lexicon semantic grammar which consists of individual words and some of their synonyms that are used in the English language grammar. While the second is a composite lexicon semantic grammar which is a combination of terminal words (terminals that exist only in the lexicon) that form phrases or sentences in a specific structure. It is designed using of UML and developed using Visual Basic.

Kanaan, Hammouri, Al-Shalabi, and Swalha presented in [10] the architecture of a question answering system. Their system depends on data redundancy rather than complicated linguistic analyses of either questions or contender answers. So, it is different from the other similar system, because a wrong answer is often worse than no answer. It receives Arabic natural language questions, and then it attempts to generate short answers. They used an existing tagger to identify proper names and other crucial lexical items and build lexical entries. They provided an analysis of Arabic question forms and attempted to formulate better kinds of more appropriated answers.

Abu Shawar introduced in [11] a method to access Arabic Web Question Answering (QA) corpus using a chatbot. This method was used properly with English and other European languages. With this method, there is no need for sophisticated NLP or logical inference. Any NLP interface to QA system is constrained to reply with the given answers, so there is no need for NLP generation to recreate well-formed answers, or for deep analysis or logical inference to map user input questions onto this logical ontology. There is simple (but large) set of pattern-template matching rules. This paper used the same chatbot to react in terms of Arabic Web QA corpus.

III. System Structure

The system receives simple requests in natural Arabic language questions as inputs from users. It is responsible of generating a final SQL command and executing it to retrieve the available answer from the database of the holy Quran. To perform that, the input sentence passes through multiple processing operations. Figure 1 presents the structure of the system.

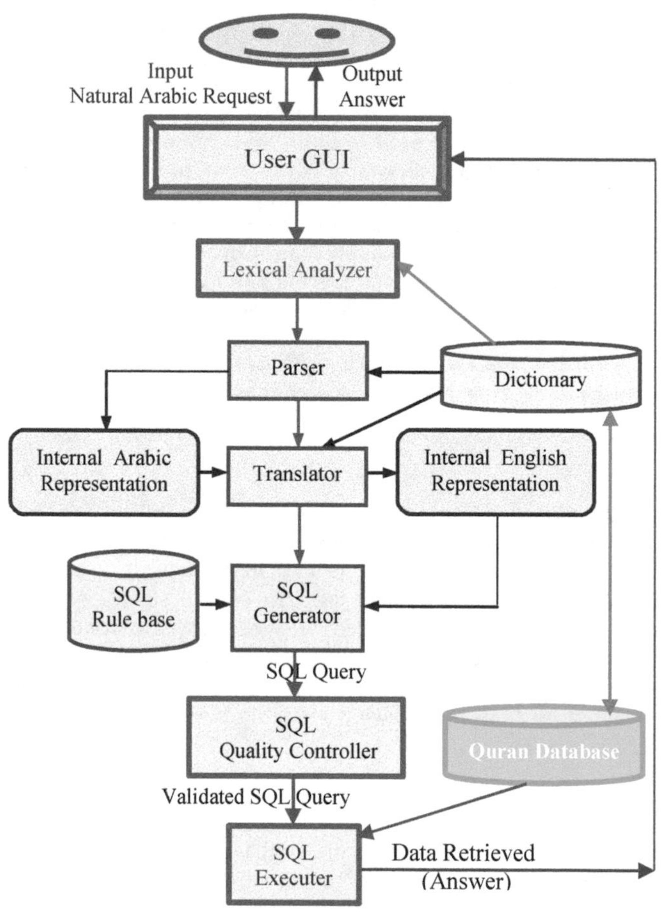

Fig. 1. Structure of the system

User GUI: the system provides an easy Graphical User Interface for easy **interaction** with its users. It is user friendly for none experts of computer or database and even for children. This is done through a menu driven and easy visual forms.

Lexical Analyzer : this stage includes performing four functions; splitting step (scan the user input character by character until recognizing a word to divide natural input into the lowest level of lexemes or words), spelling (checks the spelling of words using the dictionary, if word is not found correction is done or a new word is added to it), tokenizing (produces a token -the category or meaning – for each word), and abstracting (removes non important words that has no effect in the meaning of request and could be considered as noisy or excessive words).

Parser: The parser performs parsing according to context-free parsing techniques. The syntax of input sentences is represented in context-free grammar rules. It produces the internal Arabic representation of the input sentence according to the suitable grammar rule, using the dictionary. Its syntactic analysis is based on Augmented Transition Network (ATN), which checks if the structure of input tokens is allowed according to grammar rules.

Translator: Early research speculated that computers could be used for Machine Translation "Translation from one language to another". The translator in the proposed system uses the internal Arabic representation of the input sentence produced by the parser and uses the information stored in the dictionary to produce the corresponding internal English representation. It brings the English word corresponding to name of column or table in the applied database.

SQL Generator: The generator uses internal English representation produced by the translator and the SQL rule base to produce the SQL command. It uses the format of SQL rules stored in the SQL rule base as format or a frame and fill slots from the internal English representation.

SQL Quality Controller: The task of the SQL quality controller is to verify the generated SQL query. The query should be verified for valid names of tables, columns and format before applying to the Quran database.

SQL executor: The task of the SQL executor is to retrieve the suitable answer from the database of the holy Quran. Then, the system will consult the answer (the retrieved data) as output to the user.

IV. Database of the Holy Quran

The database used by the system was prepared to hold data about the Quran, to enable executing SQL queries retrieving its statistics data. Mainly, it keeps data about word(s), ayah(s), surah(s), and 30 Jozaa (chapter). Each Jozaa has with two Hezb (section), while each section has four quarters. Figure 2 present the Entity-Relationship diagram for the database of this system.

The entity DICTIONARY has the attributes: WordCode, WordNum, Word_Text, WordNumOfChar, Word_Meaning, Word_Root. While the entity AYAH has the attributes: AyahCode, AyahNum, Ayah_Text, AyaPageNo, and Ayah_meaning. The entity SORAH has the attributes: SurahCode, SurahNum, SurahName, Surah_Area, RevelationOrder, SurahPageNum, Quarter#, Hezb#, Jozaa#. Also there some attributes in the relationship between entities WORD and AYAH, like Word#InAyah.

Most of statistics of the words, ayah, and surah of the Holy Quran for the presented system is transferred from the database of TANZIL resources [12].

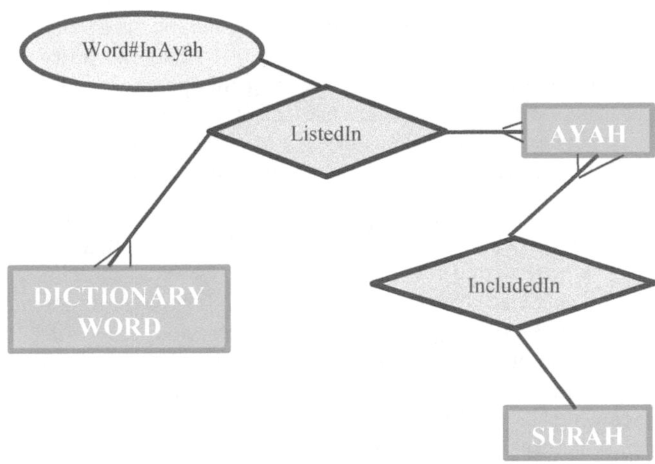

Fig. 2. Entity-Relationship Diagram of Quran Database

V. Parsing and Morpholgy

A. Parsing Process

Parsing (syntactic analysis) is the core of the proposed system where the input utterance is being checked to ensure that its syntax is correct and structured representations of the possible parses are generated. In parsing, a grammar is used to determine what sentences are legal [2].

Grammar is being applied using a parsing algorithm to produce a structure representation, or parse tree. Parser reads every input sentence, character by character, to decide what is what. It can recognize the underlying structure of a source text and checks that a token is a part of a legal pattern specified by the language grammar. It also gets some attributes of tokens from the dictionary [8].

Context-free grammar is used in parsing the Arabic request inputted to the proposed system, as will as in parsing of most of programming language because it has several advantages. It can deal with the word level and the phrase level. Also, it knows where it is in the sentence at all times. Its main disadvantage is that, it can't handle the numerous valid ways that a language can construct, due to the limitations of size and speed [8].

Context-free grammars are simply grammars consisting entirely of rules with a single symbol on the left-hand side of the BNF rules. The obvious advantage of BNF is that it is simple to define. Many of the grammars used for NLP systems are BNF, as such they have been widely studied and understood and hence highly efficient parsing mechanisms have been developed to apply them to their input.

The system uses context-free parsing technique. It begins by looking at the rules for the sentence, then it looks up the rules for the constituents of the sentence and progresses until a complete sentence structure is build up. If a sentence rule match the input sentence then the parsing process is ended, otherwise, the parser restarts again at the top level with the next rule. It performs syntax analysis recursively until firing certain rule structure or fail.

The parse tree breaks down the sentence into structured parts so that the computer can easily understand and process it. For the parsing algorithm to construct this parse tree, a set of BNF rules, which describes what tree structures are legal, must be available. These rules say that a certain symbol may be expanded in the tree by a sequence of other symbols.

Also, the system used noise disposal parsing for our application, because it is suitable for those application that concern only with a few keywords that a sentence contains, not with all associative words that make up a language. In essence, these types of applications are interested only in the information included in the sentence. This task is done by considering all unknown and un-required words as noise and discarding them. Simply, all sentences must follow a rigid format that resembles natural language.

Its main advantages are the easier implementation of extracting information from sentences, while it is not useful outside restricted situations such as the database queries. This is because it is based on two assumptions: the first is that the sentence follows a strict format, the second is that, only a few keywords or symbols are important. While in normal conversation, most words are important in some way or another.

B. Morphology and Dictionary

Morphology: NLP system doesn't always include morphological analysis. The alternative is to put all possible forms of each word into the dictionary, however storing all possible variants is inefficient and unnecessary. The terms: noun, verb, etc. are morphological but the nominal and verbal which are defined by the distribution of the forms in the sentences are syntactic analysis. Morphology process depends mainly on the dictionary entries and language grammar inflection [13]. This system will discard the prefix and postfix additional characters from a word. It is always one of those listed in Table 1.

Dictionary: a dictionary for NLP system contains the vocabularies known by that system. Its main function is to assist the parser in translating the input sentence into an internal meaning representation (IMR) to be processed.

Any word in the input sentence must be located in the dictionary, taking in consideration the necessary morphological process done by the system. It determines the capabilities of the system. The problem of the format and structure of the dictionary are closely related to the problems of the text storing. If the text is compressed to optimize storage size, the processing time is increased to compress and expand the data.

The format of each dictionary entry depends on the information stored in that entry. The most important data item within each entry is the morpheme itself called the head. Each entry has its appropriate information. The morphological algorithm is responsible of isolating the heads of the dictionary entries from the stream of the input words. Each entry contains the corresponding English meaning of each Arabic word stored in it. Actually, it has the English meaning of imperative verbs or interrogatives and the names of columns and tables of the applied database.

TABLE I. A LIST OF ADDITIONAL PREFIXES AND POSTFIXES

Addition Type		Examples and Meaning	
Pronoun	ضمير	هو, هي, هما, هم, هـ, هن	Its, his, her, their, he, she, they
Preposition	حرف جر	لـ, بـ, كـ,	for, with, as
Preposition and Pronoun	ضمير مع حرف جر	له, لها, لهما, لهم, لهن, به, ..	for him, with him, as him, ….
Definition character	أداة تعريف	ال	the
Preposition and Definition character	حرف جر مع أداة التعريف	لل, بال, كال	for the, with the, as the

VI. SQL QUERIES, INPUTS AND OUTPUTS

First of all, we should take in consideration the already exist SQL queries and their format. Then we should find out how to map between the expected inputs and the generated SQL queries, to finally generate the suitable answer.

A. SQL Queries

This system is run over MySQL database. So, it should generate complete SQL Query as usual in MySQL. Any SQL Query consists of SQL command beside names of attributes (columns) from certain tables and tables themselves and given values of some attributes as conditions if there is.

SQL commands are classified into two categories: Data Definition Language (DDL) commands like: CREATE, ALTER, DROP, DESCRIBE, etc. Data Manipulation Language (DML) commands like: SELECT UPDATE, DELETE, INSERT, etc. [14]. This system will generate SQL queries with the command SELECT only. Dependent on the natural input request, it translates the predicted request to this command. No way to generate another command.

B. Inputs and SQL Queries

The proposed system is expected to process the input sentence in the some of the following two modes [15]:

1) Imperative Mode: This sentence starts with imperative verb like:

استخرج الآيات التي بها كلمة الجنة

Retrieve the Ayah that include the word Paradise.

Table 2 shows list of some examples of imperative verbs beside their meanings and objects. As example, the first five imperative verbs (green color) are allowed from the user. But the last four imperative verbs (red color) are disallowed.

2) Question Mode: This sentence starts with interrogative question like:

ما اسم السورة التي تتضمن أحكام الصوم؟

Retrieve name of Sora include fasting rules?

Table 3 shows some examples of Questions beside their meanings and goals.

TABLE II. LIST OF SOME IMPERATIVE REQUESTS AND SQL QUERIES

Imperative Verb فعل أمر باللغة العربية	English meaning	Corresponding SQL Command
استخرج	Retrieve	SELECT
اعرض	Show	SELECT
أعطني	give me	SELECT
اذكر	List	SELECT
وضّح	illustrate	SELECT

3) Imperative and Question Mode: This sentence starts with imperative verb followed by interrogative question like:

وضح لي أين توجد آيات الحج؟

Tell me, Where is the Ayah of Pilgrim?

Similarly, the user can mix any of the allowed imperative verbs, like shown in table 2 with question like those listed in table 3.

TABLE III. LIST OF SOME QUESTIONS AND CORRESPONDING SQL

Interrogative	English meaning	Corresponding SQL Command
ما	Which	SELECT
ماذا	What	SELECT
من	Who	SELECT
كم	how many/much	SELECT
أين	Where	SELECT

C. *The Output Answer*

It is predicted the output retrieved from the database of the holy Quran, to be statistics about word (s), Sora (s), and subject(s).

VII. PROCESSING ARABIC REQUEST

All Arabic requests received from users have a common part and parcel. This common part contains all necessary information needed to build an output SQL command, and was given the name REQUEST. The REQUEST part has several forms. Each form represents certain one of the three expected input requests listed above, and were represented by some BNF grammar rules.

The REQUEST part might consists mainly from four components beside some noise data. The first part is the mode itself which included explicitly with the imperative or interrogative or declarative sentence. The second part is the TABLES component, which consists of SQL table names. The third part is the REQUIRED component, which has the names of the items of information actually needed to be retrieved (needed values of attributes of database table). The fourth part is the CONDITION component, which implies the condition to be applied on certain SQL query.

The main task of the SQL generator is to map the elements of the natural query to the elements of the SQL commands of the used databases. There is a general SQL command generated for all queries which is SELECT. So, the SQL generator should finds out the columns (attributes) to be in front of the SELECT command, table(s) to be in front of FROM clause, conditions for WHERE clause and a the method or displaying resulted data in certain order if needed.

VIII. CONCLUSIONS & FUTURE WORK

The presented system satisfies the need for accessing Quran DBs through LAN or WEB for all users, especially with no knowledge of database. It could accept natural Arabic requests such as imperative statements or questions. Then, it generated the suitable SQL command to be verified and executed. Finally it presents the answer from a database of Quran data to the user in an easy manner.

It performed parsing and little morphological processes according to a sub set of Arabic context-free grammar rules to work as an interface layer between users and Database.

In future, the database will be extended to include more tables and attributes. Also, the system will be extended to accept more complex search request and link answer with explanation of meaning of surah and ayah of the holy Quran.

REFERENCES

[1] S. J. Russell and P. Norvig, "Artificial Intelligence – A Modern Approach", 3th edition, Pearson, 2010.

[2] E. V. Provider, "Talking with Computer in Natural Language", Springer-Verlog, NY, 1986.

[3] M. Tyagi, " Natural Language Interface to Databases: A Survey", International Journal of Science and Research (IJSR), Volume 3 Issue 5, pp. 1443-1445, May, 2014. http://www.ijsr.net/archive/v3i5/MDIwMTMyMTU3.pdf

[4] A. Popescu, O. Etzioni, H. Kautz, "Towards a theory of natural language interfaces to databases", In: 8th Intl. Conf. on Intelligent User Interfaces, Miami, FL, pp. 149–157, USA, 2003. http://dl.acm.org/citation.cfm?id=604070

[5] S. N. Zawati and M. A. Muhanna, "Arabic mobile applications: Challenges of interaction design and development", 2014 International Conference (IWCMC) of Wireless Communications and Mobile Computing, Nicosia, pp 134-139, 2-8 Aug 2014. http://ieeexplore.ieee.org/xpl/login.jsp?tp=&arnumber=6906345&url

[6] D. Daoud, "Building an Arabic Application employing information extraction technology", In Proceedings of the Second International Conference on Information Technology (ICIT05), Amman, Jordan, pp. 1-9., 2005. http://icit.zuj.edu.jo/icit05/2005/Information%20Systems/193.pdf

[7] B. Al-Johar and J. McGregor, "A Logical Meaning Representation for Arabic Representation (LMRA)", . In: Proceedings of the 15th National Computer Conference, Riyadh, Saudi Arabia, pp 31-40, 1997. http://www.ccse.kfupm.edu.sa/~sadiq/proceedings/NCC1997/Pap24.doc

[8] F. Mohammad, Kh .Nasser and H. Harb "A Knowledge Based Arabic Question Answering System (AQAS)." In SIGART Bulletin, A Quarterly Publication of the ACM, Special Interest Group on Artificial Intelligence, Vol. 4, No. 4, October, 1993. dl.acm.org/citation.cfm?id=165488

[9] F. A. El-Mouadib, Z. S. Zubi, A. A. Almagrous, I. El-Feghi, "Interactive Natural Language Interface", Journal WSEAS TRANSACTIONS on COMPUTERS, Issue 4, Volume 8, pp 661-680, April 2009. http://dl.acm.org/citation.cfm?id=1558765

[10] G. Kanaan, A. Hammouri, R. Al-Shalabi, M. Swalha, "A New Question Answering System for the Arabic Language", American Journal of Applied Sciences 6: pp. 797-805, 2009. http://thescipub.com/PDF/ajassp.2009.797.805.pdf

[11] B. Abu Shawar, "A Chatbot as a Natural Web Interface to Arabic Web QA", iJET – Volume 6, Issue 1, pp. 37-43, March 2011. http://www.editlib.org/p/44956/article_44956.pdf

[12] Tanzil, "Tanzil Quran Navigator", 1/3/2015 http://tanzil.net/

[13] A. Walker, "Knowledge Systems and Prolog", IBM T.J. Watson Research Center, Addison-Wesley, 1997.

[14] R. Elmasri, and S. Navathe, "Fundamentals of Database Systems", 6th edition, Addison Wesley, 2010.

[15] F. Noama, "Summary of Arabic Grammar", Scientific Center for translation, Cairo, 1988.

33

Seamless Location Measuring System with Wifi Beacon Utilized and GPS Receiver based Systems in Both of Indoor and Outdoor Location Measurements

Kohei Arai 1
Graduate School of Science and Engineering
Saga University
Saga City, Japan

Abstract—**A seamless location measuring system with WiFi beacon utilized and GPS receiver based systems in both of indoor and outdoor location measurements is proposed. Through the experiments in both of indoor and outdoor, it is found that location measurement accuracy is around 2-3 meters for the locations which are designated in both of indoor and outdoor.**

Keywords—GPS receiver; WiFi beacon; seamless location estimation

I. Introduction

GPS based location estimation is quite popular. Mobile device, smart-phone, i-phone are utilizing GPS receivers for location estimation. It, however, does not work for indoor environments because of the fact that GPS satellite signals cannot be received in indoor environments. The location estimation accuracy of the GPS receiver based method is not so high. Furthermore, the accuracy of GPS based location estimation depends on many factors such as weather condition, the number of the acquired GPS satellites, circumstances (surrounding buildings, mountains, etc.), etc. On the other hands, WiFi-beacon can be used for location measurements in both indoor and outdoor environments. The location estimation accuracy depends on the situation of WiFi access points. Therefore, in general, WiFi beacon utilized location estimation accuracy is not good. It, however, still is somewhat useful for location estimation if the location accuracy requirement is not high. Therefore, WiFi beacon receiver based location estimation method is proposed here for both indoor and outdoor situations.

Bose and Heng classified WiFi-based positioning methods into Cell Identity (Cell-ID), Time of Arrival (TOA), Time Difference of Arrival (TDOA), Angle of Arrival (AOA), and signal strength categories [1]-[9]. Cell Identity (Cell-ID) is a basic wireless positioning system solution. It matches the target's position with its connection to an Access Point (AP). It does not require complex operations such as time synchronization and multiple APs. However, its low positional accuracy is the pitfall of its simplicity. Time of Arrival (TOA) measures a distance using the travel time of a radio signal from a transmitter to a receiver. Its application requires time synchronization of the transmitter and receiver, which is difficult to achieve for close ranges. To overcome the problem, Time Difference of Arrival (TDOA) was developed, which utilizes the time difference between receiver and two or more receivers. That is to say, whereas TOA requires time synchronization of transmitters and receivers, TDOA needs just synchronization between receivers. Angle of Arrival (AOA) determines the position of a receiver by measuring the angle to it from a transmitter. An AP must use smart antennas and be capable of mounting them under static conditions.

Signal Strength based technique uses the signal attenuation property of the radio wave Received Signal Strength Indication (RSSI) to measure the distance from a receiver to transmitter using the distance-to-signal-strength relationship. One common approach of RSSI-based system is fingerprint approach, which entails two phases: a training phase and a tracking phase. In the training phase, the received signal strength information is filtered, interpolated, and eventually stored in a database as sample points. In the tracking phase, the position is determined by comparison with the received signal strength sample points stored in the database [10]. The accuracy of this system is a function of the sample points' sampling space, an estimation method and the structure of the database. However, such a method requires the time consuming on survey procedure or calibration process.

In order to find patients in hospitals, victims in group homes, etc., GPS receivers and WiFi beacon receivers in smart-phone, i-phone and tablet terminals are used in the proposed system. The location of WiFi access points the designated hospitals and the supposed group homes are known. Also, the specific location of the hospitals and group homes, for instance, the middle of the entrance door, is known. Therefore, the locations of the patients and the victims are estimated in both of indoor and outdoor situations in seamless basis. When they are in hospitals or group homes, their locations are estimated with WiFi beacon receivers while their location is estimated with WiFi beacon receiver and GPS receivers when they are in outside of hospitals or group homes with an accuracy of a couple of meters[1]. WiFi beacon based location estimation is helpful to improve GPS based location accuracy[2]. Also, location estimation can be done with WiFi

[1] http://wiki.openstreetmap.org/wiki/Accuracy_of_GPS_data

[2] http://www.quora.com/Why-is-location-accuracy-improved-when-wi-fi-is-enabled

beacon receiver in smart-phone, i-phone, and tablet terminal[3]. The WiFi beacon based location estimation accuracy around a couple of meters in an indoor situation [11]-[12].

The location estimation method proposed here is to use both GPS based method with improvement by WiFi beacon based method (it is referred to GPS based method hereafter) in outdoor situations and WiFi beacon based method in indoor situations as well as a calibration of estimated locations at the specific location of hospitals or group homes. Because of the locations of specific positions of hospitals or group homes are known, a calibration can be done through a comparison of the estimated locations of the designated specific positions between GPS based and WiFi beacon based methods. Thus the locations of the patients in hospitals and / or victims in group homes are estimated in seamless basis. Through experiments, it is found that the proposed method does work for seamless location estimation with an acceptable accuracy, 2-3 meters for finding the patients and the victims when they are out of hospitals or group homes when a disaster occurs.

The next section describes the proposed seamless basis location estimation method followed by some experiments. Then conclusion is described together with some discussions.

II. The Proposed Method and System

The location estimation method proposed here is to use both GPS based method and WiFi beacon based method as well as a calibration of estimated locations at the specific location of hospitals or group homes. Because of the location of specific position of hospitals or group homes is known, a calibration can be done through a comparison of the estimated locations of the designated specific positions between GPS based and WiFi beacon based methods. Thus the locations of the patients in hospitals and / or victims in group homes are estimated in seamless basis.

Fig.1 shows flow chart of the proposed location estimation method. The location of smart-phone i-phone is estimated with GSP and WiFi beacon receivers equipped in the smart-phone and i-phone. The estimated locations are compared to the location of the designated known position in a prior to the location estimation such as the center of the entrance of the hospital or the group home. Then calibration of location estimation is done with the difference between the designated location and the estimated location with GPS and WiFi beacon. Thus the estimated location is calibrated. Repeat the calibration and location estimation is repeated.

The estimated locations are expressed with ISO 19155 of Place Identifier: PI. Example of the PI expression is shown below,

```
<allpi>
<placeidentifier>
<name>Kasasagi-Kaikan</name>
<latitude>33.24149339</latitude>
<longitude>130.28919659</longitude>
</placeidentifier>
<placeidentifier>
<name>Faculty Bldg. No.6</name>
<latitude>33.24139665</latitude>
<longitude>130.28864292</longitude>
</placeidentifier>
<placeidentifier>
<name>faculty Bldg. No.7</name>
<latitude>33.24118506</latitude>
<longitude>130.28842333</longitude>
</placeidentifier>
</allpi>
```

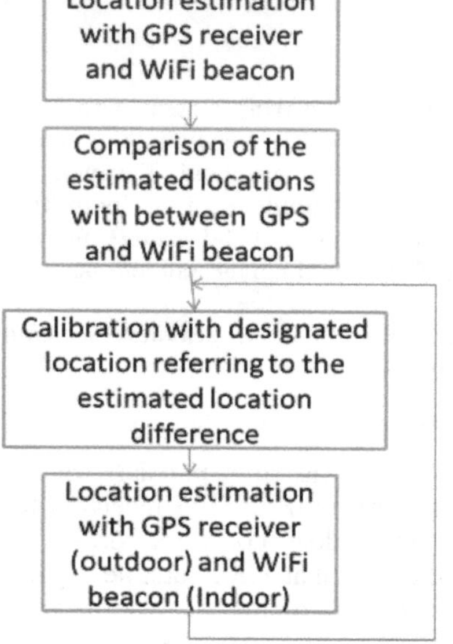

Fig. 1. Flow chart of the proposed location estimation method

The location name is located at the third line followed by latitude and longitude of the location. These name and latitude / longitude are aligned sequentially. PI expression is based on XML. Therefore, other information which is related to the location can be attached with a tag or tags. For instance, phone number, address, and the others. Therefore, these are referred crossly each other. Thus users can retrieve the location with the name, the phone number, the address, and the latitude / longitude.

III. Experiments

A. *Location estimation accuracy with GPS receiver based method (Improvement by WiFi beacon receiver based method) in outdoor situation*

Although location estimation accuracy of the GPS receivers is well reported, there are a few reports on location estimation accuracy of WiFi beacon receivers. Therefore, the following experiments are conducted at the road situated at the prefectural border between Fukuoka and Saga, Japan in night time. Test site on the map is shown in Fig.2 while the photo of the test site in day time is shown in Fig.3. At the test site on the Sazanka road, WiFi access points and mobile devices are set up as are shown in Fig.4.

In order to avoid external noises, location estimation accuracy is measured in night time. The distance between two

[3] http://engineeringblog.yelp.com/2012/08/gps-vs-wifi-the-battle-for-location-accuracy-using-yelp-check-ins.html

locations, A and B is 10 m. Distance from A and B is measured with 1m step. Received signal strength is varied with the distance between the location A and the location apart from A with 1m step. Therefore distance between both locations can be estimated.

Fig. 2. Test site location on topographic map

Fig. 3. Day time photo of the test site situated at the prefectural border between Fukuoka and Saga, Japan

Received signal strength in unit of dBm is shown in Table 1 while the estimated distance between both locations is shown in Table 2, respectively.

As the results of the experiments, it is found that location estimation error of the method of location estimation with WiFi beacon is around 5(%).

TABLE I. RECEIVED SIGNAL AT THE DIFFERENT LOCATIONS, A AND B

Distance(m)	Signal_Level_from_A(dBm)	Signal_Level_from_B(dBm)
0	-25	-58
1	-31.5	-57
2	-35.5	-56.5
3	-39	-55
4	-43	-52
5	-45	-50
6	-47;5	-47
7	-49	-45
8	-51.5	-48
9	-54	-38
10	-57.5	-27

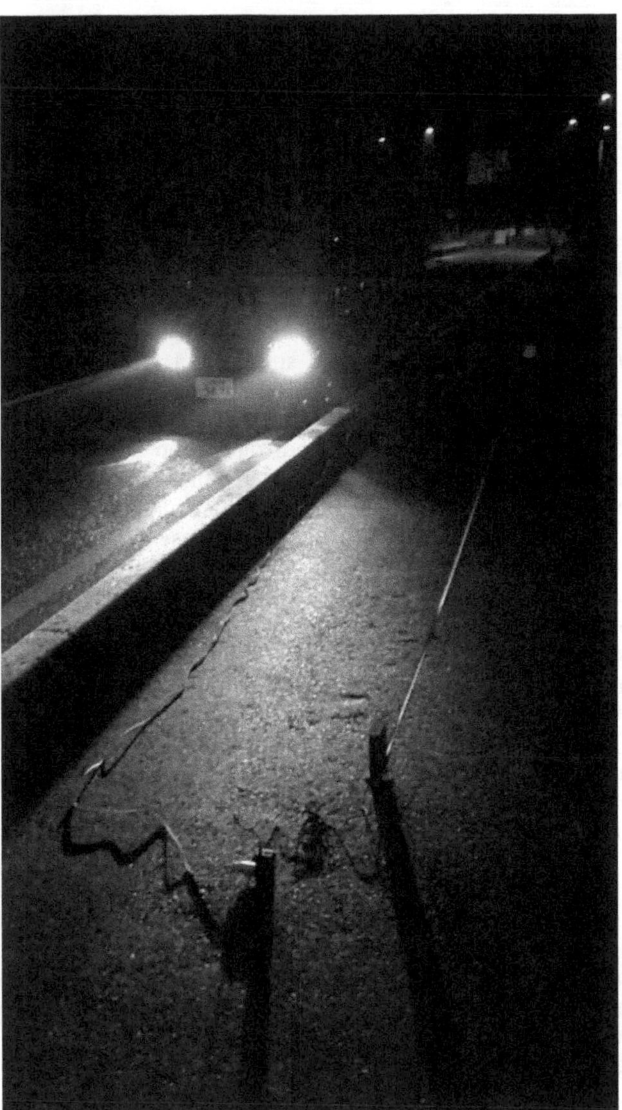

Fig. 4. WiFi access points and mobile device are set up on the Sazanka road of the test site

TABLE II. MEASURED DISTANCE ERRORS

Distance(m)	Calculated_Distance(m)	Error(%)
0	-0.28	28.0
1	1.06	5.66
2	1.89	-5.5
3	2.8	-6.67
4	4.15	3.61
5	5.0	0.0
6	6.16	2.6
7	6.84	-2.29
8	7.62	-4.99
9	8.89	-1.22
10	11.2	10.7

Location estimation accuracy of the GPS receiver based method is around 3.4 meters in average. Therefore, GPS receiver based method with the improvement by WiFi receiver based method achieved about 6 times much accurate location estimation.

B. Location estimation accuracy with WiFi beacon receiver based method in indoor situation

As described above, wireless channel signal strengths are changeful. Even immovable calibrating, the signal strengths still pulse up and down. Inconstant signal strengths on one position make estimating location difficult and inaccurate. Most research adopts mean value to solve this problem, but I think it's insufficient through inexact location estimating result. Using one mean value to stand for one position's wireless channel information is not enough. I present a Grid Segment Process to make some improvement. Assume one training process gains 100 signal strengths, I divide up these 100 signal strengths into 10 parts, each has 10 signal strengths. Then calculate mean value of each part and store them into radio map to substitute for original one mean value. Now I acquire 10 slices of mean values and have more information to estimate location. I call the divide procedure as Segment Process. Not only in offline calibration phase I do Segment Process, but in online estimation phase I execute it, as will introduce below. The mean value is used for expressing proper wireless channel characteristic. Using Segment Process to obtain more slices of mean values denotes more information to refer to and more accurate location estimation result. The received signal of each access point is converted into its color representative. This system use 3 signals information from 3 different AP's. Each access point has its basic color that different each other. The three AP (*AP1, AP2, & AP3*) use *red*, *green* and *blue* color respectively. The color map is based on signal strength information get from signal data collection process. The gradation of color is based on the *HSL (Hue Saturation Luminance)* value where *L* is a function of signal strength. Assuming S_{px} is a variable for signal strength (in percentage) for the position of x meter from the initial position. Then the color for the grid of x meter distance is measured by the formula:

$$L\text{-}Color\ AP1_x = 255 - (Hexadecimal(S_{px})*128/100)$$

where Hexadecimal(S) is function for convert the input into its hexadecimal value.

The gradation of color is based on the HSI (Hue Saturation Intensity) value where L is a function of signal strength.

The fission of the grid is half size and the color is based on the RGB combination of two adjacent colors. Then we will have a new grid in the more detail size as shown in Fig.5. The second stage of grid fusion is possible to create more detail color radio map grid. The accuracy of each color radio-map then investigates to measure the accuracy.

The effectiveness of this method is the interpolation using color grid fusion technique. Initial grid size could setup to a number that low cost on offline training. I start with grid size of 5 meters long with 2.5 widths (the corridor width). The initial color radio-map is created for this grid size that will have the initial error also 5 meters. Then too improve the accurate of the system, the initial map is interpolate in half size to determine the interpolation value (color) of middle point between two grids. Fig.6 shows the illustration of grid fusion technique. Also Fig.7 shows an example of location estimation result in the color radio-map representation.

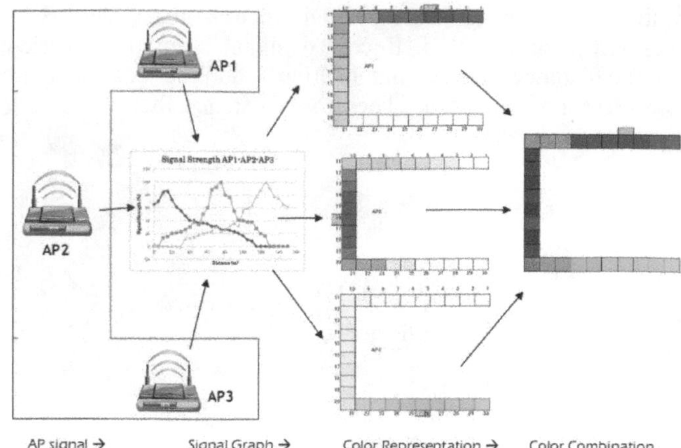

Fig. 5. Diagram flow of Color Radio Map Method

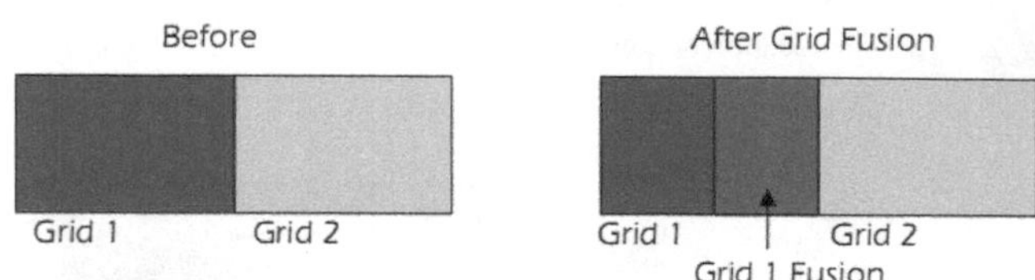

Fig. 6. Grid fusion techniques

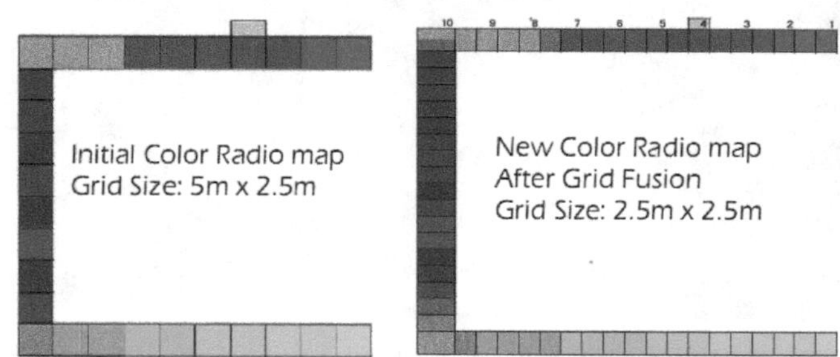

Fig. 7. Example of an estimated result

I performed our experiment in the third floor of the Science and Engineering Faculty Building No.1, Saga University. This building has a layout like show in Fig.8 with the total dimension in rectangle is 150 × 3 meters. The building is equipped with 802.11b wireless LAN environment. To form the radio map, the environment was modeled as a space of 30 locations in grid of 5 x 3 meters each. The position of AP and initial online tracking position is show in Fig.5.

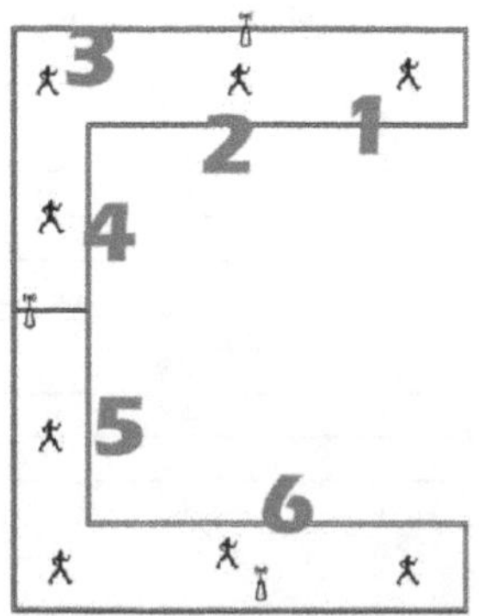

Fig. 8. Floor layout of the Bldg. in the experiment

As the result of the experiment, it is found that the location estimation accuracy is approximately 2.5m.

C. Calibration of estimated location

Application software is developed which allows location estimation with GPS and WiFi beacon receivers. The estimated location is expressed with PI. Example of the estimated locations with GPS and WiFi beacon is shown in Fig.9. Not only latitude and longitude but also direction of the target location from the current position is displayed onto smart-phone and / or i-phone as shown in Fig.10. The direction can be calculated with the following equation.

$$\theta = \mathrm{atan}(2\beta/\alpha) \qquad (1)$$

where α and β are calculated with the estimated locations.

ISO19155: Place Identifier (PI)

Fig. 9. Example of the estimated locations displayed onto smart-phone or i-phone

Example of the estimated and actual location is shown in Fig.11. In comparison of the actual location and the estimated locations with GPS and WiFi receivers are quite clearly different each other with around 5 meters. Location estimation accuracies of GPS and WiFi beacon based methods depend on circumstances, surrounding building, in this case.

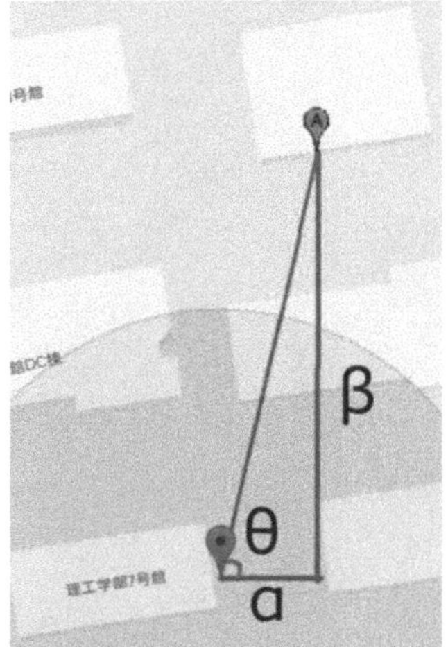

Fig. 10. Estimation of the direction of the target location from the current location

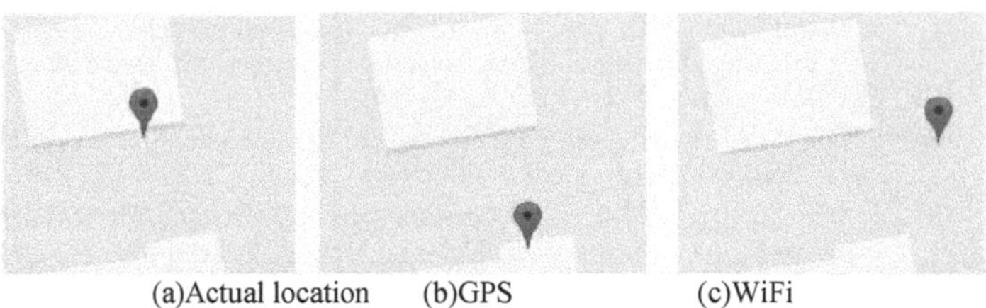

(a)Actual location (b)GPS (c)WiFi

Fig. 11. Actual and the estimated locations with GPS and WiFi beacon receivers equipped in smart-phone and i-phone

Using the difference between the actual location and the estimated locations with GPS and WiFi beacon receivers can be calibrated. Thus the estimated location is switched to GPS based method to WiFi beacon based method when the patients and the victims are getting into hospitals and or group homes from the outside and vise versa.

IV. CONCLUSION

A seamless location measuring system with WiFi beacon utilized and GPS receiver based systems in both of indoor and outdoor location measurements is proposed. Through the experiments in both of indoor and outdoor, it is found that location measurement accuracy is around 2-3 meters for the locations which are designated in both of indoor and outdoor. It is also found that GPS receiver based location estimation method with improvement by using WiFi beacon based method works well. Approximately 6 times better accuracy is achieved in comparison between GPS based method and the proposed GPS based method with improvement with WiFi beacon based method.

These GPS and WiFi beacon receivers are equipped in smart-phone and / or i-phone. Therefore, the locations of the patients in hospitals, the victims in group homes can be estimated with 2-3 meters of accuracy if they have the smart-phone and / or i-phone with the developed application software.

ACKNOWLEDGMENT

The authors would like to thank Mr. Kenji Egashira, Dr. Herman Tolle of Arai's laboratory members for their useful comments and suggestions during this research works.

REFERENCES

[1] P. Bahl, V.N. Padmanabhan, RADAR: an in-building RF-based user location and tracking system, INFOCOM 2000. 19th Annual Joint Conference of the IEEE Computer and Communications Societies, Proceedings IEEE 2 (2000) 775–784 vol.772.

[2] A. Bose, F. Chuan Heng, A Practical Path Loss Model for Indoor WiFi Positioning Enhancement, Information, Communications & Signal Processing, 2007 6th International Conference, 2007, pp. 1–5.

[3] Y. Chen, X. He, Contribution of pseudolite observations to GPS precise surveys, KSCE Journal of Civil Engineering 12 (1) (2008) 31–36.

[4] Y.K. Cho, J.-H. Youn, N. Pham, Performance tests for wireless real-time localization systems to improve mobile robot navigation in various indoor environments, in: C. Balaguer, M. Abderrahim (Eds.), Robotics and Automation in Construction, InTech Education and Publishing, 2008, pp. 355–372.

[5] Y. Fukuju,M. Minami, H.Morikawa, T. Aoyama, DOLPHIN: An Autonomous Indoor Positioning System in Ubiquitous Computing Environment, Proceedings of the IEEE Workshop on Software Technologies for Future Embedded Systems, IEEE Computer Society, 2003.

[6] S. Han, J. Kim, C.-H. Park, H.-C. Yoon, J. Heo, Optimal detection range of RFID tag for RFID-based positioning system using the k-NN algorithm, Sensors 9 (6) (2009) 4543–4558.

[7] C. Hu, W. Chen, Y. Chen, D. Liu, Adaptive Kalman filtering for vehicle navigation, Journal of Global Positioning Systems 2 (1) (2003) 6.

[8] J. Yim, C. Park, J. Joo, S. Jeong, Extended Kalman filter for wireless LAN based indoor positioning, Decision Support System 45 (4) (2008) 960–971.

[9] W.-S. Jang, M.J. Skibniewski, A wireless network system for automated tracking of construction materials on project sites, Journal of Civil Engineering and Management 14 (1) (2008) 9.

[10] H.M. Khoury, V.R. Kamat, Evaluation of position tracking technologies for user localization in indoor construction environments, Automation in Construction 18 (4) (2009) 444–457.

[11] Arai, K., Tolle Herman, Akihiro Serita, Mobile Devices Based 3D Image Display Depending on Users' Actions and Movements, International Journal of Advanced Research in Artificial Intelligence, 2, 6, 77-83, 2013.

[12] Kohei Arai, Herman Tolle, Color radio-map interpolation for efficient fingerprint WiFi-based indoor location estimation, International Journal of Advanced Research in Artificial Intelligence, 2, 3, 10-15, 2013

Hybrid Intelligent Approach for Predicting Product Compositions of a Distillation Column

Yousif Al-Dunainawi
Electronic and Computer Engineering Department
College of Engineering, Design and Physical Sciences
Brunel University London
Uxbridge, London, UK

Maysam F. Abbod
Electronic and Computer Engineering Department
College of Engineering, Design and Physical Sciences
Brunel University London
Uxbridge, London, UK

Abstract—**Compositions measurement is a vitally critical issue for the modelling and control of distillation process. The product compositions of distillation columns are traditionally measured using indirect techniques via inferring tray compositions from its temperature or by using an online analyser. These techniques were reported as inefficient and relatively slow methods. In this paper, an alternative procedure is presented to predict the compositions of a binary distillation column. Particle swarm optimisation based artificial neural network PSO-ANN is trained by different algorithms and tested by new unseen data to check the generality of the proposed method. Particle swarm optimisation is utilised, here, to choose the optimal topology of the network. The simulation results have indicated a reasonable accuracy of prediction with a minimal error between the predicted and simulated data of the column.**

Keywords—Hybrid Intelligence; Prediction; Distillation Column; Neural network; Particle swarm optimisation

I. INTRODUCTION

Over the past few decades, great developments in online analysis, monitoring and measurement of dynamic processes were made in various applications [1]–[5]. This development is partially motivated by the desire to improve quality. Unambiguously, quality is a significant indication that has a substantial impact on productivity and economy of manufacture, particularly in the field of mass production [2]. Direct measurement of the product compositions of the distillation column is a crucial issue. However, its disadvantages at stream process lie in difficulty, unreliability and high capital and operational cost. These disadvantages will have an exponentially negative effect when more than one analysis are needed to obtain a clearer picture of the different streams involved. Consequently, indirect and inferential measurement techniques are being used to design and run many distillation columns. These columns operate widely in chemical and petrochemical plants as well as refineries to separate mixtures into their individual components.

Not only non-linearity and transit behaviour make distillation, as a process, complicated to control, but the product compositions also cannot be promptly measured, nor reliable. The delay caused by the measurement and analysis of compositions will negatively affect the effectiveness and robustness of control [6]. An indirect method is proposed to monitor the products compositions of the column by using tray temperature inside the column, albeit this feature is an unreliable indicator of product compositions [7], [8]. Moreover, other considerations like consistent maintenance, regular calibrations, and high-cost equipment make composition analysers an ineffective solution for precise online measurements. Consequently, soft sensing or inferential systems have been proposed recently as practical options to replace hardware measurement systems [9].

Artificial neural networks (ANNs) is one of the most attention-grabbing branches of artificial intelligence, which has grown rapidly in the recent years as an optimal solution for the modelling, and prediction of dynamic systems. ANNs have shown outstanding performance to learn the input-output relationship of nonlinear and complex systems. This relationship could be easily, quickly and efficiently found out via reducing the error between the network output(s) and the actual output(s). After the network is trained, the output can be predicted within few seconds. ANN-based models are still being applied successfully to overcome engineering problems in different fields such as adaptive control, pattern recognition, robotics, image processing, medical diagnostics, fault detection, process monitoring, renewable and sustainable energy, laser applications and nonlinear system identification [10]–[17].

The most crucial task which faces the neural network constructer is the proper selection of the network topology to solve a particular problem. The topology means, here, the number of nodes (neurones) and the number of layers in the hidden zone. Therefore, one of the most efficient methods to determine the optimal network structure is evolutionary algorithm EA methods such as genetic algorithm GA [18] and practical swarm optimisation PSO [19] and so on.

This paper proposed a PSO-based neural network as a predictor model for estimating the product compositions of a binary distillation column.

II. DISTILLATION COLUMN MODELLING AND DESCRIPTION

Distillation is, undoubtedly, one of the most important processes in chemical and petrochemical plants. Distillation columns are used as separators of chemical compounds in petroleum, natural gas, liquid and chemical industries [20]. The major disadvantage of using those columns is that they are considered as an intensive energy process. A report from the US Department of Energy has indicated that distillation is the largest consumer of energy in the chemical industry; typically, it accounts 40% of the energy consumed by petrochemical plants. Despite its "thirst" for energy, distillation persists to be a widely utilised method for separations [21], [22].

Figure 1 is a schematic diagram of a binary distillation column, in which a feed mixture is separated into a distillate product (overhead) and a bottom product. Also, heat is transferred into the column via a reboiler (heat exchanger) to vaporise some of the liquid from the base of the column. The vapour travels up through trays inside the column to reach the top and, then, comes out to be liquefied in a condenser. Liquid from the condenser, at that point, drops into the reflux drum. Finally, the distillate is removed from this drum as a pure product. additionally, some liquid (reflux) is fed back near the top of the column while the impure product is produced at the bottom outlet.

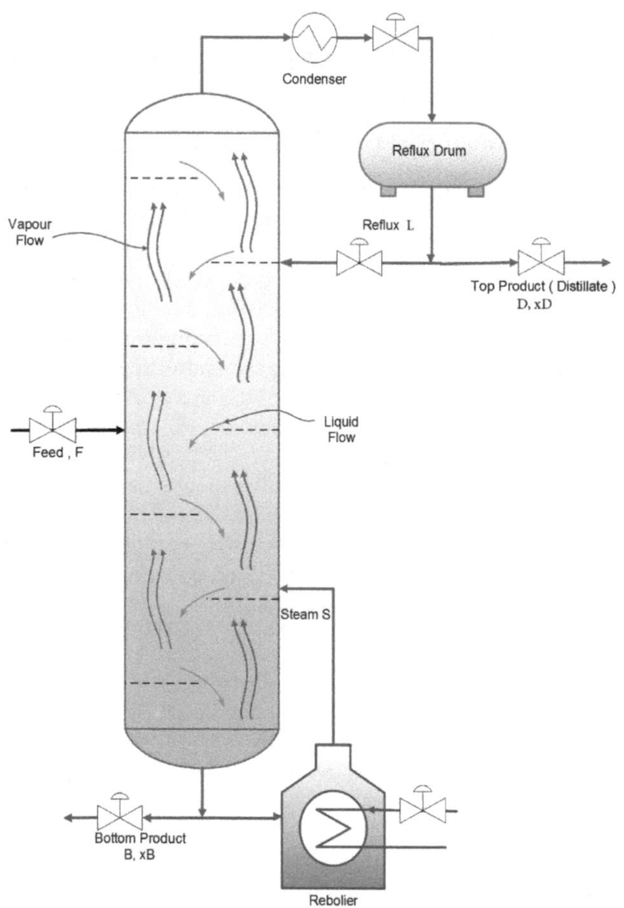

Fig. 1: Schematic diagram of a binary distillation column

The dynamic model can be simplified under the following assumptions:

- No chemical reactions occur in all stages of the column
- Constant pressure (open to atmosphere pressure)
- Binary mixture
- Constant relative volatility
- No vapour hold-up occurs
- Perfect mixing and equilibrium for vapour-liquid on all stages

While the operating conditions and technical aspects of the distillation column are detailed in the appendix at the end of this paper.

Accordingly, the mathematical expression of the model can be represented with the assumptions by the following equations: On each tray (excluding reboiler, feed and condenser stages):

- On each tray (excluding reboiler, feed and condenser stages):

$$M_i \frac{dx_i}{t} = L_{i+1}x_{i+1} + V_{i-1}y_{i-1} - L_i x_i - V_i y_i \quad (1)$$

- Above the feed stage $i = NF + 1$:

$$M_i \frac{dx_i}{t} = L_{i+1}x_{i+1} + V_{i-1}y_{i-1} - L_i x_i - V_i y_i + F_V y_F \quad (2)$$

- Below the feed stage $i = NF$:

$$M_i \frac{dx_i}{t} = L_{i+1}x_{i+1} + V_{i-1}y_{i-1} - L_i x_i - V_i y_i + F_L X_F \quad (3)$$

- In the reboiler and column base $i = 1$, $x_i = xB$:

$$M_B \frac{dx_i}{t} = L_{i+1}x_{i+1} - V_i y_i + Bx_B \quad (4)$$

- In the condenser, $i = N + 1$, $xD = xN + 1$:

$$M_D \frac{dx_i}{t} = V_{i-1}y_{i-1} - L_i x_D - Dx_D \quad (5)$$

- Vapour-liquid equilibrium relationship for each tray [23]:

$$y_i = \frac{\alpha x_i}{1 + (\alpha - 1)x_i} \quad (6)$$

- The flow rate at constant molar flow:

$$L_i = L, Vi = V + F_V \quad (7)$$

since

$$F_L = q_F \times F \quad (8)$$

$$F_v = F + F_L \quad (9)$$

- The flowrate of both condenser and reboiler as: Reboiler:

$$B = L + F_L - V \quad (10)$$

Condenser:

$$D = V + F_V - L \quad (11)$$

- The feed compositions xF and yF are found from the flash equation as:

$$F_{zF} = F_L \times x_F - F_V \times y_F \quad (12)$$

III. Hybrid Model Development and Optimisation

Recently, hybridization or combination of different learning and adaptation techniques has been employed to a large number of new intelligent system designs. The main aim of integrating these techniques is to overcome individual limitations and to achieve synergetic effects [24].

Therefore, a PSO-based artificial neural network is proposed as an estimator tool of a binary distillation column.

A. Artificial Neural Networks

Artificial neural network (ANNs) is a complicated system, which is composed of numerous neural nets. These nets fundamentally based on the principal understanding of the function, structure, and the mechanism of the brain of humankind [25]. In the last two decades or so, ANNs have been applied to a widespread range of applications due to their ability to analyse and capture the complexity and nonlinearity features of dynamic processes. One of the major applications of ANNs is a modelling or identification process of complex systems [26].

It is worth mentioning that the topology of the network is a crucial matter where as choosing the number of the neurones and layers in the hidden zone is not an easy task. So far, no systematic approach or automatic methods have been used to tackle this issue. Because the network structure depends on the nature and features of the process that would be modelled. Therefore, there are, probably, only two research methods to select from, a blind or heuristic. The blind approach, or trial and error, is an unguided and arbitrary search method, to which all possible alternatives are applied to find the optimal solution. Although this technique can eventually find the optimal ANN topology with limited search space, this method is not practical because it is considered highly expensive in terms of time and computations.

B. Particle Swarm Optimisation

Ever since particle swarm optimisation (PSO) has been proposed by Kennedy and Eberhart in 1995 [27] and 2001 [28], PSO algorithm turned to be vastly successful. The several of researchers have presented the merit of the implementation of PSO as an optimiser for various applications [29]. In PSO procedure, all individuals or particles (commonly between 10 and 100) are located at a random position and are supposed to move randomly in a defined direction in the search space. Each particle direction is then changed steadily to move assuredly along the direction of its best previous positions to discover even a new better position according to certain criteria or an objective function (fitness). The initial particle velocity and position are selected arbitrarily, and the following velocity equation can update them as

$$Vc_{i+1} = wV_i + C1R1 \times (Pb_i - x_i) + C2R2 \times (Gb - x_i) \quad (13)$$

Whereas the new particle is calculated by adding the previous one to the new velocity as shown in the following equation:

$$x_{i+1} = x_i + Vc_{i+1} \quad (14)$$

where: Vc: velocity of the particle, X: position of the particle, $R1, R2$: independent random variables uniformly distributed in [0, 1], $C1, C2$: acceleration coefficients as well as w: inertia weight. Eq. 13 is used to compute the new velocity of the particle according to its preceding velocity and the distances of its current position from its own best position (Pb) and the global best position (GB). Then, the particle moves to a new place in the search space, according to Eq. 14. The performance of each particle is measured according to a pre-defined objective function (performance index).

C. Hybrid System Design

Evolutionary-based optimisation, like PSO, can be applied by only simple mathematical operations with a few lines of code [30]. This feature provides a low-cost method concerning both memory and speed requirements. Thus, in this study, PSO is chosen to find the optimal network topology of the prediction model as depicted in Figure 2.

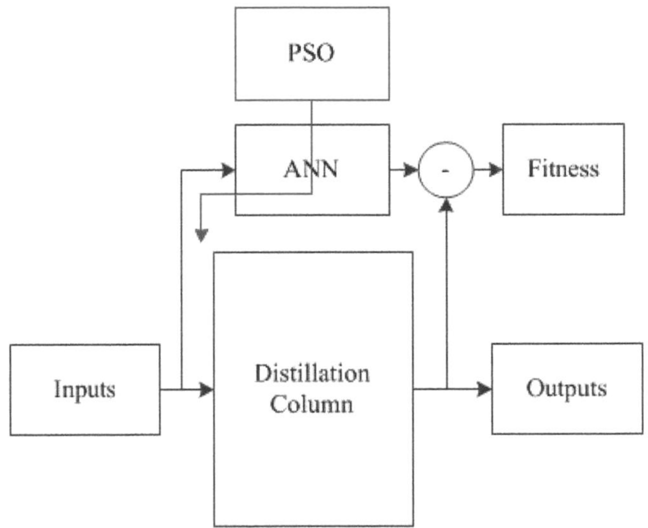

Fig. 2: Schematic diagram of the proposed hybrid system

IV. Simulation and Results

In this study, the reflux (L) and the boil-up (V) flow rates had been used as inputs to the network while distillate and bottom composition chosen as outputs. The dataset implemented for the training, validation and testing of ANN was generated by applying 40 distributed random values, each lasting 50 sampling time for (L) and (V) as shown in Figure 3. The distillate composition was approximately between 0.95 and 1 (mole fraction), while the bottom composition was around 0.005 to 0.12 (mole fraction) and a total of 2000 datasets were collected for identification, Figure 4 presents the simulated data of the column.

The dataset obtained by the simulation was randomly divided into 70%, 15% and 15% for training, validation and testing respectively. Feedforward multilayer network had been implemented to predict the product compositions of the distillation column. In addition, various backpropagation training algorithms, namely; Gradient Descent (GD), Scaled Conjugate Gradient (SCG) and LevenbergMarquardt (LM) were separately applied to decide which one performs better

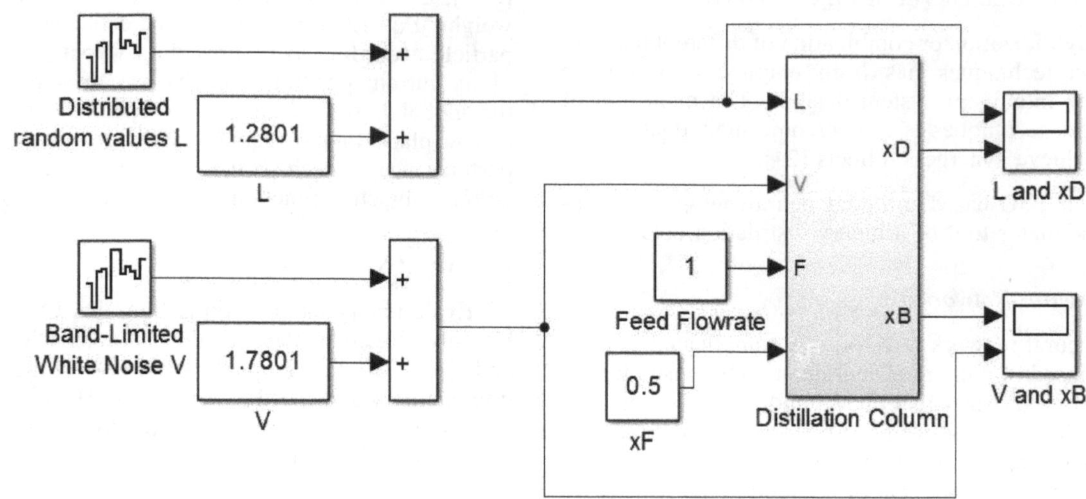

Fig. 3: MATLAB/Simulink model of the distillation column

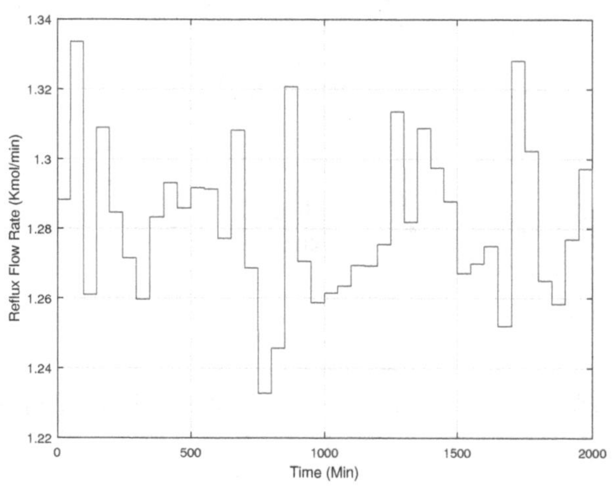

(a) Reflux folw rate

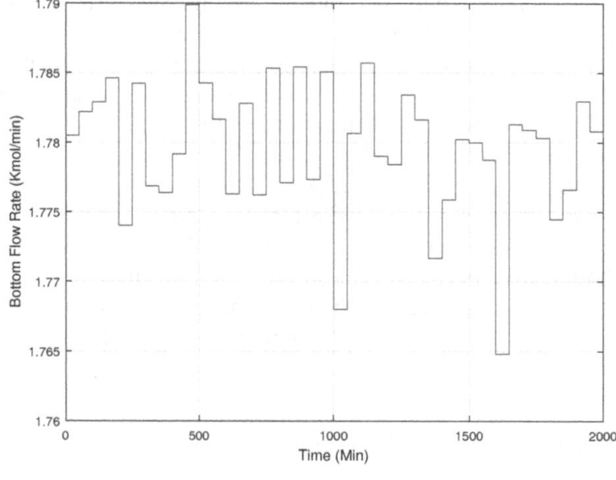

(c) Boil-up flow rate

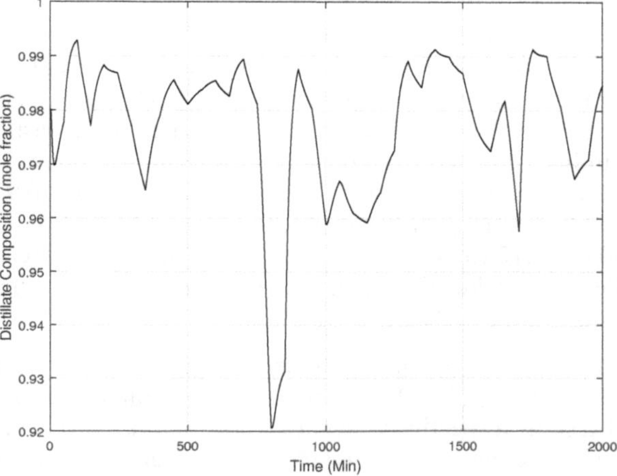

(b) Distillate composition

than the others. Moreover, Log-sigmoid activation function was embedded in the neurones of the hidden layer(s) because of its differentiability. PSO was employed to find the optimal structure of the network, the best operational parameters of PSO algorithm were chosen after extensive simulations and were set as following: For each of the network architecture,

No. of variables (dimensions)	2
Size of the swarm (no. of particles)	50
Maximum iterations (max)	100
Cognitive acceleration ($C1$)	1.2
Social acceleration ($C2$)	0.12
Momentum or inertia (w)	0.9
Minimum search space	1
Maximum search space	25

the training algorithms had been run ten times with different random initial weights and biases. After investigating the performance of different architectures using the PSO technique, a network with two hidden layers (including 23 neurons in the first and 25 in the second) trained by LM algorithm have

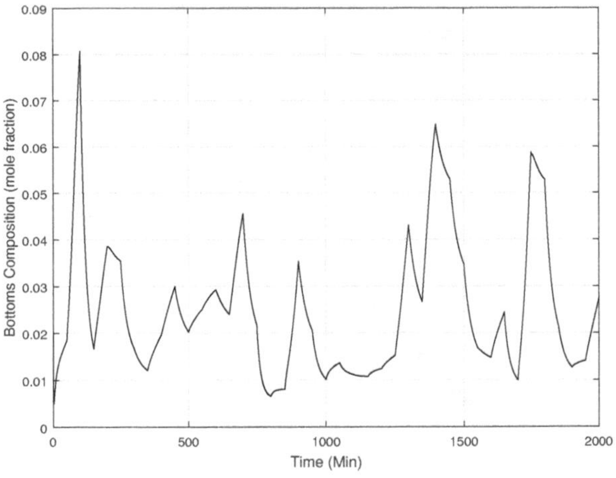

(d) Bottoms composition

Fig. 4: Inputs/outputs of the simulation of the distillation column

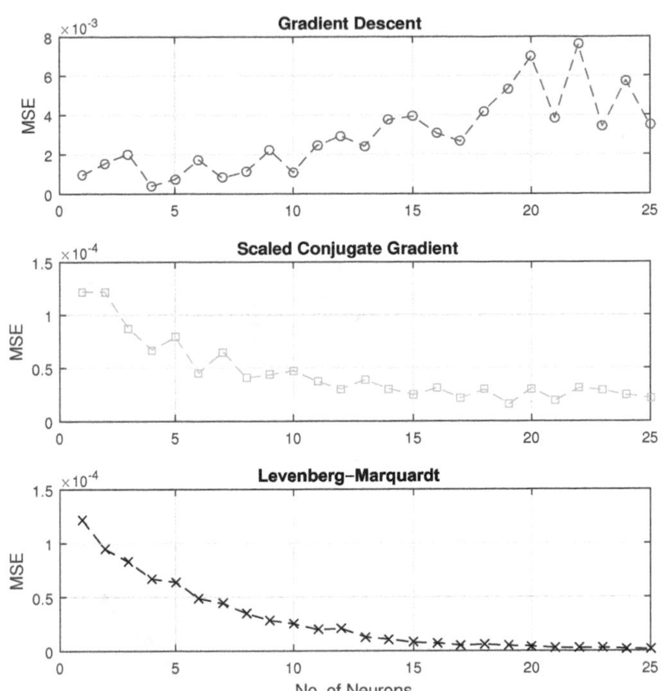

Fig. 5: Performance curves of one hidden layer PSO-ANN of different algorithms

indicated reasonably good results. Figures 5 and 6 show the performance of the network as a mean square error (MSE) versus the network architecture of the single and double hidden layer respectively. Table I demonstrates the training, testing and validation performances of different training approaches of both one and two layers in the hidden zone.

It is clearly indicated that much better results are found using LM as training algorithm with two hidden layers topology because LM uses Hessian matrix approximation as a second-order method to calculate the change in gradient. Figures 7 and 8 display regression plots of the network outputs on both compositions of training and test sets. For a perfect fit, the data must fall along a 45-degree line, where the network outputs are equal to the targets. The network with two hidden layers trained by LM algorithm, the fit, is reasonably good of both datasets, with R values in each case of 0.99 or above. Checking the test set is importantly required for examining the generalisation of the network of unseen data in the learning stage. It is worth noting that the network training and simulation was performed using MATLAB® and Simulink® platform.

Training Method	No. of Hidden Layers	No. of Neurones	Training MSE	Testing MSE	Validation MSE	Time (sec)
GD	1	4	3.972×10^{-4}	3.952×10^{-4}	3.996×10^{-4}	12.40
	2	$9-2$	2.047×10^{-4}	2.112×10^{-4}	2.015×10^{-4}	11.84
SCG	1	19	1.529×10^{-5}	1.6×10^{-5}	1.5×10^{-5}	12.39
	2	$20-25$	9.392×10^{-7}	9.429×10^{-7}	9.743×10^{-7}	38.39
LM	1	25	1.12×10^{-6}	1.225×10^{-6}	1.195×10^{-6}	18.58
	2	$23-25$	4.365×10^{-9}	3.984×10^{-9}	7.8×10^{-9}	11.37

TABLE I: Performances (MSE) of different PSO-ANN topologies

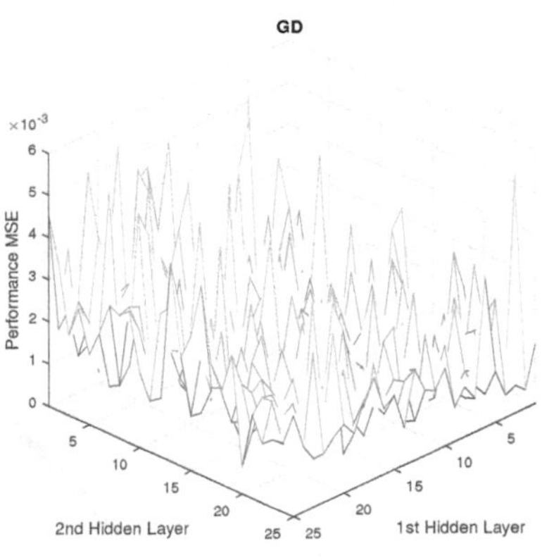

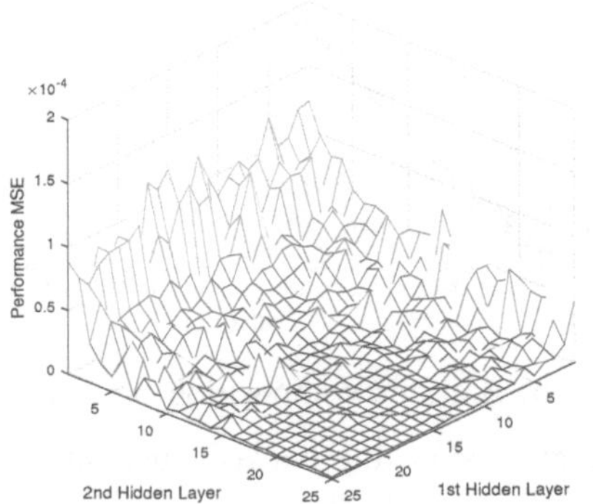

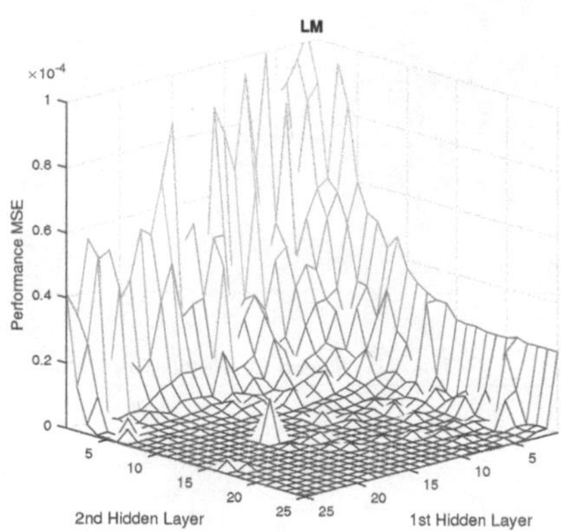

Fig. 6: Performance 3D surfaces of two hidden layers PSO-ANN of different algorithms

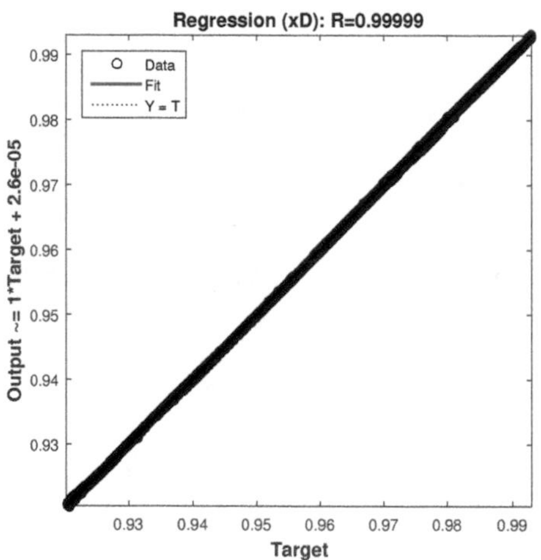

(a) Distillate composition prediction

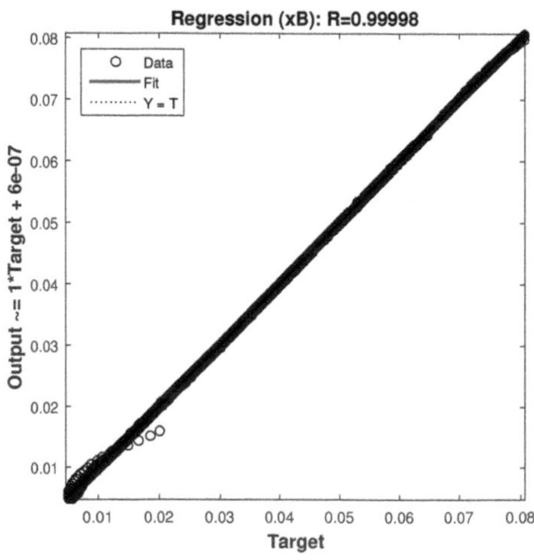

(b) Bottoms composition prediction

Fig. 7: The regression line between predicted and simulated compositions of training set

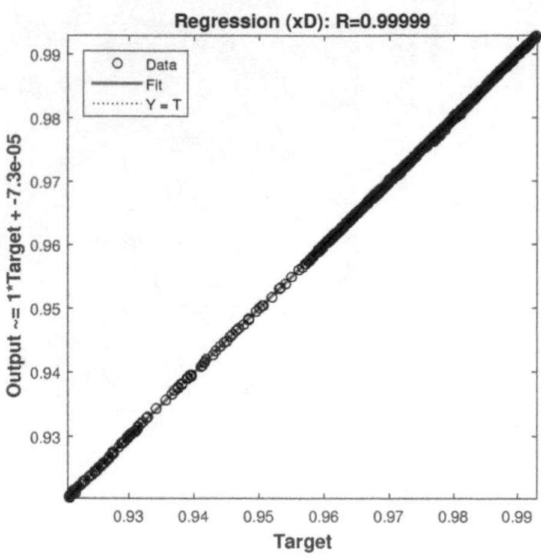

(a) Distillate composition prediction

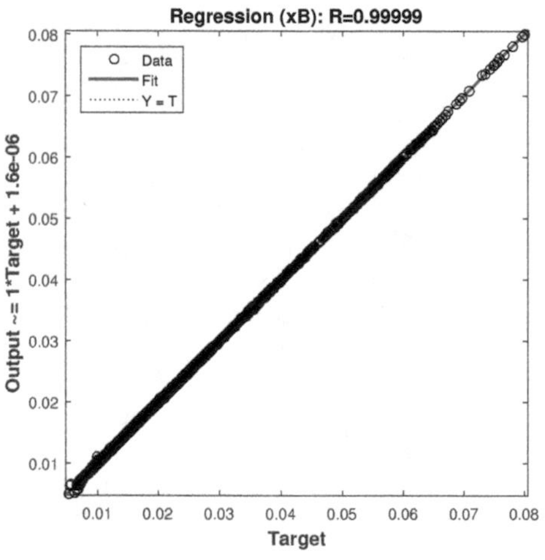

(b) Bottoms composition prediction

Fig. 8: The regression line between predicted and simulated compositions of testing set

V. CONCLUSION

A PSO-based artificial neural network has been proposed as an intelligent prediction approach to estimate product compositions of a binary distillation column; boil-up and reflux were used as inputs to the network. A double layer in the hidden zone with 23-25 neurones architecture was presented the optimal performance of the prediction model after examining different training algorithms and topologies using particle swarm optimisation. the network trained by LevenbergMarquardt algorithm gave more accurate results with less MSE compared to Gradient Descent and Scaled Conjugate Gradient. Therefore, the precision of predicted compositions of the distillation column using LM algorithm has shown to be high, and the estimated compositions have approximately been in agreement with the simulation results. The proposed ANN could be used efficiently to improve the performance of the different neural network controllers like NARMA-L2, direct inverses and NN predictive controller, which mainly depend on the prediction performance, which is to be the subject of future work.

ACKNOWLEDGMENT

The corresponding author is grateful to the Iraqi Ministry of Higher Education and Scientific Research for supporting the current research.

REFERENCES

[1] R. Teti, K. Jemielniak, G. ODonnell, and D. Dornfeld, "Advanced monitoring of machining operations," *CIRP Annals-Manufacturing Technology*, vol. 59, no. 2, pp. 717–739, 2010.

[2] R. X. Gao, X. Tang, G. Gordon, and D. O. Kazmer, "Online product quality monitoring through in-process measurement," *CIRP Annals-Manufacturing Technology*, vol. 63, no. 1, pp. 493–496, 2014.

[3] A. Vijayakumari, A. Devarajan, and N. Devarajan, "Decoupled control of grid connected inverter with dynamic online grid impedance measurements for micro grid applications," *International Journal of Electrical Power & Energy Systems*, vol. 68, pp. 1–14, 2015.

[4] H. Wang, P. Senior, R. Mann, and W. Yang, "Online measurement and control of solids moisture in fluidised bed dryers," *Chemical Engineering Science*, vol. 64, no. 12, pp. 2893–2902, 2009.

[5] A. K. Pani and H. K. Mohanta, "Online monitoring and control of particle size in the grinding process using least square support vector regression and resilient back propagation neural network," *ISA transactions*, vol. 56, pp. 206–221, 2015.

[6] A. M. F. Fileti, L. S. Pedrosa, and J. A. Pereira, "A self tuning controller for multicomponent batch distillation with soft sensor inference based on a neural network," *Computers & Chemical Engineering*, vol. 23, pp. S261–S264, 1999.

[7] W. Luyben, "Feedback control of distillation columns by double differential temperature control," *Industrial & Engineering Chemistry Fundamentals*, vol. 8, no. 4, pp. 739–744, 1969.

[8] T. Mejdell and S. Skogestad, "Estimation of distillation compositions from multiple temperature measurements using partial-least-squares regression," *Industrial & Engineering Chemistry Research*, vol. 30, no. 12, pp. 2543–2555, 1991.

[9] L. Fortuna, S. Graziani, and M. G. Xibilia, "Soft sensors for product quality monitoring in debutanizer distillation columns," *Control Engineering Practice*, vol. 13, no. 4, pp. 499–508, 2005.

[10] M. A. Hussain, "Review of the applications of neural networks in chemical process controlsimulation and online implementation," *Artificial intelligence in engineering*, vol. 13, no. 1, pp. 55–68, 1999.

[11] N. D. Ramirez-Beltran and H. Jackson, "Application of neural networks to chemical process control," *Computers & industrial engineering*, vol. 37, no. 1, pp. 387–390, 1999.

[12] P. J. Drew and J. R. Monson, "Artificial neural networks," *Surgery*, vol. 127, no. 1, pp. 3–11, 2000.

[13] J. W. Catto, D. A. Linkens, M. F. Abbod, M. Chen, J. L. Burton, K. M. Feeley, and F. C. Hamdy, "Artificial intelligence in predicting bladder cancer outcome a comparison of neuro-fuzzy modeling and artificial neural networks," *Clinical Cancer Research*, vol. 9, no. 11, pp. 4172–4177, 2003.

[14] A. Khosravi, S. Nahavandi, D. Creighton, and A. F. Atiya, "Comprehensive review of neural network-based prediction intervals and new advances," *Neural Networks, IEEE Transactions on*, vol. 22, no. 9, pp. 1341–1356, 2011.

[15] F. Amato, A. López, E. M. Peña-Méndez, P. Vaňhara, A. Hampl, and J. Havel, "Artificial neural networks in medical diagnosis," *Journal of applied biomedicine*, vol. 11, no. 2, pp. 47–58, 2013.

[16] S. Mondal, A. Bandyopadhyay, and P. K. Pal, "Application of artificial neural network for the prediction of laser cladding process characteristics at taguchi-based optimized condition," *The International Journal of Advanced Manufacturing Technology*, vol. 70, no. 9-12, pp. 2151–2158, 2014.

[17] M. Maarouf, A. Sosa, B. Galván, D. Greiner, G. Winter, M. Mendez, and R. Aguasca, "The role of artificial neural networks in evolutionary optimisation: A review," in *Advances in Evolutionary and Deterministic Methods for Design, Optimization and Control in Engineering and Sciences*. Springer, 2015, pp. 59–76.

[18] L. M. Lima de Campos, L. de Oliveira, and R. Celio, "A comparative analysis of methodologies for automatic design of artificial neural networks from the beginnings until today," in *Computational Intelligence and 11th Brazilian Congress on Computational Intelligence (BRICS-CCI & CBIC), 2013 BRICS Congress on*. IEEE, 2013, pp. 453–458.

[19] R. K. Agrawal and N. G. Bawane, "Multiobjective pso based adaption of neural network topology for pixel classification in satellite imagery," *Applied Soft Computing*, vol. 28, pp. 217–225, 2015.

[20] M. L. Darby and M. Nikolaou, "Identification test design for multivariable model-based control: an industrial perspective," *Control Engineering Practice*, vol. 22, pp. 165–180, 2014.

[21] R. W. Baker, *Membrane separation systems: recent developments and future directions*. Noyes Publications, 1991.

[22] C. L. Smith, *Distillation control: An engineering perspective*. John Wiley & Sons, 2012.

[23] W. L. Luyben, *Distillation design and control using Aspen simulation*. John Wiley & Sons, 2013.

[24] A. Abraham, "Hybrid intelligent systems: evolving intelligence in hierarchical layers," in *Do Smart Adaptive Systems Exist?* Springer, 2005, pp. 159–179.

[25] Y. Yousif, K. Daws, and B. Kazem, "Prediction of friction stir welding characteristic using neural network," *Jordan Journal of Mechanical and Industrial Engineering*, vol. 2, no. 3, 2008.

[26] G. Dreyfus, *Neural networks: methodology and applications*. Springer Science & Business Media, 2005.

[27] R. C. Eberhart, J. Kennedy *et al.*, "A new optimizer using particle swarm theory," in *Proceedings of the sixth international symposium on micro machine and human science*, vol. 1. New York, NY, 1995, pp. 39–43.

[28] J. Kennedy, J. F. Kennedy, R. C. Eberhart, and Y. Shi, *Swarm intelligence*. Morgan Kaufmann, 2001.

[29] R. Poli, "Analysis of the publications on the applications of particle swarm optimisation," *Journal of Artificial Evolution and Applications*, vol. 2008, p. 3, 2008.

[30] J. Kennedy, "Particle swarm optimization," in *Encyclopedia of machine learning*. Springer, 2011, pp. 760–766.

APPENDIX

Abbreviations, the operating conditions and technical aspects of the distillation column are detailed in following table.

Symbol	Description	Value	Unit
N	Number of trays	20	-
N_F	Feed stage location	11	-
F	Typical inlet flow rate to the column	1	kmol/min
D	Typical distillate flow rate	0.5	kmol/min
B	Typical bottoms flow rate	0.5	kmol/min
zF	Light component in the feed (mole fraction)	0.5	-
q_F	Mole fraction of the liquid in the feed	1	-
L	Typical reflux flow rate	1.28	kmol/min
V	Typical boil-up flow rate	1.78	kmol/min
α	Relative volatility	2	-
xD	Distillate composition (mole fraction)	0.98	-
xB	Bottoms composition (mole fraction)	0.02	-
i	Stage number during distillation	-	-
x	Mole fraction of light component in liquid	-	-
y	Mole fraction of light component in vapour	-	-
M	Tray hold-up liquid	0.5	kmol
MD	Condenser hold-up liquid	0.5	kmol
MB	Reboiler hold-up liquid	0.5	kmol

35

Implementation of Computer Assisted CIPP Model for Evaluation Program of HIV/AIDS Countermeasures in Bali

I Made Sundayana
Director/Lecture of Health Education
Buleleng School of Health
Bali, Indonesia

***Abstract*—One of the fact within economical development of tourism in Bali is indicated by established tourism facilities in order to support Bali tourism industry. Consquently, It has brought up effect that large numbers of new citizen search for occupation to Bali.Those people who came and settle in Bali temporaly or permanently, consequently Bali become heterogeneous.Thus, Bali become over populated. Since, over populated in Bali has risen up the economic sector and it has been spreading HIV /AIDS rapidly. As anticipation and prevention for contagious, developed and spreading HIV of Bali Provinse has regulated (regional act) Number 3 2006 concerning of prevention act fr HIV/AIDS. As the matter of fact, regional act is not properly conducted yet as, therefore it is evaluation required f0r the rule and program that have been conducted by the government.One of the technical evaluation can be applied is CIPP model. However, CIPP model is still applied in conventional way and it has not yet contributed accurate evaluational count in processing the data, therefore by using CIPP model of computer assistance. This can be proved by ending up the result of the total program percentage of HIV /AID prevention by conventional counted result as much 88.000%, meanwhile the count with computer assistance end up with 88.400% in result. It shows high category.**

***Keywords*—Evaluation; Computer Assisted CIPP Model**

I. Introduction

Tourism sector is a significant sector in order to achieve regional revenue goverment income.Tourim shall be perceived from several point of view as its complexity embedded within tourism activity. Among of those activity are tourism as resource, tourism as business, and toursm as industry. Those things indicate that tourism has potential in order to support economic sector.

One of the reality in economy base on develoving of Bali's tourism is facilities that had been established as an effort to support Bali Tourism. By Having established various of business, consequently, new comers have come to seek occupations. Pople who came to settle permanently or temporally having social interaction with local that creating heterogen society. Heterogenity is causing over populated in Bali, however it's rising up economy sector as well as spreading infected desease HIV/AIDS.

Masiive spreading of HIV/AID indicates high rate of infection. In Bali particularly HIV/AIDS infected not only in urban but also in rural. Large numbers of HIV/AIDS cases rose in rural . Until nowdays, the process of preventing HIV/AIDS structurally involves formal institutions, and traditional instutions yet socialized in rural based on geographical reason and daily activity of the traditional society.

On other side, effort to prevent HIV/AIDS consider the government policy voint of view, whereas the HIV/AIDs's subject and object is it's own. Various action of anticipation or prevention of spreading and contagious HIV/AIDs, Bali Province has Local Act Number 3 2006 regarding HIV/AIDS, however the provision is unble to well manage, therefore it is necessary to evaluate the act program which is conducted by the government.

One of the technical evaluation applied is CIPP model, However, CIPP model that has been applied conventionally yet shows accurate counted evaluation in processing its data.

It is appropriate on the results of research conducted by Dewa Gede Hendra Divayana about Program Evaluation of Management E-learning shows the model is done in the conventional CIPP still not provide an accurate evaluation calculation of the data processing[1]. From the results of these studies, the authors are interested in continuing the development of conventional CIPP model evaluation toward a computer assisted CIPP model.

II. Literature Review

A. Evaluation

In [2], Evaluation is a mean for understanding how things going.

In [3], Evaluation can be defined as the determination of conformity between the results achieved and the objectives to be achieved.

In [4], Evaluation can be defined as an activity or process to provide or specify a value above a certain object, things, institutions, and programs.

In [5], evaluation is a systematic and ongoing process to collect, describe, interpret and present information about a program to be used as a basis for making decisions.

From the opinions of the above can be concluded in general that the evaluation is an activity in collecting, analysing, and presenting information about an object of research and the results can be used to take a decision.

B. CIPP Model

In [6], the core concept of this model denoted by the CIPP acronym, which stands for the evaluation context, input, process, and product.

In [7], the CIPP evaluation there are four components that must be passed is the evaluation of the component context, the evaluation of input component, the evaluation of process components, and the evaluation of product components.

In [8], the CIPP model evaluation consists of four types, namely: context evaluation, input evaluation, process evaluation and product evaluation.

In the evaluation context is carried out to identification and assessment of the needs that underlie the program formulation. The input evaluation carried out to choose among several existing planning. In the process evaluation is carried out to access the implementation of the plan has been set. And the product evaluation conducted to identify and access the outputs and benefits of a program.

In [9], basically the CIPP evaluation model requires that a series of questions will be asked about four different elements of the model on the context, input, process, and product.

From the above opinions can be concluded in general that the CIPP model is a model in its activities through four stages of evaluation are: evaluation of the component context, input, process and product.

III. METHODOLOGY

A. Object dan Research Site

1) Research Object is HIV/AIDS countermeasures program.

2) Research Site at Health Department of Bali Province.

B. Data Type

In this research, the authors use primary data, secondary data, quantitative and qualitative data.

C. Data Collection Techniques

In this research, the authors use data collection techniques such as interviews, observation, and documentation.

D. Analysis Techniques

Analysis techniques used in this research is descriptive statistical.

E. Aspect of Evaluation

The aspects evaluated in HIV/AIDS countermeasures program can be seen in Table I bellow.

TABLE I. EVALUATION CRITERIA

No	Component	Aspects
1.	Context	Local regulations of HIV/AIDS
		The mission and purpose of program
		Readiness from Head of Health Department in implementing the regulations of HIV/AIDS
2.	Input	Guide of the program implementation
		Human resources
		Facilities and infrastructure
3.	Process	Program planning of HIV/AIDS countermeasures
		Program implementation of HIV/AIDS countermeasures
4.	Product	The impact of implementation of HIV/AIDS countermeasures program
		The expected outcome form implementation of HIV/AIDS countermeasures program

IV. RESULT AND DISCUSSION

A. Result

The research results can be seen in Table II below.

TABLE II. EVALUATION RESULTS OF HIV/AIDS COUNTERMEASURES PROGRAM WITH CIPP MODEL IN CONVENTIONAL

No	Dimension	Aspects	Respondents Score					X	%
			R1	R2	R3	R4	R5		
1.	Context	C1	5	4	5	4	4	4.4	88
		C2	5	4	4	5	5	4.6	92
		C3	5	4	4	4	5	4.4	88
Percentage of Effectiveness on Context Dimension									89
2.	Input	I1	5	5	4	5	5	4.8	96
		I2	4	5	5	4	4	4.4	88
		I3	5	4	4	5	4	4.4	88
Percentage of Effectiveness on Input Dimension									91
3.	Process	P1	4	4	4	4	5	4.2	84
		P2	4	5	5	4	4	4.4	88
Percentage of Effectiveness on Process Dimension									86
4.	Product	O1	5	4	4	5	4	4.4	88
		O2	4	5	4	4	4	4.2	84
Percentage of Effectiveness on Product Dimension									86
Total Percentage of Effectiveness									88
Category									High

Explanation :

C1 : Local regulations of HIV/AIDS
C2 : The mission and purpose of program
C3 : Readiness from Head of Health Department in implementing the regulations of HIV/AIDS
I1 : Guide of the program implementation
I2 : Human resources

I3 : Facilities and infrastructure
P1 : Program planning of HIV/AIDS countermeasures
P2 : Program implementation of HIV/AIDS countermeasures
O1 : The impact of implementation of HIV/AIDS countermeasures program
O2 : The expected outcome form implementation of HIV/AIDS countermeasures program
X : Average
% : Percentage

Category of scale effectiveness:
Highest : 90%-100%
High : 80%-89%
Sufficient : 70%-79%
Low : ≤ 69%

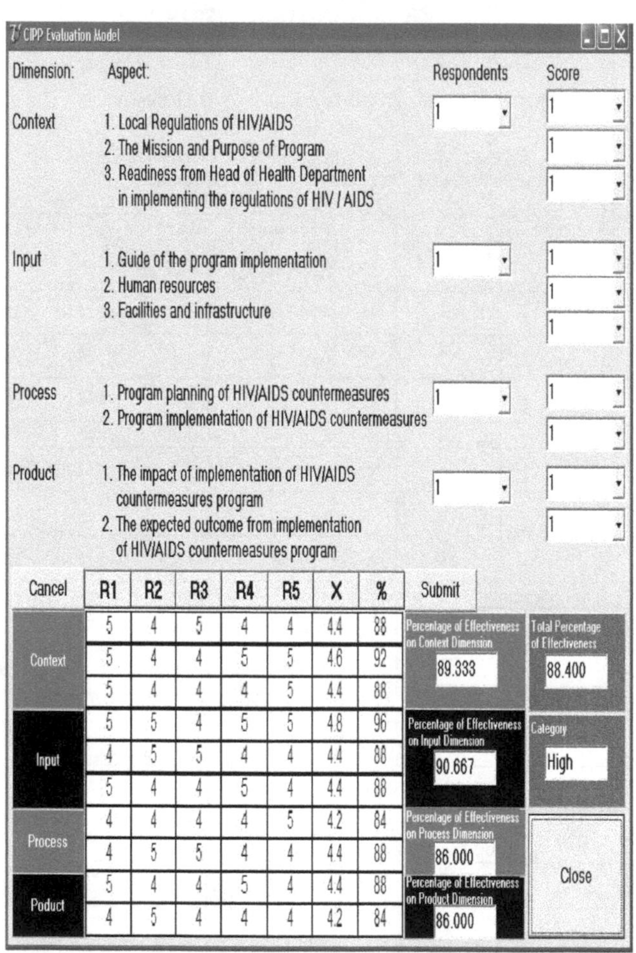

Fig. 1. Evaluation Results of HIV/AIDS Countermeasures Program With Computer Assisted CIPP Model

From the above results can be seen clearly that the results of the evaluation by Computer Assisted CIPP model calculation shows the results more accurate than using the conventional calculation method. It is seen from the results the percentage of effectiveness on context dimension with the conventional calculation shows result of 89.000%, while the computer aided calculation shows result of 89.333%. The percentage of effectiveness on input dimension with the conventional calculation shows result of 91.000%, while the computer-aided calculation shows result of 90.667%. The percentage of effectiveness on process dimension with the conventional calculation shows result of 86.000%, while the computer-aided calculation also obtained the same result of 86.000%. The percentage of effectiveness on product dimension with the conventional calculation shows result of 86.000%, while the computer-aided calculation also obtained the same result by 86.000%. The Total Percentage of Effectiveness of HIV/AIDS countermeasures program with the conventional calculation shows result of 88.000%, while the computer-aided calculation shows result of 88.400% with the higher category.

V. Conclusions

There is concusion to be drawn from this research that by applying CIPP evaluasion model based on computer assistance shall achieve more accurate and rapid counting compare to conventional way of counting. There after, decicion maker shall be quicker in order to make recomandation within decicion making whether the program shal be terminated or be proceded.

Acknowledgments

The author express their gratefulness to staff at Buleleng School of Health for all support and motivation. The author also generously thank to Dewa Gede Hendra Divayana, Ph.D. as Lecture at IT Education, Informatics Department, Ganesha University of Education.

References

[1] D.G.H. Divayana. Program Evaluation of Management E-learning (Thesis). Surabaya: YAPAN Surabaya School of Economic, 2014.

[2] L.H. Kuo, *et.al.*, "An Evaluation Model of Integrating Emerging Technology into Formal Curriculum," in The International Journal of Education and Information Technologies Vol.6, No.3, 2012,pp.250-259.

[3] D. Mardapi, Measurement, Assessment, and Evaluation of Education (1st Edition). Yogyakarta: Nuha Medika, 2012.

[4] S. Rutoto, "Observing the Guidance and Counseling Program Evaluation at School Present and Future," in Mawas Vol.1,No.1, 2010,pp.1- 15.

[5] H. Sundoyo, T. Sumaryanto, Dwijanto, "Program Evaluation of Dual System Education Based of Countenance Stake Model," in Innovative Journal of Curriculum Vol.1,No.2,2012,pp.69-73.

[6] D.L. Stufflebeam, C.L.S. Coryn, "Evaluation Theory, Models, and Applications (2nd Edition). San Francisco: Jossey-Bass,2014.

[7] Wirawan. Evaluation Theory, Model, Standards, Applications, and Profession (1st Edition). Jakarta: Rajawali Pers,2011.

[8] M. Tiantong, P. Tongchin, "A Multiple Intelligences Supported Web-based Collaborative Learning Model Using Stufflebeam's CIPP Evaluation Model," in International Journal of Humanities and Social Science Vol.3, No.7,2013,pp.157-165.

Mobile Device Based Personalized Equalizer for Improving Hearing Capability of Human Voices in Particular for Elderly Persons

Kohei Arai 1
Graduate School of Science and Engineering
Saga University
Saga City, Japan

Takuto Konishi 1
Graduate School of Science and Engineering
Saga University
Saga City, Japan

***Abstract*—Mobile device based personalized equalizer for improving the hearing capability of human voices in particular for elderly persons are proposed. Through experiments, it is found that the proposed equalizer does work well for improving hearing capability by 2 to 55 % of voice Recognition success ratio. According to the investigation of the frequency component analysis and *formant* detections, most of the voice sounds have the formant frequencies for the first to third frequencies within the range of 3445 Hz. Therefore, a nonlinear equalizing multiplier is better to enhance the frequency components for the first to third formants in particular. The experimental results with the voice above input experiments show that a good Percent Correct Recognition: PCR is required for 0 to more than 8000 Hz of frequency components. Also, 8162 Hz *cut off* frequency would be better for both noise suppressions and keeping a good PCR**

Keywords—Frequency response equalization; mobile devices; formount frequancy; hearing capability; hearing aids

I. INTRODUCTION

In general, hearing capability of human voices is getting bad for elderly persons due to the fact that a high-frequency response of elderly persons' ears is getting poor. Hearing capability is defined with the well-known averaged hearing capability level that is defined as Averaged value of hearing capability for human voices regarding frequency components ranged from 500 Hz to 4000 Hz. In accordance with the definition, 25-40 dB of loudness of human voices are difficult to hear slightly when human voice is not loud while 40-70 dB of loudness of human voices are difficult to hear when human voice is normal level.

Earlier devices, known as ear trumpets or ear horns [1], [2], were passive funnel-like amplification cones designed to gather sound energy and direct it into the ear canal. After that not so small number of methods have been proposed so far [3] – [10]

Although general purpose of frequency equalizer which allows compensation of hearing capability is used for that purpose (improvement of hearing capability), degradation of high frequency response depends on person. Therefore, customization is highly required for the frequency equalization devices. On the other hand, such customized frequency equalization devices are not so cheap and also are not so good looking. Therefore, frequency equalizing devices are not so popular.

The human voice has base frequency sounds and overtone sounds. Frequency components of human voices consist of formant[1] (peaks of frequency components which are used for characterization of personal human voices). Lower-frequency components are dominant for vowels, in general, while relatively higher-frequency components are dominant for consonants. For elderly persons, high-frequency components are getting difficult to hear which results in the consonants are getting difficult to hear. Such difficulties are depending on persons by persons. Therefore, it is not so easy to customize frequency equalizers which allow improvement of hearing capability, in particular, for elderly persons.

There are several ways of evaluating how well a hearing aid compensates for hearing loss. One approach is audiometry which measures a subject's hearing levels in laboratory conditions. The threshold of audibility for various sounds and intensities is measured in a variety of conditions. Although audiometric tests may attempt to mimic real-world conditions, the patient's experiences may differ. An alternative approach is self-report assessment of which the patient reports their experiences with the hearing aid in concern. The evaluation method proposed here is based on using Electroencephalogram : EEG sensor. Namely, in accordance with hearing quality, Peak Alpha Frequency: PAF amplitude is getting large while it is getting small when hearing quality is getting poor.

The following section describes the proposed method and implementation of the compensation filter including mobile devices followed by some experiments for a specific person. Then a conclusion is described together with some discussions.

II. PROPOSED METHOD AND IMPLEMENTATION

In order to create a new personalized frequency equalizer, mobile devices are used. Mobile devices with headsets or ear phones are getting cheap and are good looking as well. Therefore, users can carry the proposed personalized equalizer.

In order to characterize hearing capability degradation, a specific user has to try to hear some sentences which cover the

[1] http://newt.phys.unsw.edu.au/jw/formant.html

spectral range from zero (Direct Current: DC) to around 20 KHz includes all the vowels and the consonants together with their overtones. Then a spectral response of a compensation filter for the specific user is designed. The compensation filter is implemented in a mobile device such as Android tablet terminal, i-phone, smart phone, etc.

Frequency responses of ears against human voices are, in general, characterized with formants which are shown in Fig.1. Namely, human voice spectra have peaks which are named as formants.

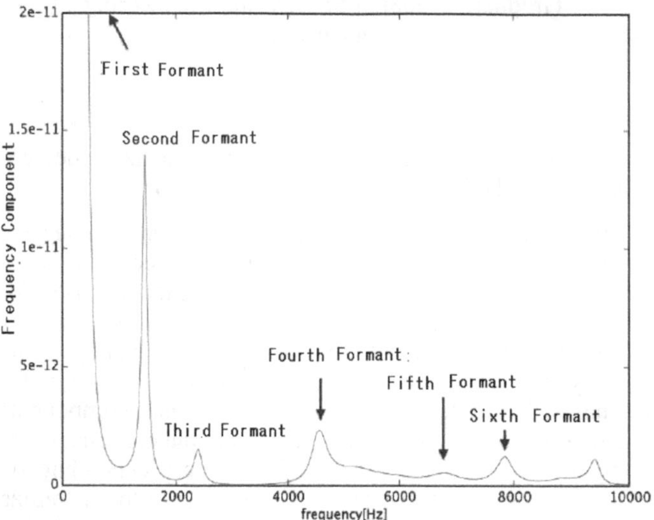

Fig. 1. Example of formants of human voices

According to the frequencies of the peaks, they are named the first formant, the second formant and so on. The proposed method detect formants through Forier Transformation first. In the mean time, input voice signals are decomposed with 32 of filter bank. Degraded formants can be found by comparing the input voice signals with the synthesized voice signals derived from Auditory Toolbox, for instance. Then the degraded formants can be compensated in accordance with the difference between actual voice and synthesized voice signals.

After that, reconstruction is applied to the degraded voice signals with 32 filter bank as shown in Fig.2. Delay time which is caused by the nonlinear equalizing multiplier can be compensated with deley element of which delay time is totally corresponding to the delay time caused by the nonlinear equalizing multiplier as shown in Fig.3

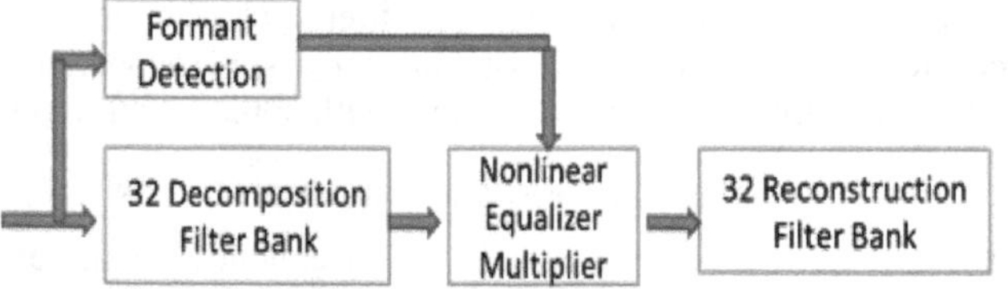

Fig. 2. Proposed method for degraded formant corrections with the consideration of formant ballunce

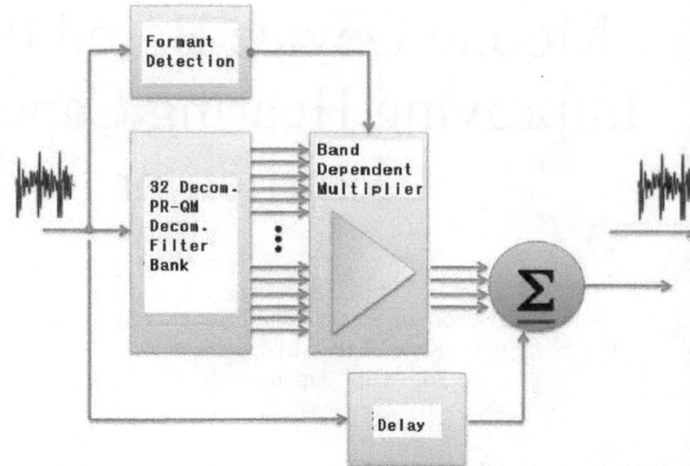

Fig. 3. Alternative of nonlinear equalizing multiplier with delay element

The correction filter is composed with hgh shelving filter of which the Frequency Transfer Function in analog filter function is expressed as follows,

$$H(s) = A\frac{As^2 + \frac{\sqrt{A}}{Q}s + 1}{S^2 + \frac{\sqrt{A}}{Q}s + A} \tag{1}$$

This can be re-written as follows,

$$H(z) = \frac{b_0 + b_1 z^{-1} + b_2 z^{-2}}{a_0 + a_1 z^{-1} + a_2 z^{-2}}$$

$$\begin{cases} b_0 = A((A+2) + (A-2)\cos\omega_0 + 2\sqrt{A}\,\alpha) \\ b_1 = 2A((A-2) + (A+2)\cos\omega_0) \\ b_2 = A((A+2) + (A-2)\cos\omega_0 - 2\sqrt{A}\,\alpha) \\ a_0 = (A+2) - (A-2)\cos\omega_0 + 2\sqrt{A}\,\alpha \\ a_1 = -2((A-2) - (A+2)\cos\omega_0) \\ a_2 = (A+2) + (A-2)\cos\omega_0 - 2\sqrt{A}\,\alpha \end{cases} \tag{2}$$

where $\alpha = \sin\omega_0/Q$、 $\omega_0 = 2\pi f_0/F_s$

The high shelving filter allows enhancement of arbitrary higher frequency components without suppression of low frequency components as shown in Fig.4. Also, the modulation transfer function of the high shelving filter is easy to design. Therefore, it is applicable for nonlinear equalizing multiplier.

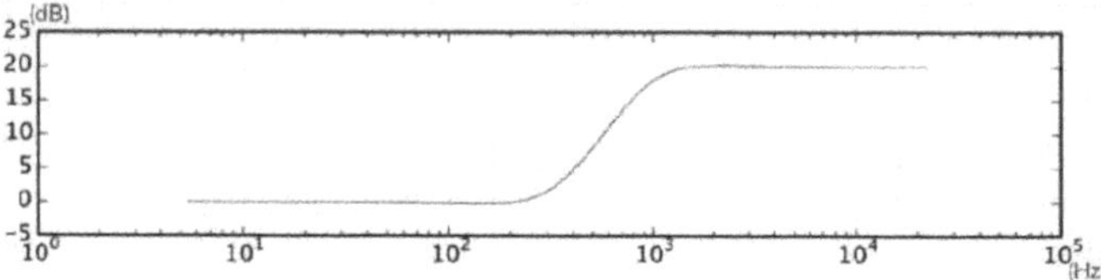

Fig. 4. Modulation Transfer Function of the high shelving filter

When the high shelving filter is applied to the input voice signals, high frequency components are enhanced as shown in Fig.5.

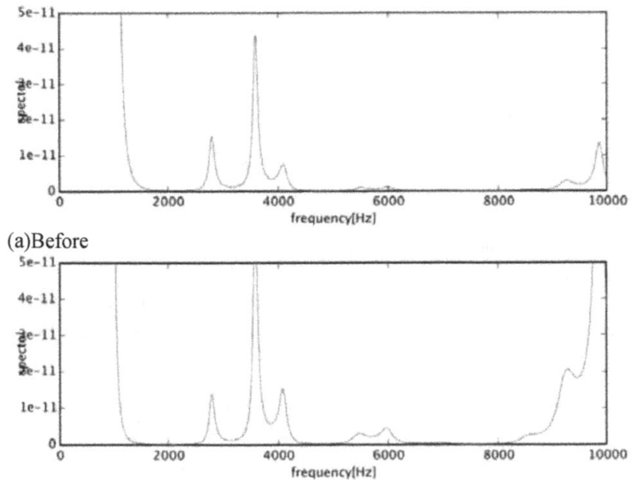

(a)Before

(b)After

Fig. 5. High frequency component enhancing nonlinear equalizing multiplier

Lastly, low pass filter is applied to the nonlinear equalizing multiplier applied voice signals for noise removal as shown in Fig.6.

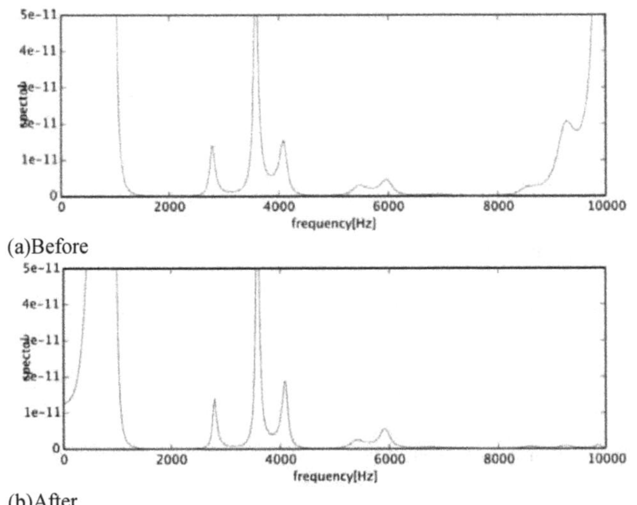

(a)Before

(b)After

Fig. 6. Before and after the low pass filter is applied to the nonlinear equalization multiplier applied voice signals

Also, digital filter featuring wavelet transformation is used for the correction filter. Haar base function of wavelet transformation is used for the first attempt. Haar wavelet transformation is illustrated in Fig.7. The original voice signals in time domain can be converted in high (H1) and low (L1) frequency components as shown in Fig.7. Then L1 component can also be converted in high (H2) and low (L2) components and so on. These components are called as wavelet coefficients (frequency components). Using the wavelet coefficients, Hn and Ln, Ln-1 can be reconstructed perfectly because Haar wavelet function is bi-orthogonal function.

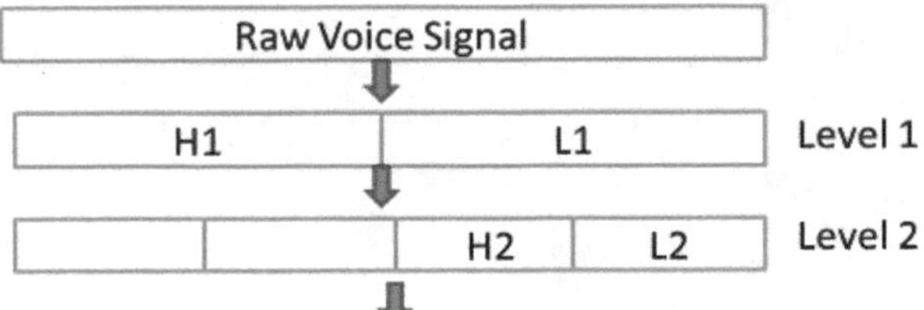

Fig. 7. Haar wavelet trasformation

Then raw input voice signal is converted through Haar wavelet trasformation with level 5 which corresponds to the third formant frequency. After that nonlinear multiplication is applied to the converted wavelet coefficients as shown in Fig.8. Also, the nonlinear multiplication is aplied to the converted wavelet coefficients with the previously designed *cut off* frequency. In particular, high frequency components sounds so noisy that the frequency components higher than *cut off* frequency is better to supressed.

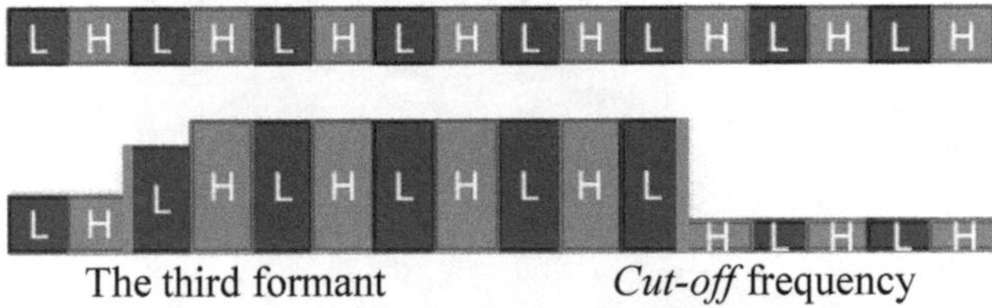

Fig. 8. Concept for the nonlinear equalizer multiplier

It is possible to constract low pass filter based on Haar wavelet transformation. Through reconstraction with the extracted low frequency component only derived from the decomposed voice signals, low pass filter can be realized as shown in Fig.9.

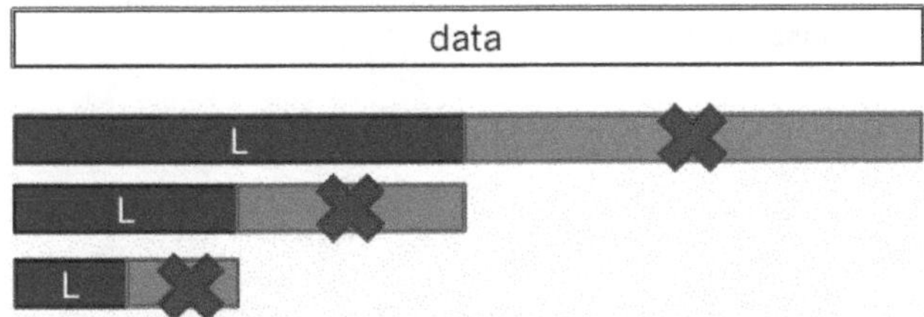

Fig. 9. Low pass filter based on Haar wavelet transformation

Therefore, arbitrary frequency components can be extracted from the level a of the wavelet coeffients. Haar wavelet transformation can be considered as filter bank which allows extraction of arbitrary frequency components. Also, it is possible to reconstruct arbitrary frequency component enhancing voice signals by adding wavelet coefficients. This is the method for frequency component equalization.

Through experiments, the following EEG sensor of ZA-9 + SleepSign-Lite manufactured by Kissei ComTec is used for evaluation of hearing quality. This EEG sensor allows measurements of EEG and electro - oculogram; EOG. Also, voice volume level meter of LM-8102 manufactured by Mother Tool Co. Ltd. is used for the experiments. Outlooks of the EEG and EOG sensor as well as voice volume level meter are shown in Fig.10 while the major specifications of the EEG and EOG sensors are shown in Table 1.

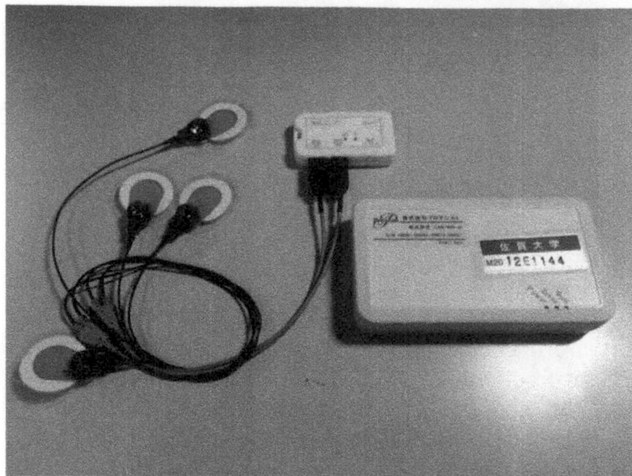

(a)EEG and EOG sensors

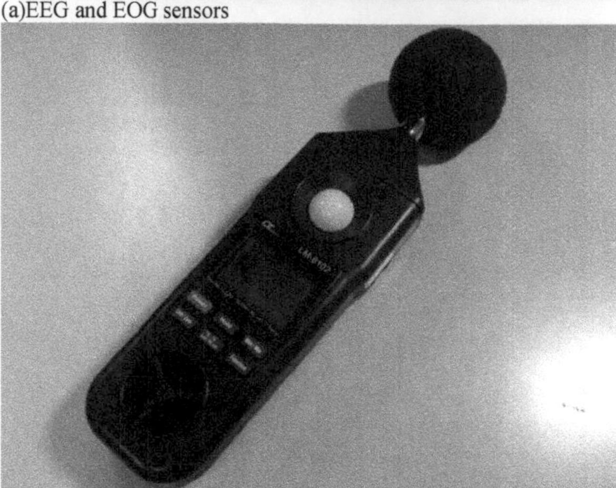

(b)Voice volume level meter

Fig. 10. Outlook of the EEG and EOG sensor as well as voice volume level meter used for the experiments

TABLE I. MAJOR SPECIFICATION OF EEG AND EOG SENSOR OF ZA-9 MANUFACTURED BY KISSEI COMTEC

Band Width	0.5～40Hz
Sampling Frequency	128Hz
AD Converter	12 bit

Meanwhile, the major specification of voice volume level meter is shown in Table 2.

TABLE II. MAJOR SPECIFICATION OF VOICE VOLUME LEVEL METER OF LM-8102 MANUFACTURED BY MOTHER TOOL CO. LTD

Auto range：30～130dB
Manual range：L=30～80dB/M=50～100dB/H=80～130dB
Resolution：0.1dB Frequency range：31.5～130dB

III. EXPERIMENTS

Example of input voice signals is shown in Fig.9. From these input voice signals and synthetic voicesignals, formant detection is performed together with creation of characteristics of nonlinear equalizer multiplier for correction of hearing capbility compensation.

Fig. 11. Example of the input voice signal

The followings are examples of frequencies of the actual and the synthetic formants for "a", "i" and "u", respectuvely.

"a"

F1 :718.75 F2 :1093.75 F3 :2437.5

(F1=730; F2=1090; F3=2440)

"i"

F1 :289.062 F2 :2296.875 F3 :3000

(F1=270; F2=2290; F3=3010)

"u"

F1 :304.688 F2 :882.812 F3 :2226.562

(F1=300; F2=870; F3=2240)

The differences between actual and synthetic formants are very small. Therefore, formants are detected almost perfectly.

In order to determine *cut off* frequency for noise suppression, the following 67 voice sounds are used.

"a", "i", "u", "e", "o"
"ka", "ki", "ku", "ke", "ko"
"sa", "si", "su", "se", "so"
"ta", "ti", "tu", "te", "to"
"na", "ni", "nu", "ne", "no"
"ha", "hi", "hu", "he", "ho"
"ma", "mi", "mu", "me", "mo"
"ya", "yu", "yo"
"ra", "ri", "ru", "re", "ro"
"wa"
"ga", "gi", "gu", "ge", "go"
"za", "zi", "zu", "ze", "zo"
"da", "di", "du", "de", "d"
"ba", "bi", "bu", "be", "bo"
"pa", "pi", "pu", "pe", "po"

Percent Correct Recognition of these voice sounds are evaluated with the different *cut off* frequencies. Nonlinear equalizing multiplier is created depending on the characteristics of hearing capabilities evaluated with EEG and EOG sensors. The results of PCR evaluation is shown in Table 3.

TABLE III. PERCENT CORRECT RECOGNITION: PCR WITH THE DIFFERENT *CUT OFF* FREQUENCIES

Frequency(Hz)	689	1378	2067	2756	3445
PCR(%)	45	79	90	98	100

If the *cut off* frequency is set at 689 Hz, then 55% of input voice sounds are not recognized. In accordance with the *cut off* frequency, PCR is increased monotonically. PCR reaches 100% at the *cut off* frequency of 3445 Hz. Therefore, the first to the third formant frequency have to be maintained their frequency components. Also it may say that 0 to 3443 Hz of frequency components is mandatory for voice recognitions. Thus it is concluded that *cut off* frequency has to be set more than 3445 Hz at least.

According to the investigation of the frequency component analysis and formant detections, most of voice sounds have the formant frequencies for the first to third frequencies within the range of 3445 Hz. Therefore, nonlinear equalizing multiplier is better to enhance the frequency components for the first to third formants in particular. The experimental results with the aforementioned voice input experiments shows that 0 to more than 8000 Hz of frequency components are required for a good PCR. Also 8162 Hz *cut off* frequency would be better for both noise suppressions and keeping a good PCR.

IV. Conclusion

Mobile device based personalized equalizer for improving hearing capability of human voices in particular for elderly persons is proposed. Through experiments, it is found that the proposed equalizer does work well for improving hearing capability by 2 to 55 % of the voice recognition success ratio.

According to the investigation of the frequency component analysis and formant detections, most of voice sounds have the formant frequencies for the first to third frequencies within the range of 3445 Hz. Therefore, the nonlinear equalizing multiplier is better to enhance the frequency components for the first to third formants, in particular. The experimental results with the voice above input experiments show that 0 to more than 8000 Hz of frequency components are required for a good PCR. Also, 8162 Hz *cut off* frequency would be better for both noise suppressions and keeping a good PCR.

Acknowledgment

The authors would like to thank all the voluntiers of saga University students who participated to the experiments.

References

[1] Bentler Ruth A., Duve , Monica R. (2000). "Comparison of Hearing Aids Over the 20th Century". *Ear & Hearing* 21 (6): 625–639. doi:10.1097/00003446-200012000-00009.

[2] Bentler, R. A.; Kramer, S. E. (2000). "Guidelines for choosing a self-report outcome measure". *Ear and hearing* 21 (4 Suppl): 37S–49S. doi:10.1097/00003446-200008001-00006. PMID 10981593. edit

[3] Jack Katz; Larry Medwetsky; Robert Burkard; Linda Hood (2009). "Chapter 38, Hearing Aid Fitting for Adults: Selection, Fitting, Verification, and Validation". *Handbook of Clinical Audiology* (6th ed.). Baltimore MD: Lippincott Williams & Wilkins. p. 858. ISBN 978-0-7817-8106-0.

[4] K. Sickel, Shortest Path Search with Constraints on Surface Models of In-ear Hearing Aids 52. IWK, Internationales Wissenschaftliches Kolloquium *(Computer science meets automation Ilmenau 10. - 13.09.2007) Vol. 2 Ilmenau : TU Ilmenau Universitätsbibliothek 2007, pp. 221-226*

[5] K. Sickel et al., Semi-Automatic Manufacturing of Customized Hearing Aids Using a Feature Driven Rule-based Framework *Proceedings of the Vision, Modeling, and Visualization Workshop 2009* (Braunschweig, Germany November 16–18, 2009), pp. 305-312

[6] Dave Fabry, Hans Mülder, Evert Dijkstra (November 2007). "Acceptance of the wireless microphone as a hearing aid accessory for adults". *The Hearing Journal* 60 (11): 32–36. doi:10.1097/01.hj.0000299170.11367.24.

[7] Hawkins D (1984). "Comparisons of speech recognition in noise by mildly-to-moderately hearing-impaired children using hearing aids and FM systems". *Journal of Speech and Hearing Disorders* 49 (4): 409. doi:10.1044/jshd.4904.409.

[8] Ricketts T., Henry P. (2002). "Evaluation of an adaptive, directional-microphone hearing aid". *International Journal of Audiology* 41 (2): 100–112. doi:10.3109/14992020209090400.

[9] Lewis M Samantha, Crandell Carl C, Valente Michael, Horn Jane Enrietto (2004). "Speech perception in noise: directional microphones versus frequency modulation (FM) systems". *Journal of the American Academy of Audiology* 15 (6): 426–439. doi:10.3766/jaaa.15.6.4.

[10] Exemption from Preemption of State and Local Hearing Aid Requirements; Applications for Exemption, Docket No. 77N-0333, 45 Fed. Reg. 67326; Medical Devices: Applications for Exemption from Federal Preemption of State and Local hearing Aid Requirements, Docket No. 78P-0222, 45 Fed. 67325 (Oct. 10, 1980).

Permissions

All chapters in this book were first published in IJARAI, by The Science and Information Organization; hereby published with permission under the Creative Commons Attribution License or equivalent. Every chapter published in this book has been scrutinized by our experts. Their significance has been extensively debated. The topics covered herein carry significant findings which will fuel the growth of the discipline. They may even be implemented as practical applications or may be referred to as a beginning point for another development.

The contributors of this book come from diverse backgrounds, making this book a truly international effort. This book will bring forth new frontiers with its revolutionizing research information and detailed analysis of the nascent developments around the world.

We would like to thank all the contributing authors for lending their expertise to make the book truly unique. They have played a crucial role in the development of this book. Without their invaluable contributions this book wouldn't have been possible. They have made vital efforts to compile up to date information on the varied aspects of this subject to make this book a valuable addition to the collection of many professionals and students.

This book was conceptualized with the vision of imparting up-to-date information and advanced data in this field. To ensure the same, a matchless editorial board was set up. Every individual on the board went through rigorous rounds of assessment to prove their worth. After which they invested a large part of their time researching and compiling the most relevant data for our readers.

The editorial board has been involved in producing this book since its inception. They have spent rigorous hours researching and exploring the diverse topics which have resulted in the successful publishing of this book. They have passed on their knowledge of decades through this book. To expedite this challenging task, the publisher supported the team at every step. A small team of assistant editors was also appointed to further simplify the editing procedure and attain best results for the readers.

Apart from the editorial board, the designing team has also invested a significant amount of their time in understanding the subject and creating the most relevant covers. They scrutinized every image to scout for the most suitable representation of the subject and create an appropriate cover for the book.

The publishing team has been an ardent support to the editorial, designing and production team. Their endless efforts to recruit the best for this project, has resulted in the accomplishment of this book. They are a veteran in the field of academics and their pool of knowledge is as vast as their experience in printing. Their expertise and guidance has proved useful at every step. Their uncompromising quality standards have made this book an exceptional effort. Their encouragement from time to time has been an inspiration for everyone.

The publisher and the editorial board hope that this book will prove to be a valuable piece of knowledge for researchers, students, practitioners and scholars across the globe.

List of Contributors

Ying Qian, Meng Li, Qingjie Wei and Xuemei Ren
The lab of Graphics and Multimedia Chongqing University of Posts and Telecommunications Chongqing, China

Dwi Mulyani
College of Informatics And Computer Management (STMIK) Banjarbaru Banjarbaru Kalsel, Indonesia

Genci Berati
Tirana University, Department of Mathematics Tirane, Albania

Indra Nugraha Abdullah
Jakarta Office, Yamaha Co. Ltd. Jakarta, Indonesia

Kensuke Kubo and Katsumi Sugawa
Fujitsu Kyushu Network Technologies, Ltd. Fukuoka Japan

Sylvia Encheva
Stord/Haugesund University College Bjørnsonsg. 45, 5528 Haugesund, Norway

Boyu Zhang, Jia Feng Liu and Xiang Long Tang
School of Computer Science and Technology, Harbin Institute of Technology, Harbin 150001, China

Adrian Brezulianu and Marius Daniel Peştină
Gheorghe Asachi Technical University of Iasi Iasi, Romania

Monica Fira
Romanian Academy Institute of Computer Science Iasi, Romania

Sylvia Encheva
Stord/Haugesund University College Bjørnsonsg. 45, 5528 Haugesund, Norway

I Putu Wisna Ariawan
Lecturer of Mathematics Education, Ganesha University of Education Bali, Indonesia

Dewa Bagus Sanjaya
Lecturer of Civics Education, Ganesha University of Education Bali, Indonesia

Dewa Gede Hendra Divayana
Lecturer of IT Education Ganesha University of Education Bali, Indonesia

Fatema Akhter
Department of Computer Science and Engineering Jatiya Kabi Kazi Nazrul Islam University Trishal, Mymensingh-2220, Bangladesh

Ernest E. Onuiri, OludeleAwodele and Sunday A. Idowu
Department of Computer Science, Babcock University, Ilishan-Remo, Ogun State, Nigeria

Jasna Hivziefendić
Faculty of Engineering and Information Technologies International Burch University Sarajevo, Bosnia and Herzegovina

Amir Hadžimehmedović
University of Tuzla Tuzla, Bosnia and Herzegovina

Majda Tešanović
Faculty of Electrical Engineering University of Tuzla Tuzla, Bosnia and Herzegovina

Amirhossein Tavanaei and Anthony S. Maida
The Center for Advanced Computer Studies University of Louisiana at Lafayette Lafayette, LA, USA

Khaled Nasser ElSayed
Computer Science Department, Umm Al-Qura University

Alaa F. Sheta
Computers and Systems Department Electronics Research Institute Giza, Egypt

Sara Elsir M. Ahmed
Computer Science Department Sudan University of Science and Technology Khartoum, Sudan

Hossam Faris
Business Information Tech. Dept. The University of Jordan Amman, Jordan

Rabab M. Ramadan
College of Computers and Information Technology University of Tabuk Tabuk, KSA

Prasun Chakrabarti
Head, Department of Computer Science and Engineering Sir Padampat Singhania University Udaipur-313601, Rajasthan, India

Prasant Kumar Sahoo
The Vice-Chancellor Utkal University Bhubaneswar - 751004,Orissa

Kohei Arai
Graduate School of Science and Engineering Saga University Saga City, Japan

Toshiya Katano
Tokyo University of Marine Science and Technology Tokyo Japan

Gahangir Hossain
Electrical & Computer Engineering Indiana University-Purdue University Indianapolis Indianapolis, IN USA

Habibah Khan
Instructional Design & Technology The University of Memphis Memphis, TN USA

Md.Iqbal Hossain
Electrical & Computer Engineering The University of Memphis Memphis, TN USA

Kenta Azuma
Cannon Electronics Inc. Tokyo, Japan

Shohei Fujise and Taka Eguchi
Graduate School of Science and Engineering Saga University Saga City, Japan

Rena Aierken and Li Xiao
The Xinjiang Technical Institute of Physics & Chemistry Academy of Sciences, Urumqi, China University of Chinese Academy of Science

Su Sha and Dawa Yidemucao
Key Laboratory of Xinjiang Multi-Language Technology, Xinjiang University, Urumqi, China Department of information science and engineering, Xinjiang University, Urumqi, China

Masanoori Sakashita
Information Science, Saga University Saga, Japan

Osamu Shigetomi and Yuko Miura
Saga Prefectural Agricultural Research Institute Saga Prefectural Government, Japan

Zhihai Yang, Zhongmin Cai and Aghil Esmaeilikelishomi
Ministry of Education Key Lab for Intelligent Networks and Network Security, Xi'an Jiaotong University, Xi'an, 710049, China

Khaled Nasser ElSayed
Computer Science Department, Umm AlQura University

Yousif Al-Dunainawi and Maysam F. Abbod
Electronic and Computer Engineering Department College of Engineering, Design and Physical Sciences Brunel University London Uxbridge, London, UK

I Made Sundayana
Director/Lecture of Health Education Buleleng School of Health Bali, Indonesia

Takuto Konishi
Graduate School of Science and Engineering Saga University Saga City, Japan

Index

Printed in the USA
CPSIA information can be obtained
at www.ICGtesting.com
JSHW051431221024
72173JS00006B/1435